Nanocosmetics

This book offers an overview of the science of cosmetics and the formulation of nanosized cosmetic products including fabrication, characterization of nanocosmetics, major challenges in the safe applications, regulatory aspects, and commercialization on a large scale. The chapters provide an understanding of the interaction of nanocarriers with skin and hair, different nanocosmetic products in the present situation, applications as well as disadvantageous toxicity associated with nanocosmetics, regulatory prospects, and future perspectives.

Features:

- Provide an explicit account on vital aspects of various nanocosmetics drug delivery approaches, thereby providing a next-generation cosmetic product
- Bring together the novel applications of nanocosmetics approaches in the biological milieu
- Explores preparation, applications, toxicity, and regulatory prospects
- Includes a dedicated chapter on Niosomal drug-delivery systems in cosmetics
- Discusses the perspectives of the technologies explored so far based upon the findings outlined in highly organized tables, illustrative figures, and flow charts

This book is aimed at researchers and professionals in nanomedicine, pharmaceuticals, biotechnology, and the health sector.

Emerging Materials and Technologies

Series Editor:
Boris I. Kharissov

The *Emerging Materials and Technologies* series is devoted to highlighting publications centered on emerging advanced materials and novel technologies. Attention is paid to those newly discovered or applied materials with potential to solve pressing societal problems and improve quality of life, corresponding to environmental protection, medicine, communications, energy, transportation, advanced manufacturing, and related areas.

The series takes into account that, under present strong demands for energy, material, and cost savings, as well as heavy contamination problems and worldwide pandemic conditions, the area of emerging materials and related scalable technologies is a highly interdisciplinary field, with the need for researchers, professionals, and academics across the spectrum of engineering and technological disciplines. The main objective of this book series is to attract more attention to these materials and technologies and invite conversation among the international R&D community.

Advanced Functional Metal-Organic Frameworks
Fundamentals and Applications
Edited by Jay Singh, Nidhi Goel, Ranjana Verma and Ravindra Pratap Singh

Nanoparticles in Diagnosis, Drug Delivery and Nanotherapeutics
Edited by Divya Bajpai Tripathy, Anjali Gupta, Arvind Kumar Jain, Anuradha Mishra and Kuldeep Singh

Functional Nanomaterials for Sensors
Suresh Sagadevan and Won-Chun Oh

Functional Biomaterials
Advances in Design and Biomedical Applications
Anuj Kumar, Durgalakshmi Dhinasekaran, Irina Savina, and Sung Soo Han

Smart Nanomaterials
Imalka Munaweera and M. L. Chamalki Madhusha

Nanocosmetics
Delivery Approaches, Applications and Regulatory Aspects
Edited by Prashant Kesharwani and Sunil Kumar Dubey

Sustainability of Green and Eco-friendly Composites
Edited by Sumit Gupta, Vijay Chaudhary, and Pallav Gupta

For more information about this series, please visit:

www.routledge.com/Emerging-Materials-and-Technologies/book-series/
CRCEMT

Nanocosmetics
Delivery Approaches, Applications and Regulatory Aspects

Edited by
Prashant Kesharwani and Sunil Kumar Dubey

CRC Press
Taylor & Francis Group
Boca Raton London New York

CRC Press is an imprint of the
Taylor & Francis Group, an **informa** business

First edition published 2024
by CRC Press
6000 Broken Sound Parkway NW, Suite 300, Boca Raton, FL 33487-2742

and by CRC Press
4 Park Square, Milton Park, Abingdon, Oxon, OX14 4RN

CRC Press is an imprint of Taylor & Francis Group, LLC

ISBN: 9781032333267 (hbk)
ISBN: 9781032333274 (pbk)
ISBN: 9781003319146 (ebk)

DOI: 10.1201/9781003319146

Typeset in Times
by codeMantra

Contents

About the Editors

Prashant Kesharwani, PhD, is an assistant professor of pharmaceutics at the School of Pharmaceutical Education and Research, Jamia Hamdard University, New Delhi, India. He has more than 300 international publications in well-reputed journals and 15 international books as an Editor. He is a recipient of several internationally acclaimed awards viz 'Ramanujan Fellowship, DST, India-2017', 'Excellence Research Award 2014', 'Young Innovator Award (Gold medal) 2012', 'International Travel Award/Grant from DST' (New Delhi) and INSA (CCSTDS, Chennai) 2012'. He has received an ICMR Senior Research Fellowship (PhD) and AICTE Junior Research Fellowship (M. Pharm.). His number of citations is = 14551; h-index = 63; i-10 index = 230. He has been invited to several talks and oral presentations organized at prestigious scientific peer conferences. He has received international acclaim and awards for research contribution, has supervised students/junior researchers and has participated actively in outreach and scientific dissemination for the service of the wider community. Four of his articles have been published in very high impact factor journals, i.e., *Progress in Polymer Sciences*, which has a high impact factor (IF 31.281) [Ref: (1) Kesharwani et al., 2014, *Progress in Polymer Science* 39 (2), 268–307; and (2) Amjad and Kesharwani et al., 2017, *Progress in Polymer Science* 64, 154–81]. Another article has been published in *Material Today*, which has a high impact factor (IF 32.072). A very recent article has been published in the journal *Progress in Materials Science*, which has a high impact factor (IF 48.165). An overarching goal of his current research is the development of nano-engineered drug delivery systems for various diseases.

Sunil Kumar Dubey, PhD, has very recently joined as General Manager, Medical Research, at Emami Ltd, Kolkata, India. Before this, he worked as an assistant professor in the Department of Pharmacy, Birla Institute of Technology and Sciences (BITS), Pilani. He has extensive research experience in the area of pharmacokinetics, nanomedicines, preclinical and clinical studies. He was also a visiting assistant research professor in the Department of Chemical and Biomolecular Engineering at University of Maryland, USA. He has about 14 years of industrial, teaching, research, and administrative experience. He has supervised several postgraduate and undergraduate students. He has published several book chapters and research articles in renowned journals and presented papers at conferences in India and abroad. During his industrial tenure, he has successfully completed many projects, including development, validation, preclinical and clinical studies. He has successfully completed various projects related to formulation development, analytical method development, validation and pharmacokinetic investigation. He is presently handling various government and industry-sponsored projects.

Contributors

Pavan Kumar Achalla
Birla Institute of Science and
  Technology, Pilani, Rajasthan, India

Waleed H. Almalki
Faculty of Pharmacy, Umm Al-Qura
  University, Makkah, Saudi Arabia

Gulin Amasya
Ankara University Faculty of Pharmacy,
  Department of Pharmaceutical
  Technology, Turkey

Marilena Antunes-Ricardo
Centro de Biotecnología FEMSA,
  Escuela de Ingeniería y Ciencias,
  Monterrey, Mexico

André Rolim Baby
Department of Pharmacy, Faculty of
  Pharmaceutical Sciences, University
  of São Paulo

Ulya Badilli
Ankara University Faculty of Pharmacy,
  Department of Pharmaceutical
  Technology, Turkey

Akansha Bisht
Department of Pharmacy, Banasthali
  Vidyapith, Banasthali, Rajasthan,
  India

Akanksha Chaturvedi
Banasthali Vidyapith, Banasthali,
  Rajasthan, India

Manisha Choudhari
Department of Pharmacy, Birla Institute
  of Science and Technology, Pilani,
  India

Eman Z. Dahmash
Faculty of Pharmacy, Isra University,
  Amman, Jordan

André Luis Máximo Daneluti
Department of Pharmacy, Faculty of
  Pharmaceutical Sciences, University
  of São Paulo

Anuradha Dey
R&D Healthcare Division, Emami Ltd,
  India

Sunil Kumar Dubey
R&D Healthcare Division, Emami Ltd,
  India

Jaya Dwivedi
Department of Chemistry, Banasthali
  Vidyapith, Banasthali, Rajasthan,
  India

Wamiq Musheer Fareed
College of Dentistry and pharmacy,
  Buraidah Private Colleges, Al
  Qassim, Saudi Arabia

Nazeer Hasan
School of Pharmaceutical Education
  and Research, Jamia Hamdard,
  India

Ozge Inal
Ankara University Faculty of Pharmacy,
  Department of Pharmaceutical
  Technology, Turkey

Affiong Iyire
Aston Pharmacy School, College of
  Health & Life Sciences, Aston
  University Birmingham

Vighnesh Jadhav
Department of Pharmacy, Birla Institute
    of Technology and Science, Pilani,
    India

Gaurav K. Jain
Delhi Pharmaceutical Sciences and
    Research University, India

Shivangi Jaiswal
Department of Pharmacy, Banasthali
    Vidyapith, Banasthali, Rajasthan,
    India

Antonio Jiménez-Rodríguez
Centro de Biotecnología FEMSA,
    Escuela de Ingeniería y Ciencias,
    Monterrey, Mexico

Yogeshvar N. Kalia
School of Pharmaceutical Sciences,
    University of Geneva

Prashant Kesharwani
Department of Pharmaceutics,
    School of Pharmaceutical Education
    and Research, Jamia Hamdard,
    India

Sanskruti Santosh Kharavtekar
Department of Pharmacy, Birla Institute
    of Technology and Science, Pilani,
    India

Lucio Martínez-Alvarado
Centro de Biotecnología FEMSA,
    Escuela de Ingeniería y Ciencias,
    Monterrey, Mexico

Motlalepula Gilbert Matsabisa
Faculty of Health Sciences, University
    of the Free State, Bloemfontein,
    South Africa

M. Mitjans
Universitat de Barcelona, Spain

Rajesh Pradhan
Department of Pharmacy, Birla Institute
    of Technology and Science, Pilani,
    India

Urushi Rehman
School of Pharmaceutical
    Education and Research, Jamia
    Hamdard, India

Shruti Richa
Department of Pharmacy, Banasthali
    Vidyapith, Banasthali, Rajasthan,
    India

Laura E. Romero-Robles
Escuela de Ingeniería y Ciencias,
    Monterrey, N.L., Mexico

Ranendra Narayan Saha
Department of Pharmacy, Birla Institute
    of Science and Technology, Pilani,
    India

Amirhossein Sahebkar
College of Pharmacy, Umm Al-Qura
    University, Saudi Arabia

Mohammad Sarwar Alam
School of Chemical and Life Sciences
    (SCLS), Jamia Hamdard, India

Sayra N. Serrano-Sandoval
The Institute for Obesity Research,
    Av. Eugenio Garza Sada 2501 Sur,
    Monterrey, NL, Mexico

Swapnil Sharma
Banasthali Vidyapith, Banasthali,
    Rajasthan, India

Afsana Sheikh
School of Pharmaceutical Education
and Research, Jamia Hamdard, India

Gautam Singhvi
Department of Pharmacy, Birla Institute
of Science and Technology, Pilani,
India

Sristi
School of Pharmaceutical Education
and Research, Jamia Hamdard, India

Rajeev Taliyan
Department of Pharmacy, Birla Institute
of Technology and Science, Pilani,
India

Nilufer Tarimci
Başkent University Faculty of
Pharmacy, Department of
Pharmaceutical Technology, Turkey

Satyajit Tripathy
Faculty of Health Sciences, University
of the Free State, Bloemfontein,
South Africa

Onyinyechi Udo-chijioke
Onyii's Cosmetics Ltd, Lagos, Nigeria

Edidiong Udofa
Young African Researcher's Academy
(YARA), Abuja, Nigeria

Kanika Verma
Banasthali Vidyapith, Banasthali,
Rajasthan, India

M.P. Vinardell
Universitat de Barcelona, Spain

Nitheesh Yanamandala
Birla Institute of Science and
Technology, Pilani, Rajasthan

1 Cosmetics science and skin care
History and concepts

Anuradha Dey and Sunil Kumar Dubey

CONTENTS

1.1 INTRODUCTION

The cosmetic industry is an ever-evolving dynamic market with an ever-expanding consumer base having a wide range of needs that keep driving the industry and the associated companies to continue to innovate (Dubey et al. 2022). Promulgation of new rules and regulations which were being set by several government agencies for controlling the market became vital as several cosmetic products and drugs had some extent of overlap which made it essential to have regulatory bodies in place (Dubey et al. 2022). One of the earliest of such instances was the terming of sunscreens as over-the-counter drugs (OTC) by the FDA in 1978 as they contained absorbents of UV rays. With the continued discovery and use of several new chemicals, polymers, etc. stringent laws assessing their safety for human use are necessary. Several other products which claim to have a medical purpose, such as antibacterial creams or lotions, anti-acne products, etc. came to being regulated by the FDA. As safety became a major concern over the years, several series of tests were developed and or

DOI: 10.1201/9781003319146-1

approved by the FDA for ensuring the same. Over the past few decades, the trend of adapting *in vitro* assays as opposed to animal testing has also been a major ongoing transformation that is advocated to curtail animal cruelty (Gerner, Liebsch, and Spielmann 2005; Takahashi et al. 2009).

Cosmetics which were once viewed as a leisurely option exclusively for the elite have evolved to become products that have vital importance in the everyday life of each individual (Bapat et al. 2020). Scrubbing different parts of the body with indigenous materials such as pumice, seashell, and limestone blocks was a common practice. Many women also used sour milk, aged wine, sugar cane juice, etc. for application on the face. Certain chemicals such as a mixture of acetic acid in vinegar or sulphuric acid in oil of vitriol were used as chemical exfoliators for countering aging (Van Scott and Yu 1989).

The public nowadays, majorly being educated, has become all the way more aware about the products which they use for personal care, and given the vastness of products to choose from, the consumers are skeptical in adapting newer products and changing their preferences, while being very mindful of the evidence being presented by the manufacturers for their products before they start using them (Pavicic et al. 2009). Cosmetics is derived from the Greek word 'Kosm-tikos' which means having the ability to arrange and decorate in order to bring harmony, thus the popular usage of the word 'kosmos' also indicates universal harmony (Halla et al. 2018). Moreover, during the Roman period, the Roman public baths had slaves called Cosmetae, who were responsible for dyeing the hair, applying ointments and make-up, and giving manicures, etc. (Blanco-Dávila 2000). In this chapter, we will go over the development that cosmetics have undergone from pre-historic times till the present age and also discuss the major mechanisms of cosmetic action.

1.2   TRACING COSMETICS THROUGH THE LENS OF HISTORY

1.2.1   Cosmetics in the Pre-historic Era up to the Middle Ages

In pre-historic times, dating back to 3000 BC, man used to color his skin as an adornment and also for camouflage while hunting. The specific colors were meant for designation, tribe identification, etc. The Patagonians were a group that smeared their faces with chalk; the Aboriginal Australians used white clay. North American Indians and the South Sea Islanders used black/red and bright colours respectively for painting their bodies ("Encyclopædia Britannica - Fourteenth Edition | Britannica" 2022). The older eastern civilizations of Egypt, China, and India have had their own practices. In early Egyptian graves pouches of malachite and galena for eye make-up and red ochre used for painting the face have been discovered. The pyramid of King Khufu built in 2600 BC was found to have a toiletry box which contained seven alabaster jars having perfume and one jar containing kohl, along with copper and gold razors, small knives, etc. all of which substantiate the practice of personal grooming and care in that era ("Poucher's Perfumes, Cosmetics and Soaps" 1993). Henna leaves were popularly used in those days as well as today for reddening nails, feet, hair, etc. For daily use and for several religious practices fragrant oils and perfumes were made using a combination of plant extracts and aromatic woods

("Encyclopædia Britannica - Fourteenth Edition | Britannica" 2022). The practice of embalming the body, from pre-historic times till the early Christian period kept undergoing revolutions with different techniques for perfecting the process (Buckley and Evershed 2001). Several fragrant oils and ointments were ubiquitously used given the extremely dry and hot climate. They were derived from seeds of saffron, linseed, pumpkin, olives, etc. along with fat obtained from cattle (Poucher and Butler 1993). Evidence of the prevalent use of face paints, fragrances, oils, etc. amongst the Jewish community, Babylonians, and Assyrians dating back to as early as 1730 BC is found in the Bible (Butler 1993).

Around 1000 BC, India already had an Ayurvedic medical code in place which gave details pertaining to the indigenous raw materials that were employed for formulating several preparations. Indus Valley excavations led to the discovery of many alabaster and ivory pots, dating back to 3000–2000 BC, which are estimated to contain perfumed oils and kohl. During the Persian conquests in 600 BC, practices of the civilizations spanning across Aegean to Indus were revamped from being mystic and superstitious to becoming more advanced after adopting several customs of the people of Medes. Chinese women used to apply a mask of tea oil and rice powder mixture as an anti-wrinkle agent.

Based on the writings of Herodotus, around 500 BC in Babylon, both men and women applied white lead and vermillion on their faces, used pumice for abrasive cleaning, and curled and perfumed their hair. The Scythian women used a mixture of cedar wood, Cyprus, and frankincense for rubbing on their skin regularly which led to smoothening of the skin and left a glossy effect (Herodotus and Godolphin 1973).

Greeks were experts at perfumery. Bottles that might have contained cosmetic preparations and were manufactured during 400–350 BC in Athens have been discovered (Lee 2015). Trade ties and exchange of goods proliferated more when Alexander the Great conquered the Persian empire, such that now under his rule connections were established as far as Uzbekistan, India, and Egypt. There are several prominent physicians or people studying science who have left behind noteworthy work which they had performed during their lifetime. Hippocrates, known as the Father of Medicine was the first person to make headway around 450 BC in separating principles of medical and cosmetic products from that of superstitions and religious beliefs. Following this, Theophrastus born in 370 BC carried out the botanical classification of plants and also left some early writings on perfumery (Green 1927).

Roman emperor Nero and his wife Poppaea used white lead and chalk for whitening their skin, kohl for darkening their eyelashes, and eyebrows, and as eye make-up, along with using barley flour, and butter for controlling pimples and moisturizing skin, etc. (Rimmel 1867). Celsus, a Roman physician who lived 7 BC–AD 53 discussed several skin and hair conditions in his medical books. Pliny the Elder (AD 23–79) also wrote on perfumes and aromatics. Dioscorides mentioned practical applications of several natural and mineral substances in Materia Medica. During AD 6–9, the Arab physicians who had undergone training at Alexandria prepared face masks known as *batikha* which was made up of a mixture of white rice, eggs, beans, limes, etc., all of which worked as dermabrasion agents. Skin cleansers known as *hemsia* and *shnouda* made of almond oil and jasmine primarily were also made by them (Blanco-Dávila 2000; Rosner 2010).

In AD 200, Galen wrote exhaustively on several topics spanning medicine, pharmacy, and hygiene, all of which were a milestone during those days, and continued to govern medical and pharmaceutical principles for the next 1500 years. The credit for formulating cooling wax (*Ceratum refrigerans*) also lies with Galen. It contained olive oil and beeswax in the ratio of 4:1 along with rose petals. This cream left a cooling sensation after application and is considered to be the forerunner of the modern era's cold creams. Galen also wrote a scientific treatise on human skin, known as "Local Remedies", which was the first of its kind (Meyerhof 1930). In AD 800–900, efforts were made towards improving experiment-related procedures for extracting essences and distilling out active constituents. From around AD 300 to 1600, China was the superpower empowering trade between the east and the west for transport of several commodities including cosmetics.

1.2.2 Cosmetics in the Modern Age and the Present Times

With the increasing demand for perfumes in Europe, in the 1500s native perfumes, creams, and toilet vinegars were produced at large scales. As the European monarchy became rich and influential, extravagant fashion styles were developed such as using iron oxide and cinnabar as rouge, lead carbonate for powdering the face, etc. According to records, Mary Queen of Scots had a particular recipe for enhancing the complexion which involved washing the face first with hot water, followed by wine and then buttermilk. The skin regime followed by the Queen became quite popular. She used a special cowslip cream for moisturizing the skin, a mix of almonds, honey, and lemon for skin brightening, and a lotion made of rosemary, camomile, sage, cloves, and thyme for strengthening the hair (Mattern 2008). The women of the European monarchy, as well as aristocracy, spent hours on getting ready for being presentable in society's eyes. They regularly applied white paint on their faces which was a mixture consisting of white lead with a small amount of mercury sublimate and grounded orris. Applying a wash of egg white was commonplace for achieving a glaze on the skin. A mixture of turpentine and sulphur followed by a layer of fresh butter was applied on acne for spot reduction. During the 16th century in America, the Aztecs had a deep understanding of the botanical sources. They had a very organized bathing system, skin care, and hair care regimes as well. For example, avocado and sugarcane were used as face masks. Both the Aztec and Mayan women used cosmetic pigments imparting reddish colour to lips, cheeks, and neck areas. They even made perfumes out of white lily, calla lily, and jasmine flowers (Blanco-Dávila 2000; Wetzel 2012).

It was only in the 17th century that disorders of the skin, hair, and nail started to be considered medical conditions rather than just being arenas for decorative remedies. Kings James I officially recognized the Apothecaries Guild in the year 1617 following which the first British Pharmacopoeia was published in 1618. All cosmetic practices were going unchecked; however, few noticed the long-term repercussions of some cosmetics, especially white lead. John Bulwer in 1653, in his writings mentioned the glaring side effects of white lead. It was found to be responsible for premature aging of the skin, leading to the formation of blackish-grey patches (Parish and Crissey 1988). In the year 1675, Robert Boyle elaborated on the Mechanical

Production of Odours and also emphasized the adoption of principles of chemistry during experimentations (Butler 1993). Towards the end of the 17th century, a pale porcelain look with rouge applied as round blobs on cheeks was adopted by men, women, and children as well. This fad lasted up till the French Revolution. Lady Coventry, a very beautiful lady, passed away at the tender age of 27 years due to excessive use of white lead paint. Similar were the cases with Lady Fortrose and Kitty Fisher amongst many others. Similarly in the US, a compound used for skin whitening, composed of lead oleate, known by the name Land's Bloom of Youth was commonly used by women. A lady in Ohio had developed severe palsy and the diagnosis was inconclusive by several doctors. Then finally, a few specialists in New York attributed the cause to lead poisoning (Witkowski and Parish 2001).

Towards the last quarter of the 18th century, foundations of chemistry laid by Robert Boyle almost a century ago were brought into the light. Work done by other scientists such as Lavoisier, Priestley, Scheele, and Bertollet also aided in the establishment of elemental chemistry, and in the early 1800s the Atomic Theory of Matter postulated by John Dalton helped in shaping more clear foundations towards the role of modern chemistry in drugs and cosmetic products (Nye 1999). The Greeks used *sulama*, a reddish pigment for reddening the cheeks, this contained mercury which had side effects as well. Materials such as beeswax, oils extracted from plants, and tallow were used for making ointments and brilliantine and white flour for powdering. During these times, in the year 1795, due to the excessive use of flour in cosmetics, there was a shortage of flour.

Advertisements of cosmetics via the distribution of handbills came into being in the 1700s. With the Industrial Revolution taking place in the 19th century, production capacities increased multi-fold in all factories. During these times, a French chemist, M.E. Chevreul, was the first person to elucidate the chemical nature of soap around the 1820s. Other observations made by C.E. Geoffroy and Scheele pertaining to fats, oil, alcohol, etc. were helpful. In 1807, William Colgate set up a factory wherein soap and candles were manufactured based on these principles. In the years to follow, further discoveries by Berzelius, Gay-Lussac, Dalton, and Avogadro helped in grasping a better understanding of the nature of matter. The work done by Thomas Graham on emulsions and colloids in the 1860s and the discovery of borax in 1856 led to improvements in the manufacturing of cold creams as now beeswax and fatty acids, borax and alkali helped in achieving more stable products. In the 1890s, V.C. Dagnet started using petroleum by-products, waxes, and refined mineral oils in place of oils from vegetable sources which gave a white appearance to the creams and also eliminated the problem of rancidity. This version of Pond's cream was launched in 1907. In 1858, another milestone was the launch of the proprietary product 'Vaseline'(Jayakumar and Micheletti 2017). Pharmacies across the US and Europe started selling several cosmetic products, even homemade ones. Eventually, several pharmaceutical companies also started manufacturing cosmetics. Notable developments were the use of collapsible tubes by Colgate for dispensing toothpaste for the first time in 1873 and also the novel use of witch hazel extracts by Pond's in their cream (Fulling 1953). H. Tetlow, in 1865, brought about the replacement of skin whitener, bismuth nitrate, with zinc oxide which proved to be much better. Moreover, a few years later he started selling finely ground magnesium silicate as talc which became widely popular (Mitsuda and Taguchi 1977). During the late 1800s, cosmetic companies underwent expansion as several parlours were set up. During the world wars, many European countries established new manufacturing units in the US given that the US had cheaper raw materials which were more readily available.

At the beginning of the 20th century, the leading giant, Gillette Company, was formed in 1901 in Boston. Gillette launched the shaving razor that is still in use today. During the 1920s and 1930s, several shaving lotions and creams were formulated that softened the beard before being shaved off. Other cosmetic products also gained momentum, for example with the advent of aerosol systems and mechanical pumps which eased dispensing, more variety of products were packaged in different forms as per the suitability or demands of the customers. Products related to hair care, foot care, nail care, etc. also started expanding. With progress in better understanding of the structure, properties, and the different conditions related to skin, etc., research started expanding to cater to all conditions. It was in the 20th century, that the term 'cosmeceutical' was coined by Dr. Albert Kligman. Now it has also been adapted in usage, generally for referring to products that are a hybrid between cosmetics and drugs (Elsner and Maibach 2000). Although this term hasn't been accepted by the FDA, it confers the meaning that cosmetic actives are capable of eliciting a pharmaceutical effect along with cosmetic benefits and this concept has revolutionized skin care. Extensive research in the field of cosmetics by several leading companies such as Beecham, L'Oreal, Gillette, and Unilever continued and in 1935 Society of Cosmetic Chemists in America was formed. The society had around 12 members in 1945, which grew to a whopping 3000 members in 1980. Similarly, the Society of Cosmetic Chemists in Great Britain was also formed in 1948. In 1969, the International Federation of Societies of Cosmetic Chemists (IFSCC) was formed and the first meeting was held in London (Sakamoto et al. 2017). Molecular identification and understanding of the role of specific plant-based constituents which are responsible for cosmetic effects were concerted efforts towards the last quarter of the 20th century.

The laying down of testing parameters that will ensure safety for use was and still is an area of continuous development and improvisations. In the 21st century, the major change has been efforts for shifting from animal testing to other alternatives which can serve the same purpose. Obtaining quality standard products that are reproducible and efficacious is the goal (Balsam and Sagarin 1972; Elsner and Maibach 2000; Sakamoto et al. 2017). The evolution of the cosmetic market in the United States reflects the growth that has occurred over time. For instance, in the late 1800s, the trading association pertaining to cosmetic products was known as Manufacturing Perfumers' Association as most of the products were either perfumes or colognes. Following this, in 1922, it was renamed as American Manufacturers of Toilet Articles and then again in the 1930s as Toilet Goods Association, as now cosmetics encompassed several types of toiletries such as soaps, shampoos, etc. Post this, in 1970 it was again revised to *The Cosmetic, Toiletry, and Fragrance Association* (CTFA) to cover a broader range of products spanning across skin care, body care, cleansing, make-up, hair care, etc. The last modification occurred in 2007 when it was renamed to *Personal Care Products Council* (PCPC), resorting to a name that is very wide ("PCPC Timeline - Personal Care Products Council" 2022).

FIGURE 1.1 Guiding principles during the development of a nanocosmetic

Cosmetic research in the 21st century is concerted towards delving deeper into and gaining a better understanding at a genomic and molecular level of what happens during the aging process and figuring out avenues for curtailing it.

With the increased knowledge of physiology and biochemistry of the skin, and of microbiological and toxicological aspects of the industry, cosmetic companies have invested heavily in molecular, genomic, and proteomic research into what causes skin cells to age, with the hope of pinpointing ways to interfere with that process (Singhvi et al. 2018). The latest tendencies of cosmetics are based on advanced research that includes the use of biotechnology-derived ingredients, adopting nanotechnology-based approaches (Waghule et al. 2020; Rapalli, Banerjee, et al. 2021), carrying out genetic profiling for individual skin care or nutritional regimes, stem cell-based products and therapies to regenerate aging tissues, or cell and tissue engineering for cosmetic purposes (Rapalli et al. 2018; Singhvi et al. 2018). Figure 1.1 highlights the guiding principles for ensuring the success of a nanocosmetic product, from the start of its formulation up till it reaches the market.

1.3 KEY AREAS OF COSMETIC CARE

There are a few major mechanisms based on which the mechanism of action of cosmetics can be classified (Zappelli et al. 2016). The first mechanism is the antioxidant action of many cosmetics, especially the ones which claim anti-aging effects. Oxidative stress is mainly due to accumulative oxidative damage caused by reactive oxygen species such as peroxides, ions, and free radicals over periods of time to DNA molecules, telomeres, and cellular membranes. This culminates in skin damage as with age our body's inherent ability to produce antioxidants as a defense mechanism reduces. Other factors and external stimuli which accelerate aging are UV rays, exposure to chemicals, either acute or chronic, electromagnetic fields, etc. Compounds such as Vitamins E and C, coenzyme Q10, superoxide dismutase, carotenoids, retinol, and different types of polyphenols are commonly incorporated in cosmetic preparations as these all are scavengers of free radicals. The second major mechanism of action of cosmetics is having a moisturizing and hydrating effect on the skin. The uppermost layer of the skin, the stratum corneum consists of dead skin cells along with a matrix of lipids such as cholesterol, ceramides, and fatty acids, all of which are responsible for working as a barrier, preventing water loss from the epidermis and providing a structural framework to the skin (Feingold 2007; Rapalli, Mahmood, et al. 2021). Ceramides make up around 50% of the lipid content

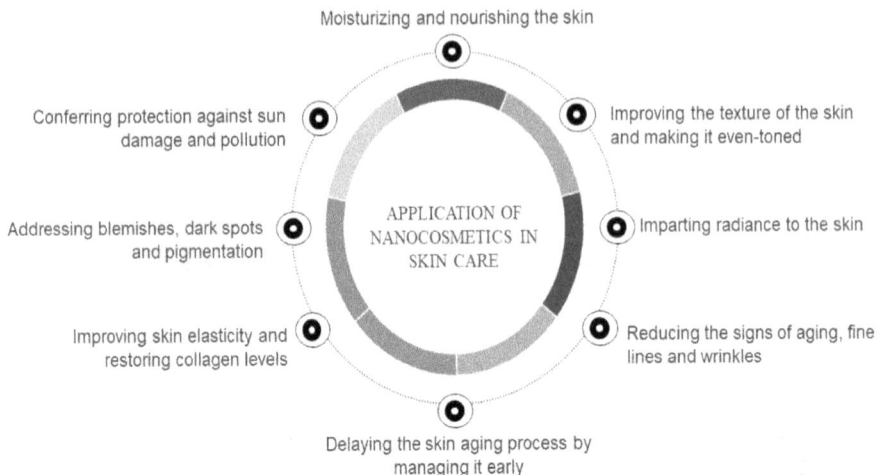

Moisturizing and nourishing the skin

Conferring protection against sun damage and pollution

Improving the texture of the skin and making it even-toned

APPLICATION OF NANOCOSMETICS IN SKIN CARE

Addressing blemishes, dark spots and pigmentation

Imparting radiance to the skin

Improving skin elasticity and restoring collagen levels

Reducing the signs of aging, fine lines and wrinkles

Delaying the skin aging process by managing it early

FIGURE 1.2   Applications of nanocosmetics in skin care

of the stratum corneum and hence several formulations such as creams, lotions, and ointments are ceramide-containing in order to maintain balance in ceramide levels. Oil extracted from plant seeds have a high content of fatty acids and hence they have also been used for moisturizing the skin (Ahmad and Ahsan 2020). With age, the skin loses its integrity and becomes thinner. This is because the components of the extracellular matrix (ECM) progressively deteriorate with age. The ECM comprises a combination of collagenous and non-collagenous compounds. The third major mechanism of action of cosmetics is restoring and maintaining the levels of ECM compounds which will facilitate cell repair and act as matrix boosters (Zappelli et al. 2016). The major areas into which cosmetics are classified are discussed here under. Figure 1.2 represents the major application of nanocosmetics in the skin care domain and the changes they bring about which makes them effective.

1.3.1   CLEANSERS AND MOISTURIZERS

Around 4000 years ago, the concept of cleansing first came into being when a group of people known as Hittites from Asia Minor started using the mixture of water and the ash of soapwort plants for cleaning their hands. Following this, the Sumerians of Ur during 2500 BC were noted to produce alkali-containing solutions for washing. Their cleansing agents were made up of soap obtained from hydrolyzing potash and animal triglycerides. A few hundred years later, the Phoenicians were the first to saponify animal fat from a goat with water and potassium carbonate. With the passage of time, when it was elucidated that bacterial infections can be warded off by cleaning our hands, Proctor along with his cousin Gamble made the famous 'creamy-white soap' in 1878. Known as 'Ivory' soap, it was hugely popular (Draelos 2018). While formulating the soap, accidentally they discovered that whipping air into the solution leads to the formation of soaps that do not sink, thus being a breakthrough given that everyone used to take baths in rivers, and soaps prior to this used to sink

(Panati 2016). Soaps that are currently available are a blend of tallow with some nut oil or suitable fatty acid, usually present in the ratio of 4:1. Modifying this ratio alters the cleansing activity of the system. For instance, if the ratio is increased, then it has higher amounts of fatty acids which makes it a milder cleanser. The cleaning formulations that are available nowadays comprise synthetic detergents which are known as 'sydnets', whose pH is in the range of 5.5–7, making them more close to human skin pH than soap which usually has a pH in the range of 9–10 (Wortzman 1991). The cleansing agents can be classified into several categories such as bar cleansers or liquid cleansers or even hand cleansers (Draelos 2018). Several of these may be mild in nature or leave a moisturizing after effect or have anti-microbial properties depending upon the ingredients it is made up of and their respective claims. Ingredients such as shea butter or petrolatum, etc. are added for moisturization.

1.3.2   FACIAL COSMETICS

Delving into historical records and scourging through cultural records have shown us that beautifying the face with the use of several chemicals or plant-based extracts, or applying layers of masks of some preparations, etc. have been a common practice in the history of mankind. As already discussed, using white lead for painting the face was a common practice that had its own horrific repercussions. During the 1600s, using face patches in order to cover facial scarring incurred due to smallpox was a common practice. 'Foundation' as it is known today, was earlier known as wet white/French white, and was made for whitening the face, hands, and neck of theatre artists as it had better adherence in comparison to loose powder (Schlossman and Feldman 1988). The breakthrough was the launch of 'cake makeup' by Max Factor in 1936, which was of great use to the common woman (Wells and Lubowe 1964). Post this breakthrough, with several other developments, the foundations available today incorporate several other functions such as providing oil control or sun protection or locking moisturization, etc., or a combination thereof based on the needs of the consumer. Cosmetic products are also used for contouring and texture improvement. Contouring is also done for addressing facial or any bodily scarring along with texture correction, thus helping to cover up blemishes and irregularities that are present on the skin surface. Pigmentation and dark spots are also addressed by different types of cosmetics (Draelos 2021).

1.3.3   SKIN LIGHTENING AGENTS

Cosmetic formulations containing azelaic acid, the improved alternative to hydroquinone, are widely used for skin lightening effect as it interferes with the metabolism of melanocytes and leads to damaging abnormal melanocytes (Baliña and Graupe 1991; Fitton and Goa 1991). Liquorice extracts comprising glycosides liquiritin and isoliquertin act as the actives to disperse melanin present in the skin and thus lighten it. Another very popular cosmetic ingredient is kojic acid which is derived from fungal species (Lim 1999). Aloesin, an active constituent obtained from aloe vera, is also used for skin lightening, but often in combination with another popular compound called arbutin. Both of them decrease the activity of tyrosinase. Arbutin is obtained from the leaves of *Vaccinium vitisidaea*. Its deoxy form, deoxyarbutin, has also been developed which has been found to have better pigment-lightening properties (Hamed et al. 2006).

1.3.4   IMPROVING SKIN TEXTURE

This entails regularizing the surface properties of the face, contouring the face, and addressing irregularities on the skin. Exfoliation is the first step in improving skin texture which involves the removal of dead skin cells. Agents such as ferulic acid, glycolic acid, lactic acid, etc, are generally incorporated. Another mechanism for improving skin texture is using products that activate retinoid receptors on the skin (Kligman, Duo, and Kligman 1984). Cosmetics containing ester forms retinyl palmitate and retinyl propionate are activated on the skin surface when the ester bond breaks and leads to the formation of retinol which is the active constituent responsible for decreasing signs of aging and improving skin texture overall (Duell et al. 1996).

1.3.5   SKIN MOISTURIZATION

Moisturizing the skin simply means keeping a check on the water loss that is occurring from the skin surface and locking it. Moisturizers act by filling the gaps that are present between the corneocytes and bring about a smooth finish. Dehydrated skin is often the underlying cause of skin wrinkling and hence the best remedy to visible wrinkles is the consistent application of moisturizers which will keep the skin hydrated. The different categories of the basis of actions of moisturizers are occlusives, humectants, and hydrophilic matrices which curtail transepidermal water loss. The first category, occlusives, being oily in nature work by sealing the skin surface to water loss. Petrolatum is an example of an occlusive as it blocks 99.9% of water. Humectants, on the other hand, are substances that have the ability to draw water from the surroundings and hold it into the skin such as glycerine, honey, urea, propylene glycol, etc. The last category is of hydrophilic matrices which form a film on the skin's surface. These agents are generally large molecular-weight protein molecules. One of the first matrices was that of colloidal oatmeal and one of the recent examples is that of hyaluronic acid which also mediates the reduction of wrinkles.

1.3.6   ENHANCING ANTI-AGING ACTION

One of the most sought-after and popular areas of skin care is that of prevention or management of the skin aging process. As discussed earlier, moisturizers do provide timely relief and smoothen the skin surface which imparts a fresh and youthful look to the skin making it softer. However, this action is only transient as it remains only as long as the moisturizer stays and is not removed by wiping or cleansing. On the contrary, compounds that claim to have anti-aging properties involve photo protection and take consistent efforts through the years to be of any impact. Several sunscreens function as anti-aging agents as they confer protection from photodamage which is one of the leading causes of aging. Compounds such as benzophenone or its several complexes are generally present in such formulations.

1.3.7 EYE-BASED COSMETICS

As early as around 4000 BC and across several centuries, kohl as loose powder or as a paste comprising of burnt almond pieces, brown clay, along with black copper oxide, and small amounts of antimony was heavily used for eye make-up and for enhancing the outline of the eyes and the eyebrows. Another common eye make-up was a green powder obtained from malachite which was applied above eyelids as it was thought to ward off infections. Usually, these were kept stored in small pots for application. Glitter made up of powdered bettle shells was also applied to the eyelids (Panati 2016). The several colored shades of eye cosmetics that are in use today were introduced in the early 1960s (Wells and Lubowe 1964). The mascara of ancient times has also undergone several refinements to reach the current mascaras which are made up of triethanolamine stearate (Rutkin 1975). The ease of application also improved drastically with the automatic dispensing tube as well as in-built brushes.

1.3.8 LIP-BASED COSMETICS

Painting the lips with colored pigments was in practice as early as 7000 BC amongst the Sumerians. The same was handed down to the Egyptians, then to the Syrians, then Babylonians, then to the Persians and Greeks, and finally down to the Romans which has continued till the present generation. Materials such as saffron, red ochre, and Brazilian wood were used for obtaining the reddish tinge. The earliest lip 'sticks' that were formulated were made from beeswax and tallow logs with the coloring pigment (Draelos 2000). The earliest version of present-day lipstick came into being in the 1920s, which had a holder system that could be pushed (Cunningham 1992). Other lip-based cosmetics include lip liners, lip balms, liquid lipsticks, etc. which have evolved in the last few decades.

1.3.9 NAIL-BASED COSMETICS

Cosmetics for the nail such as nail paints, enamels, etc. came into use in relatively more recent times. With the development of lacquer technology, which is glossy thin films of nitrocellulose, nail polish came into being in the 1920s. Before this, nails were rubbed with abrasives for imparting shine to them, and onto them oil-based pigments were applied. The first nail lacquer launched in the market was known as 'nail polish' because of the lustre it imparted. Revson was the first person to come up with the idea of adding a range of colored pigments to the lacquer. Later when Revson formed Revlon, one of the leading cosmetics manufacturers in the world, they developed nail enamels.

1.4 CONCLUSION

With the scientific advancements in the 21st century, the cosmetics that have emerged exert their action based on well-defined mechanisms of action which act at a cellular level and bring about physiological changes which lead to their rejuvenation. Cosmetology or the study of cosmetics has evolved since the start of civilizations and has continued to reflect and resonate with the customs and practices of those times respectively. Cosmetics is an ever-growing and ever-changing industry with new additions to current trends and fading of old trends as well. Cosmetics being products that are applied onto several of our body surfaces on a daily basis require serious scientific considerations and can't be afforded to be treated as just beautification gimmicks. Perceiving cosmetics as being associated with cultural customs and scientifically unchallenged to being scientifically elucidated has been a major transformation over time. Cosmetics of the present day are a confluence of science, aesthetics, and personal choices which provides a wide range of cosmetic options to choose from. Companies are more stringent regarding ensuring effectiveness and substantiating the claims of their products. Upholding the rules and regulations, adhering to the ethical guidelines, and safeguarding the patients are of paramount importance.

REFERENCES

Ahmad, Anas, and Haseeb Ahsan. 2020. "Lipid-Based Formulations in Cosmeceuticals and Biopharmaceuticals." *Biomedical Dermatology* 4 (1): 12. doi:10.1186/s41702-020-00062-9.

Baliña, L M, and K Graupe. 1991. "The Treatment of Melasma. 20% Azelaic Acid versus 4% Hydroquinone Cream." *International Journal of Dermatology* 30 (12): 893–95. doi:10.1111/j.1365-4362.1991.tb04362.x.

Balsam, Marvin S, and Edward Sagarin. 1972. *Cosmetics: Science and Technology*. Vol. 1. New York: Wiley-Interscience. https://archive.org/details/cosmeticsscience03bals .

Bapat, Ranjeet A, Tanay V Chaubal, Suyog Dharmadhikari, Anshad Mohamed Abdulla, Prachi Bapat, Amit Alexander, Sunil K Dubey, and Prashant Kesharwani. 2020. "Recent Advances of Gold Nanoparticles as Biomaterial in Dentistry." *International Journal of Pharmaceutics* 586: 119596. doi:10.1016/j.ijpharm.2020.119596.

Blanco-Dávila, F. 2000. "Beauty and the Body: The Origins of Cosmetics." *Plastic and Reconstructive Surgery* 105 (3): 1196–204. doi:10.1097/00006534-200003000-00058.

Buckley, Stephen A, and Richard P Evershed. 2001. "Organic Chemistry of Embalming Agents in Pharaonic and Graeco-Roman Mummies." *Nature* 413 (6858): 837–41. doi:10.1038/35101588.

Butler, Hilda. 1993. "Historical Background - Poucher's Perfumes, Cosmetics and Soaps: Volume 3 Cosmetics." In, edited by Hilda Butler, 639–92. Dordrecht: Springer Netherlands. doi:10.1007/978-94-011-1482-0_24.

Cunningham, J. 1992. "Color Cosmetics." In *Chemistry and Technology of the Cosmetics and Toiletries Industry*, 149–82. Dordrecht: Springer. https://link.springer.com/book/10.1007/978-94-011-2268-9

Draelos, Zoe Diana. 2000. "Cosmetics and Skin Care Products: A Historical Perspective." *Dermatologic Clinics* 18 (4): 557–59.

———. 2018. "The Science behind Skin Care: Cleansers." *Journal of Cosmetic Dermatology* 17 (1): 8–14.

———. 2021. "The Use of Cosmetic Products to Improve Self Esteem & Quality of Life." *Essential Psychiatry for the Aesthetic Practitioner*. Wiley Online Books. doi:10.1002/9781119680116.ch3.

Dubey, Sunil Kumar, Anuradha Dey, Gautam Singhvi, Murali Manohar Pandey, Vanshikha Singh, and Prashant Kesharwani. 2022. "Emerging Trends of Nanotechnology in Advanced Cosmetics." *Colloids and Surfaces B: Biointerfaces* 112440. doi:10.1016/j.colsurfb.2022.112440.

Duell, E A, F Derguini, S Kang, J T Elder, and J J Voorhees. 1996. "Extraction of Human Epidermis Treated with Retinol Yields Retro-Retinoids in Addition to Free Retinol and Retinyl Esters." *The Journal of Investigative Dermatology* 107 (2): 178–82. doi:10.1111/1523-1747.ep12329576.

Elsner, Peter, and Howard I Maibach. 2000. *Cosmeceuticals: Drugs vs. Cosmetics.* Vol. 23. Boca Raton, FL: CRC Press. https://www.taylorfrancis.com/books/mono/10.1201/9781 003002987/cosmeceuticals-peter-elsner-howard-maibach

"EncyclopædiaBritannica-FourteenthEdition|Britannica."2022.AccessedMay24.https://www. britannica.com/topic/Encyclopaedia-Britannica-English-language-reference-work/ Fourteenth-edition.

Feingold, Kenneth R. 2007. "Thematic Review Series: Skin Lipids. The Role of Epidermal Lipids in Cutaneous Permeability Barrier Homeostasis." *Journal of Lipid Research* 48 (12): 2531–46. doi:10.1194/jlr.R700013-JLR200.

Fitton, A, and K L Goa. 1991. "Azelaic Acid. A Review of Its Pharmacological Properties and Therapeutic Efficacy in Acne and Hyperpigmentary Skin Disorders." *Drugs* 41 (5): 780–98. doi:10.2165/00003495-199141050-00007.

Fulling, Edmund H. 1953. "American Witch Hazel—History, Nomenclature and Modern Utilization." *Economic Botany* 7 (4): 359–81.

Gerner, Ingrid, Manfred Liebsch, and Horst Spielmann. 2005. "Assessment of the Eye Irritating Properties of Chemicals by Applying Alternatives to the Draize Rabbit Eye Test: The Use of QSARs and in Vitro Tests for the Classification of Eye Irritation." *Alternatives to Laboratory Animals : ATLA* 33 (3): 215–37. doi:10.1177/026119290503300307.

Green, Mary L. 1927. "History of Plant Nomenclature." *Bulletin of Miscellaneous Information (Royal Botanic Gardens, Kew)* 10, 403–15. https://www.jstor.org/stable/4107555

Halla, Noureddine, Isabel P Fernandes, Sandrina A Heleno, Patrícia Costa, Zahia Boucherit-Otmani, Kebir Boucherit, Alírio E Rodrigues, Isabel C F R Ferreira, and Maria F Barreiro. 2018. "Cosmetics Preservation: A Review on Present Strategies." *Molecules.* doi:10.3390/molecules23071571.

Hamed, Saja H, Penkanok Sriwiriyanont, Mitchell A deLong, Marty O Visscher, R Randall Wickett, and Raymond E Boissy. 2006. "Comparative Efficacy and Safety of Deoxyarbutin, a New Tyrosinase-Inhibiting Agent." *Journal of Cosmetic Science* 57 (4): 291–308.

Herodotus, and Francis R B Godolphin. 1973. "Herodotus: On the Scythians." *The Metropolitan Museum of Art Bulletin* 32 (5), 129–49. https://www.jstor.org/stable/3269235

Jayakumar, Kishore L, and Robert G Micheletti. 2017. "Robert Chesebrough and the Dermatologic Wonder of Petroleum Jelly." *JAMA Dermatology* 153 (11): 1157.

Kligman, L H, C H Duo, and A M Kligman. 1984. "Topical Retinoic Acid Enhances the Repair of Ultraviolet Damaged Dermal Connective Tissue." *Connective Tissue Research* 12 (2): 139–50. doi:10.3109/03008208408992779.

Lee, Mireille M. 2015. *Body, Dress, and Identity in Ancient Greece.* Cambridge University Press.

Lim, J T. 1999. "Treatment of Melasma Using Kojic Acid in a Gel Containing Hydroquinone and Glycolic Acid." *Dermatologic Surgery: Official Publication for American Society for Dermatologic Surgery [et Al.]* 25 (4): 282–84. doi:10.1046/j.1524-4725.1999.08236.x.

Mattern, Susan P. 2008. *Galen and the Rhetoric of Healing.* Baltimore, MD: Johns Hopkins University Press.

Meyerhof, Max. 1930. "The" Book of Treasure", an Early Arabic Treatise on Medicine." *Isis* 14 (1): 55–76.

Mitsuda, Takeshi, and H Taguchi. 1977. "Formation of Magnesium Silicate Hydrate and Its Crystallization to Talc." *Cement and Concrete Research* 7 (3): 223–30.

Nye, Mary Jo. 1999. *Before Big Science: The Pursuit of Modern Chemistry and Physics, 1800–1940.* Cambridge, MA; London: Harvard University Press.

Panati, Charles. 2016. *Panati's Extraordinary Origins of Everyday Things*. New York: Chartwell Books.

Parish, Lawrence Charles, and John Thorne Crissey. 1988. "Cosmetics: A Historical Review." *Clinics in Dermatology* 6 (3): 1–4.

Pavicic, Tatjana, Stephanie Steckmeier, Martina Kerscher, and Hans Christian Korting. 2009. "Evidence-Based Cosmetics: Concepts and Applications in Photoaging of the Skin and Xerosis." *Wiener klinische Wochenschrift* 121 (13–14): 431–39. doi:10.1007/s00508-009-1204-9.

"PCPC Timeline - Personal Care Products Council." 2022. Accessed May 24. https://www.personalcarecouncil.org/pcpc-timeline/.

Poucher, W A, and H Butler. 1993. *Poucher's Perfumes, Cosmetics and Soaps: Volume 3: Cosmetics*. Dordrecht: Springer. https://link.springer.com/book/10.1007/978-94-011-1482-0

"Poucher's Perfumes, Cosmetics and Soaps." 1993. *Poucher's Perfumes, Cosmetics and Soaps*. Springer Netherlands. doi:10.1007/978-94-011-1482-0.

Rapalli, Vamshi Krishna, Saswata Banerjee, Shahid Khan, Prabhat Nath Jha, Gaurav Gupta, Kamal Dua, Md Saquib Hasnain, Amit Kumar Nayak, Sunil Kumar Dubey, and Gautam Singhvi. 2021. "QbD-Driven Formulation Development and Evaluation of Topical Hydrogel Containing Ketoconazole Loaded Cubosomes." *Materials Science and Engineering: C* 119: 111548. doi:10.1016/j.msec.2020.111548.

Rapalli, Vamshi Krishna, Arisha Mahmood, Tejashree Waghule, Srividya Gorantla, Sunil Kumar Dubey, Amit Alexander, and Gautam Singhvi. 2021. "Revisiting Techniques to Evaluate Drug Permeation through Skin." *Expert Opinion on Drug Delivery* 18 (12): 1829–42. doi:10.1080/17425247.2021.2010702.

Rapalli, Vamshi Krishna, Gautam Singhvi, Sunil Kumar Dubey, Gaurav Gupta, Dinesh Kumar Chellappan, and Kamal Dua. 2018. "Emerging Landscape in Psoriasis Management: From Topical Application to Targeting Biomolecules." *Biomedicine & Pharmacotherapy* 106: 707–13. doi:10.1016/j.biopha.2018.06.136.

Rimmel, Eugene. 1867. *The Book of Perfumes*. London: Chapman & Hall.

Rosner, Fred. 2010. "Moses Maimonides: Biographic Outlines." *Rambam Maimonides Medical Journal* 1 (1): e0002. https://www.ncbi.nlm.nih.gov/pmc/articles/PMC3721660/

Rutkin, P. (1975). Eye Make-up. In G. deNavarre (Ed.), *The Chemistry and Manufacture of Cosmetics*, 2nd. ed., 709–740. Orlando, FL: Continental Press.

Sakamoto, Kazutami, Howard Lochhead, Howard Maibach, and Yuji Yamashita. 2017. *Cosmetic Science and Technology: Theoretical Principles and Applications*. Elsevier. https://www.sciencedirect.com/book/9780128020050/cosmetic-science-and-technology#book-description

Schlossman, M L, and A J Feldman. 1988. "Fluid Foundations and Blush Make-Up." In G. deNavarre (Ed.), *The Chemistry and Manufacture of Cosmetics* 2, 741–765. Wheaton, IL: Allured Publishing.

Singhvi, Gautam, Prachi Manchanda, Vamshi Krishna Rapalli, Sunil Kumar Dubey, Gaurav Gupta, and Kamal Dua. 2018. "MicroRNAs as Biological Regulators in Skin Disorders." *Biomedicine & Pharmacotherapy* 108: 996–1004. doi:https://doi.org/10.1016/j.biopha.2018.09.090.

Takahashi, Yutaka, Takumi Hayashi, Shinichi Watanabe, Kazuhiko Hayashi, Mirei Koike, Noriko Aisawa, Shinya Ebata, et al. 2009. "Inter-Laboratory Study of Short Time Exposure (STE) Test for Predicting Eye Irritation Potential of Chemicals and Correspondence to Globally Harmonized System (GHS) Classification." *The Journal of Toxicological Sciences* 34 (6): 611–26. doi:10.2131/jts.34.611.

Van Scott, E J, and R J Yu. 1989. "Alpha Hydroxy Acids: Procedures for Use in Clinical Practice." *Cutis* 43 (3): 222–28.

Waghule, Tejashree, Srividya Gorantla, Vamshi Krishna Rapalli, Pranav Shah, Sunil Kumar Dubey, Ranendra Narayan Saha, and Gautam Singhvi. 2020. "Emerging Trends in Topical Delivery of Curcumin Through Lipid Nanocarriers: Effectiveness in Skin Disorders." *AAPS PharmSciTech* 21 (7): 284. doi:10.1208/s12249-020-01831-9.

Wells, Frederick Victor, and Irwin Irville Lubowe. 1964. *Cosmetics and the Skin*. New York; Amsterdam: Reinhold Publishing Corporation.

Wetzel, Christine L. 2012. "Permanent Cosmetics." *Plastic Surgical Nursing* 32 (3): 117–19.

Witkowski, Joseph A, and Lawrence Charles Parish. 2001. "You've Come a Long Way Baby: A History of Cosmetic Lead Toxicity." *Clinics in Dermatology* 19 (4): 367–70.

Wortzman, M S. 1991. "Evaluation of Mild Skin Cleansers." *Dermatologic Clinics* 9 (1): 35–44.

Zappelli, Claudia, Ani Barbulova, Fabio Apone, and Gabriella Colucci. 2016. "Effective Active Ingredients Obtained through Biotechnology." *Cosmetics* 3 (4): 39.

2 Transdermal drug absorption
Mathematical modelling

*Urushi Rehman, Amirhossein Sahebkar,
Nazeer Hasan, Gaurav K. Jain, Waleed
H. Almalki and Prashant Kesharwani*

CONTENTS

2.1 INTRODUCTION

Cosmetics are defined as "preparations that are meant for human applications intended for the purpose of purifying, enhancing, facilitating attractiveness, or modifying the look without affecting the human body's physiology or functions" [1,2]. Since they are widely known consumer goods in the marketplace, it is a lucrative area for various key manufacturers [3]. It is forecasted that the worldwide cosmetic real economy, valued at approximately 500 billion US dollars in 2017, is predicted to increase to 800 billion US dollars by 2023 [4]. These statistics have prompted cosmetic companies to create innovative and efficient products containing novel materials. In today's world, the application of nanoscience to enhance the performance of cosmetics is a huge hit, contributing to further breakthroughs in cosmetic science.

DOI: 10.1201/9781003319146-2

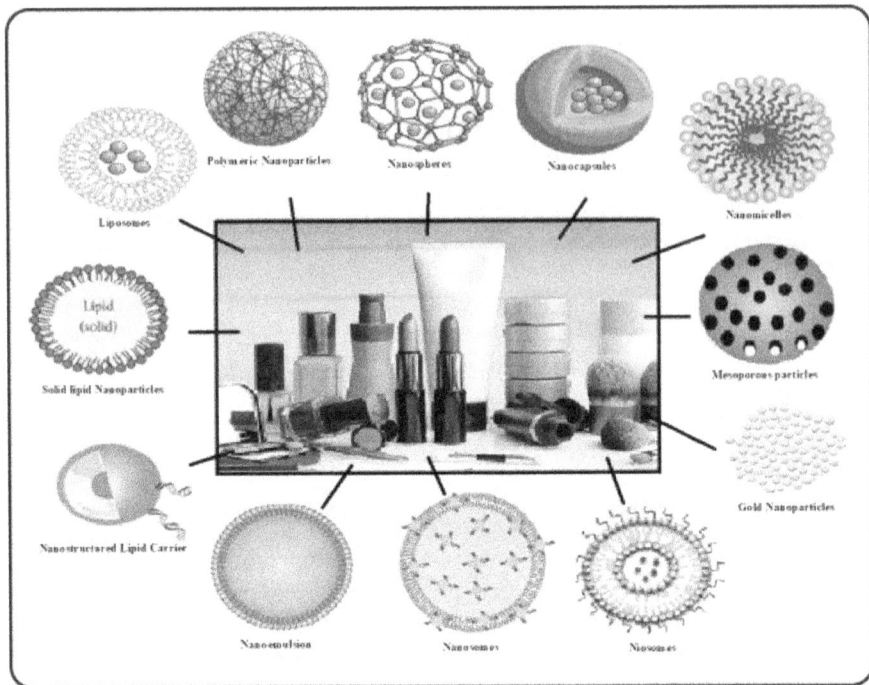

FIGURE 2.1 Types of nanoparticle systems used in cosmetic preparations. Reproduced with permission from [6]

As nano-materials attract popularity in the medical sector, the cosmetic sector is assumed to be the most expansive in integrating nano-materials into one's products. From 2006 to 2013, the number of cosmetic products based on nanotechnology grew by nearly 516% [5]. A nanocosmetic is any cosmetic preparation that contains nanoparticles. An entity including all facets in the nanometre range is considered a nanoparticle (Figure 2.1).

The particle size in the range of 1–100 nm is referred to as nanoscale [5,7–13]. Nanocosmetic applications include treatments for hair issues, wrinkles, age-related disorders, and skin problems [4]. Given its enormous surface area, human skin covers a large portion of the body. Skin is in control of safeguarding, sensory perception and regulation of other bodily functions. Human skin has an average surface area of 1.6–1.9 m² [14–20]. As a permeation barrier, the stratum corneum (SC) of the skin provides an excellent delivery route for therapeutic and cosmetic agents. In both transdermal and topical delivery, the moieties can enter the skin or subcutaneous layer and thereafter enter the portal circulation via the established routes [21,22]. The nanocarriers can either migrate to the skin without deterioration or deteriorate close to the surface of the skin, enabling the entrapped moiety to permeate through the skin layers. The delivery route for nanocosmetics depends on the interactions between nanoparticles governed by their size, surface characteristics, physicochemical properties, and drug encapsulation efficiency [22]. Diving into transdermal delivery research is crucial for pharmaceutical understanding regarding their delivery processes. It is also necessary to understand the time course of pre-pared analytes permeating through the skin in characterizing both chronic and acute exposure of the user to a potential hazardous analyte [23]. The most accurate way to explain this time course is to demonstrate the function as a mathematical model. The goal of every mathematical model is to fully reflect absorption and distribution mechanisms, summarize empirical values with mathematical functions, and predict mechanisms under different conditions. But, when it comes to explaining the mecha-nisms involved, some models are quite often too complicated to be practical [24]. We attempted to comprehend the mathematical modeling available for transdermal absorption in this chapter.

2.2 TRANSDERMAL DRUG DELIVERY (TDD): OVERVIEW

TDD is a hassle- and pain-free application of the formulation to healthy undamaged skin for systemic drug transport [25,26]. The drug infiltrates the SC first, then deeper epithelium and subcutaneous layer, with no significant accumulation in the dermal layer entering the dermal layer, the drug becomes available to systemic circulation for absorption via dermal microvasculature [27,28] (Figure 2.2). TDD offers several benefits over other traditional routes for drug delivery such as a non-invasive substi-tute for parenteral delivery [25], uniform pharmacokinetic profiles [25], reducing the risk of adverse events, improved patient compliance, and appropriate for uncoopera-tive patients (e.g. unconscious or nauseous) [29] and most importantly avoidance of hepatic-first metabolism thus enhancing bioavailability [30,31]. Skin can be consid-ered a novel location for vaccination as it is rich in dendritic cells in both the dermal and epidermal layers, which is important in immunity development, thus enabling TDD as an intriguing route for a vaccination with therapeutic peptides and proteins [32]. The demand for a cost-effective, non-invasive approach to vaccination, particu-larly in developing countries, has compelled researchers to an exhaustive investiga-tion into the creation of such TDD systems for vaccination purposes [32].

Transdermal Drug Delivery Systems

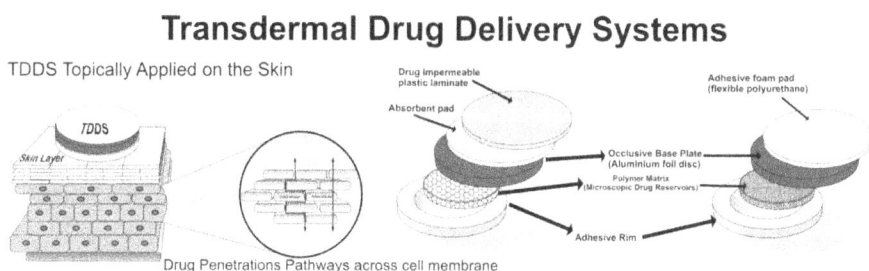

TDDS Topically Applied on the Skin

Drug impermeable plastic laminate

Absorbent pad

Adhesive foam pad (flexible polyurethane)

TDDS

Skin Layer

Occlusive Base Plate (Aluminium foil disc)

Polymer Matrix (Microscopic Drug Reservoirs)

Adhesive Rim

Drug Penetrations Pathways across cell membrane

FIGURE 2.2 Pictorial representation of a transdermal drug delivery system. Reproduced with permission from [33]

2.3 SUMMARY OF SKIN STRUCTURE

Skin with a surface area of 1.7 m², is the largest and most readily accessible organ of the body, accounting for 16% of an average individual's entire body mass [34–36]. The principal role of the skin is to shield the body from pathogens, UV rays, toxins, allergens, and dehydration [37]. The skin is differentiated into different layers starting from the outermost layer, i.e., the epidermis, containing the SC; the middle layer, i.e., the dermis and the hypodermis is the innermost layer (Figure 2.3) [38].

2.3.1 THE EPIDERMIS

The epidermis is the skin's outermost layer, varying in thickness from 0.8 mm through the entire body [37]. The epidermis' cellular content is dominated by keratinocytes, followed by other cells such as Langerhans cells, melanocytes, and Merkel cells [32]. The SC is the epidermis's surface layer that comes into direct proximity to the outside environment [39]. Its protective qualities might be explained in part by its limited hydration and exceptionally high density (1.4 g/cm³ in the dry state) [40]. The SC is primarily made up of lipid (minor-20%) and insoluble keratins (major-70%) [40].

2.3.2 THE DERMIS

The dermis has a thickness of about 2–3 mm and is made up of collagen fibres (70%) and elastin fibres, thus providing the skin with its apparent strength and flexibility [35]. Dermal blood vessels supply nutrients to both the epidermis and dermis. The dermal layer also contains phagocytic cells, nerves, and lymphatic vessels for other necessary functions [39].

2.3.3 THE HYPODERMIS

The hypodermis, also known as the subcutaneous layer, is the deepest layer beneath the skin composed of a network of adipocytes [35]. It connects with the underlying tissues within the body and thus majorly acts as a physical shock absorber, insulation, and transmitting of the vascular and neural signals [41]. Fat cells residing in hypodermis represent about 50% of total body fat, with macrophages and fibroblasts being the other major cells [42].

2.3.4 ROUTES OF DRUG PENETRATION ACROSS SKIN

There are two potential routes for drug permeation along the exposed skin, which are as follows (Figure 2.4):

1. Trans-epidermal path (across the SC)
2. Trans-follicular route (shunt pathway – Uptake through skin appendages such as hair follicles, apocrine, and eccrine glands)

The primary factors influencing the choice of the route are the drug's physicochemical properties and the type of the formulation [44].

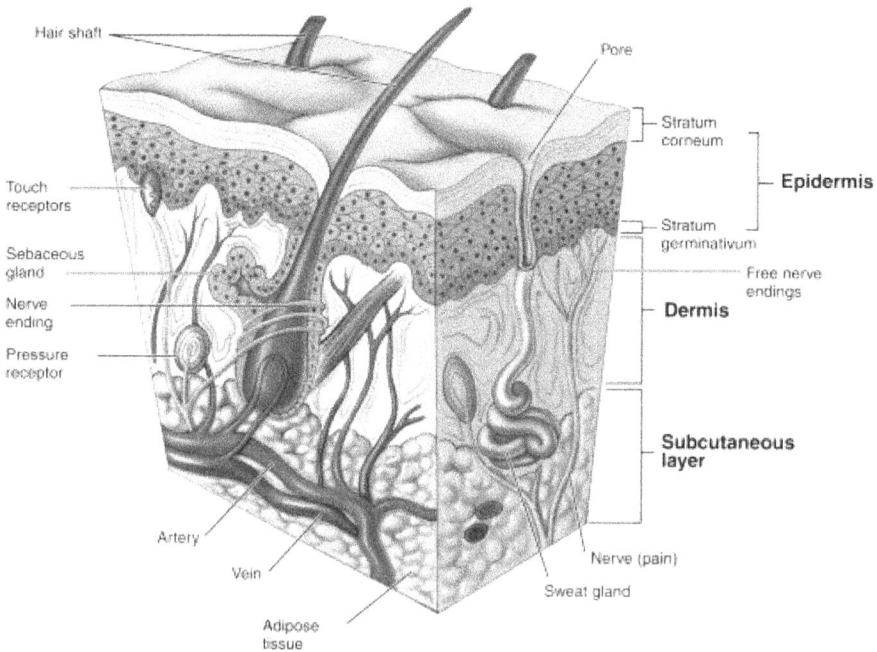

FIGURE 2.3 Structure of skin. Reproduced with permission from [43]

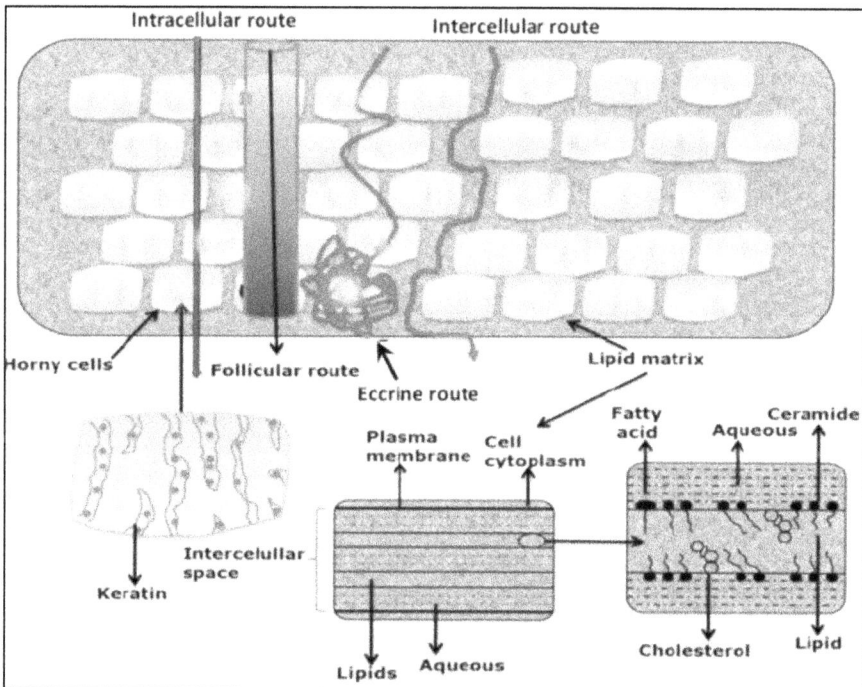

FIGURE 2.4 Different types of routes of drug penetration across human skin. Reproduced with permission from [45]

2.4   KINETICS OF TRANSDERMAL ABSORPTION

Deep knowledge of the kinetics of transdermal delivery is essential for the progression of TDD systems. A systemic drug must possess certain physicochemical characteristics that enable it to pass through the skin and gain entry to circulation to access the target tissue. The evaluation of transdermal absorption of moieties is a critical step in assessing any TDD. The infiltration of materials into multiple skin layers and penetrability across the skin into the blood circulation is referred to as percutaneous/transdermal absorption. The steps listed below are involved in transdermal absorption starting from the release of a drug molecule from a Transdermal Drug Delivery Systems (TDDS) till it gets transported to the systemic circulation (Figure 2.5) [46–48]:

  i. Dissolution from the formulation and release of the therapeutic moiety;
  ii. Penetration: The entry of therapeutic moiety into a specific layer of the skin;
  iii. Partitioning from the SC into the hydrophilic epidermis;
  iv. Diffusion: The movement of the molecule through the epidermis to the outer dermis;
  v. Permeation: The movement of the molecule between layers;
  vi. Absorption: Entry of moiety into the systemic circulation.

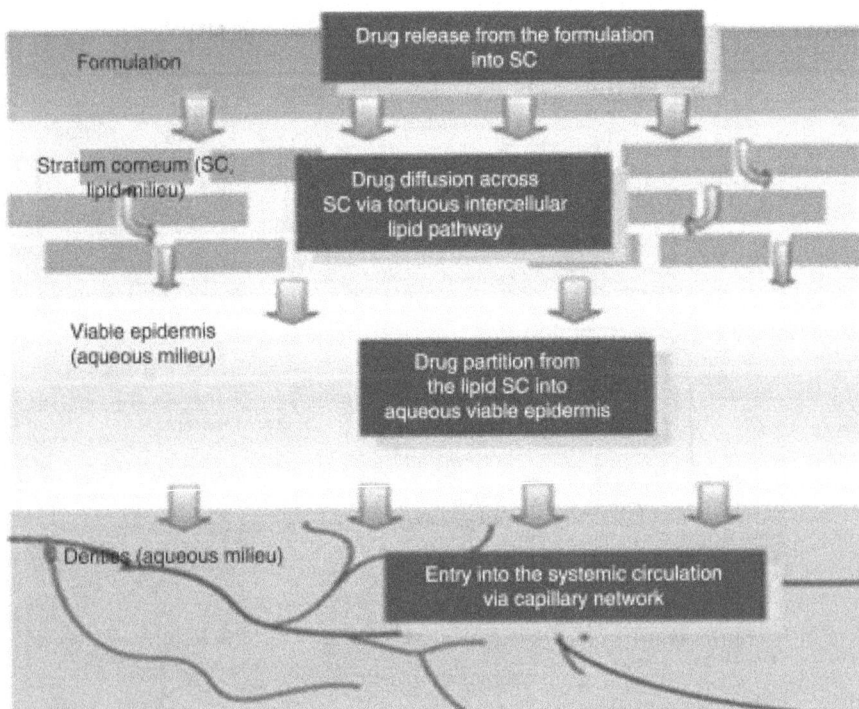

FIGURE 2.5   Stepwise events involved in absorption of drug from TDDS. Reproduced with permission from [49]

The estimation of the drug's maximum flux (J) across the skin which is usually expressed in units of g/cm²/h), is the starting point for evaluating the kinetics of any TDDS. According to Fick's law of diffusion, therapeutic moieties will be transported across the skin until the concentration gradient disappears [50]. The membrane limited flux (J) under steady state condition is explained by the equation:

$$J = \frac{DK_{o/w}C}{h}$$

Where,

    J = Amount of drug that passes through the membrane system per unit area per unit time;

    D = Membrane diffusion coefficient;

    h = Thickness of the membrane;

    K = Partition coefficient;

    C = Concentration gradient along the membrane.

2.5   MATHEMATICAL MODELLING

TDD absorption mathematical models are extremely crucial for gaining a working knowledge of bio-transport mechanisms in addition to assessing dermal exposure to industrial and ambient hazards. In the 1940s and 1970s, the framework for predictive modelling for transdermal was laid down. It was identified at this time that partitioning and solubility were significant factors in percutaneous absorption.

2.5.1   MODELS BEFORE 1990s

Most of the initial experiments relied on the analysis of similar, or closely linked, sequences of compounds, where typically merely a small number of compounds were tested, with Higuchi's research on the release of drugs from ointment being an exception [51]. Greater hydrophobicity and skin permeability were found to be inversely correlated in many of these investigations [52,53]. Although these studies demonstrated that it was possible to generate predictive modeling of percutaneous absorption, every model only applied to a particular chemical group or class of compounds. This limitation, however, was discovered much later when research concentrated on the examination of smaller datasets taken from larger datasets; this issue is covered in more detail below. Moss and co-workers recognized the issues with existing analysis, like the lack of any variability in the dataset and the difficulty of dissociating co-linear variables, such as hydrophobicity and molecular weight [54].

2.5.2   INITIAL QUANTITATIVE MODELS

Since about 20 years ago, mathematical modeling of skin penetration has helped us comprehend how substances move across the skin and the SC in particular. It is well established that a particle's physicochemical characteristics affect how effectively it penetrates and moves through the skin. Conclusions from experiments, especially those which looked at closely linked or homologous series, have been the basis of most of the research conducted in this area before 1990.

The work conducted by Brown and Rossi is one exception to the aforementioned statement [55].They devised a straightforward model that was only based on the SC's portrayal as a straightforward lipophilic barrier. They created two quantitative correlations between the penetrant's permeability and lipophilicity:

$$K_p = 0.1\left[\frac{p_{oct}^{0.75}}{120 + p_{oct}^{0.75}}\right] \tag{2.1}$$

$$I = C_w \cdot A \cdot T \cdot 0.2\left[\frac{P_{oct}^{0.75}}{120 + p_{oct}^{0.75}}\right] \tag{2.2}$$

Where,

K_p = Permeability coefficient (cm/hr);
I = Dermally absorbed intake (mg);
C_w = Penetrant's concentration in water (mg/cm³);
A = Area of skin subjected to permeation (cm²);
T = Duration of exposure (hr);
P_{oct} = Partition coefficient between octanol and water.

Two significant studies—those by Flynn [56] and Potts and Guy—were conducted after this work. These research projects represented important turning points in the effort to create a statistical model of percutaneous absorption. Except for the in vivo investigations for toluene, ethylbenzene, and styrene, Flynn presented a dataset of 97 permeability coefficients for 94 compounds across human skin in vitro from 15 different existing literature. This gave rise to the first sizable database of skin permeability values obtained in a single species—and, up until recently, the largest. Flynn's study of this data led to some findings that demonstrate a distinct correlation between a permeant's permeability and its lipophilicity (log P) and size. Flynn outlined a set of straightforward algorithms for low and high molecular weight compounds that clearly stated that different hydrophobicity-dependent QSARs might be used to estimate skin permeability for high and low molecular weight compounds and that highly hydrophilic and hydrophobic compounds had low and high skin permeability, respectively. In their groundbreaking work, Potts and Guy [57] subsequently verified this link between permeability and a molecule's physical characteristics. In Flynn's dataset, 93 compounds were observed to have the following relationship:

$$\log K_p = 0.71\log K_{ow} - 0.0061MW - 6.3$$

$$\left[n = 93; \ r^2 = 0.67; \ s \text{ not reported}; \ F \text{ not reported}\right]$$

Where,

K_p = Permeability coefficient (cm/s);
$\log K_{ow}$ = log 10 octanol-water partition coefficient;
MW = Molecular weight;

N = Number of observations;
r = Correlation coefficient;
s = Standard error of the estimate;
F = Fisher's statistic.

2.5.3 FICKIAN MODELS

Compartmental models, also widely known as pharmacokinetic (PK) skin models, are recurrently utilized for the investigation of moieties that enter or exit the body. The PK models assume the body and whole skin well-distributed compartments of homogeneous concentration that function as storage for drugs/moieties, and the transfer between compartments is represented via first-order rate constants. The PK model represents the penetration across the skin by a series of compartments imitating the partitioning and diffusion mechanisms through the SC, or as multiple compartments that differentiate the lipid-soluble SC and water-soluble epidermis separately [24]. Although the majority of the emphasis in the field of transdermal modelling has been on discussing diffusion mechanisms in the SC, it has recently been recognized that other mechanisms such as binding and metabolism as well play a key role in assessing the rate of drug uptake. Binding is particularly important since many compounds attach to keratin, affecting their permeability across the SC [58]. Roberts and co-workers discussed this binding effect on transdermal transport related to epidermal penetration, taking into account the kinetic parameters associated with the SC reservoir effect. In their study, they assumed that binding is quick, which means that equilibrium establishes at a faster rate between unbound and bound states in comparison to diffusion. The benefit of this approach is the fact that modelling is relatively easier in this case, with the diffusion coefficient D in the diffusion equation being replaced by an effective diffusion coefficient D_{eff}, where D_{eff} ¼fuD and fu is the fraction of unbound solute. Because the fraction unbound is less than one, binding causes slower diffusion and thus longer lag times [59].

2.5.4 NON-FICKIAN MODELS

The Fickian model projected distribution of the drug in the skin and vehicle frequently contradicts experimental proof, according to outcomes of experiments. Barbeiro and Ferreira recently devised a non-Fickian mathematical model for the transdermal absorption problem.

[60]. They included a non-Fickian component described with a relaxation parameter (τJ) related to the characteristics of the components, Fick's equation for the flow is altered in this new model. This value is comparable to the heat flux's relaxation time (τq) for the related heat wave diffusion problem. Hence, they got:

$$J = -D\frac{\partial c}{\partial x} - \tau_J\frac{\partial J}{\partial t}$$

A system of differential equations with a compatible condition on the boundary between the two aspects of the physical model was created by integrating the flow equation as well as the conservation of mass law. Instead of the differential PDE, a hyperbolic PDE can be obtained, as demonstrated by Haji-Sheikh and associates [61]. The continuous version of the mathematical model was introduced by Barbeiro and Ferreira to resolve it, and the difficult stability and convergent characteristics of the discrete system were investigated through the examination of numerical experiments.

2.5.5 OTHER MODELS

Tojo created a bi-layer skin/two-compartmental model which serves as the foundation for the development of a dynamic mathematical model for transdermal administration. Simulated impacts on the penetration rate time profile include those of the metabolic reaction in the viable skin, the drug binding and reservoir function in SC, and the solubility and diffusivity of the drug in the skin. An analysis was also done on how the pharmacokinetic factors affect the plasma concentration profile. The model was able to forecast the plasma concentration following transdermal drug delivery in addition to measuring the rate of skin permeability [62].

By using computational methods, particularly the Gaussian Process, on a sizable skin permeability database, Moss and associates adopted an unconventional approach for simulating skin permeability [63]. They managed to achieve this by collaborating with a diversified team of researchers with knowledge spanning all facets of the study and its methods, including statistics and mathematics, as advised by Cronin and Schultz [64]. As a result, the researchers first looked at the fundamental properties of the permeability dataset they used (which was an extension of the Flynn dataset), using both straightforward data visualization techniques and more sophisticated ones like principal and canonical component analysis. They showed that the skin database is inherently non-linear, which not only supported their choice to use non-linear methods of evaluation but also stands in sharp contrast to the vast number of studies in this field that have employed (multiple) linear regression modeling and comparable techniques.

2.6 CONCLUSION AND OUTLOOK

The estimation of transcutaneous absorption is a crucial component in the analysis of TDDS. Recognizing the variables that influence in vivo activity is a prime goal in the development and optimization of transdermal formulations. Considering the biochemical and structural ambiguity of the skin, in addition to the restrictions of in vivo and also in vitro experiments, mathematical modelling provides an alternative option that, despite assumptions that are frequently grossly simplified, could still provide some perspectives on the trend and impact of relevant factors. To that end, this chapter has given a comparatively brief and interpretive description of the field of TDD absorption modelling. The modelling of TDD absorption has dramatically improved our comprehension of the processes underlying skin permeation. The rising adoption of new techniques for the modelling of transcutaneous absorption, including Raman spectroscopy, dermato-pharmacokinetics, and mass spectrometry-based imaging technology, can provide precise and accurate endpoints unavailable to several investigators yet, and therefore will enable the possibility of modelling such anomalies in full depth in the coming years.

REFERENCES

1. Summary of Cosmetics Labeling Requirements | FDA (n.d.).
2. S.K. Dubey, A. Dey, G. Singhvi, M.M. Pandey, V. Singh, P. Kesharwani, Emerging trends of nanotechnology in advanced cosmetics, *Colloids Surf. B Biointerf.* 214 (2022) 112440. https://doi.org/10.1016/J.COLSURFB.2022.112440.
3. Cosmetic Products Market Size, Share, Growth, Trends (2022–27) (n.d.).
4. V. Dhapte-Pawar, S. Kadam, S. Saptarsi, P.P. Kenjale, Nanocosmeceuticals: Facets and aspects, *Futur. Sci. OA* 6 (2020). https://doi.org/10.2144/fsoa-2019-0109.
5. M. Ajazzuddin, G. Jeswani, A. Jha, Nanocosmetics: Past, present and future trends, *Recent Patents Nanomed.* 5 (2015). https://doi.org/10.2174/1877912305666150417232826.
6. S. Dhawan, P. Sharma, S. Nanda, Cosmetic nanoformulations and their intended use, *Nanocosmetics* (2020) 141–69. https://doi.org/10.1016/B978-0-12-822286-7.00017-6.
7. P. Kesharwani, K. Jain, N.K. Jain, Dendrimer as nanocarrier for drug delivery, *Prog. Polym. Sci.* 39 (2014) 268–307. https://doi.org/10.1016/j.progpolymsci.2013.07.005.
8. M. Fatima, A. Sheikh, N. Hasan, A. Sahebkar, Y. Riadi, P. Kesharwani, Folic acid conjugated poly(amidoamine) dendrimer as a smart nanocarriers for tracing, imaging, and treating cancers over-expressing folate receptors, *Eur. Polym. J.* 170 (2022) 111156. https://doi.org/10.1016/J.EURPOLYMJ.2022.111156.
9. V. Singh, S. Md, N.A. Alhakamy, P. Kesharwani, Taxanes loaded polymersomes as an emerging polymeric nanocarrier for cancer therapy, *Eur. Polym. J.* 162 (2022) 110883. https://doi.org/10.1016/J.EURPOLYMJ.2021.110883.
10. F. Zeeshan, T. Madheswaran, J. Panneerselvam, R. Taliyan, P. Kesharwani, Human serum albumin as multifunctional nanocarrier for cancer therapy, *J. Pharm. Sci.* (2021). https://doi.org/10.1016/j.xphs.2021.05.001.
11. S. Gorantla, G. Wadhwa, S. Jain, S. Sankar, K. Nuwal, A. Mahmood, S.K. Dubey, R. Taliyan, P. Kesharwani, G. Singhvi, Recent advances in nanocarriers for nutrient delivery, *Drug Deliv. Transl. Res.* (2021). https://doi.org/10.1007/S13346-021-01097-Z.
12. S. Kumar Dubey, R. Pradhan, S. Hejmady, G. Singhvi, H. Choudhury, B. Gorain, P. Kesharwani, Emerging innovations in nano-enabled therapy against age-related macular degeneration: A paradigm shift, *Int. J. Pharm.* 600 (2021) 120499. https://doi.org/10.1016/j.ijpharm.2021.120499.
13. A. Sheikh, S. Md, N.A. Alhakamy, P. Kesharwani, Recent development of aptamer conjugated chitosan nanoparticles as cancer therapeutics, *Int. J. Pharm.* 620 (2022) 121751. https://doi.org/10.1016/J.IJPHARM.2022.121751.
14. F. Erdo, N. Hashimoto, G. Karvaly, N. Nakamichi, Y. Kato, Critical evaluation and methodological positioning of the transdermal microdialysis technique. A review, *J. Control. Release* 233 (2016) 147–61. https://doi.org/10.1016/J.JCONREL.2016.05.035.
15. Medical definition of body surface area (n.d.).
16. T. Madheswaran, R. Baskaran, B.K. Yoo, P. Kesharwani, In vitro and in vivo skin distribution of 5α-reductase inhibitors loaded into liquid crystalline nanoparticles, *J. Pharm. Sci.* (2017). https://doi.org/10.1016/j.xphs.2017.06.016.
17. H. Kaur, P. Kesharwani, Advanced nanomedicine approaches applied for treatment of skin carcinoma, *J. Control. Release* 337 (2021) 589–611. https://doi.org/10.1016/J.JCONREL.2021.08.003.
18. A.K. Jain, S. Jain, M.A.S. Abourehab, P. Mehta, P. Kesharwani, An insight on topically applied formulations for management of various skin disorders, (2022) 1–27. https://doi.org/10.1080/09205063.2022.2103625.
19. S. Md, S.K. Bhattmisra, F. Zeeshan, N. Shahzad, M.A. Mujtaba, V. Srikanth Meka, A. Radhakrishnan, P. Kesharwani, S. Baboota, J. Ali, Nano-carrier enabled drug delivery systems for nose to brain targeting for the treatment of neurodegenerative disorders, *J. Drug Deliv. Sci. Technol.* 43 (2018) 295–310. https://doi.org/10.1016/j.jddst.2017.09.022.

20. S. Kumari, P.K. Choudhary, R. Shukla, A. Sahebkar, P. Kesharwani, Recent advances in nanotechnology based combination drug therapy for skin cancer, (2022) 1–34. https://doi.org/10.1080/09205063.2022.2054399.

21. A. Verma, A. Jain, P. Hurkat, S.K. Jain, Transfollicular drug delivery: Current perspectives, *Res. Reports Transdermal Drug Deliv.* 5 (2016) 1–17. https://doi.org/10.2147/RRTD.S75809.

22. K.S. Paudel, M. Milewski, C.L. Swadley, N.K. Brogden, P. Ghosh, A.L. Stinchcomb, Challenges and opportunities in dermal/transdermal delivery, *Ther. Deliv.* 1 (2010) 109–31. https://doi.org/10.4155/tde.10.16.

23. Y.G. Anissimov, M.S. Roberts, Mathematical models for topical and transdermal drug products, *Top. Drug Bioavail. Bioequiv. Penetr.* (2014) 249–98. https://doi.org/10.1007/978-1-4939-1289-6_15.

24. Y.G. Anissimov, J.J. Calcutt, M.S. Roberts, Mathematical models in percutaneous absorption, *Percutaneous Absorpt.* (2021) 9–52. https://doi.org/10.1201/9780429202971-2.

25. T. Han, D.B. Das, Potential of combined ultrasound and microneedles for enhanced transdermal drug permeation: A review, *Eur. J. Pharm. Biopharm.* 89 (2015) 312–28. https://doi.org/10.1016/j.ejpb.2014.12.020.

26. C.M. Schoellhammer, D. Blankschtein, R. Langer, Skin permeabilization for transdermal drug delivery: Recent advances and future prospects, *Expert Opin. Drug Deliv.* 11 (2014) 393–407. https://doi.org/10.1517/17425247.2014.875528.

27. R.F. Donnelly, T.R.R. Singh, D.I.J. Morrow, A.D. Woolfson, Microneedle-mediated transdermal and intradermal drug delivery - Ryan F. Donnelly, Thakur Raghu Raj Singh, Desmond I. J. Morrow, A. David Woolfson - Google Books, Microneedle-Mediated Transdermal Intradermal Drug Deliv. (2012).

28. K. Kretsos, G.B. Kasting, A geometrical model of dermal capillary clearance, *Math. Biosci.* 208 (2007) 430–53. https://doi.org/10.1016/J.MBS.2006.10.012.

29. M.R. Prausnitz, R. Langer, Transdermal drug delivery, *Nat. Biotechnol.* 2008 2611. 26 (2008) 1261–68. https://doi.org/10.1038/nbt.1504.

30. D. Brambilla, P. Luciani, J.C. Leroux, Breakthrough discoveries in drug delivery technologies: The next 30 years, *J. Control. Release* 190 (2014) 9–14. https://doi.org/10.1016/J.JCONREL.2014.03.056.

31. K.B. Ita, Transdermal drug delivery: Progress and challenges, *J. Drug Deliv. Sci. Technol.* 3 (2014) 245–250. https://doi.org/10.1016/S1773-2247(14)50041-X.

32. H. Suh, J. Shin, Y.-C. Kim, Microneedle patches for vaccine delivery, *Clin. Exp. Vaccine Res.* 3 (2014) 42. https://doi.org/10.7774/CEVR.2014.3.1.42.

33. A. Alexander, S. Dwivedi, Ajazuddin, T.K. Giri, S. Saraf, S. Saraf, D.K. Tripathi, Approaches for breaking the barriers of drug permeation through transdermal drug delivery, *J. Control. Release* 164 (2012) 26–40. https://doi.org/10.1016/J.JCONREL.2012.09.017.

34. G.K. Menon, New insights into skin structure: Scratching the surface, *Adv. Drug Deliv. Rev.* 54 (2002) S3. https://doi.org/10.1016/S0169-409X(02)00121-7.

35. X. Liu, P. Kruger, H. Maibach, P.B. Colditz, M.S. Roberts, Using skin for drug delivery and diagnosis in the critically ill, *Adv. Drug Deliv. Rev.* 77 (2014) 40–49. https://doi.org/10.1016/J.ADDR.2014.10.004.

36. A.C. Williams, B.W. Barry, Penetration enhancers, *Adv. Drug Deliv. Rev.* 64 (2012) 128–37. https://doi.org/10.1016/j.addr.2012.09.032.

37. H.A.E. Benson, M.E. Lane, P. Santos, A.C. Watkinson, J. Hadgraft, Passive skin penetration enhancement, *Top. Transdermal Drug Deliv.* (2012) 3–38.

38. M. Bhowmick, T. Sengodan, Mechanisms, kinetics and mathematical modelling of transdermal permeation – an updated review, *Int. J. Comprehens. Pharm.* 04 (2013) 4–7.

39. C.L. Domínguez-Delgado, I.M. Rodríguez-Cruz, M. López-Cervantes, The skin: A valuable route for administration of drugs, *Curr. Technol. Increase Transderm. Deliv. Drugs* (2010) 1–22. https://doi.org/10.2174/978160805191511001010001.

40. G.M. El Maghraby, B.W. Barry, A.C. Williams, Liposomes and skin: From drug delivery to model membranes, *Eur. J. Pharm. Sci.* 34 (2008) 203–22. https://doi.org/10.1016/j.ejps.2008.05.002.

41. A. Sherwood, J.K. Bower, J. McFetridge-Durdle, J.A. Blumenthal, L.K. Newby, A.L. Hinderliter, Age moderates the short-term effects of transdermal 17β-estradiol on endothelium-dependent vascular function in postmenopausal women, *Arterioscler. Thromb. Vasc. Biol.* 27 (2007) 1782–87. https://doi.org/10.1161/ATVBAHA.107.145383.

42. D.N. McLennan, C.J.H. Porter, S.A. Charman, Subcutaneous drug delivery and the role of the lymphatics, *Drug Discov. Today Technol.* 2 (2005) 89–96. https://doi.org/10.1016/J.DDTEC.2005.05.006.

43. J.M. Reicherter, Anatomy and physiology for polygraph examiners, *Fundam. Polygr. Pract.* (2015) 29–60. https://doi.org/10.1016/B978-0-12-802924-4.00002-5.

44. A.Z. Alkilani, M.T.C. McCrudden, R.F. Donnelly, Transdermal drug delivery: Innovative pharmaceutical developments based on disruption of the barrier properties of the stratum corneum, *Pharmaceutics* 7 (2015) 438–70. https://doi.org/10.3390/pharmaceutics7040438.

45. Y. Shahzad, R. Louw, M. Gerber, J. Du Plessis, Breaching the skin barrier through temperature modulations, *J. Control. Release* 202 (2015) 1–13. https://doi.org/10.1016/J.JCONREL.2015.01.019.

46. S.H. Curry, Novel drug delivery systems, Y. W. Chien. New York, Marcel Dekker Inc., No. of pages: 648. Price $65.00, *Biopharm. Drug Dispos.* 4 (1983) 405. https://doi.org/10.1002/BDD.2510040414.

47. V. Rastogi, P. Yadav, Transdermal drug delivery system: An overview, *Asian J. Pharm.* 6 (2012) 161–70. https://doi.org/10.4103/0973-8398.104828.

48. K.A. Walters, Dermatological and Transdermal Formulations, edited by (2002). Boca Raton: CRC Press. DOI https://doi.org/10.1201/9780824743239

49. T.A. Sonia, C.P. Sharma, Routes of administration of insulin, *Oral Deliv. Insul.* (2014) 59–112. https://doi.org/10.1533/9781908818683.59.

50. S. Wiedersberg, R.H. Guy, Transdermal drug delivery: 30+ years of war and still fighting!, *J. Control. Release* 190 (2014) 150–56. https://doi.org/10.1016/J.JCONREL.2014.05.022.

51. U.S. Denton, J.J. Quinones, Rate of release of medicaments from ointment bases containing drugs in suspension, *J. Pharm. Sci.* 50 (1961) 874–75. https://doi.org/10.1002/JPS.2600501018.

52. R.J. Scheuplein, I.H. Blank, Permeability of the skin, *Physiol. Rev.* 51 (1971) 702–47. https://doi.org/10.1152/PHYSREV.1971.51.4.702.

53. M.S. Roberts, R.A. Anderson, J. Swarbrick, Permeability of human epidermis to phenolic compounds, *J. Pharm. Pharmacol.* 29 (1977) 677–83. https://doi.org/10.1111/J.2042-7158.1977.TB11434.X.

54. G.P. Moss, J.C. Dearden, H. Patel, M.T.D. Cronin, Quantitative structure-permeability relationships (QSPRs) for percutaneous absorption, *Toxicol. Vitr.* 16 (2002) 299–317. https://doi.org/10.1016/S0887-2333(02)00003-6.

55. S.L. Brown, J.E. Rossi, A simple method for estimating dermal absorption of chemicals in water, *Chemosphere* 19 (1989) 1989–2001. https://doi.org/10.1016/0045-6535(89)90022-2.

56. T. Gerrity, C. Henry, Principles of route-to-route extrapolation for risk assessment: Proceedings of the Workshops on Principles of Route-to-Route Extrapolation for Risk Assessment held March 19–21, 1990, in Hilton Head, South Carolina and July 10–11, 1990, in Durham, North C.

57. R.O. Potts, R.H. Guy, Predicting skin permeability, *Pharm. Res.* 9 (1992) 663–69. https://doi.org/10.1023/A:1015810312465.

58. P. Liu, W.I. Higuchi, A.H. Ghanem, W.R. Good, Transport of beta-estradiol in freshly excised human skin in vitro: Diffusion and metabolism in each skin layer, *Pharm. Res.* 11 (1994) 1777–84. https://doi.org/10.1023/A:1018975602818.

59. A.C. Watkinson, K.R. Brain, Basic mathematical principles in skin permeation, in: *Dermatological and Transdermal Formulatios* (2002) 79–106. https://doi.org/10.1201/9780824743239-6.

60. S. Barbeiro, J.A. Ferreira, Coupled vehicle–skin models for drug release, *Comput. Methods in Appl. Mech.* 198 (2009) 2078–86. https://doi.org/10.1016/J.CMA.2009.02.002.

61. A. Haji-Sheikh, F. De Monte, J. V. Beck, Temperature solutions in thin films using thermal wave Green's function solution equation, *Int. J. Heat Mass Transf.* 62 (2013) 78–86. https://doi.org/10.1016/J.IJHEATMASSTRANSFER.2013.02.036.

62. K. Tojo, Mathematical modeling of transdermal drug delivery, *J. Chem. Eng. Japan* 20 (1987) 300–8. https://doi.org/10.1252/JCEJ.20.300.

63. G.P. Moss, S.C. Wilkinson, Y. Sun, G. Moss, A. Uk, Mathematical modelling of percutaneous absorption CORE view metadata, citation and similar papers at core (2012). https://doi.org/10.1016/j.cocis.2012.01.002

64. M.T.D. Cronin, T.W. Schultz, Pitfalls in QSAR, *J. Mol. Struct. Theochem.* 622 (2003) 39–51. https://doi.org/10.1016/S0166-1280(02)00616-4.

3 Nanocosmetics-principle and classification

Anuradha Dey and Sunil Kumar Dubey

CONTENTS

3.1 INTRODUCTION

Cosmetics are preparations that are used for beautifying and increasing the attractiveness of the individual. Nanocosmetics are cosmetic preparations in which the particles are nano-sized, falling in the range of 1–100 nm (Anu Mary Ealia and Saravanakumar 2017). Nanoparticles for cosmetics can be categorized into two types, the first are the ones that are biodegradable in nature, such that they undergo disintegration upon being applied to the skin. They comprise nanosystems such as liposomes or micro/nanoemulsion systems which themselves act as carriers of the cosmetic actives. The second category of nanocosmetics is those which are insoluble in nature and it is the nano-sized particles of these agents themselves which

DOI: 10.1201/9781003319146-3

confer the cosmetic action. These are titanium oxide nanoparticles and zinc oxide nanoparticles, both of which are used in sunscreen products. Other examples include fullerenes and quantum dots, etc.

3.2   WHY NANOTECHNOLOGY IN COSMETICS?

The success of any cosmetic agent is incumbent upon overcoming the most important hurdle, which is achieving efficient and effective delivery of the formulation at the desired site of action. Skin is the barrier that hinders the delivery of any agent that has been applied topically along with problems such as stability and solubility which are also dependent upon the nature of the cosmetic ingredients. Manipulating several parameters and properties such as the size of the particles, the viscosity of the system, hydrophilic or lipophilic nature of the compound, etc. are necessary for ensuring successful passage of the topical preparations. Changing the formulation composition and/or adopting newer manufacturing techniques also has a remarkable impact on fastening the delivery of cosmetic actives. The front runner in solving problems pertaining to delivery in the 21st century has been nanotechnology. Nanotechnology is the buzzword in every field and the same goes for the pharmaceutical industry. Nanotechnology has revamped the pharmaceutical industry. Efforts of formulation departments are all concerted towards developing strategic nanotechnology-based delivery systems which will aid in overcoming all the delivery-related challenges that were gripping us up till now. The most prominent nanotechnology-based cosmetic systems are nanoemulsions, nanogels, lipidic nanoparticles, fullerenes, and many more.

With the expansion of consumerism along with wide arrays of advertisements, the cosmetic industry as a whole owing to these factors has been ever-expanding. Consumers have an inclination towards specific categories of cosmetics that will address their problems related to skin aging or pigmentation or drying/flaking, blemishes, spotting, wrinkling, etc. This has given rise to a very wide range of product segments which is also specifically made depending upon the demographics of the population.

3.3   CLASSIFICATION OF NANOCOSMETICS

The majority of the companies that are leaders in manufacturing cosmetics employ nanomaterials in their several processes which lead to the formulation of the end products. Several different types of nanosystems that are of use in developing nanocosmetics have been discussed hereunder. Lipid-based nanosystems are most explored and continue to be explored further particularly for topicals as they are required to penetrate across the skin layers for effectiveness. Hence, it is needed that these topicals have a higher lipid content which is complementary to the lipid layers present on the skin surface which will ultimately aid in better blending and adherence. Broadly, the nano lipid systems are classified either as 'matrix-systems' or 'vesicles' or other newer forms. In the matrix systems, nanoparticles are present in the matrix of the cosmetic such as nanoemulsions or solid lipid nanoparticles (SLNs) or nanostructured lipid carriers (NLCs), etc. In the vesicular systems, the vesicles act as carriers, such as liposomes. Lastly, the novel systems include niosomes, novasomes,

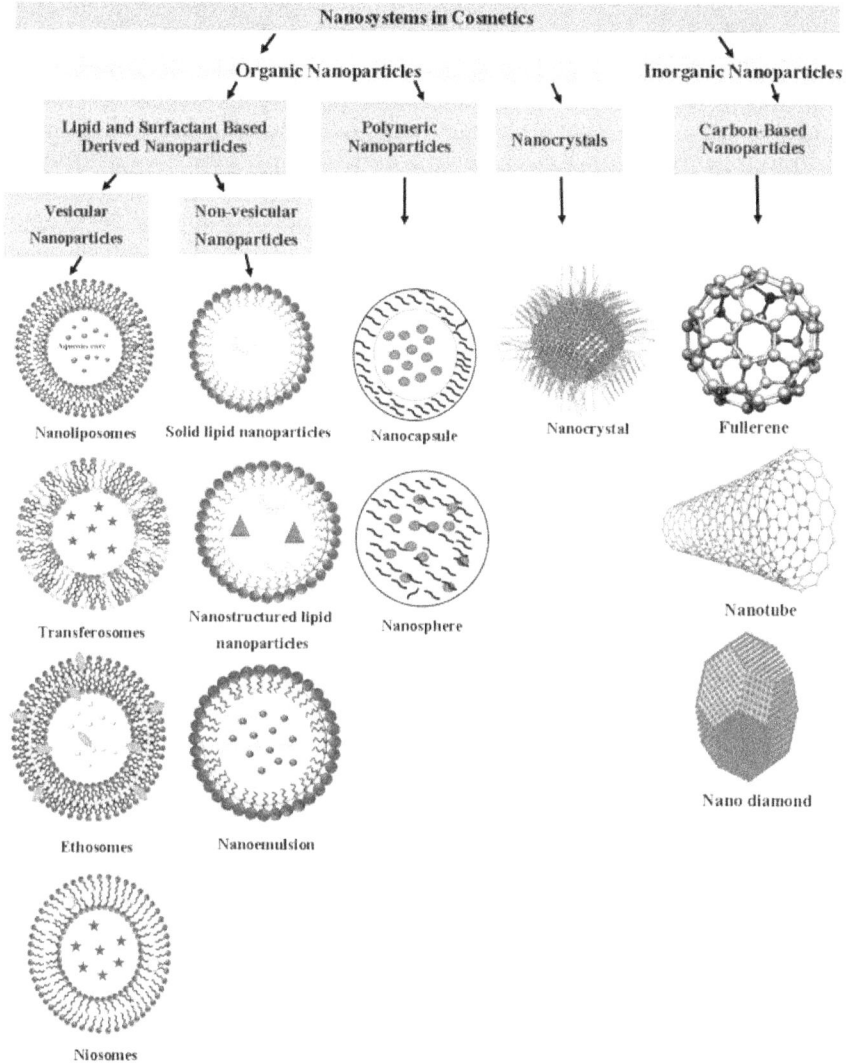

FIGURE 3.1 Different types of nanosystems that are used for developing cosmetic formulations

and cubosomes, etc. (Ahmad et al. 2018). Figure 3.1 gives a pictorial representation of the different nanocarrier systems that are employed for developing cosmetic formulations.

3.3.1 NANO-EMULSION SYSTEMS

Nanoemulsions are biphasic systems that have gained prominence in the cosmetic industry as they serve to be stable carriers of actives being thermodynamically and kinetically stable (Guglielmini 2008; Mason et al. 2006). Nanoemulsions are a very desirable delivery system of cosmetic agents because they can provision for controlled release and also help in achieving optimum dispersion of the nanocosmetic across the skin layer (Maali and Mosavian 2013). Emulsions mainly comprise of few essential components which in crude terms are oil, water, and surfactant. They are usually transparent or translucent in appearance and can be in the form of oil in water or water in oil or even having multiple layers of the same (Choudhury et al. 2019; Gorain et al. 2020). The first time the term 'nanoemulsion' came into being was in the 1990s. The size range lies within 50–200 nm which are constantly undergoing Brownian motion being in a metastable state which ensures stability and delays or prevents them from coalescing or flocculating. They have low interfacial tension and are generally produced by several techniques such as high-pressure homogenization, sonication, microfluidization, and phase inversion (Dubey et al. 2022). The nano-size of emulsion droplets imparts a blue hue. Nanoemulsions require lower concentrations of surfactants in comparison to microemulsions (5–10% vs. 20–25%) and can be prepared by both, either high or low shear techniques (Yang et al. 2012). Another advantage of nanoemulsions is that they are easily spreadable and are very successful in achieving texture merging along with acting as good hydrators. All these properties make them have an upper hand in comparison to other carriers (Sonneville-Aubrun, Simonnet, and L'Alloret 2004). Properties such as the surface charge, the globule size, and the nature of the surfactant used play key role in determining the extent of skin permeation. Cholesterol, palmitic acid, and ceramides are the three most common lipids that are present on our skin, and hence adding these to nanoemulsions is most suited, good at enhancing the elasticity of the skin and hydrating it as well (Sonneville-Aubrun, Yukuyama, and Pizzino 2018). There are several nanoemulsions-based cosmetic products such as the nanoemulsions gel Kemira and many more (Ashaolu 2021).

3.3.2 NANOSOMAL SYSTEMS

In some specific instances involving anti-aging care, it is sometimes required that the cosmetic actives have to penetrate across the deeper skin layer, however, it is made sure that they don't enter the systemic circulation. Nanosomes are vesicle-shaped systems that can achieve deeper penetration. These are self-assembled structures that encapsulate the cosmetic actives and release it upon coming in contact with the skin (Lasic 1998a; Patravale and Desai 2014). The most explored nanosones are liposomes.

3.3.2.1 Liposomes

Liposomes were the earliest nanocosmetic products to enter the market in as early as 1986 when global cosmetic giant Dior launched an anti-aging cream known as 'Capture' (Kazi et al. 2010a). Liposomes have a central aqueous core and on the periphery are surrounded by a phospholipid bilayer and their size falls in the range of nanometre and micrometer. The arrangement of this outer lipid coat further distinguishes them to be unilamellar or multilamellar vesicles. Liposomes deemed to be versatile carriers have several advantages such as increasing the solubility, being more biocompatible with the skin, and also reducing toxicity (Gorain et al. 2021; Kaur and Agrawal 2007; Sriraman and Torchilin 2014). Liposomes have been utilized for developing several types of cosmetics such as creams, moisturizers, deodorants/antiperspirants, sunscreens, hair products, etc. (Kaul et al. 2018). Liposomes restore water balance and also provide moisturization to the skin given that their lipid layers have ceramides and cholesterol (Lasic 1998b; Müller-Goymann 2004). Soyabased phospholipids, phosphatidylcholine, and different types of esterified fatty acids are often incorporated into cosmetics and other dermatological preparations. Other nanosomal structures which have developed over the last few decades are transferosomes, niosomes, cersomes, etc. (Hougeir and Kircik 2012).

3.3.2.2   Elastic liposomes

Transferosomes, as the name suggests have the ability to squeeze through the outer barrier of the skin and are quite elastic in nature, popularly called, "self-optimizing deformable structures" (Dubey et al. 2022). They comprise of two essential molecules, an edge activator and the phospholipid counterpart, both of which make it highly deformable in nature which makes possible enhanced penetration across the skin (Walve et al. 2011). Edge activators are like surfactant molecules, generally one chain long. Formulation scientists have forayed into extensive research in this area, Diractin, which is ketoprofen-loaded transferosome is a phase III candidate for osteoarthritis (Rother et al. 2007). Other actives having cosmetic significance are also being developed such as capsaicin, curcumin, and resveratrol, etc. which have anti-aging anti-oxidant properties (Opatha, Titapiwatanakun, and Chutoprapat 2020; Wu et al. 2019). Gel-based transferosomes have also been developed which mediated UV protection and treated acne as well (Pahwa et al. 2021). Another category of elastic liposomes are invasomes which are much softer than transferosomes. They are made up of a mixture of phosphatidylcholine along with terpenes and some amount of alcohol (Babaie et al. 2020). Invasomes loaded with phenylethyl resorcinol, an agent used for skin whitening, were successful in vitro in reducing the melanin content more effectively than similar liposomes (Amnuaikit et al. 2018). In another study, dapsone-loaded invasomes were formulated which have antibiotic and anti-inflammatory action (El-Nabarawi et al. 2018).

3.3.2.3   Ethosomes

Again, as the name suggests, ethosomes have a high concentration of alcohol in them which helps in disrupting the structure of lipids and consequently leads to improved penetration. Niacinamide ethosomes have been formulated, which are used for addressing pigmentation. The group of Touitou et al. was the first to patent ethosomes in the 1990s (Touitou et al. 2000). A hybrid of tranfersomes and ethosomes, known as transferosomes, has also been developed. It exhibits the combined properties of both. Curcumin-containing ethosomes have been shown to have anti-wrinkle and anti-sagging effects (Swarnlata Saraf 2014). Recently developed retinyl palmitate, which contains ethosomal hydrogel has shown efficacy in treating acne and checking peroxidation-related damages (Salem et al. 2021). In a study by Saraf et al. in 2014, curcumin extracts containing ethosomes were prepared, which exhibited promising results in vitro as a moisturizing and hydrating agent (Swarnlata Saraf 2014). In another study, lamellar liposomal vesicles containing amber extracts were prepared which make the skin appear firmer ("Cosmetic Composition Containing an Amber Extract - Patent US-7887858-B2- PubChem" 2022).

3.3.2.4 Cubosomes

As the name suggests, cubosomes are cube-like self-assembled nanostructures having a bi-continuous liquid crystalline phase, being about 100–300 nm in size and giving the appearance of dots in their 3D form (Pan et al. 2013; Rapalli et al. 2021). These are molecules that are highly stable to heat, capable of carrying amphiphilic/lipophilic/hydrophilic molecules and are generally made up of liquid crystalline particles, surfactant molecules, and water in specific ratios (Spicer 2005). They have a large surface area which imparts a high loading capacity. The cubosomal preparations developed by L'Oreal have proven to stabilize oil-in-water emulsions of cosmetic preparations which have the ability to trap pollutants (Saritha et al. 2021). Cubosomes have the unique advantage of being good carriers for dermal delivery. For example, this was substantiated by a study in which minoxidil-loaded cubosomes had much better in vitro skin permeation than the counterpart which was a solution of minoxidil-propylene glycol-water-ethanol (Kwon et al. 2010). In a recent study by Sherif et al., cubosomes loaded with 5% alpha-lipoic acid were made which had the cosmetic action of reducing wrinkles and other visible signs of aging (Sherif, Bendas, and Badawy 2014). In another study, silver sulfadiazine-loaded cubosomes were found to be effective in treating skin burns that had been infected (Morsi, Abdelbary, and Ahmed 2014). Utilizing microneedles for delivering vaccine-loaded cubosomes was found to be effective as the cubosomes were retained on the skin (Rattanapak et al. 2013).

3.3.2.5 Niosomes and novasomes

Niosomes are prepared using non-ionic surfactants and have higher drug entrapment efficiency and higher stability than normal liposomes. The non-ionic surfactants that can be employed for formulating these are brij, tween, sorbitan esters, spans, etc. (Kazi et al. 2010b). L'Oreal, in as early as 1970, had patented niosomal technology and branded the products in this space as Lancome (Nasir, Harikumar, and Kaur 2012). Novasomes are superior to liposomes and niosomes. Novasomes have a central core that is amorphous in nature and are surrounded by two to seven bi-lipid layers which makes it possible to load a much higher amount of cosmetic actives into them, be it hydrophilic or hydrophobic in nature. Moreover, novasomes make provision for loading even the actives which have a tendency of reacting with one another. This is brought about by loading a particular type of active in one lipid layer and the other active in another lipid layer, thus partitioning them (Waghmare, Patil, and Patil 2016). NOVAVAX is the laboratory that has patented this encapsulation technology of developing novasomes (Singh, Malviya, and Sharma 2011). These are very stable molecules, surviving in the pH range 2–13, temperatures 0°C–100°C along with having low manufacturing costs than liposomal systems. Some cosmetic examples of novasomes include AcneWorx (salicylic acid novasaomes) for acne, terconazole novasomes for candidiasis, etc.

3.3.2.6 Polymersomes, ultrasomes, and photosomes

All of these are novel nanosomal systems that have unique properties of their own. Polymersomes, as the name suggests, is made up of blocks of co-polymers that encapsulate a wide range of actives be it proteins or DNA/RNA moieties or small molecules, etc. (Patravale and Mandawgade 2008). Polymersomes have a rigid bilayer which makes them more stable formulations than liposomes. Patenting of several polymersome-based cosmetics are under way (Bermudez et al. 2002). Ultrasomes and photosomes are advanced liposomal systems that provide good protection from UV rays. Ultrasomes containing enzymes obtained from Micrococcus luteus have been developed which help in reversing the damages caused by exposure to light and also mediate the reduction of inflammation by reducing the levels of pro-inflammatory cytokines (Rahimpour and Hamishehkar 2012). Photosomes on the other hand contain the enzyme photolyase obtained from marine plant species, which can reverse photodamage caused to DNA (Patravale and Mandawgade 2008).

3.3.2.7 Other nanosomal systems

They include marinosomes, obtained from marine sources; cerasomes, having high amounts of ceramides, the lipids present on human skin, and phytosomes, enclosing phytoconstituents. Marinosomes have polyunsaturated fatty acids in their composition which when acted upon by human epidermal enzymes lead to the formation of several metabolites which mediate the lessening of skin irritation and inflammation. Phytosomes deliver herbal actives that are cosmetically useful such as different kinds of flavonoids, glycosides, volatile oils, etc. which rejuvenate the skin. Several other systems such as glycerosomes, having glycerine base and oleosomes having rich oil content, have found cosmetic utility for carrying cosmetic actives and moisturizing the skin as well (Dhawan, Sharma, and Nanda 2020; Shokri 2017).

3.3.3 NANOPIGMENTS

Nanopigments have the property of blocking UV rays and hence are most popularly used in sunscreens. However, they have several other avenues of application as they can be used in facial masks and different types of creams. In 2014, a novel cosmetic formulation comprising nanoparticular pigments in a vehicular system, having light-transmitting properties and capable of forming films on the skin was patented. The underlying principle of the same was optical blurring which helped in achieving reduced signs of aging (Ahmad et al. 2018). In 2008, the group of Jung et al. patented a composition of gold and silver nanopigments, which in different ratios was used for formulating a variety of coloured cosmetics ("US20090022765A1- Cosmetic Pigment Composition Containing Gold or Silver Nano-Particles - Google Patents" 2022).

3.3.4 SOLID LIPID NANOPARTICLES AND NANOSTRUCTURED LIPID CARRIERS

Lipid-based carriers have been in use for a long time. SLNs belong to the first generation and were developed in the 1990s. It comprises of solid lipid core which is dispersed in an aqueous media with aid of surfactants (Singh et al. 2021). Their occlusive nature, better biocompatibility, and physiological stability make them attractive carriers, especially for cosmetics and other topical formulations (Waghule et al. 2020). NanoLipid Repair CLR is a marketed SLN that carriers the mixture of black currant seed oil and manuka oil which have cosmetic benefits (Ahmad 2021). The second generation of lipid carriers are the NLCs. The problems associated with SLN such as low-loading capacity, chances of content expulsion, and lesser stability propelled the development of improved lipid carriers. NLCs confers the advantage of carrying out precise loading and release of the cosmetic agent along with making possible the achievement of higher loading, higher entrapping efficiency, and stability (Üner 2016). NLCs are made up of a mix of lipids, both solid and liquid in nature and they are grossly categorized into three types, amorphous NLCs or multiple NLCs or imperfect NLCs (Costa and Santos 2017). One of the earliest examples of NLC-based cosmetic products was NanoRepair Q10 cream developed in Germany (Müller et al. 2007). Recently developed softisan NLCs entrapping a mixture of volatile oils having antimicrobial, immunomodulatory properties along with acting as an antioxidant and analgesic was quite effective (Carbone et al. 2018). Overall, NLCs are more advantageous than SLNs as they have been found to be more efficient carriers of the bulk of actives, and hence being much more stable it curtails expulsion of the loaded actives during periods of storage when the formulations might undergo a polymeric transition (Müller, Radtke, and Wissing 2002). Cosmetic actives such as Coenzyme Q10 and retinol have been incorporated into NLCs (Morales et al. 2015; Vinardell and Mitjans 2015). NLCs also offer the flexibility of incorporating hydrophilic moieties into the core via conjugation with lipids. In the year 2013, a novel nanocosmetic system comprising of combined polymer-coated SLN and NLC was patented which exhibited improved penetration and higher stability against cosmetic degradation. NLCs in the form of nanobeads or nanopearls have also found cosmetic utility (Singh, Pandey, and Vishwakarma 2020).

3.3.5 POLYMERIC NANOPARTICLE-BASED COSMETICS

These are broadly classified into two types, nanospheres and nanocapsules. As the name suggests nanospheres involve spherical particles such that the cosmetic actives are dispersed all across the polymeric matrix and it is all present in a unified manner. However, nanocapsules involve the capsulation of the cosmetic actives by the polymeric coat which facilitates a controlled release (Guterres, Alves, and Pohlmann 2007). Nanocapsules are particularly preferred when the cosmetic agent being delivered is sensitive in nature and susceptible to degradation due to pH, light exposure, or surrounding contents. Nanocapsules encapsulating fragrances have been developed (El-Say and El-Sawy 2017; Frank et al. 2015). The polymer being employed for developing these polymeric nanoparticles is of crucial importance as they will have properties of their own that might influence the activity of the cosmetic active. Chitosan is one of the commonly used polymers which is very suitable for incorporation into cosmetic formulations. Hyaluronic acid-chitosan nanoparticles were found to be effective anti-wrinkle agents (Chaouat et al. 2017). Even minoxidil, a hair growth stimulator, loaded onto chitosan was seen to give better results in comparison to the conventional preparations (Matos et al. 2015). Several other polymers are utilized such as the natural ones being starch or albumin; the synthetic ones being polyacrylates or polycaprolactones, etc. Benzophenone, a UV-blocker when loaded onto polymeric nanoparticles, had a high sun protection factor (SPF) as compared to other lipid-based nanoparticles (Gilbert et al. 2016).

3.3.6 NANOFIBRES AND NANOHYDROGELS

Nanofibres are a relatively newer class of carriers that have a very high surface area, a very porous structure, and very high absorption capacity. Natural sources such as silk or even collagen can be used for making nanofibres and even synthetic fibres made of PVP or PVA can be used. Nanofibres are utilized for making facial masks, cleaning agents which can trap dirt, etc. (Zanin, Cerize, and De Oliveira, 2011). Vitamin A and E, both of which are extremely beneficial for the skin and are used extensively in cosmetic preparations when loaded onto nanofibres, exhibited sustained release (Taepaiboon, Rungsardthong, and Supaphol 2007). Nanohydrogels comprise of a polymeric matrix that swells up due to the presence of fluids. Nanohydrogels used for cosmetic care generally have water incorporated into them and the polymer used for them are either from natural sources (collagen, cellulose, alginate) or synthetic (polyvinyl alcohol) or biopolymers (xanthum gum, pectin). Hyaluronic acid nanohydrogels have shown good spreadability and moisturizing capacity, making the skin more supple (Son et al. 2017).

3.3.7 COSMETICS CONTAINING NANOGOLD AND NANOSILVER

Gold has been used for several functions ever since the inception of civilizations, be it for ornaments, vessels, or even cosmetic preparations. Gold is often used in facial masks, creams, etc. Gold has created its own niche in cosmetic preparation owing to its several advantages such as skin rejuvenation, imparting firmness, reducing visible signs of aging, along with having antimicrobial properties. Gold, as a metal, is quite inert and non-toxic in nature and even the nano-form of gold is highly stable in both solid and liquid forms. Nanogold stimulates the circulation of blood, reduces inflammation, smoothens the skin, and boosts the production of collagen by up to 200 times (Bapat et al. 2020; Singh 2021). Silver-based cosmetics and dermatological formulations have antimicrobial properties, acting against bacterial, viral, and fungal species. They are also used for dental care, hastening wound healing, and for treating burns. The cosmetic utility of silver is in treating acne, reversing photodamage, and uplifting the skin. Deodorants, creams, body sprays, foams, lip care products, etc. have silver in them for their anti-bacterial action. A gel formulation containing nanosilver was better in treating burns than silver sulfadiazine which contained 30 times more amount of silver (Dubey et al. 2022).

3.3.8 COSMETICS CONTAINING METAL OXIDES

Both zinc oxide (ZnO) and titanium oxide (TiO_2) are extensively used in sun protection products as they have the ability to absorb UV-A and UV-B rays respectively (Yang, Zhu, and Pan 2004). These metal oxides present in their nanoform, act as filters of UV light by forming a thin transparent film of the area of application. Both

ZnO and TiO$_2$ being opaque white metal oxides can adopt any colour and hence are used as colourants while making cosmetic products such as highlighters, bronzers, or brighteners, etc. Other metal oxides such as that of Cerium and Zirconium are also being developed for cosmetic applications (Bilal and Iqbal 2020).

3.3.9 DENDRIMERS AND NANOCRYSTALS

Dendrimers are densely branched tree-like polymeric organic complexes, whose inner core size generally varies from 1 to 10 nm (Mainardes and Silva 2004; Surekha et al. 2021). The extensive network of branching provisions sites for attachment by nanoparticles and thus facilitating their delivery once loaded onto these systems. Polyamidoamine (PAMAM) dendrimers are one of the most commonly used dendrimers. For example, in a study, the formulation of PAMAM dendrimers carrying indomethacin when applied onto rat skin exhibited improved in vitro penetration. Another recent study involved the formulation of cosmetic carbosiloxane dendrimer which claimed to be resistant oil and water on the skin, thus imparting a glossy finish ("US10172779B2- Copolymer Having Carbosiloxane Dendrimer Structure and Composition and Cosmetic Containing the Same - Google Patents" 2022). Many leading cosmetic companies like L'Oreal, Unilever, etc. have also been granted patents of several dendrimer-based cosmetics (Dubey et al. 2021).

Nanocrystals are clusters of the active itself which imparts enhanced solubility owing to the crystal structure which increases the surface area (Waghule et al. 2021). Nanocrystals of rutin which have antioxidant activity were found to give better cosmetic results than its water-soluble form (Li et al. 2021). Nanovital Vitanics Crystal is a nanocrystal-based cosmetic that also contains niacinamide and Vitamin C and has skin lightening and moisturizing action (Kaul et al. 2018).

3.3.10 CARBON BASED NANOPARTICLES

Fullerenes and carbon nanotubes are the two major carbon-based nanocosmetics. Fullerenes are hollow sphere-like structures whereas carbon nanotubes, as the name suggests are hollow tube-like structures having multiple/single walls of carbon atoms, generally graphene sheets (Baroli 2010). Fullerenes have the capability of working as antioxidants and hence it is used for making cosmetics that work as rejuvenators of the skin (Xiao et al. 2005). This was validated by a study in which human skin keratinocyte cultures (HaCaT) treated with fullerenes were seen to tackle and reduce the levels of reactive oxygen species (Xiao et al. 2005). Carbon nanotubes have the ability to adhere and bind with hair cells, thus making them good carriers for hair colorants (Kaul et al. 2018). Activated nano-carbon or nano-charcoal particles are also used in several masks and cleansing products wherein they are able to absorb all the dirt pollutants and oily clogs in the pores along with blackheads (Hammani et al. 2019).

3.4 CONCLUSION

Rapid scientific advancements in the sphere of nanotechnology, hand in hand with fast-paced commercialization, have spurted the prospective avenues for incorporating nanomaterials into one of the biggest commercial avenues, the cosmetic industry. Applicability of nanomaterials in developing a wide variety of cosmetic formulations is a field having substantial promise which has already begun to bear fruits as several nanocosmetics are already in the market. Given the plethora of advantages these nano-systems offers, several global leaders of the cosmetic industry are confident upon continuing to venture into research in this field. Due diligence has to be concentrated by the cosmetic companies in making sure that their nanocosmetic products are safe for human use with toxicity data backing their claims. This requires stringency on part of regulatory bodies which is yet to be developed fully and subsequently implemented. It is required that the regulatory bodies closely monitor the manufacturing of cosmetics that employ nanotechnology and lay down specific guidelines for the same. Adhering to all guidelines related to consumer safety, ethical manufacturing, and ethical sourcing, animal cruelty-free, long-term, and short-term toxicity testing are of prime importance. The world leaders of the cosmetic industry have a huge responsibility of ensuring that their nanocosmetic end-products which are reaching the consumers are safe and effective.

REFERENCES

Ahmad, Javed. 2021. "Lipid Nanoparticles Based Cosmetics with Potential Application in Alleviating Skin Disorders." *Cosmetics.* doi:10.3390/cosmetics8030084.

Ahmad, Usama, Zeeshan Ahmad, Ahmed Abdullah Khan, Juber Akhtar, Satya Prakash Singh, and Farhan Jalees Ahmad. 2018. "Strategies in Development and Delivery of Nanotechnology Based Cosmetic Products." *Drug Research* 68 (10): 545–52. doi:10.1055/a-0582-9372.

Amnuaikit, Thanaporn, Tunyaluk Limsuwan, Pasarat Khongkow, and Prapaporn Boonme. 2018. "Vesicular Carriers Containing Phenylethyl Resorcinol for Topical Delivery System; Liposomes, Transfersomes and Invasomes." *Asian Journal of Pharmaceutical Sciences* 13 (5): 472–84. doi:10.1016/J.AJPS.2018.02.004.

Anu Mary Ealia, S., and M. P. Saravanakumar. 2017. "A Review on the Classification, Characterisation, Synthesis of Nanoparticles and Their Application." *IOP Conference Series: Materials Science and Engineering* 263: 32019. doi:10.1088/1757-899x/263/3/032019.

Ashaolu, Tolulope Joshua. 2021. "Nanoemulsions for Health, Food, and Cosmetics: A Review." *Environmental Chemistry Letters* 19 (4): 3381–95. doi:10.1007/S10311-021-01216-9.

Babaie, Soraya, Azizeh Rahmani Del Bakhshayesh, Ji Won Ha, Hamed Hamishehkar, and Ki Hyun Kim. 2020. "Invasome: A Novel Nanocarrier for Transdermal Drug Delivery." *Nanomaterials* 10 (2): 341. doi:10.3390/NANO10020341.

Bapat, Ranjeet A., Tanay V. Chaubal, Suyog Dharmadhikari, Anshad Mohamed Abdulla, Prachi Bapat, Amit Alexander, Sunil K. Dubey, and Prashant Kesharwani. 2020. "Recent Advances of Gold Nanoparticles as Biomaterial in Dentistry." *International Journal of Pharmaceutics* 586: 119596. doi:10.1016/j.ijpharm.2020.119596.

Baroli, Biancamaria. 2010. "Penetration of Nanoparticles and Nanomaterials in the Skin: Fiction or Reality?" *Journal of Pharmaceutical Sciences* 99 (1): 21–50. doi:10.1002/jps.21817.

Bermudez, Harry, Aaron K. Brannan, Daniel A. Hammer, Frank S. Bates, and Dennis E. Discher. 2002. "Molecular Weight Dependence of Polymersome Membrane Structure, Elasticity, and Stability." *Macromolecules* 35 (21): 8203–8. doi:10.1021/MA020669L.

Bilal, Muhammad, and Hafiz M N Iqbal. 2020. "New Insights on Unique Features and Role of Nanostructured Materials in Cosmetics." *Cosmetics.* doi:10.3390/cosmetics7020024.

Carbone, C., C. Martins-Gomes, C. Caddeo, A. M. Silva, T. Musumeci, R. Pignatello, G. Puglisi, and E. B. Souto. 2018. "Mediterranean Essential Oils as Precious Matrix Components and Active Ingredients of Lipid Nanoparticles." *International Journal of Pharmaceutics* 548 (1): 217–26. doi:10.1016/j.ijpharm.2018.06.064.

Chaouat, C., Stéphane Balayssac, Myriam Malet-Martino, F. Belaubre, Emmanuel Questel, A. M. Schmitt, Steph Poigny, S. Franceschi, and E. Perez. 2017. "Green Microparticles Based on a Chitosan/Lactobionic Acid/Linoleic Acid Association. Characterization and Evaluation as a New Carrier System for Cosmetics." *Journal of Microencapsulation* 34 (March): 1–21. doi:10.1080/02652048.2017.1311956.

Choudhury, Hira, Nur Fadhilah B. Zakaria, Puteri Atdriann B. Tilang, Angeline S. Tzeyung, Manisha Pandey, Bappaditya Chatterjee, Nabil A. Alhakamy, et al. 2019. "Formulation Development and Evaluation of Rotigotine Mucoadhesive Nanoemulsion for Intranasal Delivery." *Journal of Drug Delivery Science and Technology* 54: 101301. doi:10.1016/j.jddst.2019.101301.

"Cosmetic Composition Containing an Amber Extract - Patent US-7887858-B2 - PubChem." 2022. Accessed May 27. https://pubchem.ncbi.nlm.nih.gov/patent/US-7887858-B2#section=Linked-Chemicals.

Costa, Raquel, and Lúcia Santos. 2017. "Delivery Systems for Cosmetics - From Manufacturing to the Skin of Natural Antioxidants." *Powder Technology* 322 (August). doi:10.1016/j.powtec.2017.07.086.

Dhawan, Surbhi, Pragya Sharma, and Sanju Nanda. 2020. "Cosmetic Nanoformulations and Their Intended Use." In, 141–69. doi:10.1016/B978-0-12-822286-7.00017-6.

Dubey, Sunil Kumar, Anuradha Dey, Gautam Singhvi, Murali Manohar Pandey, Vanshikha Singh, and Prashant Kesharwani. 2022. "Emerging Trends of Nanotechnology in Advanced Cosmetics." *Colloids and Surfaces B: Biointerfaces*, 112440. doi:10.1016/j.colsurfb.2022.112440.

Dubey, Sunil Kumar, Maithili Kali, Siddhanth Hejmady, Ranendra Narayan Saha, Amit Alexander, and Prashant Kesharwani. 2021. "Recent Advances of Dendrimers as Multifunctional Nano-Carriers to Combat Breast Cancer." *European Journal of Pharmaceutical Sciences: Official Journal of the European Federation for Pharmaceutical Sciences* 164 (September): 105890. doi:10.1016/j.ejps.2021.105890.

El-Nabarawi, Mohamed Ahmed, Rehab Nabil Shamma, Faten Farouk, and Samar Mohamed Nasralla. 2018. "Dapsone-Loaded Invasomes as a Potential Treatment of Acne: Preparation, Characterization, and In Vivo Skin Deposition Assay." *AAPS PharmSciTech* 19 (5): 2174–84. doi:10.1208/S12249-018-1025-0.

El-Say, Khalid M., and Hossam S. El-Sawy. 2017. "Polymeric Nanoparticles: Promising Platform for Drug Delivery." *International Journal of Pharmaceutics* 528 (1–2): 675–91. doi:10.1016/j.ijpharm.2017.06.052.

Frank, Luiza A., Renata V. Contri, Ruy C. R. Beck, Adriana R. Pohlmann, and Silvia S. Guterres. 2015. "Improving Drug Biological Effects by Encapsulation into Polymeric Nanocapsules." *Wiley Interdisciplinary Reviews: Nanomedicine and Nanobiotechnology* 7 (5): 623–39. doi:10.1002/WNAN.1334.

Gilbert, E., L. Roussel, C. Serre, R. Sandouk, D. Salmon, P. Kirilov, M. Haftek, F. Falson, and F. Pirot. 2016. "Percutaneous Absorption of Benzophenone-3 Loaded Lipid Nanoparticles and Polymeric Nanocapsules: A Comparative Study." *International Journal of Pharmaceutics* 504 (1–2): 48–58. doi:10.1016/j.ijpharm.2016.03.018.

Gorain, Bapi, E. Bandar Al-Dhubiab, Anroop Nair, Prashant Kesharwani, Manisha Pandey, and Hira Choudhury. 2021. "Multivesicular Liposome: A Lipid-Based Drug Delivery System for Efficient Drug Delivery." *Current Pharmaceutical Design*. doi:10.2174/1381612827666210830095941.

Gorain, Bapi, Hira Choudhury, Anroop B Nair, Sunil K Dubey, and Prashant Kesharwani. 2020. "Theranostic Application of Nanoemulsions in Chemotherapy." *Drug Discovery Today* 25 (7): 1174–88. doi:10.1016/j.drudis.2020.04.013.

Guglielmini, Giancarlo. 2008. "Nanostructured Novel Carrier for Topical Application." *Clinics in Dermatology* 26 (4): 341–46. doi:10.1016/j.clindermatol.2008.05.004.

Guterres, Sílvia S., Marta P. Alves, and Adriana R. Pohlmann. 2007. "Polymeric Nanoparticles, Nanospheres and Nanocapsules, for Cutaneous Applications." *Drug Target Insights* 2 (January): 147. /pmc/articles/PMC3155227/.

Hammani, H., F. Laghrib, A. Farahi, S. Lahrich, T. El Ouafy, A. Aboulkas, K. El Harfi, and M. A. El Mhammedi. 2019. "Preparation of Activated Carbon from Date Stones as a Catalyst to the Reactivity of Hydroquinone: Application in Skin Whitening Cosmetics Samples." *Journal of Science: Advanced Materials and Devices* 4 (3): 451–58. doi:10.1016/j.jsamd.2019.07.003.

Hougeir, Firas G., and Leon Kircik. 2012. "A Review of Delivery Systems in Cosmetics." *Dermatologic Therapy* 25 (3): 234–37. doi:10.1111/j.1529-8019.2012.01501.x.

Kaul, Shreya, Neha Gulati, Deepali Verma, Siddhartha Mukherjee, and Upendra Nagaich. 2018. "Role of Nanotechnology in Cosmeceuticals: A Review of Recent Advances." *Journal of Pharmaceutics* 2018 (March): 1–19. doi:10.1155/2018/3420204.

Kaur, I. P. and R. Agrawal. 2007. "Nanotechnology: A New Paradigm in Cosmeceuticals." *Recent Patents on Drug Delivery & Formulation* 1 (2): 171–82. doi:10.2174/187221107780831888.

Kazi, Karim Masud, Asim Sattwa Mandal, Nikhil Biswas, Arijit Guha, Sugata Chatterjee, Mamata Behera, and Ketousetuo Kuotsu. 2010a. "Niosome: A Future of Targeted Drug Delivery Systems." *Journal of Advanced Pharmaceutical Technology & Research* 1 (4). Medknow Publications: 374–80. doi:10.4103/0110-5558.76435.

———. 2010b. "Niosome: A Future of Targeted Drug Delivery Systems." *Journal of Advanced Pharmaceutical Technology & Research* 1 (4): 374–80. doi:10.4103/0110-5558.76435.

Kwon, Taek, Hyun Lee, Jong Kim, Won Shin, Seung Park, and Jin-Chul Kim. 2010. "In Vitro Skin Permeation of Cubosomes Containing Water Soluble Extracts of Korean Barberry." *Colloid Journal* 72 (April): 205–10. doi:10.1134/S1061933X10020092.

Lasic, D. D. 1998a. "Novel Applications of Liposomes." *Trends in Biotechnology* 16 (7): 307–21. doi:10.1016/S0167-7799(98)01220-7.

———. 1998b. "Novel Applications of Liposomes." *Trends in Biotechnology* 16 (7): 307–21. doi:10.1016/S0167-7799(98)01220-7.

Li, Jing, Weilong Ni, Mayinuer Aisha, Juanjuan Zhang, and Minjie Sun. 2021. "A Rutin Nanocrystal Gel as an Effective Dermal Delivery System for Enhanced Anti-Photoaging Application." *Drug Development and Industrial Pharmacy* 47 (3): 429–39. doi:10.1080/03639045.2021.1890113.

Maali, A., and M. T. Hamed Mosavian. 2013. "Preparation and Application of Nanoemulsions in the Last Decade (2000–2010)." *Journal of Dispersion Science and Technology* 34 (1): 92–105. doi:10.1080/01932691.2011.648498.

Mainardes, Rubiana M., and Luciano P. Silva. 2004. "Drug Delivery Systems: Past, Present, and Future." *Current Drug Targets* 5 (5): 449–55. doi:10.2174/1389450043345407.

Mason, T. G., J. N. Wilking, K. Meleson, C. B. Chang, and S. M. Graves. 2006. "Nanoemulsions: Formation, Structure, and Physical Properties." *Journal of Physics Condensed Matter* 18 (41): R635. doi:10.1088/0953-8984/18/41/R01.

Matos, Breno Noronha, Thaiene Avila Reis, Taís Gratieri, and Guilherme Martins Gelfuso. 2015. "Chitosan Nanoparticles for Targeting and Sustaining Minoxidil Sulphate Delivery to Hair Follicles." *International Journal of Biological Macromolecules* 75 (April): 225–29. doi:10.1016/j.ijbiomac.2015.01.036.

Morales, Javier O., Karina Valdés, Javier Morales, and Felipe Oyarzun-Ampuero. 2015. "Lipid Nanoparticles for the Topical Delivery of Retinoids and Derivatives." *Nanomedicine (London, England)* 10 (2): 253–69. doi:10.2217/nnm.14.159.

Morsi, Nadia M., Ghada A. Abdelbary, and Mohammed A. Ahmed. 2014. "Silver Sulfadiazine Based Cubosome Hydrogels for Topical Treatment of Burns: Development and In Vitro/ In Vivo Characterization." *European Journal of Pharmaceutics and Biopharmaceutics : Official Journal of Arbeitsgemeinschaft Fur Pharmazeutische Verfahrenstechnik e.V* 86 (2): 178–89. doi:10.1016/j.ejpb.2013.04.018.

Müller, R. H., R. D. Petersen, A. Hommoss, and J. Pardeike. 2007. "Nanostructured Lipid Carriers (NLC) in Cosmetic Dermal Products." *Advanced Drug Delivery Reviews* 59 (6): 522–30. doi:10.1016/J.ADDR.2007.04.012.

Müller, R. H., M. Radtke, and S. A. Wissing. 2002. "Solid Lipid Nanoparticles (SLN) and Nanostructured Lipid Carriers (NLC) in Cosmetic and Dermatological Preparations." *Advanced Drug Delivery Reviews* 54 Suppl 1 (November): S131–55. doi:10.1016/ s0169-409x(02)00118-7.

Müller-Goymann, C. C. 2004. "Physicochemical Characterization of Colloidal Drug Delivery Systems Such as Reverse Micelles, Vesicles, Liquid Crystals and Nanoparticles for Topical Administration." *European Journal of Pharmaceutics and Biopharmaceutics: Official Journal of Arbeitsgemeinschaft Fur Pharmazeutische Verfahrenstechnik e.V* 58 (2): 343–56. doi:10.1016/J.EJPB.2004.03.028.

Nasir, Ali, Harikumar Sl, and Amanpreet Kaur. 2012. "Niosomes: An Excellent Tool for Drug Delivery." *International Journal of Research in Pharmacy and Chemistry* 2 (January).

Opatha, Shakthi Apsara Thejani, Varin Titapiwatanakun, and Romchat Chutoprapat. 2020. "Transfersomes: A Promising Nanoencapsulation Technique for Transdermal Drug Delivery." *Pharmaceutics* 12 (9): 855. doi:10.3390/pharmaceutics12090855.

Pahwa, Rakesh, Shweta Pal, Kamal Saroha, Parul Waliyan, and Manish Kumar. 2021. "Transferosomes: Unique Vesicular Carriers for Effective Transdermal Delivery." *Journal of Applied Pharmaceutical Science* 11 (5): 1–8. doi:10.7324/JAPS.2021.110501.

Pan, Xin, Ke Han, Xinsheng Peng, Zhiwen Yang, Lingzhen Qin, Chune Zhu, Xintian Huang, et al. 2013. "Nanostructured Cubosomes as Advanced Drug Delivery System." *Current Pharmaceutical Design* 19 (35): 6290–97. doi:10.2174/1381612811319350006.

Patravale, Vandana B., and Preshita P. Desai. 2014. "Topical Nanointerventions for Therapeutic and Cosmeceutical Applications." In Domb, A., and Khan, W. (eds.), *Focal Controlled Drug Delivery. Advances in Delivery Science and Technology*, 535–60. Boston, MA: Springer.

Patravale, Vandana B., and S. D. Mandawgade. 2008. "Novel Cosmetic Delivery Systems: An Application Update." *International Journal of Cosmetic Science* 30 (1): 19–33. doi:10.1111/J.1468-2494.2008.00416.X.

Rahimpour, Yahya, and Hamed Hamishehkar. 2012. "Liposomes in Cosmeceutics." 9 (4): 443–55. doi:10.1517/17425247.2012.666968.

Rapalli, Vamshi Krishna, Saswata Banerjee, Shahid Khan, Prabhat Nath Jha, Gaurav Gupta, Kamal Dua, Md Saquib Hasnain, Amit Kumar Nayak, Sunil Kumar Dubey, and Gautam Singhvi. 2021. "QbD-Driven Formulation Development and Evaluation of Topical Hydrogel Containing Ketoconazole Loaded Cubosomes." *Materials Science and Engineering: C* 119: 111548. doi:10.1016/j.msec.2020.111548.

Rattanapak, Teerawan, James Birchall, Katherine Young, Masaru Ishii, Igor Meglinski, Thomas Rades, and Sarah Hook. 2013. "Transcutaneous Immunization Using Microneedles and Cubosomes: Mechanistic Investigations Using Optical Coherence Tomography and Two-Photon Microscopy." *Journal of Controlled Release: Official Journal of the Controlled Release Society* 172 (3): 894–903. doi:10.1016/j.jconrel.2013.08.018.

Rother, Matthias, Bernard J. Lavins, Werner Kneer, Klaus Lehnhardt, Egbert J. Seidel, and Stefan Mazgareanu. 2007. "Efficacy and Safety of Epicutaneous Ketoprofen in Transfersome (IDEA-033) versus Oral Celecoxib and Placebo in Osteoarthritis of the Knee: Multicentre Randomised Controlled Trial." *Annals of the Rheumatic Diseases* 66 (9): 1178–83. doi:10.1136/ard.2006.065128.

Salem, H. F., R. M. Kharshoum, S. M. Awad, M. Ahmed Mostafa, and H. A. Abou-Taleb. 2021. "Tailoring of Retinyl Palmitate-Based Ethosomal Hydrogel as a Novel Nanoplatform for Acne Vulgaris Management: Fabrication, Optimization, and Clinical Evaluation Employing a Split-Face Comparative Study." *International Journal of Nanomedicine* 16: 4251–76. doi:10.2147/IJN.S301597.

Saritha, M., Boyina Harshini, Paravastu Kamala Kumari, and Srinivasa Rao Yarraguntla. 2021. "Review on Cubosomes." *International Journal of Current Pharmaceutical Research*, November, 37–42. doi:10.22159/ijcpr.2021v13i6.1926.

Sherif, S., E. R. Bendas, and S. Badawy. 2014. "The Clinical Efficacy of Cosmeceutical Application of Liquid Crystalline Nanostructured Dispersions of Alpha Lipoic Acid as Anti-Wrinkle." *European Journal of Pharmaceutics and Biopharmaceutics : Official Journal of Arbeitsgemeinschaft Fur Pharmazeutische Verfahrenstechnik e.V* 86 (2): 251–59. doi:10.1016/J.EJPB.2013.09.008.

Shokri, Javad. 2017. "Nanocosmetics: Benefits and Risks." *BioImpacts : BI* 7 (4): 207–8. doi:10.15171/bi.2017.24.

Singh, Anupama, Rishabha Malviya, and Pramod Sharma. 2011. "Novasome-A Breakthrough in Pharmaceutical Technology a Review Article." *Advances in Biological Research* 5 (January): 184–89.

Singh, Archana. 2021. "Carbon Nanofiber in Cosmetics." In, 341–63. doi:10.1002/9781119769149.ch14.

Singh, Babita, Shivani Pandey, Mohammad Rumman, Shashank Kumar, Prem Prakash Kushwaha, Rajesh Verma, and Abbas Ali Mahdi. 2021. "Neuroprotective and Neurorescue Mode of Action of Bacopa Monnieri (L.) Wettst in 1-Methyl-4-Phenyl-1,2,3,6-Tetrahydropyridine-Induced Parkinson's Disease: An In Silico and In Vivo Study." *Frontiers in Pharmacology.* https://www.frontiersin.org/article/10.3389/fphar.2021.616413.

Singh, Suman, Satish Pandey, and Neelam Vishwakarma. 2020. "Functional Nanomaterials for the Cosmetics Industry." In, 717–30. doi:10.1016/b978-0-12-816787-8.00022-3.

Son, Seong U., Jae-woo Lim, Taejoon Kang, Juyeon Jung, and Eun-Kyung Lim. 2017. "Hyaluronan-Based Nanohydrogels as Effective Carriers for Transdermal Delivery of Lipophilic Agents: Towards Transdermal Drug Administration in Neurological Disorders." *Nanomaterials.* doi:10.3390/nano7120427.

Sonneville-Aubrun, Odile, J.-T. Simonnet, and F. L'Alloret. 2004. "Nanoemulsions: A New Vehicle for Skincare Products." *Advances in Colloid and Interface Science* 108–109 (May): 145–49. doi:10.1016/j.cis.2003.10.026.

Sonneville-Aubrun, Odile, Megumi N. Yukuyama, and Aldo Pizzino. 2018. "Application of Nanoemulsions in Cosmetics." *Nanoemulsions: Formulation, Applications, and Characterization*, January: 435–75. doi:10.1016/B978-0-12-811838-2.00014-X.

Spicer, Patrick. 2005. "Progress in Liquid Crystalline Dispersions: Cubosomes." *Current Opinion in Colloid & Interface Science* 10 (December): 274–79. doi:10.1016/j.cocis.2005.09.004.

Sriraman, Shravan Kumar, and Vladimir P. Torchilin. 2014. "Recent Advances with Liposomes as Drug Carriers." *Advanced Biomaterials and Biodevices* 9781118773635 (July): 79–119. doi:10.1002/9781118774052.CH3.

Surekha, Bhavya, Naga Sreenu Kommana, Sunil Kumar Dubey, A. V. Pavan Kumar, Rahul Shukla, and Prashant Kesharwani. 2021. "PAMAM Dendrimer as a Talented Multifunctional Biomimetic Nanocarrier for Cancer Diagnosis and Therapy." *Colloids and Surfaces B: Biointerfaces* 204: 111837. doi:10.1016/j.colsurfb.2021.111837.

Swarnlata Saraf, and Gunjan Jeswani. 2014. "Topical Delivery of Curcuma Longa Extract Loaded Nanosized Ethosomes to Combat Facial Wrinkles Research Article." *Journal of Pharmaceutics & Drug Delivery Research* 03 (01). doi:10.4172/2325-9604.1000118.

Taepaiboon, Pattama, Uracha Rungsardthong, and Pitt Supaphol. 2007. "Vitamin-Loaded Electrospun Cellulose Acetate Nanofiber Mats as Transdermal and Dermal Therapeutic Agents of Vitamin A Acid and Vitamin E." *European Journal of Pharmaceutics and Biopharmaceutics : Official Journal of Arbeitsgemeinschaft Fur Pharmazeutische Verfahrenstechnik e.V* 67 (2): 387–97. doi:10.1016/j.ejpb.2007.03.018.

Touitou, E., N. Dayan, L. Bergelson, B. Godin, and M. Eliaz. 2000. "Ethosomes - Novel Vesicular Carriers for Enhanced Delivery: Characterization and Skin Penetration Properties." *Journal of Controlled Release* 65 (3): 403–18. doi:10.1016/S0168-3659(99)00222-9.

Üner, Melike. 2016. "Characterization and Imaging of Solid Lipid Nanoparticles and Nanostructured Lipid Carriers." In, 117–41. doi:10.1007/978-3-319-15338-4_3.

"US10172779B2- Copolymer Having Carbosiloxane Dendrimer Structure and Composition and Cosmetic Containing the Same - Google Patents." 2022. Accessed May 27. https://patents.google.com/patent/US10172779B2/en.

"US20090022765A1- Cosmetic Pigment Composition Containing Gold or Silver Nano-Particles - Google Patents." 2022. Accessed May 27. https://patents.google.com/patent/US20090022765A1/en.

Vinardell, María Pilar, and Montserrat Mitjans. 2015. "Nanocarriers for Delivery of Antioxidants on the Skin." *Cosmetics.* doi:10.3390/cosmetics2040342.

Waghmare, Swapnil, Aarti Patil, and Prachi Patil. 2016. "The Pharma Innovation Journal 2016; 5(5): 34–38 Novasome: Advance in Liposome and Niosome." www.thepharmajournal.com.

Waghule, Tejashree, Srividya Gorantla, Vamshi Krishna Rapalli, Pranav Shah, Sunil Kumar Dubey, Ranendra Narayan Saha, and Gautam Singhvi. 2020. "Emerging Trends in Topical Delivery of Curcumin Through Lipid Nanocarriers: Effectiveness in Skin Disorders." *AAPS PharmSciTech* 21 (7): 284. doi:10.1208/s12249-020-01831-9.

Waghule, Tejashree, Shalini Patil, Vamshi Krishna Rapalli, Vishal Girdhar, Srividya Gorantla, Sunil Kumar Dubey, Ranendra Narayan Saha, and Gautam Singhvi. 2021. "Improved Skin-Permeated Diclofenac-Loaded Lyotropic Liquid Crystal Nanoparticles: QbD-Driven Industrial Feasible Process and Assessment of Skin Deposition." *Liquid Crystals* 48 (7): 991–1009. doi:10.1080/02678292.2020.1836276.

Walve, J. R., S. R. Bakliwal, Bhushan Rane, and S. P. Pawar. 2011. "Transfersomes: A Surrogated Carrier for Transdermal Drug Delivery System." *International Journal of Applied Biology and Pharmaceutical Technology* 2 (January): 204–13.

Wu, Pey-Shiuan, Yu-Syuan Li, Yi-Ching Kuo, Suh-Jen Jane Tsai, and Chih-Chien Lin. 2019. "Preparation and Evaluation of Novel Transfersomes Combined with the Natural Antioxidant Resveratrol." *Molecules* 24 (3). doi:10.3390/MOLECULES24030600.

Xiao, Li, Hiroya Takada, Kentaro Maeda, Mari Haramoto, and Nobuhiko Miwa. 2005. "Antioxidant Effects of Water-Soluble Fullerene Derivatives against Ultraviolet Ray or Peroxylipid through Their Action of Scavenging the Reactive Oxygen Species in Human Skin Keratinocytes." *Biomedicine & Pharmacotherapy = Biomedecine & Pharmacotherapie* 59 (7): 351–58. doi:10.1016/j.biopha.2005.02.004.

Yang, Hongying, Sukang Zhu, and Ning Pan. 2004. "Studying the Mechanisms of Titanium Dioxide as Ultraviolet-Blocking Additive for Films and Fabrics by an Improved Scheme." *Journal of Applied Polymer Science* 92 (5): 3201–10. doi:10.1002/APP.20327.

Yang, Ying, Christopher Marshall-Breton, Martin Leser, Alexander Sher, and David Mcclements. 2012. "Fabrication of Ultrafine Edible Emulsions: Comparison of High-Energy and Low-Energy Homogenization Methods." *Food Hydrocolloids* 29 (December): 398–406. doi:10.1016/j.foodhyd.2012.04.009.

Zanin, Maria Helena A., Natalia N. P. Cerize, and Adriano M. De Oliveira. (2011). Production of Nanofibers by Electrospinning Technology: Overview and Application in Cosmetics. In: Beck, R., Guterres, S., and Pohlmann, A. (eds.), *Nanocosmetics and Nanomedicines.* Berlin, Heidelberg: Springer. https://doi.org/10.1007/978-3-642-19792-5_16

4 Approaches for administration of nanocosmetics

Manisha Choudhari, Anuradha Dey, Gautam Singhvi, Ranendra Narayan Saha and Sunil Kumar Dubey

CONTENTS

DOI: 10.1201/9781003319146-4

4.1 INTRODUCTION

Since the inception of mankind and the expansion of civilizations, people across the globe have been using a variety of components on their skin to achieve cosmetic and/or therapeutic outcomes. In the current period, numerous topical delivery-based formulations have been designed to improve the appearance of the skin and for treating several skin diseases (Benson 2005; Rapalli et al. 2018). A growing market for cosmetic products provides various benefits with little effort, and the customer interest towards having healthy-looking skin has also been rising. Consumers anticipate cutting-edge technology and inventive formulations with various active ingredients that have been clinically tested. As a consequence, formulations that increase skin permeability, facilitating the delivery of actives are driving the development of new products in the cosmetics sector (Rapalli et al. 2021). The active ingredients in a therapeutic formulation must penetrate across the stratum corneum and into the viable tissues of interest after being applied onto the skin (Patravale and Mandawgade 2008). To know the transdermal and topical delivery of actives across the skin, morphology-related aspects, some aspects of understanding skin physiology, and the variables impacting transdermal and topical delivery are necessary. The challenges with skin barrier functions must be selectively overcome to effectively deliver active components into the skin cells (Kim et al. 2020).

Cosmeceuticals, particularly nanocosmeceuticals, have had an enormous increase in the market for personal care products (Brandt, Cazzaniga, and Hann 2011). Cosmeceuticals are employed for various purposes, including improving skin texture by promoting collagen formation and as antiaging preparations, because the antioxidants present bring forth their action by neutralizing the reactive oxygen species and maintaining keratin structure, resulting in healthier skin (Durazzo et al. 2019; Dubey et al. 2022). The significant benefits of nanocosmeceuticals comprise controlled release of actives, site-specific targeting, and improved occlusiveness with better hydration, all mounting to an increased skin permeability to the actives (Kaul et al. 2018). Topical administration of the active components has various advantages in nanocosmeceuticals formulation. Antiaging nanocosmeceuticals marketed as hair care items such as conditioners, shampoo, and hair

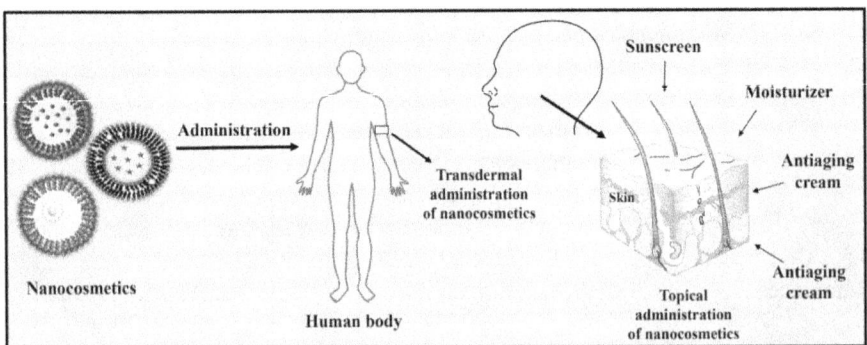

FIGURE 4.1 Different approaches for administration of nanocosmetics

growth stimulators may comprise liposomes, niosomes, and nanoemulsions (Rosen, Landriscina, and Friedman 2015). These products are intended to seal moisture around the respective cuticles and maximize the duration of contact period with the scalp and follicles by generating a coating that is protective in nature (Tripura Sundari and Anushree 2017).

Additional breakthroughs in nanomedicine-based research have resulted in the development of several transdermal delivery systems that could passively or actively distribute the necessary molecules having the desired concentration at the sites of interest on the skin (Alexander et al. 2012; Vitorino et al. 2014). In the past, creams and gels were regarded as cosmeceuticals; however, fast-paced technological advances in transdermal delivery systems such as iontophoresis, micro-needling, microdermabrasion, and electroporation were included (Swain et al. 2011). This chapter describes the different routes of administration of nanocosmetics, such as topical and transdermal delivery, and also describes the bioactive nanocarriers that might enhance the general health of the skin. Figure 4.1 illustrates the different routes of administration of nanocosmetics.

4.2 TOPICAL DELIVERY SYSTEM OF NANOCOSMETICS

The skin stands to be an exploitable interface for administering various actives due to its large surface area (Souto et al. 2020). The benefits of topical drug delivery include the potential to avoid the risks related to parenteral administration and even avoid elimination owing to first-pass metabolism, generally faced with oral therapies. Topical delivery facilitates continued therapy for longer timeframes, allowing the use of drugs with short biological half-lives, and thus reducing the associated gastrointestinal discomforts such as irritation that is usually faced with systemically administered therapies (Souto et al. 2020). Overcoming the several physiological layers, each with distinctive polarity presents to be one of the most significant barriers to delivery through the skin. The skin is known to defend the body from foreign particles, including toxic substances and pathogens, as well as pharmaceutical actives, whilst ensuring the exclusion of physiological fluids that are involved in homeostasis. The bioavailability of APIs administrated topically is lower than that of oral and intravenous methods due to the skin's tendency to impede API absorption (Mozafari 2017). Topically applied nanocosmetics comprise skin care (sunscreens, antiaging creams, moisturizers, skin cleansers), lip care, hair care, nail care, and oral care (mouthwash and toothpaste) described as follows:

4.2.1 Skin care

Depending on their intended use, skin care products fall into several categories, including moisturizers, sunscreens, cleansers, and antiaging creams. In order to enhance the effectiveness of cosmetic bases, a variety of nanocarriers, including nanoparticles, nanoemulsions, niosomes, and solid lipid nanoparticles (SLNs), have been used (Dhawan, Sharma, and Nanda 2020; Dubey et al. 2022). The list of marketed skin care nanocosmetics products showed in Table 4.1 (Kumari 2017; Kaul et al. 2018).

4.2.1.1 Sunscreens

Sunscreens act as a shielding agent over the skin, averting the skin damage induced by the UV rays. UVA (causes skin aging) and UVB (causes skin burning) rays are naturally reflected by mineral bases like zinc oxide (ZnO) and titanium dioxide (TiO$_2$), preventing their entry into the deeper layers of skin. Traditional sunscreens, after application, often lead to the formation of a chalky white residual layer on the skin's surface. In order to solve this issue, nanotechnological principles were incorporated. The sizes of the sunblocking agents were decreased to the nano levels, and effective sunscreen products with clear and natural appearances were obtained.

4.2.1.2 Antiaging creams

Aging impacts the dermal layer of the skin, which causes collagen and elastin breakdown. Since nanoparticles can reach the deeper skin layers, antiaging substances like retinol, vitamin E, vitamin C, and coenzyme Q10 should function more effectively when administered at the nanoscale. Numerous cosmeceutical companies are developing antiaging products using nanotechnology. There are various antiaging creams, such as Chantecaille's nanogold face revitalizing cream containing 24-carat gold nanoparticles and Lancome's Hydra Zen Cream containing nano encapsulated Triceramide (Kumari, Sharma, and Gupta 2017; Lohani et al. 2014).

4.2.1.3 Moisturizers

Skin dryness can be prevented and addressed by applying moisturizers that generate a thin film of compounds which are humectants. Consequently, moisturizers can minimize water loss or make the stratum corneum function normally. Liposomes, NLCs, nanoemulsions, SLNs, and niosomes are among the several nanocarriers frequently added to moisturizers because they help the skin retain moisture for a long time (Lohani et al. 2014; Dhawan, Sharma, and Nanda 2020).

4.2.1.4 Skin cleanser

The purpose of a cleanser is to remove deposited residues, makeup, oil, grime, dead skin cells, and several other impurities from the skin, usually the face. Metal nanoparticles serve as skin disinfection and decontamination agents. Cyclic cleansers that are nanosilver-based, reduce age spots and sun damage, eliminate colony-forming germs, control fungal growth, treat acne, exfoliate dead skin, and remove make-up gently (Lohani et al. 2014; Dhawan, Sharma, and Nanda 2020).

4.2.2 Lip care

Lip care products based on nanotechnology include lip balm, lipstick, lip gloss, and lip volumizer. The inclusion of nanoparticles in lip care products softens the lips by averting transepidermal water loss, stopping the migration of the concerned pigments from the lips, and prolonging the product's appearance. Lip volumizers contain liposomes that improve the hydration and volume of the lips. In lipsticks, pigments are used, which causes lead poisoning. However, it was observed that nanosilver had a yellow colour while nanogold had a red colour. In order to obtain the unique coloured pigments, silver and gold nanoparticles have been added to the lipsticks (Nanda et al.

2016). Furthermore, lipstick also contains silver nanoparticles. These nanoparticles ensure that colors are dispersed uniformly and stop the pigments from bleeding into or migrating into the lips' tiny wrinkles. Some of the marketed lip care nanocosmetics products are shown in Table 4.1.

4.2.3 Hair care

Nanotechnology has been employed in hair-care products such as shampoos, dyes/colorants, hair growth promoters, conditioners, and styling-based products. Cosmetic companies are currently investigating the use of nanoparticles in minimizing hair loss, stimulating hair growth, treating hair-associated problems, and maintaining hair's health, shine, and silkiness. The effectiveness of hair care products can be enhanced by using nanocarriers such as nanoemulsion, nanospheres, and liposomes. For example, shampoo containing Proxiphen N is used to cure alopecia, hair depilation caused by bacteria is treated with silver nanoparticles, and damaged hair cuticles are treated with sericin nanoparticles in hair products (Hu et al. 2012). Some of the marketed hair care nanocosmetics products are shown in Table 4.1.

4.2.4 Nail care

In comparison to conventional nail paints, the addition of nanotechnology to nail care products has produced more beneficial effects. Nail polishes made with nanotechnology have better toughness, more resistance, better resistance to chipping, increased durability, simplicity of application, and quick drying. Nano Labs Corp. acquired a provisional patent in 2012 for its nano-nail polish and nail lacquer, which has high flexibility, shock resistance, shock absorption, shock resistance, and the ability to dry to a highly hard condition. Furthermore, incorporating silver and metal oxide nanoparticles to nail lacquer, which have antifungal qualities, is one method of treating fungal infection of the toenails (Lohani et al. 2014). Some of the marketed nail care nanocosmetics products are shown in Table 4.1.

4.2.5 Oral care

Toothpaste and mouthwash are products that are used to improve oral hygiene. Manufacturing these items with nanotechnology has become increasingly popular to increase their performance and efficacy. Most oral care products use nanocarriers like dendrimers, hydrogels, gold nanoparticles, silver nanoparticles, quantum dots, and gold nanoparticles (Dhawan, Sharma, and Nanda 2020). Table 4.1 shows the list of marketed oral care nanocosmetics products.

4.2.5.1 Mouthwash

Mouthwash promotes oral hygiene maintenance by evading gingivitis, plaque, and tartar. In order to whiten the teeth, whitening agents are incorporated into mouthwash. Mouthwashes based on nanotechnology are widely available on the market. For example, UK dentists created "Nano Whitening Mouthwash," and also "Enamel Care Technology," which bring about teeth whitening and remineralization.

TABLE 4.1

List of marketed skin care, lip care, hair care, nail care, and oral care products

Nano-cosmetics	Product name	Company	Nanoformulation	Active ingredients	Uses
Skin care	C-vit liposomal serum	Sesderma	Liposomes	Ascorbyl glucoside is encapsulated in liposomes, mulberry extract, hyaluronic acid Syncoll, and Panthenol.	Cleanse and tone the skin
	emerginC HyperVitalizer Face cream	emerginC	Liposomes	Alpha lipoic acid, hyaluronic acid, rose extract, lutein	Antiaging facial cream helps improve the appearance of fine lines, wrinkles, and uneven skin tone.
Lip care	Fillderma Lips Lip	Sesderma	Liposomes	Low and high molecular weight of Hyaluronic acid, Urea, fermented sweet black tea, mimetic peptides, collagen, niacinamide, Centella Asiatica	Increases the lip volume, fills wrinkles in the lip contour, hydrate and outline the lips, softens and protects the lip from external aggressions
	Oriflame tender care	Kara vita	Nanosphere	Olive oil, velvet veil, Gireline, Matrixyl 3000, Green tea extract, squalene, Hyaluronic acid, Tyrostat-90, Aregirline, Pepha-Tight	Lip moisturizer
	Infracyte Luscious Lips	Luscious Lips	Nano zinc oxide	Dehydrated marine collagen, Vitamin C, E, and K, Plant extract and super fruit berries, castor oil, jojoba oil, apricot kernel oil, grape seed oil, avocado oil	Lip care
Hair Care	Identik Masque Floral Repair	Identik	Niosomes	Adenosine Punica granatum seed extract, hydrolyzed yeast extract,	Hair repair masque
	Spectral DNC-N Hair Loss Treatment	DS Laboratories, Inc.	Nanosomes	Nanoxidil 5%	Alopecia
Nail Care	Nano-In Hand and Nail Moisturizing Serum and Foot Moisturizing Serum	Nano-Infinity Nanotech	Nanoparticles	Nano-zinc oxide, natural essence, etc.	Nail moisturizer
Oral Care	NanoCare gold	Dental NanoTechnology, Poland	Nanoparticles	Nanogold or nanosilver, chlorhexidine, isopropyl alcohol.	Cavity disinfectant

4.2.5.2 Toothpaste

The preparation of toothpaste using nanotechnology aims to promote healing of the teeth on its own, along with enamel repair, and conferring protection against bacterial infections. Nanosized forms of hydroxyapatite and titanium dioxide are added to toothpaste to improve their ability to whiten teeth and provide better protection (Dhawan, Sharma, and Nanda 2020). Toothpaste is loaded with silver nanoparticles that are effective microbicides (Bapat et al. 2020). When compared to traditional toothpaste without nano additives, toothpaste containing calcium carbonate nanoparticles along with 3% nanosized sodium tri-metaphosphate enhanced the remineralization of early chronic wounds (Dhawan, Sharma, and Nanda 2020).

4.3 TRANSDERMAL DELIVERY SYSTEMS FOR NANOCOSMETICS

Over the past two decades, transdermal delivery methods have undergone extensive research to overcome the skin barrier to enable more successful implementation of pharmaceutical and cosmetic items (Brandt, Cazzaniga, and Hann 2011). The term "transdermal delivery" pertains to the site of the administration of biologically active components such as drug and cosmetic ingredients. According to the delivery mechanism, the cosmetic ingredients might be either water-soluble or insoluble, permeable or impermeable. When compared to other delivery methods, transdermal delivery has many advantages. This administration method avoids injections, lowers the dosage, and lessens the risk of infections induced by repeated needle use. Furthermore, this method has the additional benefits of being self-administered, non-invasive, offering a release for prolonged activity, and affordable (Raju et al. 2020).

Transdermal delivery contains the patches that administer the active substances via unbroken skin. Cosmetic products contain therapeutic active substances for treating dermatological disorders and maintaining skin health. In this area, transdermal delivery uses cutting-edge technologies for improving the penetration of actives across the skin. The main problem encountered herein is the ineffective delivery of creams/pastes across the dermal layers. This prompted the creation of technologies like microneedles, microdermabrasion, laser radiation, and iontophoresis, etc. These improve the permeation of the payloads across the skin and achieve the action at the desired site (Raju et al. 2020). The biologically active components and the excipients are loaded into the device and assessed for efficacy and stability. Due to their simplicity of use and lack of significant health risks, these gadgets have become popular with the general public and are now available on store shelves (Draelos 2009).

4.4 NOVEL TECHNOLOGIES FOR THE DEVELOPMENT OF TRANSDERMAL SYSTEMS

The novel technologies that are presently used for mediating successful transdermal delivery are described as follows:

4.4.1 IONTOPHORESIS

During the iontophoresis process, a low electric current is introduced directly or indirectly across the skin to increase the permeability of the biological actives (Banga, Bose, and Ghosh 1999; Narasimha Murthy, Wiskirchen, and Paul Bowers 2007). This process is based on the principle of electro-repulsion for charged solutes; for uncharged solutes, it is based on the principles of electroosmosis; and for both charge-carrying and uncharged solutes, it is based on electroperturbation. For the treatment of skin diseases, spironolactone and its corresponding metabolized form, canrenone, were successfully developed in an iontophoretic system (Ferreira-Nunes et al. 2019). The method, which intends to regulate the quality of skin care products, was shown to be effective by permeation studies due to its accuracy and precision. Iontophoresis is a method that L'Oreal has patented for delivering vitamin C to the skin. This approach includes delivering vitamin C and its numerous derivatives to the skin, combined with a polymer, using direct current (Delgadillo JC, Planard-Luong TH, inventors; L'Oreal SA 2018).

4.4.2 ELECTROPORATION

The electroporation process of applying a high-voltage pulse onto the skin causes the development of a transient pore. Here, a high voltage of approximately 100 V, for a short time of approximately a few milliseconds is used. The factors pertaining to electricity-related aspects which are examined for improving the permeability are rate, waveform, and electrical pulse. This method has been used and is profitable for molecules of various sizes and lipophilicity. Neostigmine has recently been delivered transdermally using electroporation technology, which is promising for therapeutic use because it enables the delivery of smaller dosages of hydrophilic substances (Banga, Bose, and Ghosh 1999).

4.4.3 MICRONEEDLES

Initially, the chemical was delivered via the skin using a reservoir and several projections on a microneedle device. The technological advancement either uses a reservoir or a microprojection array. Microneedles are frequently employed to transport macromolecules via the skin to increase material permeability. Typically, the length of the needle used to deliver the active ingredient is between 50 and 100 mm. According to a recent study, microneedles can be used to deliver adenosine to treat wrinkles more effectively than adenosine cream, potentially creating new cosmetic products. However, it has been demonstrated that the skin micro-needling method successfully treats atrophic acne scars (Donnelly, Raj Singh, and Woolfson 2010).

4.4.4 MICRODERMABRASION

This procedure involves rupturing the skin's top layers to achieve the permeation of the biological actives. These techniques are similar to those used by dermatologists to treat acne, scars, etc. Abrasion technology can achieve skin resurfacing, which is essential for skin rejuvenation. This method treats acne vulgaris and eliminates the complications that arise with conventional acne creams (Fernandes et al. 2014).

4.4.5 BIOACTIVE NANOCARRIERS

Nanocarriers are used to deliver the biologically active material in a manner such that they improve skin permeability and transport the actives to the affected area effectively. Nanocosmeceuticals are rapidly becoming popular for treating and/or managing diseases like acne flares, wrinkles, spots, and pigmentation. The bioactive nanocarriers have several advantages over traditional carriers, including higher epidermal penetration, the ability to facilitate sustained release of the loaded actives, increased stability, specificity, and enhanced entrapment efficiency. There are several bioactive nanocarrier-based cosmetic products that function by employing different transdermal delivery techniques as described.

4.4.5.1 Liposomes

Liposomes are vesicular structures comprised of phospholipid lipid bilayers suitable for hydrophilic and lipophilic compounds, having a size ranging from 20 nm to micrometers. Due to its softening characteristics, phosphatidylcholine, a key component of liposomes, is frequently employed in skin care-based formulations (Arora, Agarwal, and Murthy 2012; Rapalli et al. 2021). Liposomes can encapsulate active substances and are thus commonly employed in dermatological applications. Liposomes are frequently used in the cosmeceutical industry, and many researchers are working on developing cosmetic products (Sundari and Anushree 2017). Kapoor et al. have achieved transdermal administration of folate fortification in combination with liposomes in cosmetic products. No surfactant or external energy for permeation was used in this delivery (Kapoor et al. 2018). In order to give vitamin D3 transdermally, Bi et al. employed liposomes as a carrier and discovered that this increased their stability. Using this product led to the repair of the brought forth by photo-aging (Bi et al. 2019).

4.4.5.2 Nanoemulsions

Nanoemulsions are kinetically stable liquid dispersions in which the oily and aqueous phases combine with a surfactant (Shah, Bhalodia, and Shelat 2010). According to studies, nanoemulsions can be employed successfully as carriers mediating transdermal delivery of lipophilic hyaluronic acid in cosmetic applications (Patel and Joshi 2012). It was reported that the matrix comprises the nanoemulsion filled with salmon, miglyol, and rapeseed oil. According to results from the study, transdermal formulations are among the best carriers for cosmetics in terms of their turbidity, size, and stability (Kabri et al. 2011).

4.4.5.3 Solid lipid nanoparticles

Solid lipid nanoparticles consist of a single layer of shells encircling a lipoic central core with dimensions between 50 and 1000 nm. Solid lipid nanoparticles are composed of biodegradable lipids that have low toxicity profiles. The smaller size enables interaction with the stratum corneum, which promotes the penetration of active compounds. Solid lipid nanoparticles are more stable than liposomes when combined with the active material. In the recent study, the solid lipid nanoparticles incorporated idebenone, an antioxidant. The study concluded that the method might be a developing approach to obtaining transdermal formulations for a cosmetic product. Another study revealed that tazarotene was incorporated into solid lipid nanoparticles for transdermal application to treat psoriasis. This solid lipid nanoparticle formulation demonstrated greater tolerability in comparison to the different formulations in the market, demonstrating the increased potential for tazarotene transdermal delivery (Aland, Ganesan, and Rao 2019).

4.4.5.4   Nanostructured lipid carriers

Nanostructured lipid carriers are being developed to address the drawbacks of solid lipid nanoparticles. These carriers have greater entrapment efficiency than solid lipid nanoparticles (Waghule et al. 2020). Nanostructured lipid carriers have a biphasic release pattern and maintain direct contact with the stratum corneum, generating the enhanced permeability of active components via the skin. NLCs depict better stability when stored and provide improved UV protection with fewer adverse effects. An NLC-based transdermal formulation comprising isoliquiritigenin with ceremide as a solid lipid was developed, improving the efficacy of its cosmetic components. The study of NLCs focused on improving skin permeability through transdermal administration of hydroquinone (Wu et al. 2017).

4.5   CONCLUSION

The cosmeceutical sector has made noteworthy advancements in developing and manufacturing innovative and effective cosmetic actives in its products. Understanding features pertaining to skin physiology are essential for ensuring the delivery of these active ingredients. Topical and transdermal administration of nanocosmetics have multiple advantages in comparison to the other approaches of administration. Even though traditional topical formulations (such as gels, emulsions, powders, suspensions, solutions, and aerosols) are suitable for the topical delivery of cosmetic active substances; these preparations have quite a few drawbacks and can lead to a compromise or inefficiency in the safety/effectiveness of the particular activity. Various nanoformulation (liposomes, niosomes, ethosomes, nanoemulsions, solid lipid nanoparticles, polymeric particles, and nanostructured lipid carriers) have been developed to circumvent these challenges. Apart from these, transdermal delivery technology-based platforms have also been developed which facilitate overcoming the numerous barriers imposed as a function of our inherent biological barriers. Transdermal delivery of cosmeceuticals has seen a remarkable increase, both for complete products and formulations loaded with cosmetic ingredients by utilizing cutting-edge technologies like microneedles and iontophoresis. This chapter establishes the significance of continuing research works in cosmetology and pharmaceuticals by incorporating the use of these techniques. The usage of nanomaterials in the continued development of skin care products that contain active substances are innovative approaches in the health and cosmetic care segments that are advantageous to both businesses and society.

REFERENCES

Aland, Rajkumar, M. Ganesan, and P. Rajeswara Rao. 2019. "In Vivo Evaluation of Tazarotene Solid Lipid Nanoparticles Gel For Topical Delivery." *International Journal of Pharmaceutical Sciences and Drug Research* 11 (1): 45–50. https://doi.org/10.25004/ijpsdr.2019.110107.

Alexander, Amit, Shubhangi Dwivedi, Ajazuddin, Tapan K Giri, Swarnlata Saraf, Shailendra Saraf, and Dulal Krishna Tripathi. 2012. "Approaches for Breaking the Barriers of Drug Permeation through Transdermal Drug Delivery." *Journal of Controlled Release* 164 (1): 26–40. https://doi.org/10.1016/j.jconrel.2012.09.017.

Arora, N., S. Agarwal, and R. S. R. Murthy. 2012. "Latest Technology Advances in Cosmaceuticals." *International Journal of Pharmaceutical Sciences and Drug Research* 4 (3): 168–82.

Banga, Ajay K, Sagarika Bose, and Tapash K Ghosh. 1999. "Iontophoresis and Electroporation: Comparisons and Contrasts." *International Journal of Pharmaceutics* 179 (1): 1–19. https://doi.org/10.1016/S0378-5173(98)00360-3.

Bapat, Ranjeet A, Tanay V Chaubal, Suyog Dharmadhikari, Anshad Mohamed Abdulla, Prachi Bapat, Amit Alexander, Sunil K Dubey, and Prashant Kesharwani. 2020. "Recent Advances of Gold Nanoparticles as Biomaterial in Dentistry." *International Journal of Pharmaceutics* 586: 119596. https://doi.org/10.1016/j.ijpharm.2020.119596.

Benson, Heather. 2005. "Transdermal Drug Delivery: Penetration Enhancement Techniques." *Current Drug Delivery* 2 (1): 23–33. https://doi.org/10.2174/1567201052772915.

Bi, Ye, Hongxi Xia, Lianlian Li, Robert J. Lee, Jing Xie, Zongyu Liu, Zhidong Qiu, and Lesheng Teng. 2019. "Liposomal Vitamin D3 as an Anti-Aging Agent for the Skin." *Pharmaceutics* 11 (7). https://doi.org/10.3390/pharmaceutics11070311.

Brandt, Fredric S., Alex Cazzaniga, and Michael Hann. 2011. "Cosmeceuticals: Current Trends and Market Analysis." *Seminars in Cutaneous Medicine and Surgery* 30 (3): 141–43. https://doi.org/10.1016/j.sder.2011.05.006.

Delgadillo, J. C., T. H. Planard-Luong, inventors; L'Oreal SA, US. 2018. "Iontophoresis Method of Delivering Vitamin c through the Skin and Iontophoresis Device Comprising: An Electrode Assembly Including at Least One Electrode and an Aqueous Active Agent. United States Patent Application" US 16/062.

Dhawan, Surbhi, Pragya Sharma, and Sanju Nanda. 2020. "Cosmetic Nanoformulations and Their Intended Use." In, 141–69. https://doi.org/10.1016/B978-0-12-822286-7.00017-6.

Donnelly, Ryan F., Thakur Raghu Raj Singh, and A. David Woolfson. 2010. "Microneedle-Based Drug Delivery Systems: Microfabrication, Drug Delivery, and Safety." *Drug Delivery* 17 (4): 187–207. https://doi.org/10.3109/10717541003667798.

Draelos, Zoe Diana. 2009. "Cosmeceuticals: Undefined, Unclassified, and Unregulated." *Clinics in Dermatology* 27 (5): 431–34. https://doi.org/10.1016/j.clindermatol.2009.05.005.

Dubey, Sunil Kumar, Anuradha Dey, Gautam Singhvi, Murali Manohar Pandey, Vanshikha Singh, and Prashant Kesharwani. 2022. "Emerging Trends of Nanotechnology In Advanced Cosmetics." *Colloids and Surfaces B: Biointerfaces*, 112440. https://doi.org/10.1016/j.colsurfb.2022.112440.

Durazzo, Alessandra, Massimo Lucarini, Eliana B. Souto, Carla Cicala, Elisabetta Caiazzo, Angelo A. Izzo, Ettore Novellino, and Antonello Santini. 2019. "Polyphenols: A Concise Overview on the Chemistry, Occurrence, and Human Health." *Phytotherapy Research* 33 (9): 2221–43. https://doi.org/10.1002/ptr.6419.

Fernandes, Mariane, Nanci Mendes Pinheiro, Virgínia Oliveira Crema, and Adriana Clemente Mendonça. 2014. "Effects of Microdermabrasion on Skin Rejuvenation." *Journal of Cosmetic and Laser Therapy* 16 (1): 26–31. https://doi.org/10.3109/14764172.2013.854120.

Ferreira-Nunes, Ricardo, Larissa A. Ferreira, Tais Gratieri, Marcilio Cunha-Filho, and Guilherme M. Gelfuso. 2019. "Stability-Indicating Analytical Method of Quantifying Spironolactone and Canrenone in Dermatological Formulations and Iontophoretic Skin Permeation Experiments." *Biomedical Chromatography* 33 (11). https://doi.org/10.1002/bmc.4656.

Hu, Zhenhua, Meiling Liao, Yinghui Chen, Yunpeng Cai, Lele Meng, Yajun Liu, Nan Lv, Zhenguo Liu, and Weien Yuan. 2012. "A Novel Preparation Method for Silicone Oil Nanoemulsions and Its Application for Coating Hair with Silicone." *International Journal of Nanomedicine* 7: 5719–24. https://doi.org/10.2147/IJN.S37277.

Kabri, Tin Hinan, Elmira Arab-Tehrany, Nabila Belhaj, and Michel Linder. 2011. "Physico-Chemical Characterization of Nano-Emulsions in Cosmetic Matrix Enriched on Omega-3." *Journal of Nanobiotechnology* 9 (1): 1–8. https://doi.org/10.1186/1477-3155-9-41/FIGURES/3.

Kapoor, Mudra Saurabh, Anisha D'Souza, Noorjahan Aibani, Swathi Sivasankaran Nair, Puja Sandbhor, Durga Kumari, and Rinti Banerjee. 2018. "Stable Liposome in Cosmetic Platforms for Transdermal Folic Acid Delivery for Fortification and Treatment of Micronutrient Deficiencies." *Scientific Reports* 8 (1): 1–12. https://doi.org/10.1038/s41598-018-34205-0.

Kaul, Shreya, Neha Gulati, Deepali Verma, Siddhartha Mukherjee, and Upendra Nagaich. 2018. "Role of Nanotechnology in Cosmeceuticals: A Review of Recent Advances." *Journal of Pharmaceutics* 2018 (March): 1–19. https://doi.org/10.1155/2018/3420204.

Kim, Byel, Hang-Eui Cho, Sun He Moon, Hyun-Jung Ahn, Seunghee Bae, Hyun-Dae Cho, and Sungkwan An. 2020. "Transdermal Delivery Systems in Cosmetics." *Biomedical Dermatology* 4 (1): 1–12. https://doi.org/10.1186/s41702-020-0058-7.

Kumari, Karuna, Pramod Kumar Sharma, and Rahul Gupta. 2017. "Cosmeceuticals: An Emerging Novel Trend towards Dermal Care." *Adv Cosmetics Dermatol.* 1–12

Lohani, Alka, Anurag Verma, Himanshi Joshi, Niti Yadav, and Neha Karki. 2014. "Nanotechnology-Based Cosmeceuticals." In Edited by T J Ryan and T Maisch. *ISRN Dermatology* 2014: 843687. https://doi.org/10.1155/2014/843687.

Mozafari. 2017. "Enhanced Efficacy and Bioavailability of Skin-Care Ingredients Using Liposome and Nano-Liposome Technology." *Modern Applications of Bioequivalence & Bioavailability* 2 (2): 1–3. https://doi.org/10.19080/mabb.2017.02.555584.

Nanda, Sanju, Arun Nanda, Shikha Lohan, Ranjot Kaur, and Bhupinder Singh. 2016. "Nanocosmetics: Performance Enhancement and Safety Assurance." In, 47–67. https://doi.org/10.1016/B978-0-323-42868-2.00003-6.

Narasimha Murthy, S., Dora E. Wiskirchen, and Christopher Paul Bowers. 2007. "Iontophoretic Drug Delivery across Human Nail." *Journal of Pharmaceutical Sciences* 96 (2): 305–11. https://doi.org/10.1002/jps.20757.

Patel, R. P., and J. R. Joshi. 2012. "An Overview on Nanoemulsion: A Novel Approach." *International Journal of Pharmaceutical Sciences and Research* 3 (12): 4640–50.

Patravale, V. B., and S. D. Mandawgade. 2008. "Novel Cosmetic Delivery Systems: An Application Update." *International Journal of Cosmetic Science* 30 (1): 19–33. https://doi.org/10.1111/J.1468-2494.2008.00416.X.

Raju, Nikhishaa Sree, Venkateshwaran Krishnaswami, Sivakumar Vijayaraghavalu, and Ruckmani Kandasamy. 2020. *Transdermal and Bioactive Nanocarriers. Nanocosmetics.* INC. https://doi.org/10.1016/b978-0-12-822286-7.00002-4.

Rapalli, Vamshi Krishna, Arisha Mahmood, Tejashree Waghule, Srividya Gorantla, Sunil Kumar Dubey, Amit Alexander, and Gautam Singhvi. 2021. "Revisiting Techniques to Evaluate Drug Permeation through Skin." *Expert Opinion on Drug Delivery* 18 (12): 1829–42. https://doi.org/10.1080/17425247.2021.2010702.

Rapalli, Vamshi Krishna, Gautam Singhvi, Sunil Kumar Dubey, Gaurav Gupta, Dinesh Kumar Chellappan, and Kamal Dua. 2018. "Emerging Landscape in Psoriasis Management: From Topical Application to Targeting Biomolecules." *Biomedicine & Pharmacotherapy* 106: 707–13. https://doi.org/10.1016/j.biopha.2018.06.136.

Rosen, Jamie, Angelo Landriscina, and Adam J Friedman. 2015. "Nanotechnology-Based Cosmetics for Hair Care." *Cosmetics*. https://doi.org/10.3390/cosmetics2030211.

Shah, P, D Bhalodia, and Pragna Shelat. 2010. "Nanoemulsion: A Pharmaceutical Review." *Systematic Reviews in Pharmacy* 1 (January). https://doi.org/10.4103/0975-8453.59509.

Souto, Eliana B., Ana Rita Fernandes, Carlos Martins-Gomes, Tiago E. Coutinho, Alessandra Durazzo, Massimo Lucarini, Selma B. Souto, Amélia M. Silva, and Antonello Santini. 2020. "Nanomaterials for Skin Delivery of Cosmeceuticals and Pharmaceuticals." *Applied Sciences* 10 (5): 1594. https://doi.org/10.3390/APP10051594.

Sundari, P. T., and H. Anushree. 2017. "Novel Delivery Systems: Current Trend in Cosmetic Industry." *European Journal of Pharmaceutical and Medical Research* 4 (8): 617–27.

Swain, Suryakanta, Sarwar Beg, Astha Singh, Ch. Niranjan Patro, and M. E. Bhanoji Rao. 2011. "Advanced Techniques for Penetration Enhancement in Transdermal Drug Delivery System." *Current Drug Delivery* 8 (4): 456–73. https://doi.org/10.2174/156720111795767979.

Tripura Sundari, P., and H. Anushree. 2017. "Novel Delivery Systems: Current Trend in Cosmetic Industry." *European Journal of Pharmaceutical and Medical Research* 4 (8): 617–27.

Vitorino, Carla, António Almeida, João Sousa, Isabelle Lamarche, Patrice Gobin, Sandrine Marchand, William Couet, Jean-Christophe Olivier, and Alberto Pais. 2014. "Passive and Active Strategies for Transdermal Delivery Using Co-Encapsulating Nanostructured Lipid Carriers: In Vitro vs. in Vivo Studies." *European Journal of Pharmaceutics and Biopharmaceutics* 86 (2): 133–44. https://doi.org/10.1016/j.ejpb.2013.12.004.

Waghule, Tejashree, Vamshi Krishna Rapalli, Srividya Gorantla, Ranendra Narayan Saha, Sunil Kumar Dubey, Anu Puri, and Gautam Singhvi. 2020. "Nanostructured Lipid Carriers as Potential Drug Delivery Systems for Skin Disorders." *Current Pharmaceutical Design* 26 (36): 4569–79. https://doi.org/10.2174/1381612826666200614175236.

Wu, Pey Shiuan, Chih Hung Lin, Yi Ching Kuo, and Chih Chien Lin. 2017. "Formulation and Characterization of Hydroquinone Nanostructured Lipid Carriers by Homogenization Emulsification Method." *Journal of Nanomaterials* 2017. https://doi.org/10.1155/2017/3282693.

5 Fabrication of nanocosmetics

Manisha Choudhari, Anuradha Dey,
Gautam Singhvi, Ranendra Narayan
Saha and Sunil Kumar Dubey

CONTENTS

5.1 INTRODUCTION

The term cosmetics refers to "substances designed for application to the human body for cleansing, beautifying, boosting attractiveness, or changing appearance without disrupting body physiology or functions," according to the US FDA. Cosmetics is a well-known consumer product in the global marketplace; therefore, it is a profitable sector for several market players (Dhapte-Pawar et al. 2020). The demand and supply for cosmetics have increased due to the lifestyle-related changes, weather alterations, and pollution. The cosmetic market globally was worth USD 532.43 billion

DOI: 10.1201/9781003319146-5

in 2017 and is projected to reach USD 805.61 billion in 2023. The aging and elderly population are increasing worldwide due to the lowered mortality rates. The global market for anti-aging-based cosmetics has been ever-expanding given the increase in demand by both sexes to appear young and slow down aging. Therefore, older adults will have great demand for cosmetics. The growing demand for natural and herbal beauty items has exciting opportunities for companies to create innovative cosmetic products (Gautam, Singh, and Vijayaraghavan, 2011; Dhapte-Pawar et al., 2020).

Nowadays, cosmeceuticals are frequently used in several situations to avoid wrinkles, uneven skin tone, hair damage, black spots, photo aging, hyperpigmentation, and dry skin (Rapalli et al. 2021). Currently, the cosmetics sector is investigating nanotechnology's use for various potential applications. In 1986, Christian Dior introduced the term "nanocosmetics" to describe cosmetic preparation containing nanoparticles (Oberdörster, Oberdörster, and Oberdörster 2005). These products gained popularity after L'Oreal found the commercial benefits of nanocosmetics in 2005. Some of the benefits related to nanotechnology-based cosmeceuticals include increased bioavailability, prolonged activity, and enhanced aesthetic appeal of products. These products have several additional advantages over conventional cosmeceuticals, such as having a smaller size and a high surface-to-volume ratio, which renders them invaluable in cosmeceuticals (Dubey, Dey, et al. 2022).

Furthermore, incorporating nanoparticles into cosmetic compositions does not alter the properties of cosmeceuticals but promotes increased coverage, improved appearance, and skin adherence (Gupta et al. 2022). Even though they provide

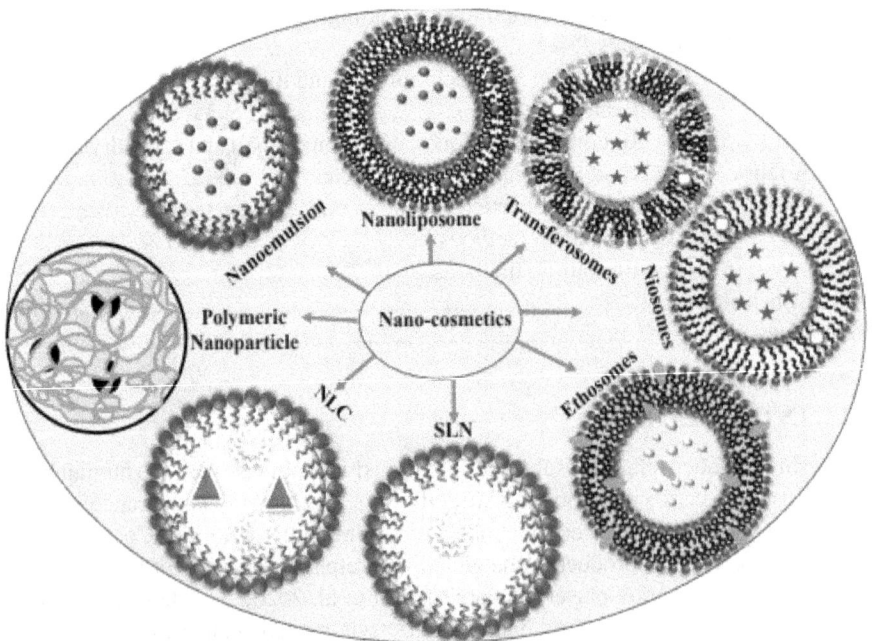

FIGURE 5.1 Different types of nanocosmaceuticals

numerous advantages, they still have drawbacks regarding scalability, cost, stability, and toxicity. Furthermore, nanomaterials' safety and toxicity characteristics are still controversial (Santos et al. 2019; Dhawan, Sharma, and Nanda 2020). Besides, they exhibit dose-dependent toxicity via various routes of administration (Gupta et al. 2022). It is usually seen that the dosage of the active ingredient has a more significant influence on its bioavailability than the physiochemical parameters of the active component (Hoag and Hussain 2001; Gupta et al. 2022).

The present chapter outlines the various nanoparticles and nano delivery systems for cosmetics, including nanoliposomes, nanostructured lipid carriers (NLCs), solid lipid nanoparticles (SLNs), niosomes, transferosomes, ethosomes, nanocapsules, and nanoemulsions, fullerenes, and carbon-based nanoparticles (Dubey, Dey, et al. 2022). Figure 5.1 illustrates the different types of nanocosmaceuticals (Gupta et al. 2022). Also, this chapter offers different fabrication methods for nanocosmetics. Each country has its own set of rules for regulating cosmetic and cosmeceutical products. It is necessary to update the guidelines to consider the use, popularity, and efficacy of cosmeceuticals driven by nanotechnology. This review also provides the current regulatory situation, vital for creating unified standards that can be widely accepted globally (Dhapte-Pawar et al. 2020).

5.2 NANOMATERIALS IN COSMETICS

Nanomaterials are substances with at least one nanoscale dimension and have unique physiochemical parameters. These components have been widely employed in the cosmetics sector for numerous years. Cosmetics containing nanomaterials have an increased number of benefits than that of microscale cosmetics (Gupta et al. 2022). The vast surface area of these particles is crucial for their absorption, transparency, bioavailability, and accessible transportation, along with the product's long-term effect (Gupta et al. 2022). A synopsis of the most popular nanomaterials that are applied in cosmetics is provided herewith the grouping of the same into two broad

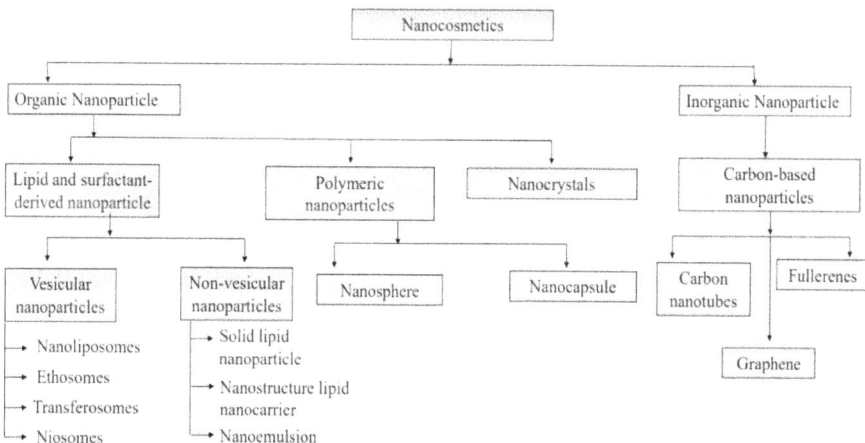

FIGURE 5.2 Classification of nanocosmetics

classes; organic and inorganic nanoparticles. The organic nanoparticle is divided into lipid and surfactant-derived nanoparticles, polymeric nanoparticles, and nanocrystals. Furthermore, the inorganic nanoparticles contain carbon-based nanoparticles. The carbon-based nanoparticles are categorized into carbon nanotubes, fullerenes, and graphene (Salvioni et al. 2021). The classification of nanocosmetics is shown in Figure 5.2.

5.3 FABRICATION METHODS OF ORGANIC NANOPARTICLES

The term "organic nanoparticles" refers to an extensive nanosized formulation comprised of either natural or synthetic components. Organic nanoparticles based on lipids, surfactants, and polymers are primarily employed in the cosmetic sector as carrier-agents of active compounds, whereas nanocrystals are solid-particulate in forms that are manufactured to increase substance solubility (van Hoogevest and Fahr 2019a; Bapat et al. 2020).

5.3.1 Lipid and surfactant-derived nanoparticles

5.3.1.1 Vesicular nanoparticles

According to the specific sizes and the number of bilayers its constitutes, vesicular nanoparticles are typically categorized as multilamellar-vesicles (MLVs), small-unilamellar vesicles (SUVs), and large-unilamellar vesicles (LUVs) (van Hoogevest and Fahr 2019b). When raw materials are dispersed in an aqueous phase, MLVs (0.5–10 m) are easily formed, and MLVs can be converted into SUVs (10–100 nm) and LUVs (100–500 nm) using a variety of processes (e.g., extrusion or sonication or high-shear homogenization, etc.) (Rahimpour and Hamishehkar 2012). Sub-micron nanocarriers are typically preferred in cosmetic science. The cosmetics sector promoted the use of such nanomaterial because of the systems resemblance to natural biological vesicles and membranes (Van Tran, Moon, and Lee 2019).

5.3.1.1.1 Nanoliposomes

Liposome was the earliest colloidal vesicular system (comprised of single or multiple lipid phosphatidylcholine bilayers) to be investigated for effective skin delivery. It is also commonly employed as a controlled release system. However, conventional liposomes are enormous and cannot enter tiny blood arteries or the skin, whereas nanoliposomes have a more remarkable ability to penetrate (Carita et al., 2017). Nanoliposomes are single/double bilayered liposomes whose sizes fall in the nanometer range. They are around 800 times smaller (about 50 nm in size) than the diameter of human hair (Fakhravar et al. 2016). Figure 5.3 shows the schematic illustration of a bilayer structure of nanoliposomes.

Nanoliposomes have multiple methods of preparation. It is to be noted that the formation of liposomes or nanoliposomes is not a process that occurs spontaneously; instead, the system requires sufficient energy to break through the energy barriers. In other words, when phospholipids like lecithin are introduced into water, lipid vesicles are created and then form bilayer structures once sufficient energy is provided (Van

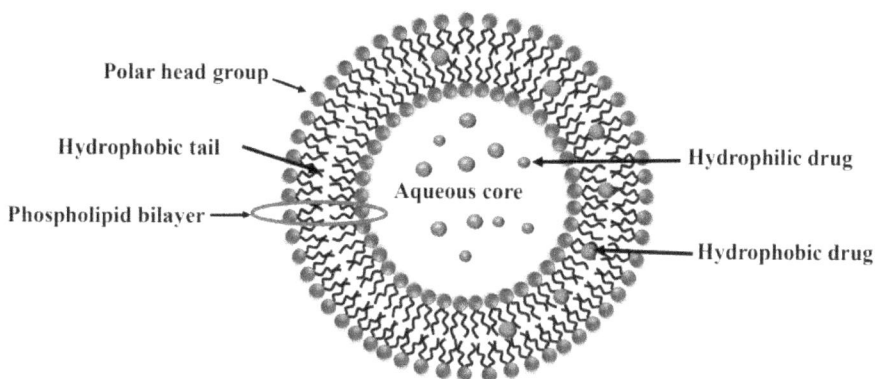

FIGURE 5.3 Schematic illustration of a bilayer structure of nanoliposomes

Tran, Moon, and Lee 2019). The lipid molecules are arranged into bilayer vesicles from energy input (such as sonication, homogenization, heating, etc.) to obatin thermodynamic equilibrium in the aqueous-phase (Mozafari 2005, 2010). The following section explains the methods for producing nanoliposomes.

5.3.1.1.1.1 Microfludization technique The microfluidization technique employing a microfluidizer is a method of producing nanoliposomes without the use of potentially hazardous solvents. This technology has been utilized in the pharmaceutical industry to make liposome-based products and emulsions of pharmaceutical relevance (Vemuri et al. 1990). Microfluidization works on the principle of splitting a pressure stream, passing both parts of the split stream via fine orifices, and directing their flows towards each other inside the microfluidizer chamber. Inside the interaction chamber, liposome particle sizes are decreased by cavitation, shear, and impact (Mahdi Jafari, He, and Bhandari 2006). Microfluidizers employ high pressures (up to 100,000 psi) to direct the flow stream through micro channels towards the region of impingement (Sorgi and Huang 1996). The fabrication method of nanoliposomes involves a few steps as follows:

Depending on the intended usage, select the nanoliposome components and the media in which they are suspended. Typically, aqueous phases like a buffer or deionized/distilled water serve as the suspension medium. A phospholipid dispersion is prepared by dissolving the nano-liposomal components into the suspension medium; post this, it is agitated with a blender or a homogenizer. The suspension of crude phospholipids is placed in the reservoir, the air regulator is set to the optimum required pressure at which it has to be operated, and post this, the dispersion is made to run through the microfluidizer. The liquid dispersion travels across a filter to the interaction chamber with an optimal configuration, where it gets divided into two streams that interact at very high speeds in dimensionally defined micro channels. Recycling the suspension is possible using the equipment. Post the sample collection, the microfluidizer cleaning can be done by recycling 95% ethanol along with distilled water across the system. Then the liposome suspension is allowed to anneal and stabilize for one hour at temperatures higher than Tc in an inert environment such as nitrogen/argon (Mahdi Jafari, He, and Bhandari 2006; Mozafari 2010).

5.3.1.1.1.2 Sonication technique   Sonication is a quite simple technique for decreasing the particle size of liposomes for producing nanoliposomes (Woodle and Papahadjopoulos 1989). The most typical laboratory-based method is treating the hydrated vesicles with a probe sonicator (titanium-tipped) for several minutes given that the temperature is well controlled during the process (Woodbury et al. 2006). A mixture of the phospholipid components has to be dissolved in either chloroform/ chloroform-methanol mixture (typically 2:1 v/v), with or without cholesterol. After that, the mixture is filtered in order to remove the small insoluble substances by filtration and reduce and diminish the pyrogens using an ultrafilter. The solution is then transferred into a round-bottomed/pear-shaped flask or container. Then a rotary evaporator is used for removing the solvents at temperatures above Tc while maintaining negative pressure and forming a thin layer of dried ingredients inside the flask. Lyophilization and spray drying are additional techniques for drying the lipid components. Using the vacuum pump, remove the residues of the organic solvent, usually overnight at a pressure less than 0.1 Pa.

An appropriate aqueous phase, for example, distilled water along with buffers, is incorporated into the flask containing the dried lipids after adding a small size of glass beads (500 mm in diameter). Handshaking or mixing is done for one to five minutes which will help the dried lipids to dissolve into the hydration fluid. At this point, micrometric liposomes of the MLV type are developed. Then the MLV-containing flask contents are transferred to a bath-type of sonicator or probe (tipped-type) sonicator and sonicated for 10–15 minutes. At this phase, nanoliposomes are generated primarily in small unilamellar vesicles (SUVs). Apart from that, nanoliposomes can be generated with the help of a bath sonicator. Firstly, the bath sonicator is filled with water and a few drops of the liquid detergent are added. Then the MLV flask is placed in the bath sonicator using a ring stand and test tube clamp. The liquid level inside and outside of the flask should be the same. Sonicate for between 20 and 40 minutes (Mozafari 2010).

5.3.1.1.1.3 Extrusion technique   By the process of extrusion, micrometric liposomes undergo structural modification; for example, the transformation of MLV to form LUV or nanoliposomes, which is dependent upon the pore size of the filters utilized. Polycarbonate filters with predetermined pore diameters are used for physically extruding vesicles under pressure (Hope et al. 1985). Firstly, liposome formulations are prepared, such as MLV, explained hitherto by the sonication method. Then the extruder filter container is stacked with one/two polycarbonate filters. Then the heating block and extruder stand are kept on a hot plate.

Then the thermometer is placed in the well present in the heating block. The hot plate is turned on and the temperature is increased above the melting point of the lipids. For minimizing the dead volume, the pre-wetted extruder parts are run once with the buffer and then the buffer is discarded. In the case of a small extruder, the liposome suspension is loaded into the donor syringe and then carefully inserted into one of the ends of the extruder by lightly twisting it. At the other end, the second syringe, the receiver syringe is inserted (MacDonald et al. 1991). Upon confirming that the plunger (receiver syringe) is at zero, the extruder device is placed into the extruder stand. Either of the opposing apexes should fall in the vertically same plane when the stainless steel hexagonal nut is inserted. The syringes are held in order to maintain good heating contact with the heating block using the swing-arm clips. Allow the liposome suspension to reach the temperature of the heating block (around 5–10 minutes). The syringe is plunged gently for transferring the entire liposome suspension to a new syringe. Repeat the extrusion operation at least seven times through the filters. The sample generally becomes more homogeneous the more it is run through the filters. The extruder is removed from the heating block with care. The nanoliposome sample is injected into a fresh new vial (Berger et al. 2001).

5.3.1.1.1.4 Heating method Most methods for making nanoliposomes either use strong shear forces or potentially hazardous solvents, for example, acetone, chloroform, methanol, and diethyl ether. The stability of the lipid vesicles may be affected by the residues formed from these hazardous solvents, which could increase potential toxicity (Vemuri and Rhodes 1995). There are multiple methods for reducing the concentration of residual solvents in liposomes, for example, performing gel filtration or carrying out dialysis, or using vacuum, etc.; however, these are complex and time-taking procedures. During the nanoliposomes preparation utilization of the high shear forces (which occur during the microfluidization), there are several reports on the destructive impacts of this method, structurally hampering the material undergoing encapsulation (Silvestri, Gabrielson, and Wu 1991; Badmeier and Chen 1993; Cencia-Rohan and Silvestri 1993; Yuh-Fun and Chung 1999). These difficulties can be avoided using alternate preparation techniques, such as the heating approach. This method allows the production of liposomes and nanoliposomes using an individual apparatus without the use of possibly harmful solvents, as mentioned in the procedure (Mozafari et al. 2002; Mortazavi et al. 2007):

An appropriate mixture of phospholipid components, either with having cholesterol or not is hydrated in an aqueous medium for one to two hours in an inert atmosphere, maintained with nitrogen/argon. Based on their solubility, the components of the nanoliposomal may be hydrated either collectively or separately. Using a pyrex beaker which is heat-resistant in nature, the lipid dispersions along with the substance to be encapsulated are combined and then glycerol is added to the final volume, maintaining the concentration of 3%. The flask containing the lipid-glycerol mixture is kept on a hot plate stirrer and agitated at 800–1000 RPM, with temperature maintained above the melting point of the lipids (~30 minutes) or until dispersion of lipids is complete. The production of nanometric vesicles can be accomplished without filtration or sonication.

5.3.1.1.1.5 Mozafari method One of the newest and easiest methods for creating liposomes and nanoliposomes was recently introduced and is termed the Mozafari method is an updated version of the heating process. Recently, the food-grade antibacterial nisin was successfully encapsulated and delivered to specific areas using the Mozafari approach (Colas et al. 2007). The Mozafari method makes provision for the production of carrier systems in a single step sans the requirement of pre-hydrating constituent material or without using hazardous solvents on small or large industrial scales. The described technique uses a simple protocol and a single vessel to produce nanoliposomes with better monodispersity and storage stability at a low cost (Mozafari 2010).

5.3.1.1.2 Transferosomes

Vesicular carrier systems known as transferosomes are produced such that there is a minimum of one inner-aqueous compartment which is surrounded by the lipid bilayer, having an edge activator molecule shown in Figure 5.4. Ultra-deformable vesicles have the property of self-optimizing and regulating themselves owing to the aqueous core that is encircled by a lipid bilayer. The elastic nature of transferosomes makes it possible for them to distort and squeeze themselves as complete vesicles through tiny pores or skin constrictions much smaller than the vesicle size (Rai, Pandey, and Rai 2017; Opatha, Titapiwatanakun, and Chutoprapat 2020).

Although there are numerous patented ways to prepare transferosomes, there is no standard preparation method or protocol for this technique. Due to this, the ideal pre-parative conditions and vesicle constituents must be determined, created, and opti-mized by carrying out specifically tailored experimental processes such that each selected therapeutic agent is coupled with the most suitable carrier for achieving the highest stability and deformability levels and drug carrying capacity. The thin-film hydration approach also called as the rotary-evaporation sonication technique is the traditional method of transferosomes preparation. Vortexing-sonication, modified handshaking procedure, suspension-homogenization, high-pressure homogeniza-tion, centrifugation, reverse-phase evaporation, and ethanol injection methods are other modified preparation techniques. The general description of each methodology is as follows:

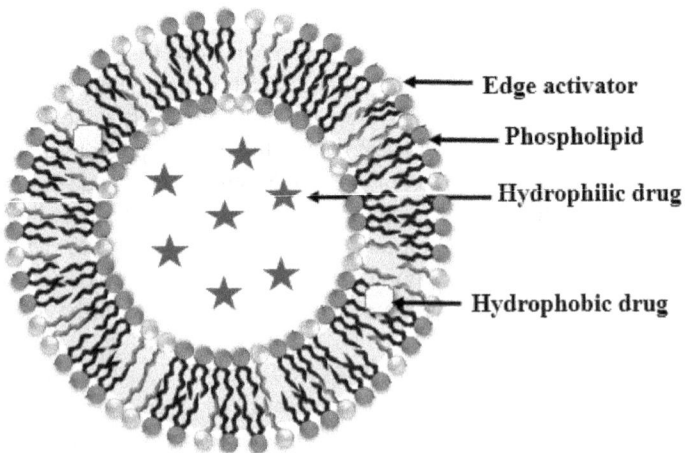

FIGURE 5.4 Schematic representation of transferosomes

5.3.1.1.2.1 Thin film hydration techniques/rotatory-evaporation sonication method In a round-bottomed flask, the phospholipids and edge activator molecules are dissolved with a volatile organic solvent mixture (chloroform-methanol). At this stage, lipophilic drugs can be incorporated. A rotary vacuum evaporator is used for evaporating the solvent at a temperature more than the lipid transition temperature with reduced pressure conditions to produce a thin layer film. The vacuum is maintained for removing the solvent residue. The deposited thin-film is then made to undergo hydration using the buffer solution with a proper pH. This stage allows for the introduction of hydrophilic drugs. The formed vesicles undergo swelling at room temperature and are then sonicated to produce smaller vesicles. The vesicles undergo homogenization via an extrusion process across a sandwich of 200–100 nm pore-sized polycarbonate membranes (Benson 2006; Dhopavkar and Kadu 2017;).

5.3.1.1.2.2 Vortexing-sonication method The drug, edge-activator, and phospholipids are mixed together using phosphate buffer as media. After that, the mixture undergoes vortexing to create a milky suspension of transferosomes. After being sonicated for the appropriate timeframe at room temperature it undergoes extrusion via polycarbonate membranes (El Zaafarany et al. 2010; Sharma, Yusuf, and Pathak 2014).

5.3.1.1.2.3 Modified handshaking method The basic principle of the modified handshaking method is similar to that of the rotatory-evaporation sonication method. The modified handshaking method involves incorporating the lipophilic API, the phospholipids, the edge activator, and the organic solvent into a flask. All components are dissolved into the organic solvent for obtaining a clear transparent solution. Instead of employing a rotating vacuum evaporator, the organic solvent is eliminated via evaporation during handshaking. The round-bottom flask is then partly submerged in a water-bath that is kept at a higher temperature (for instance, 40–60°C). After that, a thin-lipid film is formed in the interior flask wall. Post this, the flask is left for a night to allow the solvent to evaporate completely. The produced film is then made to undergo slow-hydration with buffer addition and gently shaken at a temperature slightly more than its phase transition temperature. At this stage, the hydrophilic drug can be incorporated (Dhopavkar and Kadu 2017).

5.3.1.1.2.4 Suspension homogenization method In order to produce the transferosomes, a suitable quantity of edge activators is added to an ethanolic phospholipid solution. After that, the produced suspension and buffer are combined to produce a completely concentrated lipid form. Then this formulation undergoes sonication, freezing, and thawing processes respectively (Ghai Ishan et al. 2012; Chaurasiya et al. 2019).

5.3.1.1.2.5 Centrifugation process In the first step, the phospholipid, edge-activator, and lipophilic drug molecules are dissolved in the organic solvent. The solvent is subsequently eliminated using a rotary-evaporation method with lowered pressure and appropriate temperature. The residual traces of the solvents are eliminated under vacuum. The formed lipid-film is then soaked using a suitable buffer, undergoing centrifugation at room temperature. This stage allows for incorporating hydrophilic drugs (Szoka and Papahadjopoulos 1978; Chaurasiya et al. 2019).

5.3.1.1.2.6 Reverse phase evaporation process In a round-bottomed flask, the phospholipids and edge-activator molecules are added, and the mixture of organic solvents is used to dissolve them (diethyl ether-chloroform mixture). The lipophilic active pharmaceutical ingredient (API) is added at this stage. Then, the solvent evaporation is carried out using a rotatory evaporator to produce the lipid-films. This lipid film re-dissolved in the organic-phase, mainly containing the isopropyl and diethylether. The aqueous-phase is incorporated into the organic-phase to produce the two-phase system. This stage allows for the incorporation of hydrophilic drugs. After that, a bath sonicator is used to sonicate this system till a homogenous w/o emulsion is created. The organic solvent undergoes slow evaporation using a rotary-evaporator to create a slightly viscous gel that develops into a vesicular suspension (Chen et al. 2014).

5.3.1.1.2.7 High-pressure homogenization technique The drug, edge activator, and phospholipids are evenly distributed in PBS or alcohol-containing distilled water, then swirled and ultrasonically shaken concurrently. After that, the mixture is exposed to intermittent ultrasonic shaking. A high-pressure homogenizer is subsequently used to homogenize the resulting mixture. The transferosomes are then kept in the appropriate conditions (Wu et al. 2019).

5.3.1.1.2.8 Ethanol injection method Organic phase is prepared to comprise of, dissolved phospholipids, edge-activator, and lipophilic drug molecules respectively in ethanol and agitating it using a magnetic stirrer for the appropriate time to obtain a clear solution. The water-soluble components undergo dissolution in the phosphate buffer to produce the aqueous-phase. This stage allows for introduction of hydrophilic drugs. The temperature is raised to 40–45°C. After that, the aqueous solution is continuously stirred while the ethanolic phospholipid solution is added dropwise. Removal of ethanol is carried out by transfer of the dispersion into an evaporator (under vacuum conditions) and then sonicating it for reducing the size of the particles. At normal temperatures, the resultant vesicles are enlarged. Further, sonication is performed at ambient temperature on the resulting multilamellar lipid vesicles (Yang et al. 2015; Balata et al. 2020).

5.3.1.1.3 Ethosomes

Ethosomes are unique lipid vesicular transporter systems with a high percentage of ethanol shown in Figure 5.5. This system is mainly containing ethanol, phospholipid, and water. These nanocarriers are specifically engineered to carry therapeutic drugs with varying physicochemical characteristics into the deeper skin layers. Ethosome dispersions are included in gels, patches, and creams for ease of applying and stability reasons (Verma and Pathak 2010).

There are different preparation methods of ethosomes, such as the hot, cold, and classical mechanical dispersion methods.

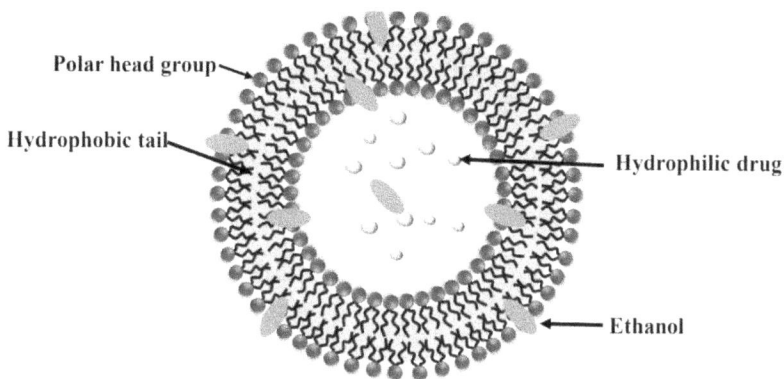

FIGURE 5.5 Schematic representation of ethosomes

5.3.1.1.3.1 Hot method Herein the API upon being dissolved in the ethanol-propylene glycol system is added to the dispersion of phospholipids present in water at around 40°C. Using a probe sonicator, the solution undergoes sonication at 4°C (three cycles; five minutes each spaced with a five minute break each). For preparing nanosized ethosomes, the formulation undergoes three cycles of homogenization at 15,000 psi with a high-pressure homogenizer (Bhalaria, Naik, and Misra 2009).

5.3.1.1.3.2 Cold method The cold method is the most popular and extensively adopted technique for preparing the ethosomes in a covered vessel incorporating phospholipids, API, and lipid components solubilized in the ethanol with vigorous stirring at room temperature. In a water bath, the mixture is warmed to 30°C. In a separate vessel, the water is warmed to 30°C and poured into the aforementioned mixture, followed by agitation for five minutes in the vessel. If required, the ethosomes formulation vesicle size can be reduced via extrusion or sonication. It has to be ensured that the formulation is appropriately stored in a refrigerator before use (Verma and Pathak 2010).

5.3.1.1.3.3 Classic mechanical dispersion method In a round-bottomed flask, soya-based phosphatidylcholine is solubilized in a mixture composed of chloroform: methanol (3:1). At temperatures higher than the lipid transition temperature, the solvents are evaporated using a rotating vacuum evaporator to create a thin lipid film on the flask wall. Finally, residues of the solvent mixture are eliminated from the deposition of lipid-coating by vacuuming the same overnight. Hydration is accomplished by spinning the flask at a proper temperature and using varied concentrations of a hydroethanolic mixture containing API (Dubey et al. 2007; Dubey, Mishra, and Jain 2007).

5.3.1.1.4 Niosomes

Niosomes are vesicular structures mainly composed of hydrated non-ionic surfactants and, in several cases, cholesterol and/or its derivatives may also be incorporated.

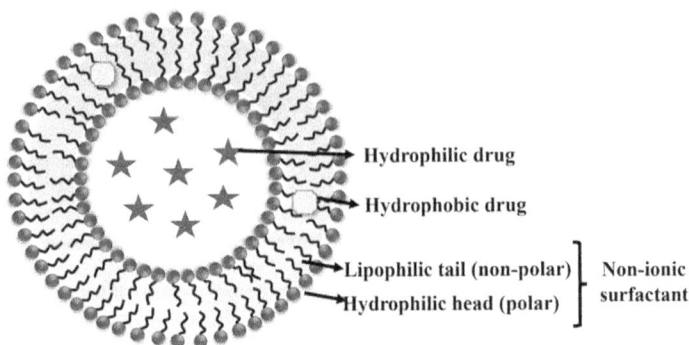

FIGURE 5.6   Schematic representation of niosomes

They can encapsulate hydrophilic and lipophilic molecules due to their distinctive structural features shown in Figure 5.6. It can be done by entrapment of hydrophilic compounds in the aqueous cores of the vesicles or by adsorbing them on the bilayers-surfaces while encapsulating the lipophilic compounds by partitioning them into the lipophilic core of the bilayers. Upon hydration, the thin lipid film and the stacks of the liquid-crystalline layer swell up and develop into liposomes. Agitation causes the hydrated lipid-sheets to separate and self-assemble into vesicular structures, which prevents the interaction of water with the bilayer's hydrocarbon-core at the borders. Production of the niosomes has already begun in the cosmetic industry, and their essential applications in drug delivery are being studied (Pardakhty and Moazeni 2012).

In the following sections, various ways are used to manufacturing niosome (shown in Figure 5.7) for drug administration and gene therapy vectors. Furthermore, they may also be coated with numerous agents such as polyethylene glycol (PEG) and hyaluronic acid (HA)-antibody conjugates for specialized applications.

5.3.1.1.4.1   Thin film hydration method   The thin-film hydration technique is a popular and easy preparation technique. In a round-bottomed flask, this method uses an organic solvent to dissolve the surfactants and some components, including cholesterol. The organic solvent removal is done by using a rotary vacuum evaporator to generate a thin film on the interior wall of the flask. An aqueous-solution containing the drug is incorporated into the water or buffered saline solution, and the dry film is made to undergo hydration at temperatures higher than the transition temperature of the surfactant. In the course of the hydration process, MLVs were formed (Uchegbu and Vyas 1998; Shilpa, Srinivasan, and Chauhan 2011). The thin film hydration method has been employed for preparing noisome entrapped in insulin (Pardakhty, Varshosaz, and Rouholamini 2007), nimesulide (Shahiwala and Misra 2002), antioxidants (Tavano et al. 2014), salicylic acid (Hao and Li 2011), etc.

5.3.1.1.4.2   Handshaking method   One approach to synthesizing MLVs is the handshaking approach which is the same as the thin film hydration method. This method involves solubilizing the surfactant and other components, such as cholesterol derivatives, in an organic solvent in a round-bottomed flask. The thin film layer was formed when the organic solvent was evaporated using a rotary evaporator. The dried film is then directly made to undergo hydration with an aqueous solution (with drug) for about an hour with mild mechanical agitation to produce niosomal dispersion having a milky texture (Verma et al. 2010).

5.3.1.1.4.3 Bubble method The "bubble" method of preparing niosomes avoids using organic solvents. This procedure involved transferring surfactants, additives, and PBS (pH 7.4), all into a glass-reactor having triple necks. The first neck contains a thermometer, the second contains nitrogen, and lastly, the third contains water-cooled reflux. Niosomes ingredients are dispersed at 70°C, and the dispersion undergoes blending for 15 seconds with high-shear homogenization. Nitrogen gas is then bubbled at 70°C immediately after the dispersion (Verma et al. 2010).

5.3.1.1.4.4 Ether injection method Surfactants, along with additives are solubilized in the organic solvent, including diethylether, and slowly injected by a needle into an aqueous-solution containing the drug kept at a constant temperature (~60°C) in the ether injection method. A rotary evaporator was used to evaporate the organic solvent. Surfactants are added to the ether vaporization process to aid in the creation of single-layers of vesicles. SUVs and LUVs are manufactured with the solvent-injection approach and have a large entrapped aqueous volume. The resulting vesicle diameter might range from 50 to 1000 nm (Verma et al. 2010; Shilpa, Srinivasan, and Chauhan 2011).

5.3.1.1.4.5 Sonication method In this method, the drug-solution (present in buffer) is incorporated into the surfactant/cholesterol mixture. For the formation of niosomes, the mixture undergoes probe sonication (at 60°C for three minutes) using a sonicator having a titanium probe (Alam et al. 2013).

5.3.1.1.4.6 Microfluidization The microfluidization approach could generate more uniform, small-sized, unilamellar vesicles, along with ensuring improved reproducibility of the same. The submerged-jet principle is applied in this process that involves two fluidized-streams interacting at extremely high velocities in precise microchannels within the interaction-chamber. The arrangement of the impingement of a thin-liquid sheet along a common frontier ensures that the energy delivered to the system stays in the region of wherein niosomes are formed (Verma et al. 2010).

5.3.1.1.4.7 Heating method Separately hydrated surfactant agents and other additives (like cholesterol) are hydrated in PBS (pH = 7.4) for around 60 minutes at ambient temperature in a nitrogen atmosphere. Post this, solution is then heated (approximately 120°C), kept on a hot-plate stirrer for about 15–20 minutes to solubilize the cholesterol. After lowering the temperature to 60°C, the remaining ingredients, such as surfactants and other additives, are mixed with the buffer wherein the cholesterol has already been dissolved whilst continuously stirring for about 15 minutes. Niosomes formed herein are maintained at 4–5°C under a nitrogen environment for 30 minutes before being used (Mortazavi et al. 2007).

5.3.1.1.4.8 Dehydration-hydration method Kirby and Gregoriadis first described the dehydration rehydration procedure in 1984 (Uchegbu and Vyas 1998). Thin film hydration prepares dehydration rehydration vesicles by freezing them in liquid nitrogen, followed by freeze-drying them for an entire night. At 60°C, niosome powders are then hydrated with PBS (pH 7.4) (Hope et al. 1986).

5.3.1.1.4.9 Freeze and thaw method This technique can produce frozen and thawed MLVs. The thin film hydration method is used to formulate the niosomal suspension which are then frozen in liquid nitrogen (one minute), then thawed at 60°C in the water bath (one minute) (Abdelkader et al. 2011).

5.3.1.1.4.10 Proniosome technology Proniosome technology was first used for niosome preparation approximately 20 years ago. Proniosomes are a unique drug-carrier manufacturing approach that has been exploited as stable precursors to the development of niosomal-carrier-systems. Proniosomes technology has been utilized to fabricate niosomes encapsulating vinpocetine, valsartan, 17-estradiol, and other APIs (Mokhtar et al. 2008).

5.3.1.2 Non-vesicular nanoparticles

5.3.1.2.1 Lipid nanoparticles

During the last few decades, lipid nanoparticles (LNPs) have drawn considerable attention. Lipid-based nanoparticles are broadly categorized into two main groups: solid lipid nanoparticles (SLNs) and nanostructured lipid carriers (NLCs) shown in Figure 5.7. In order to circumvent the drawbacks of other colloidal carriers like emulsions, liposomes, and/or polymeric nanoparticles, SLNs were developed. They have multiple advantages like improved release profiles, targeted drug delivery, and improved colloidal stability (Naseri, Valizadeh, and Zakeri-Milani 2015). The SLNs have a spherical morphology with an average size ranging across 40 and 1000 nm. SLN is made up of solid fat (~0.1–30% w/w) that is dispersed in aqueous-phase. Here, to increase the stability, surfactants are used in a concentration of about 0.5–5%. The right choice of lipids and surfactants can impact the particle size, drug loading capacity, and long-term stability associated with storage. Both hydrophilic and hydrophobic drugs may be used with SLN, being dependent on the preparation technique. SLN may have numerous advantages, including low cost, large-scale production, preparation without using organic solvents, ease of preparation, and a good release profile (Waghule, Gorantla, et al. 2020). Despite these critical features, SLN systems exhibit limitations, such as drug explosion phenomena and variable gelation tendencies (Schwarz et al. 1994; Mehnert and Mäder 2001). The second generation of LNPs, known as NLCs, overcome the above limitations. NLCs are modified SLNs are the next generation of LNPs to increase stability and capacity loadings and prevent drug expulsion from storage. The NLC innovation involves mixing solid and liquid lipids by incorporating a liquid lipid. This addition affects the formation of precise lipid crystals, reducing drug expulsion (Müller, Radtke, and Wissing 2002; Rapalli et al. 2020).

There are various methods for fabricating lipid nanoparticles. High-pressure homogenization at high/low temperatures (hot homogenization and cold

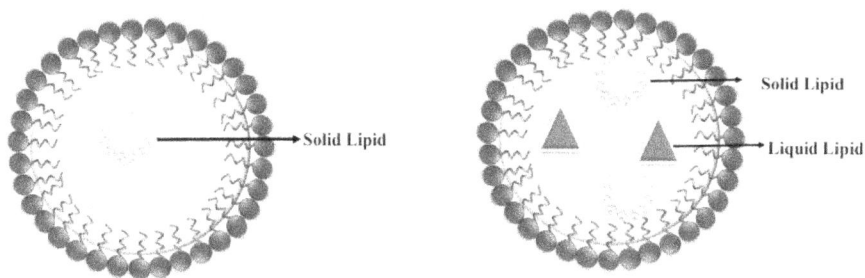

FIGURE 5.7 Schematic representation of solid lipid nanoparticles and nanostructured lipid carriers

homogenization), solvent-emulsification, evaporation or diffusion, or supercritical fluid extraction, or ultrasonication, etc. are common techniques for preparing lipid nanoparticles (Waghule, Rapalli, et al. 2020).

5.3.1.2.1.1 High-pressure homogenization High-pressure homogenization stands to be a well-established and effective method for preparing LNPs. High-pressure homogenization is capable of producing LNPs on a large scale. There are two types of homogenization processes: hot and cold homogenization. The pharmaceutical components are dissolved/dispersed in the melted lipids before high-pressure homogenization in both processes. Homogenization has several benefits, such as large-scale production, elimination of organic solvents, improved product stability, and loading capacity of API (Al Haj et al. 2008).

5.3.1.2.1.2 Hot homogenization In this method, homogenization can occur at a temperature higher than the melting point of the lipids. This method encompasses heating the lipid phase to 90°C, then dispersing the hot-lipid phase into a solution containing surfactants that are also heated to the same temperature. The pre-emulsion is homogenized at 90°C using a high-pressure homogenizer (three cycles at 5×10^7 Pa). The formed oil/water emulsion is cooled to room temperature for the solidification of the SLNs or NLCs (Souto et al. 2004).

5.3.1.2.1.3 Cold homogenization This process involves cooling the melted lipid phase until it solidifies, then grinding it to create lipid micro-particles. The generated lipid micro-particles are then dispersed in an excellent, surfactant-containing aqueous phase to create the presuspension. The presuspension is now homogenized in a high-pressure homogenizer for 5 cycles at room temperature and 1.5×10^8 Pa pressure (Pardeike, Hommoss, and Müller 2009).

5.3.1.2.1.4 Solvent emulsification/evaporation
Herein the lipid phase is dissolved in the organic phase. Post this, the organic phase is incorporated into the aqueous phase (surfactant solution in water) whilst continuously stirring at 70–80°C. The stirring has to be continued until the organic phase

has evaporated. The resulting nanoemulsion is now to be cooled (below 5°C) for solidification of the LNPs (Chen et al. 2002).

5.3.1.2.1.5 Ultrasonication/high-speed homogenization

This method dissolves the lipid phase is dissolved in an organic solvent (dichloromethane) and heating the same to a temperature of 50°C. Then, the surfactant-containing aqueous phase along with emulsifiers is heated at the same temperature. The aqueous phase is incorporated into the organic phase (at 50°C) after partially evaporating dichloromethane. The resulting emulsion is sonicated for the required time followed by cooling in an ice bath to solidify the SLNs (Luo et al. 2006).

5.3.1.2.1.6 Microemulsion formation techniques

This process involves melting the lipid at a suitable temperature and heating the aqueous phase, comprising surfactants to the same temperature. While stirring, the hot aqueous phase gets incorporated into the melted lipids. The formed hot o/w microemulsion is then dispersed in cold water at a ratio of 1:5 for solidifying the nanoparticles (Shah et al. 2014).

5.3.1.2.2 Nanoemulsion

They are a type of colloidal formulation that contains particles with sizes between 20 and 200 nm that are thermodynamically unstable shown in Figure 5.8 (Shaker et al. 2019). The finding explains that the dispersed globules associated with free energy in the continuous phase are higher than that of the immiscible liquid constituent in nanoemulsion. Oil and water are two immiscible counterparts most frequently employed in commercial applications; hence nanoemulsions are typical of w/o or o/w type. O/W type of nanoemulsion comprises the tiny oil droplets dispersed in the aqueous medium, whereas the W/O type of nanoemulsion comprises the tiny droplets of water dispersed in the oil medium (McClements and Jafari 2018; Pereira et al. 2022b).

Many different fabrication techniques can be employed to make nanoemulsions; these techniques are classified into low-energy and high-energy (McClements 2011; Gupta et al. 2016; Shams and Sahari 2016). Currently, the most prevalent techniques for formulating nanoemulsions in industrial set-ups are high-energy methods such as high-pressure valve homogenization, ultrasonication, and microfluidization (Tadros et al. 2004; McClements and Rao 2011). Low-intensity techniques depend on the formation of tiny droplets in specific surfactant-oil-water mixtures spontaneously when the system components or extrinsic factors (like temperature) are altered. The phase inversion temperature, phase inversion concentration, and spontaneous emulsification procedures are the most often utilized low-energy techniques (Komaiko and McClements 2016). The choice of the specific nanoemulsion fabrication techniques depends on the properties of the materials that are to be homogenized (especially the oil and surfactant phases) and the anticipated physicochemical and/or functional properties of the finished products.

High-energy methods comprise homogenization, ultrasonication, and microfluidization, which are discussed below

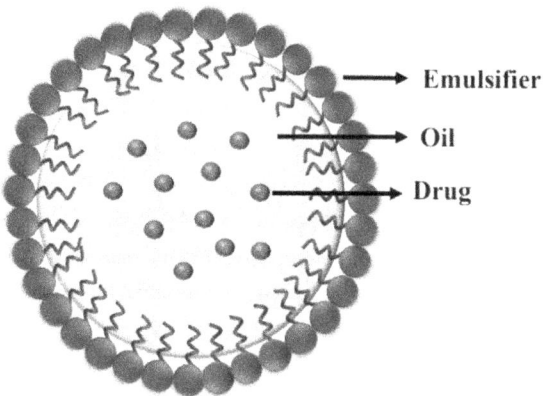

FIGURE 5.8 Schematic representation of nanoemulsion

5.3.1.2.2.1 Homogenization Recently, high-pressure homogenization (HPH) has been extensively explored for the production of nanoemulsions. When a prepared coarse emulsion is driven through a small opening, cavitation, shearing, and impact lead to the breakdown of the particle processes. The working pressures might be around 10–350 mPa, contingent on the product. There are a different number of factors that affect the droplet size, such as the number of cycles involved, the operational pressures and temperature, as well as the product composition. Given that higher temperatures tend to reduce viscosity and improve process efficiency, the process can be carried out at raised temperatures, given that the formulation contains thermally stable components. The benefits include the simplicity of scaling up and continuous batch production (Harusawa 1986; Schultz et al. 2004; Galvão, Vicente, and Sobral 2018).

5.3.1.2.2.2 Ultrasonication Droplets of the dispersed phase are broken down using this technique by cavitation, which acts as the main shearing force. Bath or probe sonicators can be employed. However, probe sonication produces more energy and is better at size reduction. The probe produces vapour bubbles under low pressure. The produced pressurized wave travels to the droplets of the interface and vanishes the energy causing the breaking of the droplets. The waves also create turbulence with high velocity, which, coupled with droplet breakage, causes agitation and mixing of the emulsion. The method is frequently utilized in lab research and is simple to regulate and alter; however, it can only be used for small batches (Pereira et al. 2022b).

5.3.1.2.2.3 Microfluidization In reference to shearing forces, a microfluidizer is analogous to a homogenizer, with the exception that a high-pressure pump propels the liquids through an inline-homogenizer to form the coarse emulsion. This emulsion is passed via microchannels in an interaction chamber. The size of the dispersed phase droplets is reduced due to the flow and impingement of the formulation, thereby producing the Nanoemulsion (Okamoto, Tomomasa, and Nakajima 2016).

Low energy methods comprise of phase-inversion temperature method, phase-inversion concentration method, and spontaneous emulsification method.

5.3.1.2.2.4   Phase-inversion temperature method This approach is reliant on temperature and entails changing an o/w emulsion at lower temperatures to a w/o emulsion at higher temperatures by adding a non-ionic surfactant whose affinity changes as a function of temperature. Because of the increased hydration of the polar groups, polyethoxylated surfactants have a large share of the surface area occupied by hydrophilic groups and hence function as an o/w surfactant. The hydrophobic 'chains' occupy a larger surface area than the hydrophilic polar groups due to the dehydration of the oxyethylene groups as the temperature rises, making the surfactant acts as a w/o emulsifier. A microemulsion or a liquid crystalline phase usually forms at an intermediary temperature when there is an equal affinity between both phases. In this stage, droplet formation starts. A quick transition from this temperature is required to create kinetically stable emulsions with smaller droplet sizes. Retinoyl palmitate nanoemulsions made with Dead Sea water have been created by Bilbao et al. Both substances are considered to treat skin conditions. The fabrication of the crude emulsion is done by mixing the oil phase (retinoyl palmitate) with the aqueous phase (7% w/w of dead sea water; 7% w/w of Brij 96; and 66% w/w distilled water) employing magnetic stirring (Pereira et al. 2022a). In order to generate a crude emulsion, the dispersion was heated at a temperature more than the phase transition temperature followed by quickly cooling in an ice bath. Post this, the emulsion was heated at 80°C, which had already been established to be the phase transition temperature using the conductivity measurement set-up. Phase inversion occurred at this temperature when the oily phase came to be continuous. After that, the resulting solution quickly cools to produce o/w nanoemulsion (Garcia-Bilbao et al. 2020).

5.3.1.2.2.5   Phase-inversion concentration method This approach encompasses preparing the formulation centered on phase inversion of the system, which occurs as a result of diluting the oil-rich phase with an aqueous phase at a constant temperature. This method, such as the phase-inversion temperature method, makes use of a surfactant whose affinity changes conditional to the dispersion vehicle as per the HLB system changes. At one concentration, a liquid crystalline phase, or microemulsion, which is thermodynamically stable, forms. The microemulsion loses its thermodynamic stability at this concentration and transforms into kinetically stable nanosized droplets. In order to create nanoemulsions made of straight-chain alkanes and esters, Zhang et al. recently developed a D-phase-emulsification technique which is based on modifying the phase-inversion-concentration approach. The protocol involves dissolving the surfactants in the polyol solvent first, then incorporating the oil and forming the clear oil surfactant gel called as O/D phase. Nanoemulsions (size of less than 200nm) were produced when this phase was diluted by incorporating the water (Zhang et al. 2021).

5.3.1.2.2.6   Spontaneous emulsification The spontaneous emulsification process uses the water-miscible solvent that contains the oil phase. When this solvent is poured into the water phase, the organic solvent diffuses and formation of oil droplets. Buzanello et al. prepared nanoemulsions using extracts from coffee beans employing the spontaneous emulsification method. The aqueous phase comprises the green coffee extract, water, and poloxamer, whereas the oily phase contains the green coffee oil, ethanol, and lecithin. A homogenous solution was developed by incorporating the aqueous phase into the organic phase. Afterward, the organic solvent was eliminated at reduced pressure to produce the colloidal dispersion and filtered under a vacuum. Herein, the nanoemulsion had an average size of 200 nm (Koroleva and Yurtov 2021).

5.3.2 POLYMERIC NANOPARTICLE

The word nanoparticles include both nanocapsules and nanospheres, which vary in morphology. Nanocapsules comprise an oil core wherein the drug is dispersed, surrounded by a polymeric shell that regulates the drug's release profile from the core. Nanospheres comprises a continuous network of polymers that allows the drug to be retained/adsorbed onto their surface. These two varieties of polymeric nanoparticles shown in Figure 5.9 are known as a matrix system (nanosphere) and a reservoir system (nanocapsule), respectively (Zielinska et al. 2020). Polymeric NPs as drug carriers have several advantages, such as the ability to protect drugs, used for controlled release, improved bioavailability, and therapeutic index (Zielinska et al. 2020).

Several preparation techniques rely on polymers that are preformed, for example, techniques such as nanoprecipitation, salting-out, emulsification solvent evaporation, emulsification solvent diffusion, and spontaneous emulsification solvent diffusion. Gelling and self-assembly are two other methods.

5.3.2.1 Solvent evaporation method

The solvent evaporation method was one of the first to be used to produce polymeric nanoparticles from the preformed polymer. Herein, an oil-in-water (o/w) emulsion must first be prepared before producing nanospheres (Jose et al. 2014). Firstly, the organic phase comprises the polar organic solvents into which polymer and drug

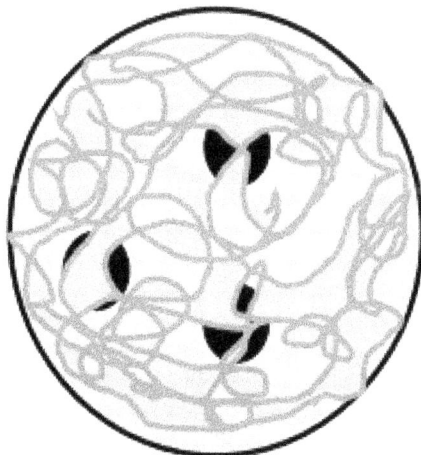

FIGURE 5.9 Schematic representation of polymeric nanoparticles

are dissolved. The aqueous phase containing surfactant (Polyvinyl acetate, PVA) is prepared next. For the production of the nanodroplets dispersion, the organic solution undergoes emulsification in the aqueous phase containing the surfactant and then is processed with high-pressure homogenization (HPH). Upon evaporating the polymer solvent, it is allowed to diffuse through the continuous emulsion phase, resulting in the formation of a suspension of NPs. The solvent is either slowly evaporated under reduced pressure (as with dichloromethane and chloroform) or by continuously employing magnetic stirring. The solidified nanoparticles can now be washed and collected by centrifugation after the solvent has been evaporated, followed by freeze-drying for long-term storability. This method enables the fabrication of nanospheres (Bohrey, Chourasiya, and Pandey 2016).

5.3.2.2 Nanoprecipitation

This technique is as well known as the solvent displacement method and involves the use of miscible solvents. The internal phase comprises a polymer dispersed in organic solvents like acetone/acetonitrile. They are easily removed by evaporating the same owing to their immiscible nature in water. Herein, the technique is centered on the interfacial deposition of the polymer that occurs after the organic solvents have been displaced from the lipophilic solution across to the aqueous phase. The polymer is dissolved into a water-miscible solvent, having intermediate polarity. This solution is incorporated gradually into the aqueous solution while stirring dropwise or at a controlled rate of addition. The nanoparticles form instantaneously as spontaneous diffusion of the polymer solution happens across the aqueous phase. The polymer now precipitates as nanocapsules and nanospheres once the solvent is diffused out of the nanodroplets (Jelvehgari et al. 2017).

5.3.2.3 Emulsification-solvent diffusion method

This procedure involves producing an o/w emulsion comprising an aqueous solution with the surfactant and a partly miscible solvent with the polymer and API. This internal emulsion phase is comprised of a partially hydro-miscible organic solvent, like benzyl alcohol or ethyl acetate, that has already undergone saturation using water to create an initial thermodynamic equilibrium between the two phases at room temperature. Post this, diluting with a larger quantity of water, the solvent diffuses from the dispersion of droplets into the surrounding external phase, resulting in colloidal particle formation (Quintanar-Guerrero et al. 1998; Jelvehgari et al. 2017).

5.3.2.4 Emulsification reverse-salting-out

The previously mentioned emulsification solvent diffusion method can be considered to be an alteration of the emulsification reverse salting-out method. The salting-out method is dependent upon the salting-out effect to bring about the separation of a hydro-miscible solvent from the aqueous solution, which may result in the formation of nanospheres. In this method, the composition of the o/w emulsion is prepared from the water-miscible solvent and an aqueous solution comprising the gel, colloidal stabilizer, and salting-out agents. The miscibility of acetone and water is reduced by saturation of the aqueous phase, culminating in forming of an o/w emulsion from the other miscible counterparts. At room temperature, the o/w emulsion is made with vigorous stirring. The emulsion then undergoes dilution with the apposite amount of deionized water/aqueous solution to permit the organic solvent's diffusion across to the external phase, the polymer's precipitation, and ultimately the formation of nanospheres (Vauthier and Bouchemal 2009).

5.3.3 Nanocrystals

Nanocrystal preparation has numerous potential applications in cosmetics, pharmaceuticals, and biomedical areas. Nanocrystals are used in cosmetics such as moisturizers, toothpaste, sunscreen, makeup, and sunscreen. The products made from nanocrystals are assumed to be the next generation of cosmetic delivery systems that are promising for improving skin hydration, skin maintenance, bioavailability, and stability. The nanocrystals could be used as new building blocks in producing APIs with improved aqueous solubility (Kulkarni and Myerson 2017).

There are numerous methods for the fabrication of nanocrystals. These methods are classified into top-down and bottom-up approaches. The two primary top-down methods for reducing particle size are milling and high-pressure homogenization. At the same time, bottom-up methods for reducing the particle size are control flow cavitation, spray drying, supercritical fluid, impinging jet crystallization, and emulsion methods.

5.3.3.1 Top-down approach: milling and high-pressure homogenization

The most important technique for producing nanocrystals is the top-down approach. Milling and high-pressure homogenization are two of the most common top-down size reduction techniques. Wet milling has been used for developing the nanocrystals products that have been launched in the market. In wet milling, which uses mechanical attrition, the particles are wetted with an aqueous surfactant solution before being sheared and ground by using balls in a milling container. Although the particle size undergoes a reduction to a few hundred micrometers, modified conventional milling can produce nanosized crystals (Peltonen and Hirvonen 2010).

Furthermore, in the high-pressure homogenization process, two fluid streams usually contain particle suspensions that collide in a chamber at high pressure, causing a particle collision and subsequent particle rupture. In piston-gap homogenizers, solid nanosized particles are created by applying high pressure to a piston to push a suspension of drug particles through a small gap. The main drawbacks of high-pressure homogenization include the need for complex equipment, a high process temperature, high energy input, potential component degradation, and lower yields than wet milling (Junghanns and Müller 2008).

5.3.3.2 Bottom-up techniques: emulsion method and spray drying

The method of producing nanocrystals through emulsion is becoming increasingly popular in fundamental research and various industrial fields. A three-step process is required to produce organic nanocrystals using the emulsion method. The first step is rapidly adding a compound solution in the organic phase directly to the aqueous phase at high temperatures to prepare the emulsion. The stable emulsion is created using a high, stirring speed and ultrasound irradiation. Solutes are crystallized in the second step by gradually carrying out the cooling of the dispersion at low temperatures. An antifoaming agent is also added in the third step to break the emulsion and carry out the separation of the organic solvent. In an aqueous phase, nanocrystals are obtained as a stable form of dispersion (Ujiiye-Ishii et al. 2008; Malik, Wani, and Hashim 2012).

Spray drying is the one-step method that converts the emulsion, suspension, solution, pastes, and slurries into powder form. It also enables the development of particles such that their size and morphological characteristics are controlled (Ré 2006). The conventional spray drying method has a limitation, that only particles of 2–5 μm in size can be developed because of the poor collection efficiency of the cyclone separator. These limitations seem to be circumvented by a spray dryer equipped with a piezoelectrically driven vibrating mesh atomizer and an electrostatic powder collector (Wan, Heng, and Chia 2008). Another novel technology is the dissolution of drug molecules and polymeric dispersant systems using a suitable solvent. The powder containing the drug is formed after spray drying the resulting solution either in the form of a molecular dispersion system in the polymer matrix such that it forms a solid solution or by carrying out dispersion as submicron particles resulting in a solid suspension.

5.4   A FABRICATION METHOD OF INORGANIC NANOPARTICLES

Inorganic nanoparticles are more hydrophilic, biocompatible, safer, and significantly more stable than natural nanoparticles. These nanoparticles are made of inorganic materials (Ag, Au, Ti, etc.), whereas natural nanoparticles are made of polymers; they might be very different. Inorganic nanoparticles are widely employed in cosmetic products. A few fundamental particles are described below:

5.4.1   CARBON-BASED NANOPARTICLES

Carbon-based nanoparticles are wholly made of carbon atoms. Herein, classification can be done as carbon nanotubes (CNT), graphene, fullerenes, and activated carbons, all present as nanosize (Anu Mary Ealia and Saravanakumar 2017). Fullerenes (C60) are sphere-shaped carbon molecules of carbon atoms, bonded with one another owing to their sp^2 hybridization. Structurally, fullerene comprises of about 28–1500 carbon atoms, with a single layer having a diameter of up to 8.2 nm and multilayer fullerenes having around 4–36 nm diameter size. Graphene being allotropic of carbon has a two-dimensional planar hexagonal honeycomb-like lattice network made of carbon atoms. Thickness of a graphene sheets is typically 1 nm. Carbon nanotubes (CNT) are formed by winding a nanofoil of graphene with a honeycomb-shaped lattice of carbon atoms resulting in forming hollow cylinders and finally nanotubes with small diameter sizes ~0.7 nm for single-layer CNTs and 100 nm for multi-layer CNTs and lengths ranging up to millimeters (Bhaviripudi et al. 2007).

There are various approaches for fabricating carbon-based nanoparticles. These methods are classified into top-down and bottom-up approaches. The top-down or destructive method involves separating large material into nanometric scale particles. Some of the most common top-down methods for nanoparticle fabrication are mechanical milling, laser ablation, thermal decomposition, and nanolithography.

At the same time, the bottom-up or constructive methods assemble materials from the atom to clusters to nanoparticles. Bottom-up methods for nanoparticle production include sol-gel, spinning, chemical vapor deposition (CVD), and pyrolysis (Anu Mary Ealia and Saravanakumar 2017).

5.4.1.1 Top-down methods

Mechanically milling, thermally decomposing, nanolithography, using lasers for ablation, etc. are the different top-down methods for fabricating carbon-based nanoparticles. They are discussed hereunder.

Mechanical milling being one of the most widely used top-down methods, can be used for manufacturing different types of nanoparticles. Different elements undergo milling in an inert atmospheric condition during the synthesis process using mechanical milling. Various influencing factors, such as plastic deformation, which determines particle shape, a fracture that reduces particle size, and cold-welding, which increases particle size, directly impact the mechanical milling (Prasad Yadav, Manohar Yadav, and Pratap Singh 2012).

The process of developing nanometric scale structures, such that, they have a minimum of one dimension and sizes ranging from 1 to 100 nm is known as nanolithography. Nanolithographic processes comprise optical, multi-photon, electron-beam, nano-imprint, and scanning probe lithography. In general, lithography is a method via which imprinting of a necessary shape or structure on a light-sensitive compound is done such that it selectively eliminates a part of the compound to develop the required shape and structure (Hulteen et al. 1999; Pimpin and Srituravanich 2012).

Laser ablation synthesis-solution is a popular technique for developing nanoparticles from various solvents. A laser beam brings forth condensation of the plasma plume that produces nanoparticles when it strikes a metal that is present immersed in a liquid solution. It is a "green" process because laser ablation synthesis in solution offers stable nanoparticle synthesis using organic solvents and water, without using chemical stabilizers or other chemicals (Amendola and Meneghetti 2009).

Thermal decomposition is an endothermic process in which chemical decomposition takes place which is caused by heat. This breaks the compound's chemical bond. The decomposition-temperature is the specific temperature at which a particular element chemically decomposes. Thus, nanoparticles are created by a breakdown of the metal at the above-discussed specific temperatures, and this is followed by a chemical reaction that generates secondary products (Salavati-Niasari, Davar, and Mir 2008).

5.4.1.2 Bottom-up methods

Bottom-up methods include sol-gel, spinning, chemical vapour deposition, and pyrolysis, which are employed for synthesizing carbon-based nanoparticles.

When a colloidal solution containing solids is dispersed in a liquid phase it is referred to as sol. A gel is a group of solid macromolecules immersed in a solvent. Amongst all, the most popular bottom-up method is the sol-gel method due to its ease of use and the fact that most nanoparticles can be synthesized. Sol-gel is a wet chemical method comprising a chemical solution that functions as a precursor to an integrated system of discrete particles. In this process, metal oxides and chlorides are the common precursors. The precursor is then suspended into the host liquid via stirring, followed by sonication and shaking, resulting in the formation of a solid and liquid-phase system. The nanoparticles are then recovered through phase separation employing different methods, including sedimentation, filtration, and centrifugation, and the moisture content is removed further by drying (Mann et al. 1997).

A spinning disc reactor (SDR) is also used for developing nanoparticles by spinning them. It comprises the rotating disc within a chamber/reactor wherein physical parameters such as temperature is under control. The reactor is typically filled with inert gases to bring about the removal of oxygen in order to prevent unwanted chemical reactions. Herein the liquid precursor along with water is pumped into the disc while rotating at various speeds. The process of spinning induces the fusion of atoms or molecules, which causes them to precipitate, collect, and dry. The characteristics of the nanoparticles produced using SDR are then determined (Mohammadi, Harvey, and Boodhoo 2014).

The process wherein a thin film of gaseous reactants is deposited onto a substrate is known as chemical vapor deposition (CVD). Deposition takes place in a reactor vessel at room temperature upon the combination of gas molecules. A chemical reaction ensues when any heated substrate comes into contact with this combined gas. The reaction leads to the forming of a thin film of product at the substrate surface, which is further recovered and reused. In CVD, the influential variable affecting the process is substrate temperature. CVD produces mostly pure, uniform, complex, and potent nanoparticles. The shortcomings of CVD include the need for specialized equipment and the highly toxic gaseous byproducts (Adachi, Tsukui, and Okuyama 2003).

Pyrolysis is the process that is most frequently used in industries to produce nanoparticles on a large scale. It implies using a flame to burn a precursor. The precursors can be liquid/vapour, are introduced into the furnace which is under high pressure, via a small hole, where it burns. The combustion or byproduct gases are then air-classified to recover the nanoparticles. Some furnaces produce high temperatures for simple evaporation using laser and plasma rather than flame. Pyrolysis is a simple, performance-efficient, cost-effective, and continual process having a high yield (Wegner and Pratsinis 2004).

5.5   CONCLUSION

The demand for cosmetics at a global scale is ever-rising. Continuous expansion in cosmetics product categories and the inclusion of nanotechnology have led to the development of cosmeceuticals into nanocosmeceuticlas. Nanotechnology-based cosmetics are now becoming popular because of their numerous advantages. Understanding the properties of nanocarriers is complex and challenging that needed professional expertise in nanotechnology and significant financial investments in the fabrication of nanotechnology-based products. Currently, novel nanocarriers used in the formulation of various cosmetics and cosmeceuticals with enhanced effectiveness include liposomes, ethosomes, NLC, SLNs, nanoemulsions, niosomes, transferosomes, and carbon-based nanoparticles. These formulations are carried across the skin by nanosystems using a variety of delivery mechanisms, and they serve various purposes, such as sun protection, moisturization, and wrinkle reduction. Although the market value of these nanomaterial products is increasing significantly, there is a great deal of controversy surrounding their safety and toxicity in humans, necessitating further research.

REFERENCES

Abdelkader, Hamdy, Sayed Ismail, Amal Kamal, and Raid G Alany. 2011. "Design and Evaluation of Controlled-Release Niosomes and Discomes for Naltrexone Hydrochloride Ocular Delivery." *Journal of Pharmaceutical Sciences* 100 (5): 1833–46. doi:10.1002/jps.22422.

Adachi, Motoaki, Shigeki Tsukui, and Kikuo Okuyama. 2003. "Nanoparticle Synthesis by Ionizing Source Gas in Chemical Vapor Deposition." *Japanese Journal of Applied Physics, Part 2: Letters* 42 (1 A/B): 4–7. doi:10.1143/jjap.42.l77.

Al Haj, Nagi A, Rasedee Abdullah, Siddig Ibrahim, and Ahmad Bustamam. 2008. "Tamoxifen Drug Loading Solid Lipid Nanoparticles Prepared by Hot High Pressure Homogenization Techniques." *American Journal of Pharmacology and Toxicology* 3 (3): 219–24. doi:10.3844/ajptsp.2008.219.224.

Alam, Maroof, Swaleha Zubair, Mohammad Farazuddin, Ejaj Ahmad, Arbab Khan, Qamar Zia, Abida Malik, and Owais Mohammad. 2013. "Development, Characterization and Efficacy of Niosomal Diallyl Disulfide in Treatment of Disseminated Murine Candidiasis." *Nanomedicine: Nanotechnology, Biology and Medicine* 9 (2): 247–56. doi:10.1016/j.nano.2012.07.004.

Amendola, Vincenzo, and Moreno Meneghetti. 2009. "Laser Ablation Synthesis in Solution and Size Manipulation of Noble Metal Nanoparticles." *Physical Chemistry Chemical Physics* 11 (20): 3805–21. doi:10.1039/b900654k.

Anu Mary Ealia, S, and M P Saravanakumar. 2017. "A Review on the Classification, Characterisation, Synthesis of Nanoparticles and Their Application." *IOP Conference Series: Materials Science and Engineering* 263: 32019. doi:10.1088/1757-899x/263/3/032019.

Badmeier, Roland, and Huagang Chen. 1993. "Hydrolysis of Cellulose Acetate and Cellulose Acetate Butyrate Pseudolatexes Prepared by a Solvent Evaporation-Microfluidization Method." *Drug Development and Industrial Pharmacy* 19 (5): 521–30. doi:10.3109/03639049309062964.

Balata, Gehan F, Mennatullah M Faisal, Hanaa A Elghamry, and Shereen A Sabry. 2020. "Preparation and Characterization of Ivabradine HCl Transfersomes for Enhanced Transdermal Delivery." *Journal of Drug Delivery Science and Technology* 60: 101921. doi:10.1016/j.jddst.2020.101921.

Bapat, Ranjeet A, Tanay V Chaubal, Suyog Dharmadhikari, Anshad Mohamed Abdulla, Prachi Bapat, Amit Alexander, Sunil K Dubey, and Prashant Kesharwani. 2020. "Recent Advances of Gold Nanoparticles as Biomaterial in Dentistry." *International Journal of Pharmaceutics* 586: 119596. doi:10.1016/j.ijpharm.2020.119596.

Benson, H A E. 2006. "Transfersomes for Transdermal Drug Delivery." *Expert Opinion on Drug Delivery* 3 (6): 727–37.

Berger, N, A Sachse, J Bender, R Schubert, and M Brandl. 2001. "Filter Extrusion of Liposomes Using Different Devices: Comparison of Liposome Size, Encapsulation Efficiency, and Process Characteristics." *International Journal of Pharmaceutics* 223 (1): 55–68. doi:10.1016/S0378-5173(01)00721-9.

Bhalaria MK, Naik S, Misra AN. 2009. "Ethosomes: A Novel Delivery System for Antifungal Drugs in the Treatment of Topical Fungal Diseases." *Indian Journal of Experimental Biology* 47 (5): 368–75.

Bhaviripudi, Sreekar, Ervin Mile, Stephen A Steiner, Aurea T Zare, Mildred S Dresselhaus, Angela M Belcher, and Jing Kong. 2007. "CVD Synthesis of Single-Walled Carbon Nanotubes from Gold Nanoparticle Catalysts." *Journal of the American Chemical Society* 129 (6): 1516–17. doi:10.1021/ja0673332.

Bohrey, Sarvesh, Vibha Chourasiya, and Archna Pandey. 2016. "Polymeric Nanoparticles Containing Diazepam: Preparation, Optimization, Characterization, In-Vitro Drug Release and Release Kinetic Study." *Nano Convergence* 3 (1): 3–9. doi:10.1186/s40580-016-0061-2.

Carita, Amanda C, Josimar O Eloy, Marlus Chorilli, Robert J Lee, and Gislaine Ricci Leonardi. 2017. "Recent Advances and Perspectives in Liposomes for Cutaneous Drug Delivery." *Current Medicinal Chemistry* 25 (5): 606–35. doi:10.2174/0929867324666171009120154.

Cencia-Rohan, Lisa, and Shawn Silvestri. 1993. "Effect of Solvent System on Microfluidization-Induced Mechanical Degradation." *International Journal of Pharmaceutics* 95 (1): 23–28. doi:10.1016/0378-5173(93)90386-T.

Chaurasiya, Priyanka, Eisha Ganju, Neeraj Upmanyu, Sudhir Kumar Ray, and Prabhat Jain. 2019. "Transfersomes: A Novel Technique for Transdermal Drug Delivery." *Journal of Drug Delivery and Therapeutics* 9 (1): 279–85. doi:10.22270/jddt.v9i1.2198.

Chen, Da Bing, Tian Zhi Yang, Wang Liang Lu, and Qiang Zhang. 2002. "In Vitro and in Vivo Study of Two Kinds of Long-Circulating Solid Lipid Nanoparticles Containing Paclitaxel." *Yaoxue Xuebao* 37 (1): 54–58.

Chen, Guanyu, Danhui Li, Ye Jin, Weiyu Zhang, Lirong Teng, Craig Bunt, and Jingyuan Wen. 2014. "Deformable Liposomes by Reverse-Phase Evaporation Method for an Enhanced Skin Delivery of (+)-Catechin." *Drug Development and Industrial Pharmacy* 40 (2): 260–65. doi:10.3109/03639045.2012.756512.

Colas, Jean-Christophe, Wanlong Shi, V S N Malleswara Rao, Abdelwahab Omri, M Reza Mozafari, and Harjinder Singh. 2007. "Microscopical Investigations of Nisin-Loaded Nanoliposomes Prepared by Mozafari Method and Their Bacterial Targeting." *Micron* 38 (8): 841–47. doi:10.1016/j.micron.2007.06.013.

Dhapte-Pawar, Vividha, Shivajirao Kadam, Shai Saptarsi, and Prathmesh P. Kenjale. 2020. "Nanocosmeceuticals: Facets and Aspects." *Future Science OA* 6 (10). doi:10.2144/fsoa-2019-0109.

Dhawan, Surbhi, Pragya Sharma, and Sanju Nanda. 2020. "Cosmetic Nanoformulations and Their Intended Use." In, 141–69. doi:10.1016/B978-0-12-822286-7.00017-6.

Dhopavkar, Shreya, Pramod Kadu. 2017. "Transfersomes – A Boon for Transdermal." *Indo American Journal of Pharmaceutical Sciences* 4 (9): 2908–19.

Dubey, Sunil Kumar, Anuradha Dey, Gautam Singhvi, Murali Manohar Pandey, Vanshikha Singh, and Prashant Kesharwani. 2022. "Emerging Trends of Nanotechnology in Advanced Cosmetics." *Colloids and Surfaces B: Biointerfaces*, 112440. doi:10.1016/j.colsurfb.2022.112440.

Dubey, Vaibhav, Dinesh Mishra, Tathagata Dutta, Manoj Nahar, D K Saraf, and N K Jain. 2007. "Dermal and Transdermal Delivery of an Anti-Psoriatic Agent via Ethanolic Liposomes." *Journal of Controlled Release* 123 (2): 148–54. doi:10.1016/j.jconrel.2007.08.005.

Dubey, Vaibhav, Dinesh Mishra, and N K Jain. 2007. "Melatonin Loaded Ethanolic Liposomes: Physicochemical Characterization and Enhanced Transdermal Delivery." *European Journal of Pharmaceutics and Biopharmaceutics* 67 (2): 398–405. doi:10.1016/j.ejpb.2007.03.007.

El Zaafarany, Ghada M, Gehanne A S Awad, Samar M Holayel, and Nahed D Mortada. 2010. "Role of Edge Activators and Surface Charge in Developing Ultradeformable Vesicles with Enhanced Skin Delivery." *International Journal of Pharmaceutics* 397 (1): 164–72. doi:10.1016/j.ijpharm.2010.06.034.

Fakhravar, Zohreh, Pedram Ebrahimnejad, Hadis Daraee, and Abolfazl Akbarzadeh. 2016. "Nanoliposomes: Synthesis Methods and Applications in Cosmetics." *Journal of Cosmetic and Laser Therapy* 18. doi:10.3109/14764172.2015.1039040.

Galvão, K C S, A A Vicente, and P J A Sobral. 2018. "Development, Characterization, and Stability of O/W Pepper Nanoemulsions Produced by High-Pressure Homogenization." *Food and Bioprocess Technology* 11 (2): 355–67. doi:10.1007/s11947-017-2016-y.

Garcia-Bilbao, Amaia, Paloma Gómez-Fernández, Liraz Larush, Yoram Soroka, Blanca Suarez-Merino, Marina Frušić-Zlotkin, Shlomo Magdassi, and Felipe Goñi-de-Cerio. 2020. "Preparation, Characterization, and Biological Evaluation of Retinyl Palmitate and Dead Sea Water Loaded Nanoemulsions toward Topical Treatment of Skin Diseases." *Journal of Bioactive and Compatible Polymers* 35 (1): 24–38. doi:10.1177/0883911519885970.

Gautam, A, D Singh, and R Vijayaraghavan. 2011. "Dermal Exposure of Nanoparticles an Understanding." *Journal of Cell and Tissue Research* 11 (1): 2703–8.

Ghai Ishan, Chaudhary Hema, Ghai Shashank, Kohli Kanchan and Kr Vikash. 2012. "A Review of Transdermal Drug Delivery Using Nano-Vesicular Carriers: Transfersomes." *Recent Patents on Nanomedicine* 2 (2). doi:10.2174/1877912311202020164.

Gupta, Ankur, H Burak Eral, T Alan Hatton, and Patrick S Doyle. 2016. "Nanoemulsions: Formation, Properties and Applications." *Soft Matter* 12 (11): 2826–41. doi:10.1039/C5SM02958A.

Gupta, Vaibhav, Sradhanjali Mohapatra, Harshita Mishra, Uzma Farooq, Keshav Kumar, Mohammad Javed Ansari, Mohammed F Aldawsari, Ahmed S Alalaiwe, Mohd Aamir Mirza, and Zeenat Iqbal. 2022. "Nanotechnology in Cosmetics and Cosmeceuticals-A Review of Latest Advancements." *Gels (Basel, Switzerland)* 8 (3). doi:10.3390/gels8030173.

Hao, Yong-Mei, and Ke'an Li. 2011. "Entrapment and Release Difference Resulting from Hydrogen Bonding Interactions in Niosome." *International Journal of Pharmaceutics* 403 (1): 245–53. doi:10.1016/j.ijpharm.2010.10.027.

Harusawa, Fuminori. 1986. "Formation and Stability of Emulsions." *Journal of Japan Oil Chemists' Society* 35 (1): 50–54. doi:10.5650/jos1956.35.50.

Hoag, Stephen W, and Ajaz S Hussain. 2001. "The Impact of Formulation on Bioavailability." *Journal of Nutrition* 131 (January): 1389–91.

Hope, Michael J, Marcel B Bally, Lawrence D Mayer, Andrew S Janoff, and Pieter R Cullis. 1986. "Generation of Multilamellar and Unilamellar Phospholipid Vesicles." *Chemistry and Physics of Lipids* 40: 89–107.

Hope, Michael J, Marcel B Bally, G Webb, and P R Cullis. 1985. "Production of Large Unilamellar Vesicles by a Rapid Extrusion Procedure. Characterization of Size Distribution, Trapped Volume and Ability to Maintain a Membrane Potential." *Biochimica et Biophysica Acta (BBA) - Biomembranes* 812 (1): 55–65. doi:10.1016/0005-2736(85)90521-8.

Hulteen, John C, David A Treichel, Matthew T Smith, Michelle L Duval, Traci R Jensen, and Richard P Van Duyne. 1999. "Nanosphere Lithography: Size-Tunable Silver Nanoparticle and Surface Cluster Arrays." *Journal of Physical Chemistry B* 103 (19): 3854–63. doi:10.1021/jp9904771.

Jelvehgari, M, Sara Salatin, Jaleh Barar, Mohammad Barzegar-Jalali, Khosro Adibkia, Farhad Kiafar, and Mitra Jelvehgari. 2017. "Development of a Nanoprecipitation Method for the Entrapment of a Very Water Soluble Drug into Eudragit RL Nanoparticles." *Research in Pharmaceutical Sciences* 12 (1): 1–14.

Jose, S, S Sowmya, T A Cinu, N A Aleykutty, S Thomas, and E B Souto. 2014. "Surface Modified PLGA Nanoparticles for Brain Targeting of Bacoside-A." *European Journal of Pharmaceutical Sciences* 63: 29–35. doi:10.1016/j.ejps.2014.06.024.

Junghanns, Jens Uwe A H, and Rainer H Müller. 2008. "Nanocrystal Technology, Drug Delivery and Clinical Applications." *International Journal of Nanomedicine* 3 (3): 295–309.

Komaiko, Jennifer S, and David Julian McClements. 2016. "Formation of Food-Grade Nanoemulsions Using Low-Energy Preparation Methods: A Review of Available Methods." *Comprehensive Reviews in Food Science and Food Safety* 15 (2): 331–52. doi:10.1111/1541-4337.12189.

Koroleva, Marina Yu, and Evgeny V Yurtov. 2021. "Ostwald Ripening in Macro- and Nanoemulsions." *Russian Chemical Reviews* 90 (3): 293–323. doi:10.1070/rcr4962.

Kulkarni, Samir A, and Allan S Myerson. 2017. "Methods for Nano-Crystals Preparation Chapter 16 Methods for Nano-Crystals Preparation," no. November 2019. doi:10.1007/978-94-024-1117-1.

Luo, YiFan, DaWei Chen, LiXiang Ren, XiuLi Zhao, and Jing Qin. 2006. "Solid Lipid Nanoparticles for Enhancing Vinpocetine's Oral Bioavailability." *Journal of Controlled Release* 114 (1): 53–59. doi:10.1016/j.jconrel.2006.05.010.

MacDonald, Robert C, Ruby I MacDonald, Bert Ph M Menco, Keizo Takeshita, Nanda K Subbarao, and Lan-Rong Hu. 1991. "Small-Volume Extrusion Apparatus for Preparation of Large, Unilamellar Vesicles." *Biochimica et Biophysica Acta (BBA) - Biomembranes* 1061 (2): 297–303. doi:10.1016/0005-2736(91)90295-J.

Mahdi Jafari, Seid, Yinghe He, and Bhesh Bhandari. 2006. "Nano-Emulsion Production by Sonication and Microfluidization - A Comparison." *International Journal of Food Properties* 9 (3): 475–85. doi:10.1080/10942910600596464.

Malik, Maqsood Ahmad, Mohammad Younus Wani, and Mohd Ali Hashim. 2012. "Microemulsion Method: A Novel Route to Synthesize Organic and Inorganic Nanomaterials: 1st Nano Update." *Arabian Journal of Chemistry* 5 (4): 397–417. doi:10.1016/j.arabjc.2010.09.027.

Mann, Stephen, Sandra L Burkett, Sean A Davis, Christabel E Fowler, Neil H Mendelson, Stephen D Sims, Dominic Walsh, and Nicola T Whilton. 1997. "Sol-Gel Synthesis of Organized Matter." *Chemistry of Materials* 9 (11): 2300–10. doi:10.1021/cm970274u.

McClements, David Julian. 2011. "Edible Nanoemulsions: Fabrication, Properties and Functional Performance." *Soft Matter* 7 (6): 2297–316. doi:10.1039/C0SM00549E.

McClements, David Julian, and Seid Mahdi Jafari. 2018. *General Aspects of Nanoemulsions and Their Formulation. Nanoemulsions: Formulation, Applications, and Characterization.* Elsevier Inc. doi:10.1016/B978-0-12-811838-2.00001-1.

McClements, David Julian, and Jiajia Rao. 2011. "Food-Grade Nanoemulsions: Formulation, Fabrication, Properties, Performance, Biological Fate, and Potential Toxicity." *Critical Reviews in Food Science and Nutrition* 51 (4): 285–330. doi:10.1080/10408398.2011.559558.

Mehnert, Wolfgang, and Karsten Mäder. 2001. "Solid Lipid Nanoparticles: Production, Characterization and Applications." *Advanced Drug Delivery Reviews* 47 (2): 165–96. doi:10.1016/S0169-409X(01)00105-3.

Mohammadi, Somaieh, Adam Harvey, and Kamelia V.K. Boodhoo. 2014. "Synthesis of TiO2 Nanoparticles in a Spinning Disc Reactor." *Chemical Engineering Journal* 258: 171–84. doi:10.1016/j.cej.2014.07.042.

Mokhtar, Mahmoud, Omaima A Sammour, Mohammed A Hammad, and Nagia A Megrab. 2008. "Effect of Some Formulation Parameters on Flurbiprofen Encapsulation and Release Rates of Niosomes Prepared from Proniosomes." *International Journal of Pharmaceutics* 361 (1): 104–11. doi:10.1016/j.ijpharm.2008.05.031.

Mortazavi, S Moazam, M Reza Mohammadabadi, Kianoush Khosravi-Darani, and M Reza Mozafari. 2007. "Preparation of Liposomal Gene Therapy Vectors by a Scalable Method without Using Volatile Solvents or Detergents." *Journal of Biotechnology* 129 (4): 604–13. doi:10.1016/j.jbiotec.2007.02.005.

Mozafari, M Reza. 2005. "Liposomes: An Overview of Manufacturing Techniques." *Cellular & Molecular Biology Letters* 10: 711–19.

Mozafari, M Reza. 2010. "Nanoliposomes: Preparation and Analysis." *Methods in Molecular Biology (Clifton, N.J.)* 605: 29–50. doi:10.1007/978-1-60327-360-2_2.

Mozafari M Reza, C J Reed, C Rostron, C Kocum, and E Piskin. 2002. "Construction of Stable Anionic Liposome-Plasmid Particles Using the Heating Method: A Preliminary Investigation." *Cellular & Molecular Biology Letters* 7(3):923–27.

Müller, R H, M Radtke, and S A Wissing. 2002. "Nanostructured Lipid Matrices for Improved Microencapsulation of Drugs." *International Journal of Pharmaceutics* 242 (1): 121–28. doi:10.1016/S0378-5173(02)00180-1.

Naseri N, H Valizadeh, and P Zakeri-Milani. 2015. "Solid Lipid Nanoparticles and Nanostructured Lipid Carriers: Structure, Preparation and Application."*Advanced Pharmaceutical Bulletin* Sep;5(3): 3. doi:10.15171/apb.2015.043.

Oberdörster, Günter, Eva Oberdörster, and Jan Oberdörster. 2005. "Nanotoxicology: An Emerging Discipline Evolving from Studies of Ultrafine Particles." *Environmental Health Perspectives* 113 (7): 823–39. doi:10.1289/ehp.7339.

Okamoto, Toru, Satoshi Tomomasa, and Hideo Nakajima. 2016. "Preparation and Thermal Properties of Fatty Alcohol/Surfactant/Oil/Water Nanoemulsions and Their Cosmetic Applications." *Journal of Oleo Science* 65 (1): 27–36. doi:10.5650/jos.ess15183.

Opatha, Shakthi Apsara Thejani, Varin Titapiwatanakun, and Romchat Chutoprapat. 2020. "Transfersomes: A Promising Nanoencapsulation Technique for Transdermal Drug Delivery." *Pharmaceutics* 12 (9): 855. doi:10.3390/pharmaceutics12090855.

Pardakhty, Abbas, and Esmaeil Moazeni. 2012. "Nano-Niosomes in Drug, Vaccine and Gene Delivery: A Rapid Overview." *Nanomedicine Journal* 1: 1–13.

Pardakhty, Abbas, Jaleh Varshosaz, and Abdolhossein Rouholamini. 2007. "In Vitro Study of Polyoxyethylene Alkyl Ether Niosomes for Delivery of Insulin." *International Journal of Pharmaceutics* 328 (2): 130–41. doi:10.1016/j.ijpharm.2006.08.002.

Pardeike, Jana, Aiman Hommoss, and Rainer H Müller. 2009. "Lipid Nanoparticles (SLN, NLC) in Cosmetic and Pharmaceutical Dermal Products." *International Journal of Pharmaceutics* 366 (1–2): 170–84. doi:10.1016/j.ijpharm.2008.10.003.

Peltonen, Leena, and Jouni Hirvonen. 2010. "Pharmaceutical Nanocrystals by Nanomilling: Critical Process Parameters, Particle Fracturing and Stabilization Methods." *Journal of Pharmacy and Pharmacology* 62 (11): 1569–79. doi:10.1111/j.2042-7158.2010.01022.x.

Pereira, Galvina, Clara Fernandes, Vivek Dhawan, and Vaishali Dixit. 2022a. "3- Preparation and Development of Nanoemulsion for Skin Moisturizing." In *Micro and Nano Technologies*, edited by Siti Hamidah Mohd Setapar, Akil Ahmad, and Mohammad B T - Nanotechnology for the Preparation of Cosmetics Using Plant-Based Extracts Jawaid, 27–47. Elsevier. doi:10.1016/B978-0-12-822967-5.00008-4.

———. 2022b. *Nanoemulsion for Skin Moisturizing. Nanotechnology for the Preparation of Cosmetics Using Plant-Based Extracts.* INC. doi:10.1016/B978-0-12-822967-5.00024-2.

Pimpin, Alongkorn, and Werayut Srituravanich. 2012. "Reviews on Micro- and Nanolithography Techniques and Their Applications." *Engineering Journal* 16 (1): 37–55. doi:10.4186/ej.2012.16.1.37.

Prasad Yadav, Thakur, Ram Manohar Yadav, and Dinesh Pratap Singh. 2012. "Mechanical Milling: A Top Down Approach for the Synthesis of Nanomaterials and Nanocomposites." *Nanoscience and Nanotechnology* 2 (3): 22–48. doi:10.5923/j.nn.20120203.01.

Quintanar-Guerrero D, E Allémann, E Doelker, H Fessi, D Quintanar-Guerrero, E Allémann, E Doelker, H Fessi. 1998. "Preparation and Characterization of Nanocapsules from Preformed Polymers by a New Process Based on Emulsification-Diffusion Technique" Jul;15(7). doi:10.1023/a:1011934328471.

Rahimpour, Yahya, and Hamed Hamishehkar. 2012. "Liposomes in Cosmeceutics." *Expert Opinion on Drug Delivery* 9 (4): 443–55. doi:10.1517/17425247.2012.666968.

Rai, Shubhra, Vikas Pandey, and Gopal Rai. 2017. "Transfersomes as Versatile and Flexible Nano-Vesicular Carriers in Skin Cancer Therapy: The State of the Art." *Nano Reviews & Experiments* 8 (1): 1325708. doi:10.1080/20022727.2017.1325708.

Rapalli, Vamshi Krishna, Vedhant Kaul, Tejashree Waghule, Srividya Gorantla, Swati Sharma, Aniruddha Roy, Sunil Kumar Dubey, and Gautam Singhvi. 2020. "Curcumin Loaded Nanostructured Lipid Carriers for Enhanced Skin Retained Topical Delivery: Optimization, Scale-up, in-Vitro Characterization and Assessment of Ex-Vivo Skin Deposition." *European Journal of Pharmaceutical Sciences : Official Journal of the European Federation for Pharmaceutical Sciences* 152 (September): 105438. doi:10.1016/j.ejps.2020.105438.

Rapalli, Vamshi Krishna, Arisha Mahmood, Tejashree Waghule, Srividya Gorantla, Sunil Kumar Dubey, Amit Alexander, and Gautam Singhvi. 2021. "Revisiting Techniques to Evaluate Drug Permeation through Skin." *Expert Opinion on Drug Delivery* 18 (12): 1829–42. doi:10.1080/17425247.2021.2010702.

Ré, Maria. 2006. "Formulating Drug Delivery Systems by Spray Drying." *Drying Technology* 24: 433–46. doi:10.1080/07373930600611877.

Salavati-Niasari, Masoud, Fatemeh Davar, and Noshin Mir. 2008. "Synthesis and Characterization of Metallic Copper Nanoparticles via Thermal Decomposition." *Polyhedron* 27 (17): 3514–18. doi:10.1016/j.poly.2008.08.020.

Salvioni, Lucia, Lucia Morelli, Evelyn Ochoa, Massimo Labra, Luisa Fiandra, Luca Palugan, Davide Prosperi, and Miriam Colombo. 2021. "The Emerging Role of Nanotechnology in Skincare." *Advances in Colloid and Interface Science* 293: 102437. doi:10.1016/j.cis.2021.102437.

Santos, Ana Cláudia, Francisca Morais, Ana Simões, Irina Pereira, Joana A D Sequeira, Miguel Pereira-Silva, Francisco Veiga, and António Ribeiro. 2019. "Nanotechnology for the Development of New Cosmetic Formulations." *Expert Opinion on Drug Delivery* 16 (4): 313–30. doi:10.1080/17425247.2019.1585426.

Schultz, Stefan, Gerhard Wagner, Kai Urban, and Joachim Ulrich. 2004. "High-Pressure Homogenization as a Process for Emulsion Formation." *Chemical Engineering and Technology* 27 (4): 361–68. doi:10.1002/ceat.200406111.

Schwarz, C, W Mehnert, J S Lucks, and R H Müller. 1994. "Solid Lipid Nanoparticles (SLN) for Controlled Drug Delivery. I. Production, Characterization and Sterilization." *Journal of Controlled Release* 30 (1): 83–96. doi:10.1016/0168-3659(94)90047-7.

Shah, Rohan M, François Malherbe, Daniel Eldridge, Enzo A Palombo, and Ian H Harding. 2014. "Physicochemical Characterization of Solid Lipid Nanoparticles (SLNs) Prepared by a Novel Microemulsion Technique." *Journal of Colloid and Interface Science* 428: 286–94. doi:10.1016/j.jcis.2014.04.057.

Shahiwala A, Misra A. 2002. "Studies in Topical Application of Niosomally Entrapped Nimesulide." *Journal of Pharmacy and Pharmaceutical Sciences* 5(3):220–25.

Shaker, Dalia S, Rania A H Ishak, Amira Ghoneim, and Muaeid A Elhuoni. 2019. "Nanoemulsion: A Review on Mechanisms for the Transdermal Delivery of Hydrophobic and Hydrophilic Drugs." *Scientia Pharmaceutica* 87 (3). doi:10.3390/scipharm87030017.

Shams, Najmeh, and Mohammad Ali Sahari. 2016. "Nanoemulsions: Preparation, Structure, Functional Properties and Their Antimicrobial Effects." *Applied Food Biotechnology* 3 (3): 138–49. doi:10.22037/afb.v3i3.11773.

Sharma, Vijay, Mohd Yusuf, and Kamla Pathak. 2014. "Nanovesicles for Transdermal Delivery of Felodipine: Development, Characterization, and Pharmacokinetics." *International Journal of Pharmaceutical Investigation* 4 (3): 119. doi:10.4103/2230-973x.138342.

Shilpa, S, Bhartur Srinivasan, and Meenakshi Chauhan. 2011. "Niosomes as Vesicular Carriers for Delivery of Proteins and Biologicals." *International Journal of Drug Delivery* 3: 14–24. doi:10.5138/ijdd.2010.0975.0215.03050.

Silvestri, Shawn, Gustave Gabrielson, and Li Li Wu. 1991. "Effect of Terminal Block on the Microfluidization Induced Degradation of a Model A-B-A Block Copolymer." *International Journal of Pharmaceutics* 71 (1): 65–71. doi:10.1016/0378-5173(91)90068-Y.

Sorgi, Frank L, and Leaf Huang. 1996. "Large Scale Production of DC-Chol Cationic Liposomes by Microfluidization." *International Journal of Pharmaceutics* 144 (2): 131–39. doi:10.1016/S0378-5173(96)04733-3.

Souto, E B, S A Wissing, C M Barbosa, and R H Müller. 2004. "Development of a Controlled Release Formulation Based on SLN and NLC for Topical Clotrimazole Delivery." *International Journal of Pharmaceutics* 278 (1): 71–77. doi:10.1016/j.ijpharm.2004.02.032.

Szoka, F, and D Papahadjopoulos. 1978. "Procedure for Preparation of Liposomes with Large Internal Aqueous Space and High Capture by Reverse-Phase Evaporation." *Proceedings of the National Academy of Sciences of the United States of America* 75 (9): 4194–98. doi:10.1073/pnas.75.9.4194.

Tadros, Tharwat, P Izquierdo, J Esquena, and C Solans. 2004. "Formation and Stability of Nano-Emulsions." *Advances in Colloid and Interface Science* 108–109 (May): 303–18. doi:10.1016/j.cis.2003.10.023.

Tavano, Lorena, Rita Muzzalupo, Nevio Picci, and Bruno de Cindio. 2014. "Co-Encapsulation of Antioxidants into Niosomal Carriers: Gastrointestinal Release Studies for Nutraceutical Applications." *Colloids and Surfaces B: Biointerfaces* 114: 82–88. doi:10.1016/j. colsurfb.2013.09.058.

Uchegbu, Ijeoma F, and Suresh P Vyas. 1998. "Non-Ionic Surfactant Based Vesicles (Niosomes) in Drug Delivery." *International Journal of Pharmaceutics* 172 (1): 33–70. doi:10.1016/ S0378-5173(98)00169-0.

Ujiiye-Ishii, Kento, Eunsang Kwon, Hitoshi Kasai, Hachiro Nakanishi, and Hidetoshi Oikawa. 2008. "Methodological Features of the Emulsion and Reprecipitation Methods for Organic Nanocrystal Fabrication." *Crystal Growth and Design* 8 (2): 369–71. doi:10.1021/cg700708g.

van Hoogevest, Peter, and Alfred Fahr. 2019a. "Phospholipids in Cosmetic Carriers." In *Nanocosmetics*. doi:10.1007/978-3-030-16573-4_6.

———. 2019b. "Phospholipids in Cosmetic Carriers." In *Nanocosmetics: From Ideas to Products*, edited by Jean Cornier, Cornelia M Keck, and Marcel de Voorde, 95–140. Cham: Springer International Publishing. doi:10.1007/978-3-030-16573-4_6.

Van Tran, Vinh, Ju-Young Moon, and Young-Chul Lee. 2019. "Liposomes for Delivery of Antioxidants in Cosmeceuticals: Challenges and Development Strategies." *Journal of Controlled Release* 300: 114–40. doi:10.1016/j.jconrel.2019.03.003.

Vauthier, Christine, and Kawthar Bouchemal. 2009. "Methods for the Preparation and Manufacture of Polymeric Nanoparticles." *Pharmaceutical Research* 26 (5): 1025–58. doi:10.1007/s11095-008-9800-3.

Vemuri, Sriram, and C T Rhodes. 1995. "Preparation and Characterization of Liposomes as Therapeutic Delivery Systems: A Review." *Pharmaceutica Acta Helvetiae* 70 (2): 95–111. doi:10.1016/0031-6865(95)00010-7.

Vemuri, Sriram, Cheng Der Yu, Vuthichai Wangsatorntanakun, and Niek Roosdorp. 1990. "Large-Scale Production of Liposomes by a Microfluidizer." *Drug Development and Industrial Pharmacy* 16 (15): 2243–56. doi:10.3109/03639049009043797.

Verma, Poonam, and K Pathak. 2010. "Therapeutic and Cosmeceutical Potential of Ethosomes: An Overview." *Journal of Advanced Pharmaceutical Technology & Research* 1 (3): 274. doi:10.4103/0110-5558.72415.

Verma, S, Shailendra Singh, N Syan, Pooja Mathur, and V Valecha. 2010. "Nanoparticle Vesicular Systems: A Versatile Tool for Drug Delivery." *Journal of Chemical and Pha rmaceutical Research* 2: 496–509.

Waghule, Tejashree, Srividya Gorantla, Vamshi Krishna Rapalli, Pranav Shah, Sunil Kumar Dubey, Ranendra Narayan Saha, and Gautam Singhvi. 2020. "Emerging Trends in Topical Delivery of Curcumin Through Lipid Nanocarriers: Effectiveness in Skin Disorders." *AAPS PharmSciTech* 21 (7): 284. doi:10.1208/s12249-020-01831-9.

Waghule, Tejashree, Vamshi Krishna Rapalli, Srividya Gorantla, Ranendra Narayan Saha, Sunil Kumar Dubey, Anu Puri, and Gautam Singhvi. 2020. "Nanostructured Lipid Carriers as Potential Drug Delivery Systems for Skin Disorders." *Current Pharmaceutical Design* 26 (36): 4569–79. doi:10.2174/1381612826666200614175236.

Wan, Lucy, Paul Heng, and Cecilia Chia. 2008. "Spray Drying as a Process for Microencapsulation and the Effect of Different Coating Polymers." *Drug Development and Industrial Pharmacy* 18: 997–1011. doi:10.3109/03639049209069311.

Wegner, Karsten, and Sotiris E. Pratsinis. 2004. "Flame Synthesis of Nanoparticles." *Chimica Oggi* 22 (9): 27–29. doi:10.1205/cerd.82.11.1444.52025.

Woodbury, Dixon J, Eric S Richardson, Aaron W Grigg, Rodney D Welling, and Brian H Knudson. 2006. "Reducing Liposome Size with Ultrasound: Bimodal Size Distributions." *Journal of Liposome Research* 16 (1): 57–80. doi:10.1080/08982100500528842.

Woodle, Martin C, and Demetrios Papahadjopoulos. 1989. "[9] Liposome Preparation and Size Characterization." In *Biomembranes Part R*, 171: 193–217. Methods in Enzymology. Academic Press. doi:10.1016/S0076-6879(89)71012-0.

Wu, Pey Shiuan, Yu Syuan Li, Yi Ching Kuo, Suh Jen Jane Tsai, and Chih Chien Lin. 2019. "Preparation and Evaluation of Novel Transfersomes Combined with the Natural Antioxidant Resveratrol." *Molecules* 24 (3): 1–12. doi:10.3390/molecules24030600.

Yang, Yan, Rujing Ou, Shixia Guan, Xiaoling Ye, Bo Hu, Yi Zhang, Shufan Lu, et al. 2015. "A Novel Drug Delivery Gel of Terbinafine Hydrochloride with High Penetration for External Use." *Drug Delivery* 22 (8): 1086–93. doi:10.3109/10717544.2013.878856.

Yuh-Fun, Maa, and C Hsu Chung. 1999. "Performance of Sonication and Microfluidization for Liquid–Liquid Emulsification." *Pharmaceutical Development and Technology* 4 (2): 233–40.

Zhang, Wanping, Yubo Qin, Shaonian Chang, Haiyang Zhu, and Qianjie Zhang. 2021. "Influence of Oil Types on the Formation and Stability of Nano-Emulsions by D Phase Emulsification." *Journal of Dispersion Science and Technology* 42 (8): 1225–32. doi:10.1080/01932691.2020.1737538.

Zielinska, Aleksandra, Filipa Carreiró, Ana M Oliveira, Andreia Neves, Bárbara Pires, D Nagasamy Venkatesh, Alessandra Durazzo, et al. 2020. "Polymeric Nanoparticles: Production, Characterization, Toxicology and Ecotoxicology." *Molecules* 25 (16): 3731. doi:10.3390/molecules25163731.

6 In vitro and in vivo characterization of nano-cosmetics

Satyajit Tripathy, Wamiq Musheer Fareed and Motlalepula Gilbert Matsabisa

CONTENTS

6.1 INTRODUCTION

Products developed with nanoparticles have several significant advantages over cosmetics produced on a lower scale. To obtain greater stability and lasting effects, the cosmetic industry employs nanoparticles (NMs). The compounds can be carried

DOI: 10.1201/9781003319146-6

through the skin more successfully due to the large surface areas of nanomaterials. One of the main purposes of using nanomaterials in cosmetics is to increase the effectiveness of substances penetrating the skin for increased product delivery, new color components (such as in lipsticks and nail polishes), transparency, and long-lasting effects (e.g., in makeup). NMs are currently most widely employed in skincare products, such as sunscreens, that act as UV filters (Raj et al., 2012).

The use of nanomaterials, which are made up of particles smaller than 0.1 mm, in consumer goods, including cosmetics, is rising. Risks to consumer health that may be presented using nanomaterials in cosmetics must be assessed and managed by both regulators and producers. This risk assessment is crucial because it will reveal whether steps need to be done to mitigate and/or reduce the risks that have been identified. The content of *in vitro* and *in vivo* characterization of nano-cosmetics, the risk assessment, is covered in this chapter. The physicochemical characterization of the nanomaterials, any potential risks (toxicity) they may cause, and finally the risk assessment itself must all be covered within the field of nano-cosmetics (Jong et al., 2016).

The Scientific Committee on Emerging and Newly Identified Risks' (SCENIHR's) conclusions underline the importance of adequate nanomaterial characterization (SCENIHR 2006, 2007; Hansen et al., 2008). It is crucial that the particulate sample is representative of the material, and that the particle size and shape properties be assessed in the most appropriate dispersion state and under settings that simulate potential exposure to humans and the environment. For nanoparticles, mass is probably not the most important parameter. Size and surface area could be important additional factors. In these chapters, the characterization techniques are covered. The dermal skin penetration of substances has been studied and quantified using a variety of ways (Sekkat and Guy, 2001). However, established assessment techniques have come under scrutiny due to the usage of topical preparations incorporating liposomes and other nanomaterials. New test methods will be required because of the unique properties of nanoparticles to identify the potential harm pathways they may induce (Nel et al., 2006).

The Safety Assessment of Nanomaterials in Cosmetics (SCCS 2019) emphasizes that risk evaluation of cosmetic in nanomaterials (NMs) may be influenced by exposure concerns, with a focus on in-depth characterization of the NMs and NM-related issues during toxicological evaluation. *In vivo* testing of cosmetic chemicals is no longer allowed, and the approved *in vitro* methodologies now available only cover specific toxicological endpoints, leading to the advocacy of exposure-based risk assessment.

According to additional SCCS guidance, the hazard information provided should be in relation to the same NM that is intended for use in the finished product, but even more crucially, it should be in relation to the NM released from the product since this is the substance to which a consumer is exposed. The substance emitted from the product may vary (Mitrano et al., 2015).

6.2 MICROSCOPIC TECHNIQUES

Microscopic analysis of the treated skin can provide additional relevant information from the *in vitro* method (or even from some *in vivo* methods). Visualizing the tissue that an active component in a vector has been applied to can give important information even though absolute quantification might not be attainable. This problem was tackled using laser scanning confocal microscopy (LSCM), which produced cross-sectional pictures as well as optical "slices" of the treated skin (Alvarez-Román et al., 2004). The benefit of this technique is that it allows for the creation of three-dimensional representations of the skin using reasonably thick tissue samples with minimal or no tissue artifacts.

However, the method depends on the availability of suitable fluorophores, which ideally enable separate tracking of the active and the vector (an objective not yet attained). But by employing this technique, researchers were able to demonstrate the impact of a nanoparticle formulation on the movement of a model active (AlvarezRomán et al., 2004) and observe how particulate vectors bind to follicular openings. The imaging of individual particles in exceedingly thin tissue sections is possible using alternative techniques like high-resolution transmission electron microscopy.

X-ray analysis can then be used to determine the chemical makeup of the visible vector. The limitations of this method cast doubt on the representativeness of any given image due to the small field of vision; and (ii) the possibility of artifacts due to the lengthy sample preparation process. A few ion-beam methods are also being used in the field, including Rutherford back-scatter spectrometry, scanning transmission ion microscopy, and particle induced X-ray emission (PIXE), which generates elemental maps.

Large fields of view, simple sample preparation, and easy artifact removal are advantages of these techniques; individual particle visualization is not one of them. Radiolabelling with the positron emitter 48V (half-life: 16 days) is a final method. Thin slices and nuclear microemulsions are used in this autoradiographic technique. The method is extremely sensitive, rather simple to use, has a wide field of view, and can display individual positron tracks but not actual particles. It seems to be helpful for directing particles to certain skin structures, such hair follicles and "furrows" in the skin.

6.3 IN VITRO TESTS

6.3.1 Validated in vitro tests

In vitro toxicology has advanced greatly in recent years, with 3Rs strategy (refinement, reduction, and replacement) serving as its guiding concept in Europe (Russell and Burch, 1959). While there are various available techniques and technologies for examining the molecular processes underlying a compound's biological activity, only a small number of these techniques are relevant for essential mechanisms, particularly the evaluation of a substance's risk. Only proven techniques are permitted for substances and products used in cosmetics.

When testing is necessary for the safety assessment of cosmetic ingredients, as mandated by the Cosmetic Directive 78/768/EEC, certain validated procedures must be applied. Acute and short-term toxicity can be measured using methods that have been validated, but repeated-dose toxicity and long-term toxicity have not yet been done so. Only a small portion of the verified methods include replacement tests:

6.3.1.1   Challenges of skin irritation test

Episkin is a human skin reconstruction model that was approved by the European Union Reference Laboratory for alternatives to animal testing (EURL ECVAM) Scientific Advisory Committee (ESAC) in April 2007. Its endpoint is the 3-(4,5)-dimethyl-2-thiazolyl-2,5-dimethyl-2H-bromide reduction. By adding IL-1α (interleukine-1α) measurement as a second endpoint, a better sensitivity is achieved without sacrificing specificity.

6.3.1.2   Skin corrosion testing via TER (transcutaneous electrical resistance)

In a two-compartment test setup where the skin discs serve as the separation between the compartments, the test chemical is administered for up to 24 hours to the epidermal surfaces of the skin discs. Rats aged 28–30 days were killed humanely to produce the skin discs. When a TER decreases below a threshold level, a substance is considered corrosive if it can cause a loss of normal stratum corneum integrity and barrier function. Based on substantial data for a wide range of chemicals, a cut-off value of 5 kΩ has been chosen for rat skin TER, where most readings were either clearly considerably above (often >10 kΩ), or well below (commonly 3 kΩ).

6.3.1.3   Neutral red uptake phototoxicity test (NRPT)

The term "photo irritation" refers to a hazardous reaction that occurs when the skin is first exposed to specific chemicals and then exposed to light, or when the skin is exposed to light following systemic injection (oral, intravenous) of a chemical agent. This test is intended to determine whether the test chemical is cytotoxic in the presence or absence of UV radiation. In this situation, cytotoxicity is employed as a stand-in for acute phototoxicity, which is a chemically induced skin irritant that needs light to start acting. The procedure is a cell-based assay that use 3T3 cells. After exposure to the test material, cytotoxicity is evaluated using neutral red absorption (and light or not).

6.3.1.4   Implementation of Franz cell for dermal absorption measurements

An internal technique utilizing Static Franz diffusion cells and dialysis membranes was developed for the evaluation of possible components for skincare products. L-ascorbic acid, -tocopherol, and benzoic acid were chosen as model substances. The cell environment was designed to promote transmembrane diffusion.

Additionally, the following tests are also considered:

• Genotoxicity/mutagenicity testing via a set of three recommended tests
• Bacterial reverse mutation test – *in vitro* mammalian cell gene mutation test
• *In vitro* micronucleus test or *in vitro* mammalian chromosome aberration test
• Embryotoxicity testing via three tests EST (embryonal stem cell test), MM (micromass assay), and WEC (whole embryo culture)

These assays have been approved for the safety evaluation of conventional cosmetic components, but they have not been approved for the safety evaluation of nanomaterials. Therefore, it is unknown if these established assays can be used for nanoparticles.

6.3.2   NON-VALIDATED IN VITRO APPROACHES

There are different non-validated methods available. These are not appropriate for quantitative risk assessment but are particularly concerned with the identification of chemical hazards under REACH (Regulation (EC) No 1907/2006). Current unvalidated tests include:

6.3.2.1   Screening of eye corrosives and severe irritants by

6.3.2.1.1   BCOP (bovine cornea opacity permeability test)

BCOP is an *in vitro* experiment that gauges how newly acquired bovine corneas respond to potentially irritating substances in terms of permeability and opacity (a by-product of abattoirs). When exposed to a test material or chemical, corneal opacity is directly assessed using an opacimeter. Changes in optical density are used to quantify corneal permeability spectrophotometrically. Corneal permeability is tested with sodium fluorescein, a dye that frequently cannot pass through corneal epithelial cells. To compare the potential for irritation of different test substances/ chemicals, these data are used to create an endpoint irritation score that can be used to compare the potential for irritancy of various test substances/chemicals.

6.3.2.1.2   ICE (isolated chicken eye test)

The ICE test's endpoints are modifications in corneal thickness, opacity, and permeability. Each of these endpoints is thought to be a possible standalone sign of eye irritancy. Corneal thickness increases are a symptom of edema, or excessive fluid retention, between the different layers of the cornea. Since the cornea oversees a large portion of the light refraction required for proper eye focus, increases in corneal thickness might result in hazy vision. The corneal opacity tells us how much light can enter the eye. It is brought on by the denaturation or precipitation of corneal proteins, basically changing their transparency to opacity. Any alterations in opacity signify a problem with the eye's health. Corneal permeability is brought on by damage to the corneal epithelial cells, which normally serve as a barrier for the inside of the eye. Any permeability implies the risk of injury or functional loss from foreign objects, substances, or chemicals entering the eye.

6.3.2.1.3   IRE (isolated rabbit eye test)

An organotypic model known as the isolated rabbit eye test (IRE) is used to assess the impacts of substances on the cornea of an isolated rabbit's eye *in vitro*. To assess a substance's irritancy, variations in corneal thickness, clarity, and permeability are considered.

6.3.2.1.4   HET-CAM (hen's egg test - chorioallantoic membrane)

The HET-CAM assay (OECD TG 405, 2012; Method B.5, Annex to Commission Regulation 440/2008/EC, EU 2008) measures the acute irritative effects of a chemical on mucosal ocular tissue. The Draize rabbit eye irritation test has been replaced by the in ovo HET-CAM (HET Hühnerei Test), which was first developed by Luepke (1986). A vascular fetal membrane known as the CAM is created by joining the allantois's surrounding wall with the chorion, the embryo's outermost sac. The CAM is made of three-cell-thick ectodermal, mesodermal, and endodermal layers that also contain blood vessels, connective tissue, and ground material (composed of squamous cells). Most studies show coagulation, vascular lysis, and hemorrhage as results. Some authors additionally make use of vasoconstriction, hyperemia, and vascular injection (minor bleeding) (https://ntp.niehs.nih.gov/iccvam/docs/annrpt/biennialrpt2007-508.pdf). It is believed that the chorioallantoic membrane of the avian can be used as a model to examine how compounds affect the conjunctivae when they are detected *in vivo* (in mammals). Thus, it is thought that the CAM's small blood vessels and proteins are adversely affected in a manner like how irritants cause conjunctival reactions. The hen's egg test is unable to feel any pain since nerve tissue and the ability to perceive pain have not yet evolved during this stage of development (Rosenbruch 1997).

6.3.2.2 Tests for sensitivity screening via a reduced LLNA (local lymph node assay)

The conception and approval of the murine local lymph node assay (LLNA) for the identification of medicines with low molecular weight sensitizing properties. It differs from other tests for sensitization in that it provides a quantitative endpoint, dose-responsive information, and potency prediction. Predictability, degrees of false positive reactions, and vehicle-related variability are only a few of the problems raised by this test. The basic tenet of the conventional LLNA is that chemical sensitizers induce initial proliferation of lymphocytes in the lymph nodes draining the site of chemical administration, also known as the induction phase. The amount of radiolabeled thymidine integrated into the lymph nodes' cellular DNA can be used to quantify this initial proliferation. Low molecular weight (LMW) chemical sensitizers known as haptens are too small to be allergenic on their own; instead, they must bind to a protein to elicit an allergic response. It has been discovered that keratinocytes, Langerhans cells, and T lymphocytes are all essential for the induction phase of Allergic contact dermatitis (ACD).

Predictomics, ReProTect, Sensit-iv, AcuTox, Liintop, Carcinogenomics, and other FP6 Research projects are developing a number of *in vitro* tests and methods. It must be highlighted that these tests do not necessarily apply to nanomaterials, just to conventional chemicals. Therefore, approved in vitro techniques created especially for the use of nanomaterials as cosmetic ingredients are required. The literature, however, indicates that significant efforts are being made to develop *in vitro* testing methods that could be employed for the pertinent toxicological assessment of nanomaterials. According to a report created by NRCG Task Force 3 in August 2006, the following relevant toxicological endpoints are significant for nanomaterials:

   i. infiltration
  ii. Accumulation and transposition
 iii. Cytocompatibility
 iv. oxidative stress and inflammation
  v. mutagenicity/genotoxicity

6.3.3 INFILTRATION

Skin: Human skin's stratum corneum serves as a superior barrier. Cosmetic substances may penetrate the skin when they are applied to it. In general, for testing cosmetics, animal skins are not appropriate... It is understood that the skin of hairy rodents causes the penetration of human skin to be overestimated. However, these models may magnify any impacts of any skin penetration enhancers in the formulation, making hairless species more advantageous. Tests of systems using nanomaterials may be affected by the significant variations in follicular density between hairy species and humans. An extra danger of barrier function disruption exists when hairy skin is shaved or depilated prior to treatment, further complicating the challenge of accurately measuring nanoparticle absorption.

Although the efficacy of pig skin *ex vivo* has been established in various applications and it adequately approximates human absorption, the question of whether the follicular characteristics in this model are reasonably comparable to those of human skin remains. *In vivo* skin absorption is described in OECD Guideline 427 (OECD 2004a), which is why it won't be permitted for cosmetic compounds in the future. Dermal absorption can also be tested in a diffusion cell using exfoliated pig or human skin (Franz cell). OECD Guideline 428 describes the technique, which can be applied to conventional cosmetic components (SCCNFP/0750/03; SCCP/0970/06), but not necessarily to nanoparticles. According to OECD Guideline 428, skin divides a donor compartment, which contains the active's formulation, from a receptor phase, which normally consists of a physiological buffer (OECD 2004b).

By sampling the receiver chamber over time, it is possible to determine the substance's permeation. Instead, the testing can be stopped at a certain point, the tissue removed, and the ingredient then examined in several compartments. However, the method is not well suited for observing the interaction between an ingredient and a carrier. By examining the substance in the skin or receptor phase, it may be feasible to determine whether the carrier affects the transfer of the chemical, but the mechanism underlying this cannot be inferred. A nanoparticle carrier's ability to be detected in the receiving medium is also exceedingly unlikely. Furthermore, TER (Transcutaneous Electrical Resistance) or TEWL (Trans Epidermal Water Loss) tests must be used to accurately evaluate the skin's integrity.

Lung: A significant entry barrier is made up of the bronchial and alveolar epithelia. For bronchial epithelium, a few human cell lines have been suggested. They might not accurately depict the *in vivo* environment (Forbes and Ehrhardt 2005). For cell types that imitate alveolar epithelium, the same is true. It would be preferable to employ isolated primary bronchial and alveolar cells as a monolayer.

Mouth: Only those components and products, such as those found in lipstick, toothpaste, etc., which may or may not include nanomaterials, can penetrate the gastro-intestinal system are significant when it comes to the oral route for cosmetics. The principal application of immortalized cell lines is the investigation of penetration through the intestinal epithelium. In specifically, human colon adenocarcinoma-derived Caco-two cell lines are employed in culture. Standardization of the various cell line types is a difficulty, but an *in vitro* system also does not accurately represent the actual *in vivo* environment.

6.4   ACCUMULATION AND TRANSPOSITION

It is crucial to take into account how a nanomaterial's size, chemical composition, and surface reactivity will affect how well it is absorbed by macrophages. According to research by Möller et al. (2005), ultrafine carbon black particles disrupt cytoskeletal function and hinder phagosome trafficking while briefly raising intracellular calcium levels. These effects appeared to be significantly influenced by the particles' particular surface area. Moss and Wong (2006) created a technique that makes use of the cumulative predicted area of the particles involved in macrophage absorption. It has been proposed that non-endocytotic mechanisms allow nanoparticles to pass through cellular membranes.

Various particle types can have different sub-cellular distributions, such as endolySOS(bacterial genotoxicity test)omal, cytoplasmic, nuclear, and mitochondrial particles (Rothen-Rutishauser et al., 2006). Various *in vitro* models, such as excised human or pig skin and rebuilt epidermis, have been employed for translocation research (Chen et al., 2006). There are not many *in vitro* investigations on translocation through different types of obstacles. In order to assess the transcytosis and toxicity of nanoparticles at the endothelial tight junction at the blood brain barrier (BBB), Lu et al. (1998) developed a co-culture paradigm using rat brain capillary endothelial cells and astrocytes.

The processes and kinetics of particle translocation remain unknown, despite the fact that numerous investigations have demonstrated that inhaled ultrafine particles can enter the bloodstream (Kreyling et al., 2006; Nemmar et al., 2002a,b; Oberdörster et al., 2002). Recently, a method for measuring the translocation of nanoparticles over a layer of lung cells *in vitro* was created (Geys et al., 2006).

6.5   CYTOCOMPATIBILITY

Depending on the target tissue being studied, cytocompatibility can be assessed *in vitro* utilizing a number of cell lines and original cells in culture. Apoptosis, intracellular metabolic alterations, and membrane integrity are the three main factors that are often quantified in measurements. Here are a few current possibilities:

Membrane integrity is checked on:

- passive dye uptake, e.g. tryphan blue, by damaged and dead cells and light microscopic counting
- spectrophotometric measurement of intracellular enzyme release from lysis of injured and dead cells with, such as lactate dehydrogenase (LDH) (LDH kits are commercially available)
- spectrophotometric analysis of active dye uptake, e.g. neutral red, by living cells

Intracellular alterations are quantified:

- spectrophotometric, mass spectrometric techniques, radioactivity, GC, HPLC, or other methodologies for enzyme activity in phase I and phase II biotransformation

- MTT assay to test the impaired mitochondrial reductive activity by the reduction of tetrazolium salts to purple colored formazan products
- the ATP content of cells and energy failure in dead cells (enzymatic assay)

Induction of programmed cell death results biochemical and morphological changes and checked by:

- caspase activity (in particular caspase 3)
- expression of Apaf-1, pro-apoptotic Bcl-2 proteins Bax and Bid, tumor suppressor p53
- annexin V-labelling of phosphatidyl serine of plasma membranes
- TUNEL assay: DNA fragmentation analysis by gel electrophoresis and labelling of DNA ends

There are some instances of nanomaterials being used in current scientific literature. The use of BRL 3A immortalized rat liver cells (ATCC, CRL1442) was suggested by Hussain et al. (2005) to assess the acute toxic effects of various metal/metal oxide nanomaterial sizes. Under control and exposed circumstances, morphology, mitochondrial function (MTT assay), membrane leakage (LDH assay), reduced glutathione (GSH) levels, reactive oxygen species (ROS), and mitochondrial membrane potential (MPP) were evaluated for toxicity. Modifications in morphology, LDH leakage, and mitochondrial dysfunction were caused by nanoparticles. Different metals showed varying degrees of toxicity, such as silver, which showed an unique toxicity associated to oxidative stress with a marked decrease in GSH level, a reduction in mitochondrial membrane potential, and an increase in ROS level (Hussain et al., 2005).

Other research on the cytotoxicity of nanomaterials in human lung cancer cell lines have used oxidative stress as an endpoint for cytotoxicity (Han et al., 2014). Additionally, effectively employed cell types include human neuronal astrocytes, human hepatoma (HepG2) cells, and human dermal fibroblasts in culture (Sayes et al., 2005). This *in vitro* methodology is only acceptable as a screening method, not a replacement for *in vivo* investigations, as it was discovered from cytotoxicity experiments with industrial nanomaterials that nanoparticle dispersion is a critical component impacting cytotoxic reactions (Ayres et al., 2008).

With relation to cytotoxicity, the behavior of nanoparticles in suspension needs to be taken into account. In cell culture media, nanoparticles may diffuse, settle, and clump together depending on the surrounding conditions (medium, ionic strength, pH, and viscosity), as well as the particle's characteristics (size, shape, and density) (Teeguarden et al., 2007). Consequently, cellular dose is impacted because the factors mentioned above control the delivery rate to the cultured cells. Because a simple calculation of the nanomaterial's concentration in the culture medium can significantly misinterpret reactions and uptake data obtained *in vitro*, nanoparticle kinetics in cell culture systems must be studied (Teeguarden et al., 2007).

6.6 IN VITRO STUDY ON CELLULAR STRESS

6.6.1 OXIDATIVE STRESS

The ability of nanoparticles to produce reactive oxygen species (ROS), which cause oxidative stress, has been shown in numerous investigations. Since oxidative stress may be examined in both cell-free (Beck-Speier et al., 2001; Brown et al., 2001; Xia et al., 2006) and cellular systems, it has been proposed as a suitable endpoint (Oberdörster et al., 2005) for a variety of tissue types. Examples include brain microglia (Gurr et al., 2005), lung cells (Xia et al., 2006), bronchial epithelial cells, and macrophages (Beck-Spier, 2005).

The mechanisms of oxidative stress were examined in the majority of these research. Regarding cellular absorption, subcellular localization, and the capacity to catalyze the generation of reactive oxygen species under biotic and abiotic circumstances, particles varied. Comparative tests have been done to see how well manufactured and ambient nanoparticles can cause oxidative stress (Xia et al., 2006). ROS can be measured in a variety of methods.

Measurements of the following variables are currently used to quantify ROS:

- Testing of reduced glutathione (GSH) level
- Measurement of glutathione in oxidized form (GSSG)
- Calculation of ratio of oxidized and reduced form of glutathione in cell
- generation of free radicals using the colorimeter's thiobarbituric acid approach, and in more detail, using spin trapping agents and measurements of the produced stable adducts' electron spin resonance
- Assay of adduct production of hydroxyl radicals with 8-OH-deoxyguanosine

6.6.2 INFLAMMATION

It is possible to track the release of pro-inflammatory mediators from various cellular sources, including cytokines, chemokines, nitric oxide, and the up-regulation of transcription factors like nuclear factor kappa (NF KB) and activator protein one (all known to play a role in the initiation of inflammatory responses). The epithelium of entrance pathways, local neutrophils and macrophages, as well as cells from target organs like the liver, kidney, and nervous system, are all possible cellular targets for nanoparticles. Although primary cells from multiple species can be employed, interspecies reactions may differ from those of humans.

While working with cell lines produced from non-cancerous tissue is simpler, comparison to the *in vivo* environment might be challenging. There is a clear link between nanoparticle surface area, ROS-generating ability, and pro-inflammatory effects (Oberdörster et al., 2005). These investigations have shown that the transcription of pro-inflammatory cytokines like IL8, IL1, and GM-CSF (GranulocyteMacrophage-Colony Stimulating Factor) is dependent on oxidative stress for nuclear factor NF-KB activation.

It has been demonstrated that lung cells exposed to ultrafine carbon black particles express more NF-KB-related genes (Kim et al., 2003). These particles not only caused cytotoxic damage and inflammation, but also prevented vascular endothelial cells from growing (Stone et al., 1998). Similar outcomes were previously attained utilizing human dermal endothelial cells (Chang et al., 2005).

6.6.3 MUTAGENICITY/GENOTOXICITY

Using *in vitro* experiments in mammalian cells, the genotoxic potential of nanoparticles can theoretically be evaluated. The tests must, however, be timed so that the nanomaterial can enter the nucleus. It would be required to look at the second post-treatment metaphase in addition to the first one in the *in vitro* chromosomal aberration test. In order to assess the cells after the second post-treatment mitosis, the cells in the cytokinesis-block micronucleus test *in vitro* should be exposed for one cell cycle without cytochalasin B (Cyt-B), followed by another cycle in its presence.

In the same way as other particle materials, it is not anticipated that bacterial genotoxicity testing will be helpful. Simple *in vitro* experiments might not be sufficient to demonstrate the genotoxic potential of nanoparticles if those effects are related to inflammation. Inflammation and genotoxic consequences are generally poorly known, and many other forms of poorly soluble bigger (non-nano) particles pose further questions in this regard. Depending on the endpoint under study, it may be necessary to conduct somewhat lengthy tests to demonstrate a link between inflammation and genotoxic consequences. However, it is not obvious if such investigations are feasible in the context of genotoxicity evaluation.

It is reported that, 15 months following treatment, rats given intratracheal instillations of nanoscale carbon black as well as fine titanium dioxide (anatase) and fine quartz experienced an increase in hprt mutations in alveolar type II cells (Driscoll et al., 1997). Before conclusions about the significance of researching inflammation-related genotoxicity of nanoparticles can be reached, it appears that fundamental studies on the relationship between inflammation and genotoxicity are necessary. Since mutations may only be corrected when a cell divides, actively dividing cells are anticipated to be the main targets of genotoxic effects related to carcinogenesis *in vivo*.

Type II pneumocytes, for example, are cells of importance because they can create new lung cells. The basal layer of the epidermis is the part of the skin that is most important. The same applies to gastrointestinal tract cells. The commonly used *in vivo* genotoxicity tests, the liver UDS (Unscheduled DNA Synthesis) test and the bone marrow micronucleus test, both identify genotoxic substances that enter the bone marrow and liver, respectively, or that have systemic genotoxic effects. For evaluating genotoxicity in the anticipated target tissues *in vivo*, there are currently no validated standard methods available, however procedures like the comet assay, micronucleus test, and gene mutation analyses in transgenic mice might likely be used.

6.7 IN VIVO METHODS

In vivo investigations of hazard identification are crucial for learning about the biodistribution, transport, accumulation, and clearance of nanomaterials. However, for acute and local toxicity testing after 2009 and for repeated dosage toxicity and long-term testing after 2013, the use of animals for safety testing of cosmetic compounds, including nanomaterials, will no longer be permitted. To develop verified *in vitro* procedures, effort is needed. This is especially true when it comes to strategies for validating nanomaterials. If the validation is done out utilizing nanomaterials, the creation of human skin substitutes such as EpiskinTM, EpidermTM, SkinethicTM, and other instances of recreated human skin could give instruments for safety assessment of cosmetic chemicals and final products.

As stated under 3.5.4.2(i), the *in vitro* test (OECD 428) must be used in place of the *in vivo* skin absorption test for nanomaterials, which is a fundamental test in safety evaluation (OECD 427). Sequential tape-stripping of the stratum corneum after a preparation treatment gives information on the uptake of substances in the skin's upper layers (Hoppel et al., 2014).

However, persistent formulation in skin 'furrows' and/or hair follicles may make estimates difficult. Furthermore, the stratum corneum barrier may have breaches that allow particles to enter (Menon et al., 2007). Even though stratum corneum tape-stripping is a tried-and-true technique for learning about the contents of the stratum corneum, it is insufficient to detect nanoparticle translocation to the deeper layers of the skin, necessitating the use of more sophisticated visualization techniques like confocal and high-resolution electron microscopy, ion beam technologies, and autoradiography.

It is necessary to be able to detect both carrier particles and active substances as well as individually label in a stable manner. For conventional chemicals, the EC and OECD have created recommendations; however, nanoparticles have not yet received this treatment. It is unclear whether the regulatory toxicological tests that are already in place are sufficient or applicable for assessing nanomaterials, or whether particular changes will be needed and new techniques developed. As a result, more study must be done and a case-by-case analysis of the nanomaterials' current hazards is required.

6.8 CONCLUSION

Cosmetic nanoparticles may be used for several things (e.g., UVA and UVB filters in sunscreens, nano-preservatives). The peculiar qualities of any nanomaterial that could produce the intended function or feature of the cosmetic product could also put consumers at risk. Given this, a comprehensive evaluation of the safety of all nanomaterials is necessary, which should include testing addressing the nano-characteristics (such as their ability to penetrate viable skin layers due to their small size as well as inhalation trials in the case of sprays and powders). The most recent guidance states that the FDA should consider the physicochemical parameters, NM aggregation and size distribution, shape, solubility, density, porosity, stability, and contaminants among other things. Additionally, it is important to determine the potential exposure pathways for NMs and gather *in vitro* and *in vivo* toxicological data, such as research on dermal penetration and potential inhalation, genotoxicity, and potential skin and eye irritation. The absence of information provided by the applicants is what makes the results inconclusive. The exposure routes of the NMs are among the most crucial factors to consider. The stratum corneum, the first layer of the epidermis, is exposed via the skin. There are still some questions about whether NMs could penetrate the stratum corneum into viable levels, where toxicological issues might surface. Even though NMs that are very small nevertheless have considerably larger molecular weights compared to known molecules that permeate the skin, additional studies should be carried out on each NM before it is employed in a cosmetic recipe. Sprays and aerosols that could contain NMs should be subject to a more thorough safety assessment because inhalation exposure is a possibility. A non-exhaustive list of the parameters needed for an exposure scenario was published in the SCCS Notes of Guidance (SCCS/1602/18). For NMs, the concentration should be stated in terms of particle number concentration and surface area in addition to the weight-based concentration of the NM.

REFERENCES

Alvarez-Román R, Naik A, Kalia YN, Guy RH, Fessi H. Skin penetration and distribution of polymeric nanoparticles. *Journal of Controlled Release.* 2004 Sep 14;99(1):53–62.

Ayres JG, Borm P, Cassee FR, Castranova V, Donaldson K, Ghio A, Harrison RM, Hider R, Kelly F, Kooter IM, Marano F. Evaluating the toxicity of airborne particulate matter and nanoparticles by measuring oxidative stress potential—a workshop report and consensus statement. *Inhalation Toxicology.* 2008 Jan 1;20(1):75–99.

Beck-Speier I, Dayal N, Karg E, Maier KL, Roth C, Ziesenis A, Heyder J. Agglomerates of ultrafine particles of elemental carbon and TiO2 induce generation of lipid mediators in alveolar macrophages. *Environmental Health Perspectives.* 2001 Aug;109(suppl 4):613–8.

Brown DM, Wilson MR, MacNee W, Stone V, Donaldson K. Size-dependent proinflammatory effects of ultrafine polystyrene particles: A role for surface area and oxidative stress in the enhanced activity of ultrafines. *Toxicology and Applied Pharmacology.* 2001 Sep 15;175(3):191–9.

Chang CC, Chiu HF, Wu YS, Li YC, Tsai ML, Shen CK, Yang CY. The induction of vascular endothelial growth factor by ultrafine carbon black contributes to the increase of alveolar-capillary permeability. *Environmental Health Perspectives.* 2005 Apr;113(4):454–60.

Chen J, Tan M, Nemmar A, Song W, Dong M, Zhang G, Li Y. Quantification of extrapulmonary translocation of intratracheal-instilled particles *in vivo* in rats: Effect of lipopolysaccharide. *Toxicology.* 2006 May 15;222(3):195–201.

De Jong WH, Delmaar C, Gosens I, Nijkamp M, Quik JT, Vandebriel RJ, van Kesteren PC, Visser MJ, Park MV, Wijnhoven SW. Description of a nanocosmetics tool for risk assessment. 2016. https://rivm.openrepository.com/handle/10029/595315

Driscoll KE, Deyo LC, Carter JM, Howard BW, Hassenbein DG, Bertram TA. Effects of particle exposure and particle-elicited inflammatory cells on mutation in rat alveolar epithelial cells. *Carcinogenesis.* 1997 Feb 1;18(2):423–30.

Forbes B, Ehrhardt C. Human respiratory epithelial cell culture for drug delivery applications. *European Journal of Pharmaceutics and Biopharmaceutics.* 2005 Jul 1;60(2):193–205.

Geys J, Coenegrachts L, Vercammen J, Engelborghs Y, Nemmar A, Nemery B, Hoet PH. *In vitro* study of the pulmonary translocation of nanoparticles: A preliminary study. *Toxicology Letters.* 2006 Jan 25;160(3):218–26.

Gurr JR, Wang AS, Chen CH, Jan KY. Ultrafine titanium dioxide particles in the absence of photoactivation can induce oxidative damage to human bronchial epithelial cells. *Toxicology.* 2005 Sep 15;213(1–2):66–73.

Han JW, Gurunathan S, Jeong JK, Choi YJ, Kwon DN, Park JK, Kim JH. Oxidative stress mediated cytotoxicity of biologically synthesized silver nanoparticles in human lung epithelial adenocarcinoma cell line. *Nanoscale Research Letters.* 2014 Dec;9(1):1–4.

Hansen SF, Michelson ES, Kamper A, Borling P, Stuer-Lauridsen F, Baun A. Categorization framework to aid exposure assessment of nanomaterials in consumer products. *Ecotoxicology.* 2008 Jul;17(5):438–47.

Hoppel M, Baurecht D, Holper E, Mahrhauser D, Valenta C. Validation of the combined ATR-FTIR/tape stripping technique for monitoring the distribution of surfactants in the stratum corneum. *International Journal of Pharmaceutics*. 2014 Sep 10;472(1–2):88–93. http://ec.europa.eu/health/ph_risk/committees/04_scenihr/docs/scenihr_o_003b.pdf, http://ec.europa.eu/health/ph_risk/committees/04_scenihr/docs/scenihr_o_004c.pdf

Hussain SM, Hess KL, Gearhart JM, Geiss KT, Schlager JJ. *In vitro* toxicity of nanoparticles in BRL 3A rat liver cells. *Toxicology In Vitro*. 2005 Oct 1;19(7):975–83.

Kim H, Liu X, Kobayashi T, Kohyama T, Wen FQ, Romberger DJ, Conner H, Gilmour PS, Donaldson K, MacNee W, Rennard SI. Ultrafine carbon black particles inhibit human lung fibroblast-mediated collagen gel contraction. *American Journal of Respiratory Cell and Molecular Biology*. 2003 Jan;28(1):111–21.

Kreyling WG, Semmler-Behnke M, Möller W. Health implications of nanoparticles. *Journal of Nanoparticle Research*. 2006; 8:543–62.

Lu PJ, Ho IC, Lee TC. Induction of sister chromatid exchanges and micronuclei by titanium dioxide in Chinese hamster ovary-K1 cells. *Mutation Research/Genetic Toxicology and Environmental Mutagenesis*. 1998 May 11;414(1–3):15–20.

Luepke NP, Kemper FH. The HET-CAM test: An alternative to the Draize eye test. *Food and Chemical Toxicology*. 1986 Jun 1;24(6–7):495–6.

Menon GK, Brandsma JL, Schwartz PM. Particle-mediated gene delivery and human skin: Ultrastructural observations on stratum corneum barrier structures. *Skin Pharmacology and Physiology*. 2007;20(3):141–7.

Mitrano DM, Motellier S, Clavaguera S, Nowack B. Review of nanomaterial aging and transformations through the life cycle of nano-enhanced products. *Environment International*. 2015 Apr 1;77:132–47.

Möller W, Brown DM, Kreyling WG, Stone V. Ultrafine particles cause cytoskeletal dysfunctions in macrophages: Role of intracellular calcium. *Particle and Fibre Toxicology*. 2005 Dec;2(1):1–2.

Moss OR, Wong VA. When nanoparticles get in the way: Impact of projected area on *in vivo* and *in vitro* macrophage function. *Inhalation Toxicology*. 2006 Jan 1;18(10):711–6.

Nel A, Xia T, Madler L, Li N. Toxic potential of materials at the nanolevel. *Science*. 2006 Feb 3;311(5761):622–7.

Nemmar A, Hoet PM, Vanquickenborne B, Dinsdale D, Thomeer M, Hoylaerts MF, Vanbilloen H, Mortelmans L, Nemery B. Passage of inhaled particles into the blood circulation in humans. *Circulation*. 2002a Jan 29;105(4):411–4.

Nemmar A, Hoylaerts MF, Hoet PH, Dinsdale D, Smith T, Xu H, Vermylen J, Nemery B. Ultrafine particles affect experimental thrombosis in an *in vivo* hamster model. *American Journal of Respiratory and Critical Care Medicine*. 2002b Oct 1;166(7):998–1004.

Oberdörster G, Maynard A, Donaldson K, Castranova V, Fitzpatrick J, Ausman K, Carter J, Karn B, Kreyling W, Lai D, Olin S. Principles for characterizing the potential human health effects from exposure to nanomaterials: Elements of a screening strategy. *Particle and Fibre Toxicology*. 2005 Dec;2(1):1–35.

Oberdörster G, Sharp Z, Atudorei V, Elder A, Gelein R, Lunts A, Kreyling W, Cox C. Extrapulmonary translocation of ultrafine carbon particles following whole-body inhalation exposure of rats. *Journal of Toxicology and Environmental Health, Part A*. 2002 Oct 12;65(20):1531–43.

OECD Guideline for testing chemicals - Guideline 427: Skin absorption: *In vivo* method.

OECD Guideline for testing chemicals - Guideline 428: Skin absorption: *In vitro* method. Organization for Economic Cooperation and Development, Paris, adopted 13 April 2004a.

OECD (2004b). OECD TG 430. OECD Guideline for the Testing of Chemicals: *In Vitro* Skin Corrosion: Transcutaneous Electrical Resistance Test Method (TER). Adopted 13 April 2004, last updated 28 Jul 2015.

Raj S, Jose S, Sumod US, Sabitha M. Nanotechnology in cosmetics: Opportunities and challenges. *Journal of Pharmacy & Bioallied Sciences*. 2012 Jul;4(3):186.

Rosenbruch M. The sensitivity of chicken embryos in incubated eggs [Article in German]. *ALTEX-Alternatives to Animal Experimentation*. 1997 Aug 1;14(3):111–3.

Rothen-Rutishauser BM, Schürch S, Haenni B, Kapp N, Gehr P. Interaction of fine particles and nanoparticles with red blood cells visualized with advanced microscopic techniques. *Environmental Science & Technology*. 2006 Jul 15;40(14):4353–9.

Russell WM, Burch RL. *The principles of humane experimental technique*. Methuen and Co. Ltd, UK (reprinted by the Universities Federation for Animal Welfare, Potters Bar, UK); 1959.

Sayes CM, Gobin AM, Ausman KD, Mendez J, West JL, Colvin VL. Nano-C60 cytotoxicity is due to lipid peroxidation. *Biomaterials*. 2005 Dec 1;26(36):7587–95.

SCCS (2019). Guidance on the safety assessment of nanomaterials in cosmetics. https://health.ec.europa.eu/system/files/2020-10/sccs_o_233_0.pdf

SCENIHR (Scientific Committee on Emerging and Newly Identified Health Risks), 10 March 2006: The appropriateness of existing methodologies to assess the potential risks associated with engineered and adventitious products of nanotechnologies, modified opinion after public consultation.

SCENIHR (Scientific Committee on Emerging and Newly Identified Health Risks), 29 March 2007: Approved for public consultation: The appropriateness of the risk assessment methodology in accordance with the Technical Guidance Documents for new and existing substances for assessing the risks of nanomaterials.

Sekkat N, Guy RH. Biological models to study skin permeation. *Pharmacokinetic Optimization in Drug Research*. 2001 Feb 23:155–72.

Stone V, Shaw J, Brown DM, MacNee W, Faux SP, Donaldson K. The role of oxidative stress in the prolonged inhibitory effect of ultrafine carbon black on epithelial cell function. *Toxicology In Vitro*. 1998 Dec 1;12(6):649–59.

Teeguarden JG, Hinderliter PM, Orr G, Thrall BD, Pounds JG. Particokinetics *in vitro*: Dosimetry considerations for *in vitro* nanoparticle toxicity assessments. *Toxicological Sciences*. 2007 Feb 1;95(2):300–12.

Xia T, Kovochich M, Brant J, Hotze M, Sempf J, Oberley T, Sioutas C, Yeh JI, Wiesner MR, Nel AE. Comparison of the abilities of ambient and manufactured nanoparticles to induce cellular toxicity according to an oxidative stress paradigm. *Nano Letters*. 2006 Aug 9;6(8):1794–807.

7 Liposomal drug-delivery system in cosmetics

Antonio Jiménez-Rodríguez, Lucio Martínez-Alvarado, Sayra N. Serrano-Sandoval, Laura E. Romero-Robles and Marilena Antunes-Ricardo

CONTENTS

7.1 INTRODUCTION

The skin is the largest organ in the body, it plays multiple essential roles to to maintain body homeostasis, that go from a protective barrier function to its immunological, endocrine, sensitive, metabolic, and thermoregulatory properties; without forgetting that it can also be the target of local diseases or even express an underlying systemic condition. It is also very important because it has to do with personal presentation, and its disruption can interfere with proper psychosocial development. This majestic organ is mainly organized into three complex levels: epidermis, dermis, and hypodermis (Jevtić et al., 2020) (Figure 7.1A). The protecting function of the skin is provided mainly by the *stratum corneum* (SC) the most outer of the skin tissue (Okasaka et al. 2019). SC constitutes the major obstacle limiting the rate of drug percutaneous absorption (Zhou et al. 2018). Drug permeability through the skin is strongly influenced by the physical state and organization of the intercellular lipids

DOI: 10.1201/9781003319146-7

matrix in SC and the interaction of drug with this (Barba et al. 2019). Therefore, drugs with improved partition into the SC lipid layer shows high skin permeability (Nan et al. 2018).

Delivery of active molecules across the skin can occur by transdermal or topical pathways. Transdermal delivery occurs when the molecule crosses a layer of skin and reaches the bloodstream, affecting the whole body, while topical delivery has a local effect on the skin with minimal systemic absorption (Castañeda-Reyes et al. 2020).

Regarding transdermal skin delivery have been described three different routes in which the molecules can be permeated across the skin: (1) intercellular, (2) transcellular, and (3) appendageal pathways through hair follicles or sweat glands (Figure 7.1B). Transcellular pathway, also known as intracellular route, occurs through the SC constituted by corneocytes and intercellular lipid bilayer matrix arranged as the "Brick and Mortar". This physiological disposition becomes the skin the principal barrier for the entry of large active molecules primarily due to their molecular weight (>500 Da), hydrophobicity, and charged state (Chaturvedi and Garg 2021). Because of SC composition, routing active molecules or ingredients need to exhibit an appropriate partition coefficient between hydrophilic and lipophilic environments, such as a molecular weight lower than 500 Da, high potency, and a low melting point (Chaturvedi and Garg 2021).

The transcellular pathway is considered the most important route for percutaneous absorption of drug molecules since it allows drugs to arrive at the lower levels of skin levels, the epidermis, and the dermis. The intercellular pathway

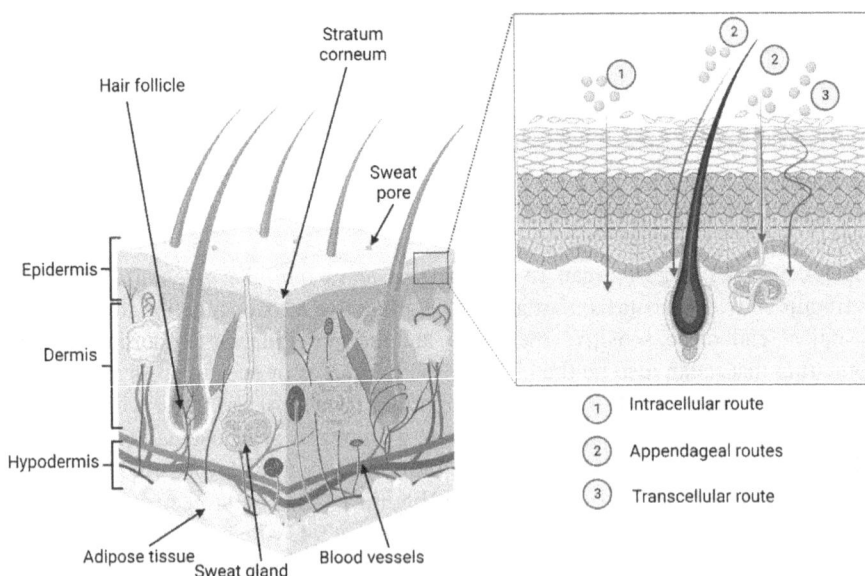

FIGURE 7.1 Schematic diagram of transcellular and intercellular routes of drug delivery via the skin

occurs through the spaces between the lipid bilayer matrix and the membranes of their adjacent cells. The hydrophilic molecules migrate mainly laterally along surfaces of less abundant water-filled interlamellar spaces or across these pockets (Jiménez-Rodríguez, Guardado-Félix, and Antunes-Ricardo 2022). Finally, the follicular pathways should be considered the best way for large molecules to enter, owing to the apparent preferential pathway along the follicular channels. Recent data demonstrate that the pilosebaceous route shows lower resistance to permeation of nanocarriers, therefore it could be considered an effective method of percutaneous drug delivery, mainly for cosmetic formulations (Elmowafy 2021).

7.2 LIPOSOMES AS DELIVERY SYSTEMS OF COSMETIC ACTIVE MOLECULES

Considering the skin composition and structure, research on the skin delivery systems has been focused on liposomes due to their similitude to biological membranes. Liposomes are lipid carriers with a size that ranged from micro to nanometers and constituted by one or multiple lipid bilayers surrounded by an aqueous core (Guimarães, Cavaco-Paulo, and Nogueira 2021). These lipid bilayers are composed by combinations of natural or synthetic phospholipids, and cholesterol, this combinations confer to the structure of an amphiphilic nature to carry out active molecules of different polarity and flexibility. This amphiphilicity nature confers to liposome self-assembly, emulsifying, and wetting properties. Also, liposomes can exhibit negative or positive charges, and this determines their chemical stability and potential cell permeability (Ibaraki et al. 2019; Lin et al. 2018; Maione-Silva et al. 2019).

Liposomes can be classified based on their composition, particle size, and lipid-bilayer structure. According to the size, liposomes can be classified into multilamellar vesicles (MLV) which are composed of many lipid bilayers (0.5–5 μm size), and unilamellar vesicles (ULV) composed of a single spherical phospholipid bilayer. Likewise, unilamellar vesicles can be also divided into (a) giant unilamellar vesicles (GULV) (1–50 μm size), (b) large unilamellar vesicles (LUV) (larger than 50 nm size), and (c) small unilamellar vesicles (SUV) (less than 50 nm size) (Salehi et al. 2020; Tadros 2018).

On the other hand, according to their constituent's composition, liposomes can be also classified into five categories: (*1*) *Conventional liposomes, (2) Cationic liposomes, (3) Long circulatory liposomes, (4) pH-sensitive liposomes, and (5) Fusogenic liposomes*. The main characteristics of each one are presented below.

Conventional liposomes are the most versatile assemblies in drug delivery systems. These are constituted by self-forming enclosed bilayer phospholipids (neutral or negatively charged) in contact with the aqueous solution. These phospholipids behave like the skin lipid matrix, and by themselves can act as moisturizing agents (Ahmed et al. 2019; Wang et al. 2022). Conventional liposomes have been extensively used in different applications including pharmaceutical, cosmetic, food industry, and recently in the functionalization of cosmetic textiles

(Amjadi et al. 2018; Fang et al. 2018; Lee et al. 2020; Maione-Silva et al. 2019; Sayıt et al. 2021).

Cationic liposomes: These are liposomes positively charged with cationic lipids that are widely used for the delivery of nucleic acids (DNA/RNA) for gene therapy or vaccines and the delivery of topical or transdermal active molecules. This delivery system has shown high efficiency in drug delivery, due to an increased diffusion and retention time into epithelial cells, facts attributed to the strong electrostatic attraction generated between the positive charge of the cationic lipids and the negative charge of proteins present in cell membranes (Kuznetsova, Vasileva, Gaynanova, Pavlov, et al. 2021; Kuznetsova, Vasileva, Gaynanova, Vasilieva, et al. 2021; Lee et al. 2021).

Long circulatory liposomes: These are used to prolong the plasma half-life of drugs provided by the inclusion of steric stabilizing amphiphiles such as polyethylene glycol (PEG)-modified phospholipids (Duan et al. 2016; Sesarman et al. 2018; Wang et al. 2019).

pH-sensitive liposomes: These are lipidic vesicles that become unstable under acidic conditions leading to destabilization, fusion with endosomal membrane, and content release. These delivery systems are often constituted by a neutral cone-shaped lipid dioleoylphosphatidyl-ethanolamine (DOPE) and a weakly acidic amphiphile (de Oliveira Silva et al. 2019; Paliwal, Paliwal, and Vyas 2015). These delivery systems have shown to be uptaken efficiently by cells allowing them to deliver their content into the cytosol (Yamazaki et al. 2017). Besides, it has the opportunity to develop a target intelligent drug delivery system by modifying the structure of the liposome so that it can become a response to different stimuli of specific cellular microenvironments (Seong, Yun, and Park 2018; Wang et al. 2020).

Fusogenic liposomes: These are unilamellar lipidic vesicles that easily fuse with the cellular plasma membrane releasing the drug (usually with lipophilic characteristics) into the cellular membrane. There are different techniques for fusogenic liposomes preparation including the incorporation of special lipids (e.g. dioleoylphosphatidylethanolamine (DOPE), 1, 2-dioleoyl-3-trimethylammonium-propane (DOTAP), 3,5-didodecyloxybenzamidine (TRX), 3-methylglutarylated hyperbranched poly(glycidol) (MGlu-HPG)), contain inactivated envelope components from the Sendai virus or incorporate fusogenic peptides (Akbarzadeh et al. 2013; Daudey et al. 2021; Farid et al. 2020; Khalil and Harashima 2018; Kim et al. 2016; Wiedenhoeft et al. 2019; Yoshizaki et al. 2017).The rigid structure of liposomes causes them to be retained in the SC reducing their ability to penetrate deep skin layers (Kuznetsova, Vasileva, Gaynanova, Vasilieva, et al. 2021; Sun et al. 2022). To overcome this limitation, flexible or liposomes known as ultradeformable liposomes (UDL) have been developed (Ramanunny et al. 2021). This new generation of liposome carriers exhibits high deformability, high trapping efficiency, and an increased transdermal permeability rate (Calienni et al. 2021). UDL exhibit lower elastic modulus compared to conventional liposomes, and this confers them the ability to achieve a high deformation and permeate across SC using in their favor the transdermal humidity gradient (Calienni et al. 2018; Chen et al. 2017; Izquierdo

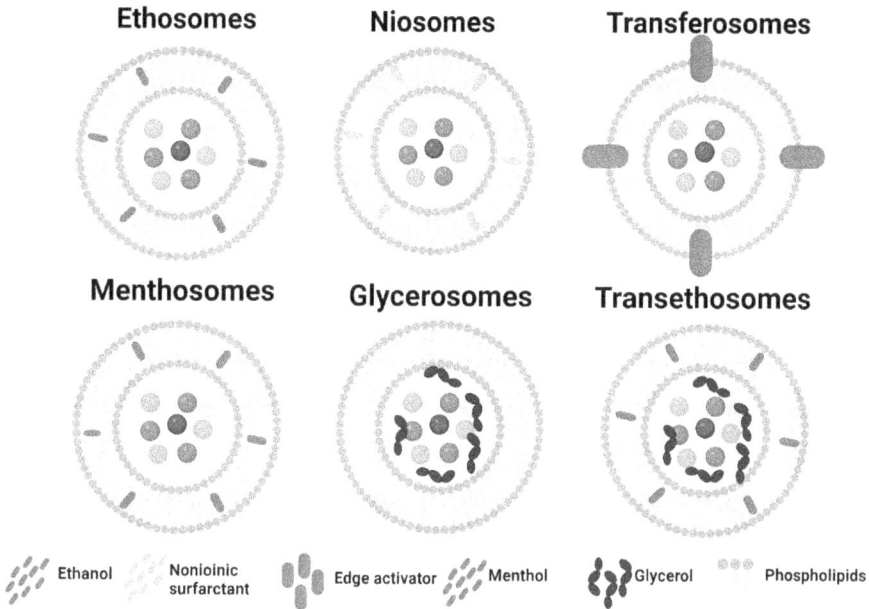

FIGURE 7.2 Schematic illustration of the structures of ultradeformable liposomes (UDL)

et al. 2020; Rangsimawong et al. 2018). Many types of UDL have been successfully developed for different applications such as transferosomes, niosomes, ethosomes, menthosomes, glycerosomes, and more recently, transethosomes (Figure 7.2). The general characteristics of UDL and their more recent cosmetic applications are presented in Table 7.1.

Different compounds have been encapsulated in UDL, mainly phytochemicals, herbal extracts, vitamins, and co-enzymes. Likewise, the most investigated in these delivery systems include antioxidant, anti-inflammatory, anti-aging, and skin-whitening agents, among others. Depending on the lipid-based delivery system type and its composition, different mechanisms of drug transdermal penetration can occur (Figure 7.3). Conventional liposomes are usually transported through hair follicles and sometimes through their adsorption or fusion with membrane cells releasing the drug into the dermis. Regarding transferosomes and niosomes, these have in their composition edge activator or nonionic surfactant agents that confer them high deformable abilities and allow them to be transported through the keratinocytes layer reaching the epidermis (Abd El-Alim et al. 2019; Niu et al. 2019). On the other side, ethosomes and transethosomes disturb the lipid matrix of the SC enhancing its fluidity and allowing the drug to penetrate easily into deeper layers of the skin and deliver the drug inside the cells. This effect is mainly attributed to their flexibility and the ethanol present in their composition, about 20–50% (Asadi et al. 2021; Durga and Veera 2020; Nastiti et al. 2017; Ruan, Hu, and Chen 2015).

TABLE 7.1

Composition, loaded drug, physical parameters, and biological activities of different types of ultradeformable liposomes (UDL) used in cosmetics

UDL type	Composition	Drug loaded	Size (nm)	Encapsulation Efficiency (%)	Activity	References
Ethosomes	L-α-phosphatidylcholine (phospholipids) + ethanol	*Rosa canina* L. extracts	196	92.3	Antioxidant	(Sallustio et al. 2022)
	Soybean lecithin (phospholipids) + ethanol	Niacinamide	616	10.8	Nutricosmetic	(Yang and Kim 2018)
	Soybean phosphatidylcholine (phospholipids) + ethanol	Rosmarinic acid	138	55	Reduces wrinkles (antioxidant, anticollagenase, anti-elastase inhibition)	(Yücel, Şeker Karatoprak, and Değim 2019a)
Ehtosome hydrogel	Soybean phosphatidylcholine (phospolipids) + ethanol + stearylamine + propyleneglycol	Resveratrol	196.8	65.78	Antioxidant	(Arora and Nanda 2019)
Transfersomes	Soybean phosphatidylcholine (phospholipids) + sodium deoxycholate (edge activator)	Ascorbic palmitate (AP)	110	91.3	Skin whitening agent (tyrosinase inhibitor)	(J. Li et al. 2021)
	Epikuron 200 (phospholipids) + Cholesterol + Tween 80 (edge activator)	Catechin	76.23	>90	Skin whitening agent	(Hsieh et al. 2021)

(Continued)

TABLE 7.1 (Continued)

UDL type	Composition	Drug loaded	Size (nm)	Encapsulation Efficiency (%)	Activity	References
	Soybean lecithin (phospholipids) + sodium deoxycholate (edge activator)	Eosin	305.5	33	Palmar hyperhidrosis treatment (photosensitizer)	(Fadeel et al. 2022)
	Soybean lecithin (phospholipids) + Tween 80 (edge activator)	Coenzyme Q10	146	97.63	Androgenic alopecia	(El-Zaafarany et al. 2021)
Pro-transferosoma	L-α-Phosphatidylcholine (phospholipids) + Tween 80 (edge activator) + Oleic Acid	Coenzyme Q10	201.5	45.64	Anti-aging	(Ayunin et al. 2022)
Liposomes, transferosomes, and ethosomes	Liposomes: Phosphatidylcholine from egg yolk (phospholipids)+ cholesterol; Transferosomes: Phosphatidylcholine from egg yolk + sodium cholate; Ethosome: Phosphatidylcholine from egg yolk + ethanol	*Sambucus ebulus* L. leaves extract	Liposomes: 123 Transferosomes: 155 Ethosomes: 190	Liposomes: 80.05 Transferosomes:75.10 Ethosomes: 85.10	Antioxidant	(Păvăloiu et al. 2019)
Transethosomes	Sunflower phospholipid (phospholipids) + Tween 80 (edge activator) + ethanol	Cycloastragenol (plant saponin)	100 to 400	31.5	Anti-aging supplement (Telomerase activator)	(F. C. Wang et al. 2022)

(Continued)

TABLE 7.1 (Continued)

UDL type	Composition	Drug loaded	Size (nm)	Encapsulation Efficiency (%)	Activity	References
Ethosomes and Transethosomes	Ethosomes: phosphatidylcholine (phospholipids) + ethanol Transethosomes: phosphatidylcholine (phospholipids) + Tween 80 (edge activator) + ethanol	Mangiferin	Ethosomes: 189.8 Transferosomes: 169.3	Ehosomes: 68 Transferosomes: 63	Anti-inflammatory and antioxidant	(Sguizzato et al. 2021)
Niosome	Neutral niosome: Cholesterol + Brij 52 (nonionic surfactant); Cationic niosome: Cholesterol + Brij 52 + CTAB (cationic surfactant)	Gallic acid (3,4,5-trihydroxybenzoic acid)	Neutral niosome: 131.23 Cationic niosome: 143.77	Neutral niosome: 4.48 Cationic niosome: 10.94	Antioxidant	(Chaikul et al. 2019)
	Cholesterol + Tween 20 (nonionic surfactant) + Span 20 (nonionic surfactant)	Arbutin	114.76	35.55	Skin whitening agent	(Radmard et al. 2021)
Glycethosomes	Glycerol + ethanol	Glycyrrhetinic acid	94.5	99.80	Anti-inflammatory and antioxidant	(Zhang et al. 2022)
Hybrid vesicular nanosystem	Collagen, niosomes (cholesterol + Span 60 (nonionic surfactant)) + zinc oxide nanoparticles	Zinc oxide nanoparticles	94	Not reported	Antibacterial	(Malathi et al. 2021)

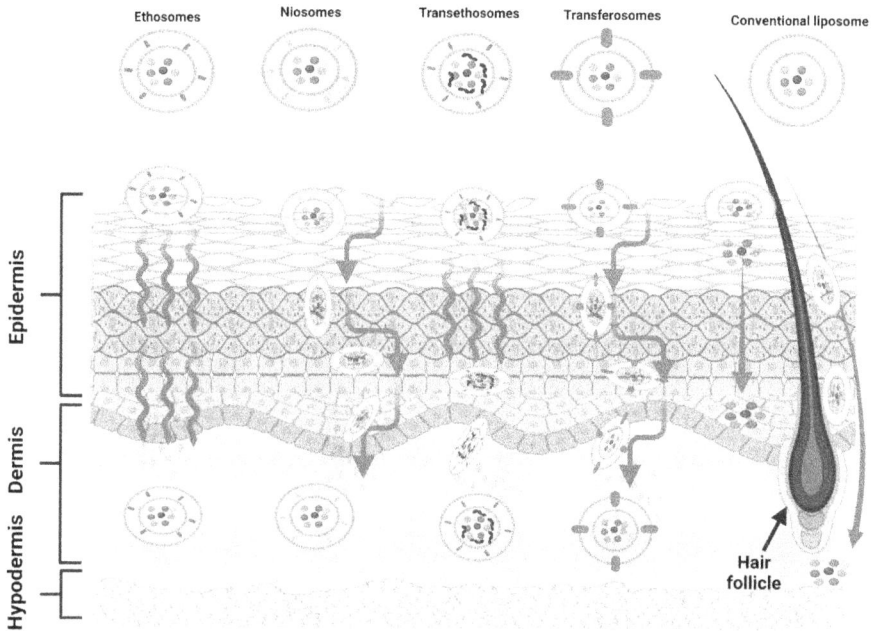

FIGURE 7.3 Some mechanisms of drug transdermal penetration exhibited by lipid-based delivery systems

7.3 LIPID-BASED DELIVERY SYSTEMS FOR COSMECEUTICALS

In recent years, the cosmetic market has focused its interest on the "natural cosmetic" taking as the main element the incorporation of ingredients from natural sources including herbal extracts. Cosmetic products that contain herbal components are usually called cosmeceuticals since these phytoconstituents play a role in the health or beauty of the skin. Nevertheless, the incorporation of these natural molecules into cosmetic formulations represents many challenges because of their environmental and chemical instability, poor solubility, and limited absorption (Nishimoto-Sauceda, Romero-Robles, and Antunes-Ricardo 2022). To overcome these limitations, an emerging type of lipid nanocarriers known as phytosomes has been developed, a vesicle system that is able to encapsulate polar and nonpolar molecules by reducing their surface tension with the solvent (Susilawati, Chaerunisa, and Purwaningsih 2021). This lipid-based delivery system constitutes a phospholipid matrix, mainly phosphatidylcholine, in which herbal extracts or phytochemicals demonstrating beneficial effects on skin health are incorporated.

Different cosmeticeutical products available in the market contain herbal components-loaded phytosomes for their use as anti-wrinkle, skin conditioner, anti-aging, moisturizer, and skin brightening products (Table 7.2). Some of the biological activities of these herbal components-loaded phytosomes include antioxidant, anti-inflammatory, antielastase, and anticollagenase activities, tyrosinase inhibitors, sun protectors, among others (Alharbi et al. 2021).

TABLE 7.2

Phytosomes-containing commercial cosmetic products

Commercial product	Botanical Origen	Cosmetic use
Ginselect®	*Panax ginseng* root extract	Hair care/ Elasticizing agent
Silymarin Phytosome®	*Silybum marianum* extract	Antioxidant/ Anti-aging
Boswellia Phytosome®	*Boswellia serrata* extract	Antioxidant/ Anti-aging
Life Extension® Bio-Quercetin Phytosome®	*Japanese sophora* concentrate (flower bud)	Nutricosmetic
Ginkgoselect® Phytosome®	*Ginkgo biloba* L. - Leaf	Antioxidant/ Anti-aging
Greenselect® Phytosome®	*Camellia sinensis* (L.) Kuntze	Antioxidant/ Anti-aging
Leucoselect® Phytosome®	*Vitis vinifera* L.	Antioxidant/ Anti-aging
Quercefit® Quercetin Phytosome®	*Sophora japonica* L.	Antioxidant/ Anti-aging
Siliphos® Silybin Phytosome®	*Silybum marianum* (L.) Gaertn.	UV protection
Curcumin Phytosome	*Curcuma longa L.* extract	Nutricosmetic

Other herbal components that have been encapsulated in these lipid nanocarriers for cosmetic applications include *Moringa olifera* seed oil (Hartini, Sumaiyah, and Hasibuan 2022), geranium essential oil from *Pelargonium graveolens* leaves, and calendula essential oil from *Calendula officinalis* flowers (Lohani, Mishra, and Verma 2019; Lohani et al. 2021) *Intsia bijuga* (Colebr.) wood (Sari et al. 2021), *Daemonorops acehensis* resin extracts (Sari et al. 2022), and *Artocarpus nobilis* (Liyanaararchchi et al. 2022), among others.

The selection of a suitable delivery system for cosmetic ingredients depends on the physical and chemical characteristics of the active ingredient and the expected cosmetic effects. Liposomes as nanocarriers have advantages and disadvantages that need to be considered before selection.

7.4 LIPOSOMAL DELIVERY SYSTEMS IN COSMETIC: ADVANTAGES AND DISADVANTAGES

Since liposomes are amphiphilic structures, they offer the opportunity to incorporate hydrophilic and lipophilic unstable molecules, protecting them from environmental or biological degradation. In some cases, liposomes can act as a sustained drug release system improving the drug effect at the target site and reducing the side effects. Besides, phosphatidylcholine, one of the majoritarian constituents of liposomes, is widely used as a softener in cosmetics exerting moisturizing effects on the skin (Castañeda-Reyes et al. 2020; Oliveira et al. 2022;). Liposomes have been recognized as non-toxic and biodegradable lipid-based formulations. These lipidic vehicles increase skin drug retention because of their lipid composition which mimics biological membranes improving the bioavailability and efficacy of loaded compounds (Castañeda-Reyes et al. 2020). Another important advantage of these delivery systems is the feasibility to produce them at an industrial scale. In fact, nowadays, there are many companies dedicated to the design and industrial production of lipid-based vesicles as drug carriers.

Despite their multiple advantages and applications, liposomes have some limitations to consider when selecting them as a delivery system. One of the most important challenges to overcome is their low physical and chemical stability which conducts liposome degradation during storage (Figueroa-Robles, Antunes-Ricardo, and Guajardo-Flores 2021; Salvioni et al. 2021). Other important issues to solve include their limited solubility in aqueous solutions, reduced body half-life time, limited cellular uptake, and leakage and fusion of the loaded drugs. These limitations coupled with a low loading capacity, low reproducibility, and high production cost limited the incorporation of liposomes in commercial cosmetic products (Nakhaei et al. 2021).

7.5 LIPID-BASED DELIVERY SYSTEMS FOR COSMETIC APPLICATIONS AND OTHER SKIN ALTERATIONS

7.5.1 ANTI-AGING

Skin aging is a process driven by environmental factors [ultraviolet (UV) light exposure] and time. It is usually associated with aesthetic changes in the skin, such as an increase in wrinkles and cutis laxa (Bi et al. 2019). Skin exposure to UV light promotes skin aging due to the damage it causes to the skin (Gilchrest, Garmyn, and Yaar 1994). Long-term exposure to UV induces the creation of reactive oxygen species (ROS), which can destroy DNA, membrane lipids, and the extracellular matrix of skin cells; causing the appearance of wrinkles, age spots, loss of elasticity, and skin thinning (Bi et al. 2019; Bondock, Khalifa, and Fadda 2007; Fisher et al. 2000; Yücel, Şeker Karatoprak, and Değim 2019a; Zouboulis et al. 2019).

Several strategies to prevent and treat skin aging have been developed (Table 7.3). However, one of the main challenges for the delivery of antioxidant compounds in anti-aging treatments is their stability and efficient penetration into the skin. For this reason, liposomes are suitable for the development of anti-aging products (Rahimpour and Hamishehkar 2012).

In addition, the use of liposomes in anti-aging products may enhance the penetration, solubility, or stability of the main compound. Besides enhancing the longevity of the anti-aging effect, the specificity of the ingredient to the desired site of action reduce toxicity, increase control over pharmacokinetics and pharmacodynamics, and make the product more cost-effective compared to other delivery carrier systems (Ahmadi Ashtiani et al. 2016).

In a study made by Bi et al. (2019), vitamin D3 liposomes enhanced the antiphotoaging effect versus a vitamin D3 solution. These liposomes had skin retention of 1.65 times higher compared to the normal vitamin D3 solution; meaning that vitamin D3 liposomes could repair skin after photoaging and promote the production of new collagen fibers. Another study (Kwon et al. 2015) showed the potential anti-aging effect of *Polygonum aviculare* L. extract (which has superior antioxidative and cellular membrane protective activity) loaded onto cell-penetrating peptide (CPP) conjugated liposomes. These liposomes effectively promoted depigmentation and inhibited wrinkle formation in hairless HR-1 mice. Moreover, Desai and Mallya (2021) developed

TABLE 7.3

Different approaches for anti-aging skin prevention and treatment. Adapted from (Zouboulis et al. 2019).

Type of strategy	Common treatments
Cosmetology care	Daily skincare
	Correct sun protection
	Esthetic noninvasive procedures
Topical medical agents	Antioxidants
	Cell regulators
Invasive procedures	Chemical peelings
	Intense pulsed light
	Radiofrequency
	Injectable skin biostimulation and rejuvenation
	Redistribution of fat and volume skin loss, skin augmentation, contouring
Systemic agents	Hormone replacement therapy
Preventive medicine (avoiding exogenous factors of aging, correction of lifestyle and habits)	Smoking
	Pollution
	UV irradiation
	Stress
	Nutrition and diet
	Physical activity
	General health control

a chlorogenic acid (obtained from green coffee beans) liposomal gel which had an anti-elastase activity and a prolonged release of 12 hours. Liposomal anti-aging and antioxidant formulations reduce the water loss of the skin, which reduces wrinkles and lines of expression, restoring the normal balance of the skin (Souto et al. 2020).

Different anti-aging products loaded into lipid-based delivery systems are available in the market, some examples of those are Decorin Cream® (ethosomes) manufactured by Genome Cosmetics, Niosome + Perfected Age Treatment® (niosomes) by Lancome, Revitalift® (nanosomes) by L'Oreal, Liposome Face and Neck Lotion (liposomes) by Clinicians Complex, Royal Jelly Lift Concentrate (liposomes) by Jafra Cosmetics, Capture Totale® (liposome) by Christian Dior, among many others (Dubey et al. 2022; Ferraris et al. 2021).

7.5.2 HAIR LOSS/ANDROGENIC ALOPECIA

Hair loss is associated with a variety of diseases, including malnutrition, aging, hormonal imbalances, and oxidative stress. Although this alteration is not limiting to the lives of patients, it can generate self-esteem problems and affect patients' social interactions (Zhou et al. 2021). An interesting approach to treating hair loss derived from androgenetic alopecia is a liposomal formulation co-loaded with a combination of minoxidil and sodium bicarbonate (Ramos et al. 2020). Minoxidil is an FDA-approved synthetic pro-drug used to stimulate hair growth by increasing the blood flow to the hair follicles promoting their growth and strength. Since minoxidil is a pro-drug, it requires the action of the sulfotransferases enzymes (SULT1A1) to convert it into the active drug (Dhurat et al. 2022). Sulfotransferases activity increases at high pH values, thus the addition of sodium bicarbonate increases the pH of the follicular cells, increasing SULT1A1 activity and therefore converting efficiently the minoxidil into the active drug. Results indicated that liposomal minoxidil (booster) produced an increase SULT1A1 activity of hair follicles in 53% of the patients with androgenetic alopecia after 14 days (Ramos et al. 2020). Similar results were obtained when a combination of minoxidil and tretinoin hydrogel, a comedolytic agent, were incorporated into a liposomal formulation. This combination enhanced the permeation of minoxidil through the skin and did not show irritant effects on the skin (Kochar et al. 2020).

Another factor involved in the etiology of androgenic alopecia is oxidative stress because of the presence of molecules susceptible to oxidation and chronic inflammation present in the hair of patients with androgenic alopecia. Coenzyme Q10 (CoQ10) is a liposoluble antioxidant present in cell membranes and mitochondria. Transethosomes have shown to be the ideal carrier for CoQ10 since these displayed a small particle size (146 nm), high zeta potential, and high entrapment efficiency (97.63%). Likewise, about 95% of the CoQ10 transethosomes evaluated showed high disposition in the different skin layers. Finally, patients suffering from androgenic alopecia and treated with CoQ10 transethosomes exhibited better clinical responses after dermoscopic examination than those treated with CoQ10 topical solution (El-Zaafarany et al. 2021). Recent developments to treat hair loss or androgenic alopecia that have boarded oxidative stress as an important causal factor, have focused on the synergic action between existing commercial drugs and natural extracts or phytochemicals recognized by their antioxidant character. *In vivo* studies have demonstrated that when finasteride, the only oral synthetic drug for the treatment of hair loss connected to androgenic alopecia, and baicalin were co-loaded into glycerol-hyalurosomes, hair follicle dermal papilla cells were stimulated to proliferate and for consequence, the number of follicles increased and hair growth was accelerated (Mir-Palomo et al. 2020). Liposomal honokiol, a majoritarian biphenolic compound isolated from the Magnolia tree, was able to stimulate the hair growth in the shaving area of C57BL/6N mice by the activation of the Wnt3a/β-catenin pathway and downregulating the transforming growth factor-β1 (TGF- β1) (S. Li et al. 2021). This beneficial effect on hair growth has been also observed in patients with alopecia after chemotherapy. Also, a combination of butterfly pea (*Clitoria ternatea*) extract and clove oil (*Eugenia caryophyllus*) was loaded into liposomes and showed to prevent hair loss and promote hair growth on eyebrows in 15 volunteers (Choochuen and Jimtaisong 2022). The evaluation of this effect was based on photos and the satisfaction assessment of the volunteers scoring from one to five (very poor-excellent). Results indicated that after 60 days of application, about 33.33% of the participants felt that the enhancement of eyebrow thickness and darkness was good while 66.7% considered it excellent. Another controversial product used for treating hair loss is the deer antler velvet (DAV) extract, which loaded into niosomes improved skin permeation with deposition into the deepest skin layers compared with other formulations. After 14–30 days of application of DAV niosomes on human scalp a significant hair elongation and an increase in the melanin content was observed compared to non-treated control. In addition, an increased skin hydration without observing skin irritation (Tansathien et al. 2021).

A more biotechnological approach to treating hair loss is using growth factors, proteins that act as potent inducers of cell proliferation and differentiation. Fibroblast growth factor (FGF-2) has shown to be effective to promote hair growth. Nerveless, free FGF-2 exhibit poor stability and low permeability across the skin. To overcome these challenges, FGF-2 was first encapsulated into liposomes (FGF-2-LIP) and then functionalized by the incorporation of silk fibroin (SF) to produce a novel FGF-2-LIP-SF hydrogel. After treatment with FGF-2-LIP-SF hydrogel, the hair of testosterone (TES)-treated mice was rapidly regrown and hair follicles were also recovered to the anagen (growth) phase. Moreover, the expression of inflammation-associated cytokines such as tumor necrosis factor (TNF)-α and interleukin (IL)-6 was significantly inhibited after treatment with FGF-2-LIP-SF hydrogel (Xu et al. 2018).

It has been established that hair loss is a consequence of disorders affecting the hair follicles and the hair follicle cycle that can be caused by genetic, inflammatory, hormonal, and environmental factors, usually acting in combination. Thus, response to treatments can be different in males and females depending on the active components and the action mechanisms activated. A study, evaluated the effects of the capillary lotions, containing dihomo-γ-linolenic acid, S-equol, and propionyl-L-carnitine, applied on 30 men and 30 women. Dihomo-γ-linolenic acid is a precursor of the prostaglandin (PG)-E1, which acts by improving microcirculation; S-equol inhibits 5α-reductases, thus preventing the transformation of testosterone into dihydrotestosterone; and propionyl-L-carnitine promotes lipid metabolism, stimulating energy production. These three molecules were loaded into liposomes for their effective transdermal delivery. Daily topical application of the lotions resulted in a hair count that significantly increased for women and marginally increased for men after six months of treatment. Furthermore, a significant increase in anagen hair and a significant decrease in telogen hair were observed starting from three months in male and one month in female patients (Brotzu et al. 2019).

There are on the market different liposomes-based products for the treatment of hair loss, some examples are Lip-44® by NanoVec loaded with guarana, yerba mate, and *Centella asiatica* extracts, together with panthenol, biotin, caffeine, and vitamin E to promote hair growth by having positive effects on the hair follicle. LIP-5® by NanoVec is formed by soybean phospholipids and loaded only with D-panthenol y vitamin E, Seskavel Mulberry® anti-hair loss foam by Sederma loaded with Curcuma longa, Morus alba, glycyrrhetinic acid, nicotinic acid, panthenol, zinc, vitamin B6, retinyl propionate, L-carnitine. Rinfoltil® liposomal serum anti-hair loss effect for intensive hair growth by Pharmalife Research loaded with vitamin E (alpha-tocopherol acetate), vitamin A (retinol palmitate), vitamin C (ascorbic acid), vitamin B1 (thiamine hydrochloride), vitamin B3 (nicotinamide), vitamin B5 (calcium pantothenate), B6 (pyridoxine hydrochloride), among other.

7.5.3 Antiacne

Acne vulgaris is a common disorder [caused by *Propionibacterium acnes* (*P. acnes*)] of the pilosebaceous unit, which ranks 8th in overall disease prevalence in the world, affecting around 80% of individuals in their whole life (Madan et al. 2019; Zaenglein 2018). *P. acnes* infection causes an increase in the production of sebum, distorted follicular keratinization, inflammation, and infection, which leads to blockage of hair follicles and accumulation of sebum (Graham et al. 2004).

There has been a development of anti-inflammatory and antimicrobial products to treat acne disorders in humans, there, which have different effects against it. Topical retinoids (e.g. adapalene, tretinoin, and tazarotene); antibiotics such as topical benzoyl peroxide, hormone-based therapies, procedural therapies (e.g. light and laser therapies), and isotretinoin; are the main treatments for this disorder (Habeshian and Cohen 2020; Tripathi et al. 2012). Nevertheless, the application of these treatments may have side effects for the patients, such as skin irritation, dryness, peeling, burning, stinging, skin itching, and erythema. Moreover, oral acne treatments can have side effects related to the gastrointestinal system, including gastrointestinal distress, diarrhea, nausea, esophagitis, abdominal cramping, and vomiting (Tripathi et al. 2012). In the search for treatments without or with fewer side effects for acne patients, there has been an interest in the use of compounds that have a lower side effect impact with better specificity, and that are associated with novel delivery systems, such as liposomes.

Liposomes have been widely studied in the application and formulation of cosmetics with anti-acne effects. For example, positively charged liposomes can penetrate easily into the dermis through skin appendages by efficiently binding to the negatively charged SC which improves the drug disposition at the site of action (Desai, Patlolla, and Singh 2010; Madan et al. 2019; Shanmugam et al. 2009;). These structures can be used as the main delivery systems for cosmetics products related to acne-solving problems because these mechanisms increase drug bioavailability and improve an unfavorable risk-benefit ratio (Zhao, Hu, and Xue 2018).

A study made by (Honzak and Šentjurc 2000) demonstrated that liposomal encapsulation of clindamycin showed better results (33% of mice were without open comedons) after six weeks of treatments compared to classic clindamycin (8.33%). In addition, these liposomes had no reports of side effects. Table 7.4 shows different liposomes and their effect on the type of acne treated.

TABLE 7.4
Different liposomal applications for acne treatment

Active agent	Results	Reference
Myoinositol and trehalose-loaded liposomes	The ready-to-use peel-off facial mask improved the cosmetic appearance of adult female acne. By de reduction of cutaneous androgen content and the promotion of skin autophagy.	(Fabbrocini et al. 2017)
Clindamycin loaded liposomes	After 6-week treatment, with clindamycin-loaded liposomes, 33.3% of hairless mice were without open comedons, while only 8.33% were without open comedons compared to a classic clindamycin drug (Klimicin® T, Lek).	(Honzak and Šentjurc 2000)
Clindamycin and green tea extract-loaded liposomes	Topical formulation of liposome-containing Clindamycin and green tea had a greater antibacterial activity compared to the non-liposomal marketed formulation.	(Sankar et al. 2019)
Benzoyl peroxide & adapalene loaded liposomes	Enhancement of dermal bioavailability (Adapalene 2.1, 5.4; benzoyl peroxide 3.0, 7.83-fold) and reduction in skin irritation and papule density were observed compared with free drugs and Epiduo, respectively.	(Jain et al. 2018)
Tretinoin loaded liposomes	Tretinoin-loaded liposomes had an advantage over marketed formulations in the enhancement of tretinoin against non-inflammatory acne lesions	(Rahman et al. 2016)

As it is shown, liposomes are a great tool to develop anti-acne products, which can act like toolboxes that can be manipulated, tuned, and manufactured depending on the condition of the skin (Bayat et al. 2020). Acnel Lotion N® (liposomes) by Dermaviduals, and Clearly It!® Complexion Mist (liposomes) by Kara Vita are other examples of cosmetic products available in the market for acne treatment that use lipid-based nanocarriers for the delivery of their active ingredients (Ferraris et al. 2021).

7.5.4 ATOPIC DERMATITIS, PSORIASIS, AND SKIN INFLAMMATION

Atopic dermatitis (AD) is a common skin disease characterized by an intense inflammatory response with the consequent release of cytokines and chemokines associated with innate immunity (Magnifico et al., 2020). An alteration in lipid organization and skin injury is also observed in this disease because the skin barrier function is disrupted. Intense itching, rash, edema, hemorrhage, and erosion of the skin surface are the main characteristics of AD (Goudarzi et al. 2021).

AD therapy has been focused on the reduction of inflammatory biomarkers to regulate the immune response such as the signal transducer and activator of transcription three (STAT3) and nuclear factor (NF)-kB, critical transcription factors that regulate inflammatory responses. Some studies have established the relationship between the action of various antioxidants and the substantially decreased skin alterations in patients with AD compared to healthy controls. These effects were related to an increase in the activity of antioxidant enzymes such as superoxide dismutase (SOD), glutathione peroxidase (GPx), and glutathione (GSH). Lee et al. (2020) demonstrated that a liposomal formulation of astaxanthin (3,3′-dihydroxy-β, β-carotene-4,4′-dione), a carotenoid usually found in microalgae and crustaceans, significantly reduced the levels of inflammatory cytokines and oxidative stress in an animal model of AD in comparison with free astaxanthin.

Likewise, a study reported that ROCEN®, a liposomal formulation containing 2% arthroncen, an oil extract of avocado and soybean, increased the expression of transforming growth factor-beta (TGF-β) and production of collagen resulting in wound healing acceleration and pain alleviation in a short period which can be an excellent therapeutic alternative to treat AD. The liposomal formulation including a combination of zedoary turmeric oil and tretinoin significantly improved the penetration of drugs into the hair follicles and retained higher drugs concentration in the skin compared with conventional formulations (Goudarzi et al. 2021). Likewise, the liposomal gel was more effective than conventional gel in treating skin chronic inflammation related to psoriasis and had a significant dose-dependent effect (Chen et al. 2021).

Tofacitinib citrate has recently gained interest in treating skin disorders such as psoriasis, atopic dermatitis, and baldness. Unfortunately, oral administration shows side effects, such as decreased neutrophil counts. The liposomes based on propylene glycol, named proposomes, carrying tofacitinib, were evaluated observing an increase of 4–11 folds in skin retention of tofacitinib was reported. Also, these proposomes were stable in storage for at least six months (Cárcamo-Martínez et al. 2021).

Some vitamins have also demonstrated a potential effect in controlling the inflammatory symptoms associated with skin dermatitis. Cyanocobalamin (vitamin B12) loaded into liposomes or transferosomes showed a skin permeability about

3.5–5-times higher than aqueous solutions or even ethosomes-loaded vitamin B12. It was also demonstrated that lipidic vesicles such as transferosomes can permeate through the SC until deep skin layers reach the dermis (>25 µm) (Guillot et al. 2021).

Skin inflammation can be caused by different extrinsic and intrinsic factors (Wu et al. 2021). An example of these external factors is the effects of PM2.5, very small (≤2.5 µm) particulate matter, considered a dangerous air pollutant related to an increase in oxidative stress, affecting the respiratory system, aging- and inflammation-related damage to the skin. Farnesol, a natural benzyl semiterpene, possesses anti-inflammatory, antioxidative, and antibacterial properties. But has poor water solubility limiting its use for biomedical applications. Liposomal formulation of farnesol at 2 mm or 4 mm showed the restoration of injured hair follicles and reduced by about 1.5-times the acute and chronic inflammation induced by PM2.5 in comparison with free farnesol solution.

7.5.5 VITILIGO

Vitiligo is a depigmenting skin disorder, characterized by white patches on the skin due to partial or complete loss of melanocytes (Doppalapudi, Mahira, and Khan 2017). It is the most common depigmenting skin disorder (0.5–2% prevalence of the population) in adults and children worldwide (Picardo et al. 2015). There have been several proposals of different mechanisms for the melanocyte's destruction in vitiligo (e.g. genetic, autoimmune responses, oxidative stress, generation of inflammatory mediators, and melanocyte detachment mechanisms) (Ezzedine et al. 2015). However, these theories do not support enough the cause of different vitiligo phenotypes that currently exist (Bergqvist and Ezzedine 2020).

Several treatments have been developed to minimize or reduce this skin disorder, such as topical calcineurin inhibitors, topical corticosteroids, phototherapy, psoralen plus ultraviolet A light phototherapy (UVA), narrowband ultraviolet B-light (UVB) phototherapy, targeted UVB phototherapy, oral immunosuppression, vitamin D analogs, and topical prostaglandin analogs; among others (Karagaiah et al. 2020; Rodrigues et al. 2017). These treatments can be applied in conjunction with them to have better results on vitiligo patients, and a way to improve the efficacy of the combination of treatments is to use loaded liposomes due to their high selectivity and specificity in the desired site of action.

A recent study (Doppalapudi, Mahira, and Khan 2017) formulated ultra-deformable liposomes loaded with resveratrol and psoralen for the evaluation of psoralen ultraviolet A light and its antioxidant properties in vitiligo. These liposomes had an entrapment efficiency of 74.9% and 76.91% for psoralen and resveratrol respectively. Results showed a better stimulation of melanin, tyrosinase, and antioxidant activity, which meant that the combination of psoralen and resveratrol-loaded liposomes has therapeutic potential for the treatment of vitiligo. Moreover (Mir-Palomo et al. 2019) developed baicalin and berberine (polyphenols that have been used for skin protection and the treatment of skin disorders) ultradeformable liposomes, which showed antioxidant and photoprotective activity *in vitro* with keratinocytes and fibroblasts. These activities are correlated with the stimulation of melanin production and tyrosinase activity which can lead to a novel treatment for vitiligo patients.

In addition De Leeuw et al. (2003) performed a study where they evaluated khellin encapsulated in L-phenylalanine stabilized phosphatidylcholine liposomes with the combination of UVA/UVB light therapy in patients with vitiligo. After 12 months of treatment, the re-pigmentation response was 50%–100% in 72% of the treated locations; re-pigmentation of 75–100% was achieved on the face in 63%, the back in 59%, the arms in 58%, the trunk in 57%, the legs in 56% and on the hands in 4% of the patients. In addition, no side effects were seen and the control patients group (only treated with UVA/B light) did not show re-pigmentation.

The use of liposomes in vitiligo treatment facilitates the ability to develop drugs to possess sustained or controlled release behavior, enhances therapeutic efficiency, and reduces side effects.

7.6   GREEN COSMETICS: THE TENDENCY IN COSMETIC FORMULATIONS

The current consumer, in addition to demanding efficient products, is also concerned about the safety of the ingredients and the responsible use of natural resources without affecting the environment. From this genuine interest were born natural cosmetics that are made with natural or ecological ingredients and produced sustainably. This new tendency in the cosmetic market has promoted more research and product development focused on green ingredients and the processes to produce them, the use of natural preservatives and colorants, and even sustainable packaging (COSMOS 2020; Dini and Laneri 2021). In this context, the valorization of waste or by-products from agro-industrial activities has become a valuable source of active ingredients for cosmetic use, setting the standard for the responsible use of natural resources (Agudelo et al. 2021; Hoss et al. 2021; Kewlani et al. 2022; Pinto et al. 2021).

7.7   SAFETY AND REGULATORY ASPECTS OF LIPOSOMES IN COSMETIC PRODUCTS

A key element to consider when developing new delivery systems for cosmetic applications is the proper processing and regulatory aspects.

Since there is a wide variety of materials and types of nanocarriers, it is rightly important to control to assure their safety for human or animal application. Among the parameters to control in lipid-based nanocarriers are physiochemical properties such as size, distribution, surface range and charge, topography, purity, composition, hydrophobicity, solubility, degrees of scattering, concentration, as well as toxicity, biological properties, and biological effects (Shukla et al. 2017). Differences in lipidic vesicles may affect the performance, quality, safety, and/or effectiveness, if applicable, of a product that incorporates that nanomaterial.

Recently, the United States Food and Drug Administration (FDA) presented guidance for industry safety of nanomaterials in cosmetic products (FDA 2014, 2022). Nanomaterials are used in a variety of FDA-regulated products because of their unique properties, imparting potential advantages to products considered for development. The application of nanotechnology may result in product attributes that differ from those of conventionally manufactured products, and thus may merit examination. Such materials, due to their nanoscale size, chemical, physical, and biological properties that differ from those of their larger counterparts. For example, the small particle size of a nanomaterial has the potential to alter the distribution and bioavailability of that material compared to a larger scale material with the same chemical composition. The small size leads to increased surface area relative to the mass of the particle, which could result in increased biological interactions. In addition, the uptake, absorption, and biodistribution of the material may be altered, leading to potential systemic exposure.

Likewise, the European Union, in the Regulation (EC) No 1223/2009 of the European Parliament and of the Council of 30 November 2009 on cosmetic products (European Parliament and of the Council 2009), sets the standards that must follow commercialized cosmetic products, including those containing nanomaterials to ensure the functioning of the internal market and achieve a high level of protection of human health. Article 16 of this legislation speaks specifically of nanomaterials in the cosmetic industry covering the identification of the nanomaterial including its chemical name (IUPAC), the specification of the nanomaterial including size of particles, physical, and chemical properties, an estimate of the quantity of nanomaterial contained in cosmetic products, the toxicological profile of the nanomaterial, the safety data of the nanomaterial relating to the category of cosmetic product, and the reasonably foreseeable exposure conditions.

7.8 CONCLUSIONS

This chapter highlights the potential uses of lipid-based nanocarriers for cosmetic and cosmeceutical applications. There is a great diversity of nanocarriers but those based on liquids have shown greater efficiency in the delivery and greater retention of drugs through the different layers of the skin due to their similarity to biological membranes. Within the wide variety of these systems, those classified as ultra-deformable have been the trend in the investigation of new ingredients and/or products. Lipid-based delivery systems have been widely developed for hair loss or androgenic alopecia, vitiligo, anti-aging, -acne, -dermatitis, -psoriasis, and -inflammation proposes. New market trends move to natural cosmetics using non-systematic ingredients not only as active components, but also as a substitute for preservatives, excipients, antioxidants, or stabilizers. Finally, adherence to regulations is essential to assure the safety of these delivery systems on human health and to establish clear guidelines for new developments.

REFERENCES

Abd El-Alim, S.H., A.A. Kassem, M. Basha, and A. Salama. 2019. Comparative Study of Liposomes, Ethosomes and Transfersomes as Carriers for Enhancing the Transdermal Delivery of Diflunisal: In Vitro and in Vivo Evaluation. *International Journal of Pharmaceutics* 563 (May): 293–303. https://linkinghub.elsevier.com/retrieve/pii/S0378517319302571.

Agudelo, C., K. Bravo, A. Ramírez-Atehortúa, D. Torres, L. Carrillo-Hormaza, and E. Osorio. 2021. Chemical and Skincare Property Characterization of the Main Cocoa Byproducts: Extraction Optimization by RSM Approach for Development of Sustainable Ingredients. *Molecules* 26, no. 24 (December 7): 7429. https://www.mdpi.com/1420-3049/26/24/7429.

Ahmadi Ashtiani, H.R., P. Bishe, N.-A. Lashgari, M.A. Nilforoushzadeh, and S. Zare. 2016. Liposomes in Cosmetics. *Journal of Skin and Stem Cell* 3, no. 3: 65815.

Ahmed, K.S., S.A. Hussein, A.H. Ali, S.A. Korma, Q. Lipeng, and C. Jinghua. 2019. Liposome: Composition, Characterisation, Preparation, and Recent Innovation in Clinical Applications. *Journal of Drug Targeting* 27, no. 7 (August 9): 742–61. https://www.tandfonline.com/doi/full/10.1080/1061186X.2018.1527337.

Akbarzadeh, A., R. Rezaei-Sadabady, S. Davaran, S.W. Joo, N. Zarghami, Y. Hanifehpour, M. Samiei, M. Kouhi, and K. Nejati-Koshki. 2013. Liposome: Classification, Preparation, and Applications. *Nanoscale Research Letters* 8, no. 1 (December): 102. https://nanoscalereslett.springeropen.com/articles/10.1186/1556-276X-8-102.

Alharbi, W.S., F.A. Almughem, A.M. Almehmady, S.J. Jarallah, W.K. Alsharif, N.M. Alzahrani, and A.A. Alshehri. 2021. Phytosomes as an Emerging Nanotechnology Platform for the Topical Delivery of Bioactive Phytochemicals. *Pharmaceutics* 13, no. 9 (September 15): 1475. https://www.mdpi.com/1999-4923/13/9/1475.

Amjadi, S., M. Ghorbani, H. Hamishehkar, and L. Roufegarinejad. 2018. Improvement in the Stability of Betanin by Liposomal Nanocarriers: Its Application in Gummy Candy as a Food Model. *Food Chemistry* 256 (August): 156–62. https://linkinghub.elsevier.com/retrieve/pii/S0308814618303522.

Arora, D., and S. Nanda. 2019. Quality by Design Driven Development of Resveratrol Loaded Ethosomal Hydrogel for Improved Dermatological Benefits via Enhanced Skin Permeation and Retention. *International Journal of Pharmaceutics* 567 (August): 118448. https://linkinghub.elsevier.com/retrieve/pii/S037851731930482X.

Asadi, P., A. Mehravaran, N. Soltanloo, M. Abastabar, and J. Akhtari. 2021. Nanoliposome-Loaded Antifungal Drugs for Dermal Administration: A Review. *Current Medical Mycology* (May 26). https://publish.kne-publishing.com/index.php/CMM/article/view/6247.

Ayunin, Q., A. Miatmoko, W. Soeratri, T. Erawati, J. Susanto, and D. Legowo. 2022. Improving the Anti-Ageing Activity of Coenzyme Q10 through Protransfersome-Loaded Emulgel. *Scientific Reports* 12, no. 1 (December): 906. https://www.nature.com/articles/s41598-021-04708-4.

Barba, C., C. Alonso, M. Martí, V. Carrer, I. Yousef, and L. Coderch. 2019. Selective Modification of Skin Barrier Lipids. *Journal of Pharmaceutical and Biomedical Analysis* 172 (August): 94–102. https://linkinghub.elsevier.com/retrieve/pii/S0731708519302304.

Bayat, F., R. Hosseinpour-Moghadam, F. Mehryab, Y. Fatahi, N. Shakeri, R. Dinarvand, T.L.M. Ten Hagen, and A. Haeri. 2020. Potential Application of Liposomal Nanodevices for Non-Cancer Diseases: An Update on Design, Characterization and Biopharmaceutical Evaluation. In *Advances in Colloid and Interface Science.* Elsevier.Bergqvist, C., and K. Ezzedine. 2020. Vitiligo: A Review. In *Dermatology.* Karger Publishers.

Bi, Y., H. Xia, L. Li, R.J. Lee, J. Xie, Z. Liu, Z. Qiu, and L. Teng. 2019. Liposomal Vitamin D3 as an Anti-Aging Agent for the Skin. *Pharmaceutics* 11, no. 7 (July 3): 311.

Bondock, S., W. Khalifa, and A.A. Fadda. 2007. Synthesis and Antimicrobial Evaluation of Some New Thiazole, Thiazolidinone and Thiazoline Derivatives Starting from 1-Chloro-3,4-Dihydronaphthalene-2-Carboxaldehyde. *European Journal of Medicinal Chemistry* 42, no. 7 (July 1): 948–54.

Brotzu, G., A.M. Fadda, M.L. Manca, T. Manca, F. Marongiu, M. Campisi, and F. Consolaro. 2019. A Liposome-Based Formulation Containing Equol, Dihomo-γ-Linolenic Acid and Propionyl-1-Carnitine to Prevent and Treat Hair Loss: A Prospective Investigation. *Dermatologic Therapy* 32, no. 1 (January): e12778. https://onlinelibrary.wiley.com/doi/10.1111/dth.12778.

Calienni, M.N., D. Maza Vega, C.F. Temprana, M.C. Izquierdo, D.E. Ybarra, E. Bernabeu, M. Moretton, et al. 2021. The Topical Nanodelivery of Vismodegib Enhances Its Skin Penetration and Performance In Vitro While Reducing Its Toxicity In Vivo. *Pharmaceutics* 13, no. 2. https://www.mdpi.com/1999-4923/13/2/186.

Calienni, M.N., M.J. Prieto, V.M. Couto, E. de Paula, S. del V. Alonso, and J. Montanari. 2018. 5-Fluorouracil-Loaded Ultradeformable Liposomes for Skin Therapy. In, 020024. Solapur, India. http://aip.scitation.org/doi/abs/10.1063/1.5047778.

Cárcamo-Martínez, Á., B. Mallon, Q.K. Anjani, J. Domínguez-Robles, E. Utomo, L.K. Vora, I.A. Tekko, E. Larrañeta, and R.F. Donnelly. 2021. Enhancing Intradermal Delivery of Tofacitinib Citrate: Comparison between Powder-Loaded Hollow Microneedle Arrays and Dissolving Microneedle Arrays. *International Journal of Pharmaceutics* 593: 120152. https://www.sciencedirect.com/science/article/pii/S0378517320311376.

Castañeda-Reyes, E.D., M. de J. Perea-Flores, G. Davila-Ortiz, Y. Lee, and E. Gonzalez de Mejia. 2020. Development, Characterization and Use of Liposomes as Amphipathic Transporters of Bioactive Compounds for Melanoma Treatment and Reduction of Skin Inflammation: A Review. *International Journal of Nanomedicine* Volume 15 (October): 7627–50. https://www.dovepress.com/development-characterization-and-use-of-liposomes-as-amphipathic-trans-peer-reviewed-article-IJN.

Chaikul, P., N. Khat-udomkiri, K. Iangthanarat, J. Manosroi, and A. Manosroi. 2019. Characteristics and In Vitro Anti-Skin Aging Activity of Gallic Acid Loaded in Cationic CTAB Niosome. *European Journal of Pharmaceutical Sciences* 131 (April): 39–49. https://linkinghub.elsevier.com/retrieve/pii/S0928098719300612.

Chaturvedi, S., and A. Garg. 2021. An Insight of Techniques for the Assessment of Permeation Flux across the Skin for Optimization of Topical and Transdermal Drug Delivery Systems. *Journal of Drug Delivery Science and Technology* 62 (April): 102355. https://linkinghub.elsevier.com/retrieve/pii/S1773224721000356.

Chen, J., Y. Ma, Y. Tao, X. Zhao, Y. Xiong, Z. Chen, and Y. Tian. 2021. Formulation and Evaluation of a Topical Liposomal Gel Containing a Combination of Zedoary Turmeric Oil and Tretinoin for Psoriasis Activity. *Journal of Liposome Research* 31, no. 2 (April 3): 130–44. https://www.tandfonline.com/doi/full/10.1080/08982104.2020.1748646.

Chen, R., R. Li, Q. Liu, C. Bai, B. Qin, Y. Ma, and J. Han. 2017. Ultradeformable Liposomes: A Novel Vesicular Carrier for Enhanced Transdermal Delivery of Procyanidins: Effect of Surfactants on the Formation, Stability, and Transdermal Delivery. *AAPS PharmSciTech* 18, no. 5 (July): 1823–32. http://link.springer.com/10.1208/s12249-016-0661-5.

Choochuen, N., and A. Jimtaisong. 2022. Physical Stability and Subjective Efficacy Study of Liposome Loaded with Clitoria Ternatea (Butterfly Pea) Flower Extract and Eugenia Caryophyllus (Clove) Oil. *Pharmaceutical Sciences Asia* 49, no. 1: 51–58. https://pharmacy.mahidol.ac.th/journal/journalabstract.php?jvol=49&jpart=1&jconnum=7.

COSMOS. 2020. COSMOS-Standard. *COSMOS-Standard AISBL.* https://www.cosmos-standard.org/.

Daudey, G.A., M. Shen, A. Singhal, P. van der Est, G.J.A. Sevink, A.L. Boyle, and A. Kros. 2021. Liposome Fusion with Orthogonal Coiled Coil Peptides as Fusogens: The Efficacy of Roleplaying Peptides. *Chemical Science* 12, no. 41: 13782–92. http://xlink.rsc.org/?DOI=D0SC06635D.

De Leeuw, J., N. Van Der Beek, G. Maierhofer, and W.D. Neugebauer. 2003. A Case Study to Evaluate the Treatment of Vitiligo with Khellin Encapsulated in L-Phenylalanin Stabilized Phosphatidylcholine Liposomes in Combination with Ultraviolet Light Therapy. *European Journal of Dermatology* 13, no. 5: 474–77.

de Oliveira Silva, J., R.S. Fernandes, C.M. Ramos Oda, T.H. Ferreira, A.F. Machado Botelho, M. Martins Melo, M.C. de Miranda, et al. 2019. Folate-Coated, Long-Circulating and PH-Sensitive Liposomes Enhance Doxorubicin Antitumor Effect in a Breast Cancer Animal Model. *Biomedicine & Pharmacotherapy* 118 (October): 109323. https://linkinghub.elsevier.com/retrieve/pii/S0753332219329774.

Desai, J., and R. Mallya. 2021. Development of Green Coffee Beans Extract Loaded Anti-Aging Liposomal Gel. *Indian Journal of Pharmaceutical Education and Research* 55, no. 4: 979–88.

Desai, P., R.R. Patlolla, and M. Singh. 2010. Interaction of Nanoparticles and Cell-Penetrating Peptides with Skin for Transdermal Drug Delivery. In *Molecular Membrane Biology*. Taylor & Francis.

Dhurat, R., S. Daruwalla, S. Pai, M. Kovacevic, J. McCoy, J. Shapiro, R. Sinclair, S. Vano-Galvan, and A. Goren. 2022. SULT1A1 (Minoxidil Sulfotransferase) Enzyme Booster Significantly Improves Response to Topical Minoxidil for Hair Regrowth. *Journal of Cosmetic Dermatology* 21, no. 1 (January): 343–46. https://onlinelibrary.wiley.com/doi/10.1111/jocd.14299.

Dini, I., and S. Laneri. 2021. The New Challenge of Green Cosmetics: Natural Food Ingredients for Cosmetic Formulations. *Molecules* 26, no. 13 (June 26): 3921. https://www.mdpi.com/1420-3049/26/13/3921.

Doppalapudi, S., S. Mahira, and W. Khan. 2017. Development and in Vitro Assessment of Psoralen and Resveratrol Co-Loaded Ultradeformable Liposomes for the Treatment of Vitiligo. *Journal of Photochemistry and Photobiology B: Biology* 174 (September 1): 44–57.

Duan, Y., L. Wei, J. Petryk, and T. Ruddy. 2016. Formulation, Characterization and Tissue Distribution of a Novel PH-Sensitive Long-Circulating Liposome-Based Theranostic Suitable for Molecular Imaging and Drug Delivery. *International Journal of Nanomedicine* Volume 11 (November): 5697–708. https://www.dovepress.com/formulation-characterization-and-tissue-distribution-of-a-novel-ph-sen-peer-reviewed-article-IJN.Dubey, S.K., A. Dey, G. Singhvi, M.M. Pandey, V. Singh, and P. Kesharwani. 2022. Emerging Trends of Nanotechnology in Advanced Cosmetics. *Colloids and Surfaces B: Biointerfaces* 214 (June): 112440. https://linkinghub.elsevier.com/retrieve/pii/S0927776522001230.

Durga, B., and L. Veera. 2020. Recent Advances of Non-Ionic Surfactant-Based Nano-Vesicles (Niosomes and Proniosomes): A Brief Review of These in Enhancing Transdermal Delivery of Drug. *Future Journal of Pharmaceutical Sciences* 6, no. 1 (December): 100. https://fjps.springeropen.com/articles/10.1186/s43094-020-00117-y.

Elmowafy, M. 2021. Skin Penetration/Permeation Success Determinants of Nanocarriers: Pursuit of a Perfect Formulation. *Colloids and Surfaces B: Biointerfaces* 203 (July): 111748. https://linkinghub.elsevier.com/retrieve/pii/S0927776521001922.

El-Zaafarany, G. M., Abdel-Aziz, R. T., Montaser, M. H. A., & Nasr, M. (2021). Coenzyme Q10 phospholipidic vesicular formulations for treatment of androgenic alopecia: *ex vivo* permeation and clinical appraisal. *Expert Opinion on Drug Delivery*, 18(10), 1513-1522.10.1080/17425247.2021.1936497

European Parliament and of the Council. 2009. Regulation (EC) No 1223/2009 of the European Parliament and of the Council. *EUR-Lex*. https://eur-lex.europa.eu/eli/reg/2009/1223/2019-08-13.

Ezzedine, K., V. Eleftheriadou, M. Whitton, and N. Van Geel. 2015. Vitiligo. In *The Lancet*, 386:74–84.

Fabbrocini, G., C. Capasso, M. Donnarumma, M. Cantelli, M. Le Maître, G. Monfrecola, and E. Emanuele. 2017. A Peel-off Facial Mask Comprising Myoinositol and Trehalose-Loaded Liposomes Improves Adult Female Acne by Reducing Local Hyperandrogenism and Activating Autophagy. *Journal of Cosmetic Dermatology* 16, no. 4 (December): 480–84. https://onlinelibrary.wiley.com/doi/10.1111/jocd.12340.

Fadeel, D.A.A., M. Fadel, A. Tawfik, and Y. Omar. 2022. Transfersomal Eosin Topical Delivery Assisted by Fractional CO2 Laser for Photodynamic Treatment of Palmar Hyperhidrosis: Case Study. *Drug Delivery and Translational Research* (April 20). https://link.springer.com/10.1007/s13346-022-01164-z.

Fang, C.-L., Y. Wang, K.H.-Y. Tsai, and H.-I. Chang. 2018. Liposome-Encapsulated Baicalein Suppressed Lipogenesis and Extracellular Matrix Formation in Hs68 Human Dermal Fibroblasts. *Frontiers in Pharmacology* 9 (March 6): 155. http://journal.frontiersin.org/article/10.3389/fphar.2018.00155/full.

Farid, M., T. Faber, D. Dietrich, and A. Lamprecht. 2020. Cell Membrane Fusing Liposomes for Cytoplasmic Delivery in Brain Endothelial Cells. *Colloids and Surfaces B: Biointerfaces* 194 (October): 111193. https://linkinghub.elsevier.com/retrieve/pii/S092777652030549X.

FDA. 2014. Guidance for Industry: Safety of Nanomaterials in Cosmetic Products. *Food and Drug Administration.* https://www.fda.gov/regulatory-information/search-fda-guidance-documents/guidance-industry-safety-nanomaterials-cosmetic-products.

———. 2022. Cosmetics Nanotechnology. *Food and Drug Administration.* https://www.fda.gov/cosmetics/cosmetics-science-research/cosmetics-nanotechnology#:~:text=FDA%20monitors%20the%20use%20of, a%20legal%20definition%20for%20nanotechnology.

Ferraris, C., C. Rimicci, S. Garelli, E. Ugazio, and L. Battaglia. 2021. Nanosystems in Cosmetic Products: A Brief Overview of Functional, Market, Regulatory and Safety Concerns. *Pharmaceutics* 13, no. 9 (September 5): 1408. https://www.mdpi.com/1999-4923/13/9/1408.

Figueroa-Robles, A., M. Antunes-Ricardo, and D. Guajardo-Flores. 2021. Encapsulation of Phenolic Compounds with Liposomal Improvement in the Cosmetic Industry. *International Journal of Pharmaceutics* 593 (January): 120125. https://linkinghub.elsevier.com/retrieve/pii/S0378517320311108.

Fisher, G.J., S. Datta, Z.Q. Wang, X.Y. Li, T. Quan, J.H. Chung, S. Kang, and J.J. Voorhees. 2000. C-Jun–Dependent Inhibition of Cutaneous Procollagen Transcription Following Ultraviolet Irradiation Is Reversed by All-Trans Retinoic Acid. *The Journal of Clinical Investigation* 106, no. 5 (September 1): 663–70.

Gilchrest, B.A., M. Garmyn, and M. Yaar. 1994. Aging and Photoaging Affect Gene Expression in Cultured Human Keratinocytes. *Archives of Dermatology* 130, no. 1 (January 1): 82–86.

Goudarzi, R., M. Eskandarynasab, A. Muhammadnejad, A.R. Dehpour, and A. Partoazar. 2021. Beneficial Effects of ROCEN (Topical Nano-Arthrocen) on Atopic Dermatitis in Mice. *BMC Complementary Medicine and Therapies* 21, no. 1 (December): 226. https://bmccomplementalternmed.biomedcentral.com/articles/10.1186/s12906-021-03393-0.

Graham, G.M., M.D. Farrar, J.E. Cruse-Sawyer, K.T. Holland, and E. Ingham. 2004. Proinflammatory Cytokine Production by Human Keratinocytes Stimulated with Propionibacterium Acnes and P. Acnes GroEL. *British Journal of Dermatology* 150, no. 3 (March 1): 421–28.

Guillot, A.J., E. Jornet-Mollá, N. Landsberg, C. Milián-Guimerá, M.C. Montesinos, T.M. Garrigues, and A. Melero. 2021. Cyanocobalamin Ultraflexible Lipid Vesicles: Characterization and In Vitro Evaluation of Drug-Skin Depth Profiles. *Pharmaceutics* 13, no. 3 (March 20): 418. https://www.mdpi.com/1999-4923/13/3/418.

Guimarães, D., A. Cavaco-Paulo, and E. Nogueira. 2021. Design of Liposomes as Drug Delivery System for Therapeutic Applications. *International Journal of Pharmaceutics* 601: 120571. https://www.sciencedirect.com/science/article/pii/S0378517321003768.

Habeshian, K.A., and B.A. Cohen. 2020. Current Issues in the Treatment of Acne Vulgaris. *Pediatrics* 145, no. Supplement_2 (May 1): S225–30.

Hartini, P.T., Sumaiyah, and P.A.Z. Hasibuan. 2022. Formulation and Evaluation of Liposome Moringa Oleifera Seed Oil (Moringa Oleifera L.) as Anti-Aging. *International Journal of Science, Technology & Management* 3, no. 4 (July 22): 1179–83. https://ijstm.inarah.co.id/index.php/ijstm/article/view/546.

Honzak, L., and M. Šentjurc. 2000. Development of Liposome Encapsulated Clindamycin for Treatment of Acne Vulgaris. *Pflügers Archiv - European Journal of Physiology* 440, no. S1 (January): R044–45. http://link.springer.com/10.1007/s004240000000.

———. 2016. Development of Liposome Encapsulated Clindamycin for Treatment of Acne Vulgaris. *Pflügers Archiv - European Journal of Physiology 2000 440:1* 440, no. 1 (October 27): R044–45.

Hoss, I., H.N. Rajha, R. El Khoury, S. Youssef, M.L. Manca, M. Manconi, N. Louka, and R.G. Maroun. 2021. Valorization of Wine-Making By-Products' Extracts in Cosmetics. *Cosmetics* 8, no. 4 (November 18): 109. https://www.mdpi.com/2079-9284/8/4/109.

Hsieh, W.-C., C.-W. Fang, M. Suhail, Q. Lam Vu, C.-H. Chuang, and P.-C. Wu. 2021. Improved Skin Permeability and Whitening Effect of Catechin-Loaded Transfersomes through Topical Delivery. *International Journal of Pharmaceutics* 607 (September): 121030. https://linkinghub.elsevier.com/retrieve/pii/S037851732100836X.

Ibaraki, H., T. Kanazawa, C. Oogi, Y. Takashima, and Y. Seta. 2019. Effects of Surface Charge and Flexibility of Liposomes on Dermal Drug Delivery. *Journal of Drug Delivery Science and Technology* 50 (April): 155–62. https://linkinghub.elsevier.com/retrieve/pii/S1773224718310141.

Izquierdo, M.C., C.R. Lillo, P. Bucci, G.E. Gómez, L. Martínez, S. del V. Alonso, M.N. Calienni, and J. Montanari. 2020. Comparative Skin Penetration Profiles of Formulations Including Ultradeformable Liposomes as Potential Nanocosmeceutical Carriers. *Journal of Cosmetic Dermatology* 19, no. 11 (November): 3127–37. https://onlinelibrary.wiley.com/doi/10.1111/jocd.13410.

Jain, S., D.P. Kale, R. Swami, and S.S. Katiyar. 2018. Codelivery of Benzoyl Peroxide & Adapalene Using Modified Liposomal Gel for Improved Acne Therapy. *Nanomedicine* 13, no. 12 (June): 1481–93. https://www.futuremedicine.com/doi/10.2217/nnm-2018-0002.

Jevtić, M., A. Löwa, A. Nováčková, A. Kováčik, S. Kaessmeyer, G. Erdmann, K. Vávrová, and S. Hedtrich. 2020. Impact of Intercellular Crosstalk between Epidermal Keratinocytes and Dermal Fibroblasts on Skin Homeostasis. *Biochimica et Biophysica Acta (BBA) - Molecular Cell Research* 1867, no. 8 (August): 118722. https://linkinghub.elsevier.com/retrieve/pii/S016748892030080X.

Jiménez-Rodríguez, A., D. Guardado-Félix, and M. Antunes-Ricardo. 2022. Challenges and Strategies for Topical and Transdermal Delivery of Bioactive Peptides. *Critical Reviews[TM] in Therapeutic Drug Carrier Systems* 39, no. 1: 1–31. http://www.dl.begellhouse.com/journals/3667c4ae6e8fd136,6ed0f7a902a1b2f0,08cc8a443e810f96.html.

Karagaiah, P., Y. Valle, J. Sigova, N. Zerbinati, P. Vojvodic, D. Parsad, R.A. Schwartz, S. Grabbe, M. Goldust, and T. Lotti. 2020. Emerging Drugs for the Treatment of Vitiligo. *Expert Opinion on Emerging Drugs* 25, no. 1: 7–24.

Kewlani, P., L. Singh, B. Singh, and I.D. Bhatt. 2022. Sustainable Extraction of Phenolics and Antioxidant Activities from Prinsepia Utilis Byproducts for Alleviating Aging and Oxidative Stress. *Sustainable Chemistry and Pharmacy* 29 (October): 100791. https://linkinghub.elsevier.com/retrieve/pii/S2352554122001954.

Khalil, I.A., and H. Harashima. 2018. An Efficient PEGylated Gene Delivery System with Improved Targeting: Synergism between Octaarginine and a Fusogenic Peptide. *International Journal of Pharmaceutics* 538, no. 1–2 (March): 179–87. https://linkinghub.elsevier.com/retrieve/pii/S0378517318300073.

Kim, J.S., M.W. Kim, H.Y. Jeong, S.J. Kang, S.I. Park, Y.K. Lee, H.S. Kim, K.S. Kim, and Y.S. Park. 2016. Sendai Viroplexes for Epidermal Growth Factor Receptor-Directed Delivery of Interleukin-12 and Salmosin Genes to Cancer Cells: Cancer-Targeted Transfection by Sendai Viroplexes. *The Journal of Gene Medicine* 18, no. 7 (July): 112–23. https://onlinelibrary.wiley.com/doi/10.1002/jgm.2884.

Kochar, P., K. Nayak, S. Thakkar, S. Polaka, D. Khunt, and M. Misra. 2020. Exploring the Potential of Minoxidil Tretinoin Liposomal Based Hydrogel for Topical Delivery in the Treatment of Androgenic Alopecia. *Cutaneous and Ocular Toxicology* 39, no. 1 (January 2): 43–53. https://www.tandfonline.com/doi/full/10.1080/15569527.2019.1694032.

Kuznetsova, D.A., L.A. Vasileva, G.A. Gaynanova, R.V. Pavlov, A.S. Sapunova, A.D. Voloshina, G.V. Sibgatullina, et al. 2021. Comparative Study of Cationic Liposomes Modified with Triphenylphosphonium and Imidazolium Surfactants for Mitochondrial Delivery. *Journal of Molecular Liquids* 330 (May): 115703. https://linkinghub.elsevier.com/retrieve/pii/S0167732221004281.

Kuznetsova, D.A., L.A. Vasileva, G.A. Gaynanova, E.A. Vasilieva, O.A. Lenina, I.R. Nizameev, M.K. Kadirov, K.A. Petrov, L.Ya. Zakharova, and O.G. Sinyashin. 2021. Cationic Liposomes Mediated Transdermal Delivery of Meloxicam and Ketoprofen: Optimization of the Composition, In Vitro and In Vivo Assessment of Efficiency. *International Journal of Pharmaceutics* 605 (August): 120803. https://linkinghub.elsevier.com/retrieve/pii/S0378517321006086.

Kwon, S.S., S.Y. Kim, B.J. Kong, K.J. Kim, G.Y. Noh, N.R. Im, J.W. Lim, J.H. Ha, J. Kim, and S.N. Park. 2015. Cell Penetrating Peptide Conjugated Liposomes as Transdermal Delivery System of Polygonum Aviculare L. Extract. *International Journal of Pharmaceutics* 483, no. 1–2 (April 10): 26–37.

Lee, J., S.I. Park, S.H. Heo, M. Kim, and M.S. Shin. 2020. Enhancing the Moisturizing Ability of the Skin Softener Using Nanoemulsion Based on Phospholipid Liposome. *International Journal of Advanced Culture Technology* 8, no. 1 (March 31): 236–42. https://doi.org/10.17703/IJACT.2020.8.1.236.

Lee, M.S., J.W. Lee, S.J. Kim, O. Pham-Nguyen, J. Park, J.H. Park, Y.M. Jung, J.B. Lee, and H.S. Yoo. 2021. Comparison Study of the Effects of Cationic Liposomes on Delivery across 3D Skin Tissue and Whitening Effects in Pigmented 3D Skin. *Macromolecular Bioscience* 21, no. 5 (May): 2000413. https://onlinelibrary.wiley.com/doi/10.1002/mabi.202000413.

Lee, Y.S., S.H. Jeon, H.J. Ham, H.P. Lee, M.J. Song, and J.T. Hong. 2020. Improved Anti-Inflammatory Effects of Liposomal Astaxanthin on a Phthalic Anhydride-Induced Atopic Dermatitis Model. *Frontiers in Immunology* 11 (December 1): 565285. https://www.frontiersin.org/articles/10.3389/fimmu.2020.565285/full.

Li, J., N. Duan, S. Song, D. Nie, M. Yu, J. Wang, Z. Xi, et al. 2021. Transfersomes Improved Delivery of Ascorbic Palmitate into the Viable Epidermis for Enhanced Treatment of Melasma. *International Journal of Pharmaceutics* 608 (October): 121059. https://linkinghub.elsevier.com/retrieve/pii/S0378517321008656.

Li, S., J. Chen, F. Chen, C. Wang, X. Guo, C. Wang, Y. Fan, Y. Wang, Y. Peng, and W. Li. 2021. Liposomal Honokiol Promotes Hair Growth via Activating Wnt3a/β-Catenin Signaling Pathway and down Regulating TGF-B1 in C57BL/6N Mice. *Biomedicine & Pharmacotherapy* 141 (September): 111793. https://linkinghub.elsevier.com/retrieve/pii/S0753332221005758.

Lin, H., Q. Xie, X. Huang, J. Ban, B. Wang, X. Wei, Y. Chen, and Z. Lu. 2018. Increased Skin Permeation Efficiency of Imperatorin via Charged Ultradeformable Lipid Vesicles for Transdermal Delivery. *International Journal of Nanomedicine* Volume 13 (February): 831–42. https://www.dovepress.com/increased-skin-permeation-efficiency-of-imperatorin-via-charged-ultrad-peer-reviewed-article-IJN.

Liyanaararchchi, G.D., A.S. Perera, J.K.R.R. Samarasekera, K.R.R. Mahanama, K.D.P. Hemalal, S. Dlamini, H.D.S.M. Perera, Q. Alhadidi, Z.A. Shah, and L.M.V. Tillekeratne. 2022. Bioactive Constituents Isolated from the Sri Lankan Endemic Plant Artocarpus Nobilis and Their Potential to Use in Novel Cosmeceuticals. *Industrial Crops and Products* 184 (September): 115076. https://linkinghub.elsevier.com/retrieve/pii/S0926669022005593.

Lohani, A., A.K. Mishra, and A. Verma. 2019. Cosmeceutical Potential of Geranium and Calendula Essential Oil: Determination of Antioxidant Activity and In Vitro Sun Protection Factor. *Journal of Cosmetic Dermatology* 18, no. 2 (April): 550–57. https://onlinelibrary.wiley.com/doi/10.1111/jocd.12789.

Lohani, A., A. Verma, G. Hema, and K. Pathak. 2021. Topical Delivery of Geranium/Calendula Essential Oil-Entrapped Ethanolic Lipid Vesicular Cream to Combat Skin Aging. Ed. Dr. Abdul Ahad. *BioMed Research International* 2021 (September 11): 1–13. https://www.hindawi.com/journals/bmri/2021/4593759/.

Madan, S., C. Nehate, T.K. Barman, A.S. Rathore, and V. Koul. 2019. Design, Preparation, and Evaluation of Liposomal Gel Formulations for Treatment of Acne: In Vitro and In Vivo Studies. *Drug Development and Industrial Pharmacy* 45, no. 3 (March 4): 395–404.

Magnifico, I., G. Petronio Petronio, N. Venditti, M.A. Cutuli, L. Pietrangelo, F. Vergalito, K. Mangano, D. Zella, and R. Di Marco. 2020. Atopic Dermatitis as a Multifactorial Skin Disorder. Can the Analysis of Pathophysiological Targets Represent the Winning Therapeutic Strategy? *Pharmaceuticals* 13, no. 11 (November 22): 411. https://www.mdpi.com/1424-8247/13/11/411.

Maione-Silva, L., E.G. de Castro, T.L. Nascimento, E.R. Cintra, L.C. Moreira, B.A.S. Cintra, M.C. Valadares, and E.M. Lima. 2019. Ascorbic Acid Encapsulated into Negatively Charged Liposomes Exhibits Increased Skin Permeation, Retention and Enhances Collagen Synthesis by Fibroblasts. *Scientific Reports* 9, no. 1 (December): 522. http://www.nature.com/articles/s41598-018-36682-9.

Malathi, S., P. Balashanmugam, T. Devasena, and S.N. Kalkura. 2021. Enhanced Antibacterial Activity and Wound Healing by a Novel Collagen Blended ZnO Nanoparticles Embedded Niosome Nanocomposites. *Journal of Drug Delivery Science and Technology* 63 (June): 102498. https://linkinghub.elsevier.com/retrieve/pii/S1773224721001787.

Mir-Palomo, S., A. Nácher, M.A. Ofelia Vila Busó, C. Caddeo, M.L. Manca, M. Manconi, and O. Díez-Sales. 2019. Baicalin and Berberine Ultradeformable Vesicles as Potential Adjuvant in Vitiligo Therapy. *Colloids and Surfaces B: Biointerfaces* 175 (March 1): 654–62.

Mir-Palomo, S., A. Nácher, M.A. Ofelia Vila-Busó, C. Caddeo, M.L. Manca, A.R. Saurí, E. Escribano-Ferrer, M. Manconi, and O. Díez-Sales. 2020. Co-Loading of Finasteride and Baicalin in Phospholipid Vesicles Tailored for the Treatment of Hair Disorders. *Nanoscale* 12, no. 30: 16143–52. http://xlink.rsc.org/?DOI=D0NR03357J.

Nakhaei, P., R. Margiana, D.O. Bokov, W.K. Abdelbasset, M.A. Jadidi Kouhbanani, R.S. Varma, F. Marofi, M. Jarahian, and N. Beheshtkhoo. 2021. Liposomes: Structure, Biomedical Applications, and Stability Parameters With Emphasis on Cholesterol. *Frontiers in Bioengineering and Biotechnology* 9 (September 9): 705886. https://www.frontiersin.org/articles/10.3389/fbioe.2021.705886/full.

Nan, L., C. Liu, Q. Li, X. Wan, J. Guo, P. Quan, and L. Fang. 2018. Investigation of the Enhancement Effect of the Natural Transdermal Permeation Enhancers from Ledum Palustre L. Var. Angustum N. Busch: Mechanistic Insight Based on Interaction among Drug, Enhancers and Skin. *European Journal of Pharmaceutical Sciences* 124 (November): 105–13. https://linkinghub.elsevier.com/retrieve/pii/S0928098718303907.

Nastiti, C., T. Ponto, E. Abd, J. Grice, H. Benson, and M. Roberts. 2017. Topical Nano and Microemulsions for Skin Delivery. *Pharmaceutics* 9, no. 4 (September 21): 37. http://www.mdpi.com/1999-4923/9/4/37.

Nishimoto-Sauceda, D., L.E. Romero-Robles, and M. Antunes-Ricardo. 2022. Biopolymer Nanoparticles: A Strategy to Enhance Stability, Bioavailability, and Biological Effects of Phenolic Compounds as Functional Ingredients. *Journal of the Science of Food and Agriculture* 102, no. 1: 41–52.

Niu, X.-Q., D.-P. Zhang, Q. Bian, X.-F. Feng, H. Li, Y.-F. Rao, Y.-M. Shen, et al. 2019. Mechanism Investigation of Ethosomes Transdermal Permeation. *International Journal of Pharmaceutics: X* 1 (December): 100027. https://linkinghub.elsevier.com/retrieve/pii/S2590156719300416.

Okasaka, M., K. Kubota, E. Yamasaki, J. Yang, and S. Takata. 2019. Evaluation of Anionic Surfactants Effects on the Skin Barrier Function Based on Skin Permeability. *Pharmaceutical Development and Technology* 24, no. 1 (January 2): 99–104. https://www.tandfonline.com/doi/full/10.1080/10837450.2018.1425885.

Oliveira, C., C. Coelho, J.A. Teixeira, P. Ferreira-Santos, and C.M. Botelho. 2022. Nanocarriers as Active Ingredients Enhancers in the Cosmetic Industry—The European and North America Regulation Challenges. *Molecules* 27, no. 5 (March 3): 1669. https://www.mdpi.com/1420-3049/27/5/1669.

Paliwal, S.R., R. Paliwal, and S.P. Vyas. 2015. A Review of Mechanistic Insight and Application of PH-Sensitive Liposomes in Drug Delivery. *Drug Delivery* 22, no. 3 (April 3): 231–42. http://www.tandfonline.com/doi/full/10.3109/10717544.2014.882469.

Păvăloiu, R.-D., F. Sha'at, C. Bubueanu, M. Deaconu, G. Neagu, M. Sha'at, M. Anastasescu, et al. 2019. Polyphenolic Extract from Sambucus Ebulus L. Leaves Free and Loaded into Lipid Vesicles. *Nanomaterials* 10, no. 1 (December 25): 56. https://www.mdpi.com/2079-4991/10/1/56.

Picardo, M., M.L. Dell'Anna, K. Ezzedine, I. Hamzavi, J.E. Harris, D. Parsad, and A. Taieb. 2015. Vitiligo. In *Nature Reviews Disease Primers*. Nature Publishing Group.

Pinto, D., M. de la L. Cádiz-Gurrea, J. Garcia, M.J. Saavedra, V. Freitas, P. Costa, B. Sarmento, C. Delerue-Matos, and F. Rodrigues. 2021. From Soil to Cosmetic Industry: Validation of a New Cosmetic Ingredient Extracted from Chestnut Shells. *Sustainable Materials and Technologies* 29 (September): e00309. https://linkinghub.elsevier.com/retrieve/pii/S2214993721000646.

Radmard, A., M. Saeedi, K. Morteza-Semnani, S.M.H. Hashemi, and A. Nokhodchi. 2021. An Eco-Friendly and Green Formulation in Lipid Nanotechnology for Delivery of a Hydrophilic Agent to the Skin in the Treatment and Management of Hyperpigmentation Complaints: Arbutin Niosome (Arbusome). *Colloids and Surfaces B: Biointerfaces* 201 (May): 111616. https://linkinghub.elsevier.com/retrieve/pii/S0927776521000606.

Rahimpour, Y., and H. Hamishehkar. 2012. Liposomes in Cosmeceutics. 9, no. 4 (April): 443–55. http://dx.doi.org/10.1517/17425247.2012.666968.

Rahman, S.A., N.S. Abdelmalak, A. Badawi, T. Elbayoumy, N. Sabry, and A. El Ramly. 2016. Tretinoin-Loaded Liposomal Formulations: From Lab to Comparative Clinical Study in Acne Patients. *Drug Delivery* 23, no. 4 (May 3): 1184–93. https://www.tandfonline.com/doi/full/10.3109/10717544.2015.1041578.

Ramanunny, A.K., S. Wadhwa, M. Gulati, S.K. Singh, B. Kapoor, H. Dureja, D.K. Chellappan, et al. 2021. Nanocarriers for Treatment of Dermatological Diseases: Principle, Perspective and Practices. *European Journal of Pharmacology* 890: 173691. https://www.sciencedirect.com/science/article/pii/S0014299920307834.

Ramos, P.M., J. McCoy, C. Wambier, J. Shapiro, S. Vañó-Galvan, R. Sinclair, and A. Goren. 2020. Novel Topical Booster Enhances Follicular Sulfotransferase Activity in Patients with Androgenetic Alopecia: A New Strategy to Improve Minoxidil Response. *Journal of the European Academy of Dermatology and Venereology* 34, no. 12 (December). https://onlinelibrary.wiley.com/doi/10.1111/jdv.16645.

Rangsimawong, W., Y. Obata, P. Opanasopit, T. Ngawhirunpat, and K. Takayama. 2018. Enhancement of Galantamine HBr Skin Permeation Using Sonophoresis and Limonene-Containing PEGylated Liposomes. *AAPS PharmSciTech* 19, no. 3 (April): 1093–104. http://link.springer.com/10.1208/s12249-017-0921-z.

Rodrigues, M., K. Ezzedine, I. Hamzavi, A.G. Pandya, and J.E. Harris. 2017. Current and Emerging Treatments for Vitiligo. *Journal of the American Academy of Dermatology* 77, no. 1: 17–29. doi: 10.1016/j.jaad.2016.11.010.

Ruan, Y., K. Hu, and H. Chen. 2015. Autophagy Inhibition Enhances Isorhamnetin-Induced Mitochondria-Dependent Apoptosis in Non-Small Cell Lung Cancer Cells. *Molecular Medicine Reports* 12, no. 4 (October): 5796–806. https://www.spandidos-publications.com/10.3892/mmr.2015.4148.

Salehi, B., A.P. Mishra, M. Nigam, F. Kobarfard, Z. Javed, S. Rajabi, K. Khan, et al. 2020. Multivesicular Liposome (DepoFoam) in Human Diseases. *Iranian Journal of Pharmaceutical Research* no. Online First (June). https://doi.org/10.22037/ijpr.2020.112291.13663.

Sallustio, V., I. Chiocchio, M. Mandrone, M. Cirrincione, M. Protti, G. Farruggia, A. Abruzzo, et al. 2022. Extraction, Encapsulation into Lipid Vesicular Systems, and Biological Activity of Rosa Canina L. Bioactive Compounds for Dermocosmetic Use. *Molecules* 27, no. 9 (May 8): 3025. https://www.mdpi.com/1420-3049/27/9/3025.

Salvioni, L., L. Morelli, E. Ochoa, M. Labra, L. Fiandra, L. Palugan, D. Prosperi, and M. Colombo. 2021. The Emerging Role of Nanotechnology in Skincare. *Advances in Colloid and Interface Science* 293 (July): 102437. https://linkinghub.elsevier.com/retrieve/pii/S0001868621000786.

Sankar, C., S. Muthukumar, G. Arulkumaran, S. Shalini, R. Sundaraganapathy, and S. joji samuel. 2019. Formulation and Characterization of Liposomes Containing Clindamycin and Green Tea for Anti Acne. *Research Journal of Pharmacy and Technology* 12, no. 12: 5977. http://www.indianjournals.com/ijor.aspx?target=ijor:rjpt&volume=12&issue=12&article=061.Sari, R.K., Y.H. Prayogo, S.A. Rozan, M. Rafi, and I. Wientarsih. 2022. Antioxidant Activity, Sun Protection Activity, and Phytochemical Profile of Ethanolic Extracts of Daemonorops Acehensis Resin and Its Phytosomes. *Scientia Pharmaceutica* 90, no. 1 (February 3): 10. https://www.mdpi.com/2218-0532/90/1/10.

Sari, R.K., Y.H. Prayogo, R.A.L. Sari, N. Asidah, M. Rafi, I. Wientarsih, and W. Darmawan. 2021. Intsia Bijuga Heartwood Extract and Its Phytosome as Tyrosinase Inhibitor, Antioxidant, and Sun Protector. *Forests* 12, no. 12 (December 17): 1792. https://www.mdpi.com/1999-4907/12/12/1792.

Sayıt, G., S.T. Tanrıverdi, Ö. Özer, and E. Özdoğan. 2021. Preparation of Allantoin Loaded Liposome Formulations and Application for Cosmetic Textile Production. *The Journal of The Textile Institute* (March 23): 1–12. https://www.tandfonline.com/doi/full/10.1080/00405000.2021.1903197.

Seong, J.S., M.E. Yun, and S.N. Park. 2018. Surfactant-Stable and PH-Sensitive Liposomes Coated with N-Succinyl-Chitosan and Chitooligosaccharide for Delivery of Quercetin. *Carbohydrate Polymers* 181 (February): 659–67. https://linkinghub.elsevier.com/retrieve/pii/S014486171731384X.

Sesarman, A., L. Tefas, B. Sylvester, E. Licarete, V. Rauca, L. Luput, L. Patras, M. Banciu, and A. Porfire. 2018. Anti-Angiogenic and Anti-Inflammatory Effects of Long-Circulating Liposomes Co-Encapsulating Curcumin and Doxorubicin on C26 Murine Colon Cancer Cells. *Pharmacological Reports* 70, no. 2 (April): 331–39. https://linkinghub.elsevier.com/retrieve/pii/S1734114017301007.

Sguizzato, M., F. Ferrara, S.S. Hallan, A. Baldisserotto, M. Drechsler, M. Malatesta, M. Costanzo, et al. 2021. Ethosomes and Transethosomes for Mangiferin Transdermal Delivery. *Antioxidants* 10, no. 5 (May 12): 768. https://www.mdpi.com/2076-3921/10/5/768.

Shanmugam, S., C.K. Song, S. Nagayya-Sriraman, R. Baskaran, C.S. Yong, H.G. Choi, D.D. Kim, J.S. Woo, and B.K. Yoo. 2009. Physicochemical Characterization and Skin Permeation of Liposome Formulations Containing Clindamycin Phosphate. *Archives of Pharmacal Research* 32, no. 7 (July 31): 1067–75.

Shukla, S., Y. Haldorai, S.K. Hwang, V.K. Bajpai, Y.S. Huh, and Y.-K. Han. 2017. Current Demands for Food-Approved Liposome Nanoparticles in Food and Safety Sector. *Frontiers in Microbiology* 8 (December 5): 2398. http://journal.frontiersin.org/article/10.3389/fmicb.2017.02398/full.

Souto, E.B., A.R. Fernandes, C. Martins-Gomes, T.E. Coutinho, A. Durazzo, M. Lucarini, S.B. Souto, A.M. Silva, and A. Santini. 2020. Nanomaterials for Skin Delivery of Cosmeceuticals and Pharmaceuticals. *Applied Sciences* 10, no. 5 (February 27): 1594.

Sun, M., O.O. Osipitan, E.K. Sulicz, and A.J. Di Pasqua. 2022. Preparation and Optimization of an Ultraflexible Liposomal Gel for Lidocaine Transdermal Delivery. *Materials* 15, no. 14 (July 14): 4895. https://www.mdpi.com/1996-1944/15/14/4895.

Susilawati, Y., A. Chaerunisa, and H. Purwaningsih. 2021. Phytosome Drug Delivery System for Natural Cosmeceutical Compounds: Whitening Agent and Skin Antioxidant Agent. *Journal of Advanced Pharmaceutical Technology & Research* 12, no. 4: 327. http://www.japtr.org/text.asp?2021/12/4/327/328622.

Tadros, T.F. 2018. 12. Formulation of Liposomes and Vesicles in Cosmetic Formulations. In *Pharmaceutical, Cosmetic and Personal Care Formulations*, 243–50. De Gruyter. https://www.degruyter.com/document/doi/10.1515/9783110587982-014/html.

Tansathien, K., P. Chareanputtakhun, T. Ngawhirunpat, P. Opanasopit, and W. Rangsimawong. 2021. Hair Growth Promoting Effect of Bioactive Extract from Deer Antler Velvet-Loaded Niosomes and Microspicules Serum. *International Journal of Pharmaceutics* 597 (March): 120352. https://linkinghub.elsevier.com/retrieve/pii/S0378517321001563.

Tripathi, S. V., C.J. Gustafson, K.E. Huang, and S.R. Feldman. 2012. Side Effects of Common Acne Treatments. 12, no. 1 (January): 39–51. http://dx.doi.org/10.1517/14740338.2013.740456

Wang, F.C., P.L. Hudson, K. Burk, and A.G. Marangoni. 2022. Encapsulation of Cycloastragenol in Phospholipid Vesicles Enhances Transport and Delivery across the Skin Barrier. *Journal of Colloid and Interface Science* 608 (February): 1222–28. https://linkinghub.elsevier.com/retrieve/pii/S0021979721018129.

Wang, Q., W. Liu, J. Wang, H. Liu, and Y. Chen. 2019. Preparation and Pharmacokinetic Study of Daidzein Long-Circulating Liposomes. *Nanoscale Research Letters* 14, no. 1 (December): 321. https://nanoscalereslett.springeropen.com/articles/10.1186/s11671-019-3164-y.

Wang, Y., B. Lei, J. Xu, M. Sun, S. Xu, and H. Liu. 2020. A PH/Reduction Dual-Sensitive Copolymer Inserted in Liposomal Bilayer Acts as a Protective "Umbrella." *Colloids and Surfaces A: Physicochemical and Engineering Aspects* 602 (October): 125128. https://linkinghub.elsevier.com/retrieve/pii/S0927775720307214.

Wiedenhoeft, T., S. Tarantini, Á. Nyúl-Tóth, A. Yabluchanskiy, T. Csipo, P. Balasubramanian, A. Lipecz, et al. 2019. Fusogenic Liposomes Effectively Deliver Resveratrol to the Cerebral Microcirculation and Improve Endothelium-Dependent Neurovascular Coupling Responses in Aged Mice. *GeroScience* 41, no. 6 (December): 711–25. http://link.springer.com/10.1007/s11357-019-00102-1.

Wu, Y.-C., W.-Y. Chen, C.-Y. Chen, S.I. Lee, Y.-W. Wang, H.-H. Huang, and S.-M. Kuo. 2021. Farnesol-Loaded Liposomes Protect the Epidermis and Dermis from PM2.5-Induced Cutaneous Injury. *International Journal of Molecular Sciences* 22, no. 11 (June 4): 6076. https://www.mdpi.com/1422-0067/22/11/6076.

Xu, H.-L., P.-P. Chen, L. Wang, W. Xue, and T.-L. Fu. 2018. Hair Regenerative Effect of Silk Fibroin Hydrogel with Incorporation of FGF-2-Liposome and Its Potential Mechanism in Mice with Testosterone-Induced Alopecia Areata. *Journal of Drug Delivery Science and Technology* 48 (December): 128–36. https://linkinghub.elsevier.com/retrieve/pii/S1773224718308219.

Yamazaki, N., T. Sugimoto, M. Fukushima, R. Teranishi, A. Kotaka, C. Shinde, T. Kumei, et al. 2017. Dual-Stimuli Responsive Liposomes Using PH- and Temperature-Sensitive Polymers for Controlled Transdermal Delivery. *Polymer Chemistry* 8, no. 9: 1507–18. http://xlink.rsc.org/?DOI=C6PY01754A.

Yang, J., and B. Kim. 2018. Synthesis and Characterization of Ethosomal Carriers Containing Cosmetic Ingredients for Enhanced Transdermal Delivery of Cosmetic Ingredients. *Korean Journal of Chemical Engineering* 35, no. 3 (March): 792–97. http://link.springer.com/10.1007/s11814-017-0344-2.

Yoshizaki, Y., E. Yuba, N. Sakaguchi, K. Koiwai, A. Harada, and K. Kono. 2017. PH-Sensitive Polymer-Modified Liposome-Based Immunity-Inducing System: Effects of Inclusion of Cationic Lipid and CpG-DNA. *Biomaterials* 141 (October): 272–83. https://linkinghub.elsevier.com/retrieve/pii/S0142961217304519.

Yücel, Ç., G. Şeker Karatoprak, and İ.T. Değim. 2019a. Anti-Aging Formulation of Rosmarinic Acid-Loaded Ethosomes and Liposomes. *Journal of Microencapsulation* 36, no. 2 (February 17): 180–91. https://www.tandfonline.com/doi/full/10.1080/02652048.2019.1617363.

Zaenglein, A.L. 2018. Acne Vulgaris. Ed. Caren G. Solomon. 379, no. 14 (October 3): 1343–52. https://doi.org/10.1056/NEJMcp1702493.

Zhang, Y., R. Liang, C. Liu, and C. Yang. 2022. Improved Stability and Skin Penetration through Glycethosomes Loaded with Glycyrrhetinic Acid. *International Journal of Cosmetic Science* 44, no. 2 (April): 249–61. https://onlinelibrary.wiley.com/doi/10.1111/ics.12771.

Zhao, G., C. Hu, and Y. Xue. 2018. In Vitro Evaluation of Chitosan-Coated Liposome Containing Both Coenzyme Q10 and Alpha-Lipoic Acid: Cytotoxicity, Antioxidant Activity, and Antimicrobial Activity. *Journal of Cosmetic Dermatology* 17, no. 2 (April 1): 258–62.

Zhou, H., D. Luo, D. Chen, X. Tan, X. Bai, Z. Liu, X. Yang, and W. Liu. 2021. Current Advances of Nanocarrier Technology-Based Active Cosmetic Ingredients for Beauty Applications. *Clinical, Cosmetic and Investigational Dermatology* 14 (July): 867–87. https://www.dovepress.com/current-advances-of-nanocarrier-technology-based-active-cosmetic-ingre-peer-reviewed-fulltext-article-CCID.

Zhou, X., Y. Hao, L. Yuan, S. Pradhan, K. Shrestha, O. Pradhan, H. Liu, and W. Li. 2018. Nano-Formulations for Transdermal Drug Delivery: A Review. *Chinese Chemical Letters* 29, no. 12 (December): 1713–24. https://linkinghub.elsevier.com/retrieve/pii/S1001841718304248.

Zouboulis, C.C., R. Ganceviciene, A.I. Liakou, A. Theodoridis, R. Elewa, and E. Makrantonaki. 2019. Aesthetic Aspects of Skin Aging, Prevention, and Local Treatment. *Clinics in Dermatology* 37, no. 4 (July 1): 365–72.

8 Niosomes as nanocarrier systems in cosmetics

Affiong Iyire, Edidiong M. Udofa, Onyinyechi Udo-chijioke and Eman Z. Dahmash

CONTENTS

DOI: 10.1201/9781003319146-8

8.1 INTRODUCTION

Niosomes (Figure 8.1) are generally bilayered, self-assembled, lamellar vesicles comprised of non-ionic surfactants (Figure 8.1), with the capacity to entrap both hydrophilic and hydrophobic substances within the formed vesicle. Basically, niosomes are formed by the hydration of non-ionic surfactant, cholesterol and/or other amphiphilic molecules, at temperatures above the gel transition temperature of the lipids. The organisation of the components into a bilayer results in the enclosing of an aqueous solution of solute/hydrophilic components within the aqueous core while lipophilic components are embedded within the lipid bilayer [1]. Although the vesicles are self-assembled, they require energy such as physical agitation or heat to aid their formation and this often results in a bilayered structure; with the hydrophobic portions oriented away from the aqueous solvent while the hydrophilic head groups remain in contact with the aqueous solvent [2]. The amphiphilic nature of the non-ionic surfactants is primarily responsible for the formation of niosomes. More specifically, the opposing forces generated by the polar and non-polar regions are responsible for this self-assembly. Hydrophobic interactions in the non-polar (aliphatic chain) groups alongside the high interfacial tension between these groups and water, cause them to associate while the hydrophilic repulsion and steric hindrance in the polar regions alongside their affinity for the aqueous medium cause them to point outwards and remain in contact with the aqueous medium [3, 4]. The size of the polar head group, the volume of the hydrocarbon chain as well as the critical chain length determine the self-assembly. This, in addition to the input of energy, results in the formation of the closed bilayered vesicle called the noisome [3]. Structurally, niosomes differ from their predecessors – liposomes, in that lipids used to fabricate liposomes consist of a double hydrophobic tail while nonionic surfactants have a single hydrophobic tail.

Niosomes present a novel delivery system with the potential to improve the bioavailability of cosmetic agents. General advantages of niosomes as a delivery system include [2, 3, 5–7];

1. Application for a variety of molecules. With niosomes, hydrophobic agents, as well as hydrophilic and amphiphilic agents, can be accommodated. Hydrophilic drugs can either be enclosed within the aqueous core of the

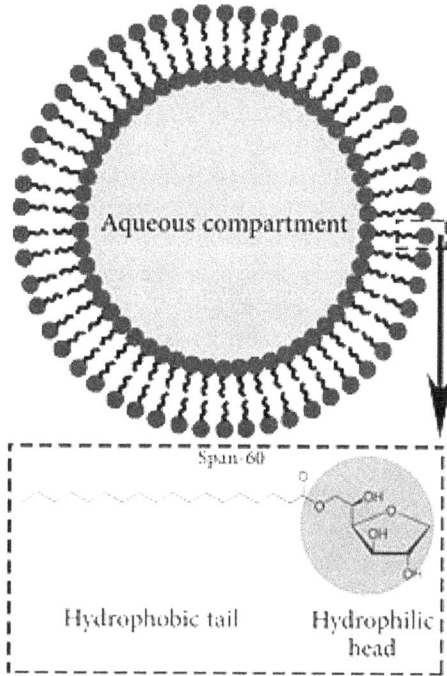

FIGURE 8.1 Structure of a noisome showing the bilayer structure of the vesicle. The hydrophilic heads of the non-ionic surfactant orient towards the aqueous environment, while the hydrophobic tails shield from the aqueous medium, thereby forming the lipid bilayer. Insert reveals the structure of a surfactant commonly used for preparing niosomes; highlighting the hydrophilic head and lipophilic tail (redrawn from Moghassemi and Hadjizadeh [23])

noisome or attached to the surface of the vesicle (prone to active ingredient loss), while hydrophobic agents would be embedded within the lipid bilayer.

2. Niosomes can apply to various routes of administration. This goes on to increase the bioavailability of ingredients via the different routes.

3. Potential to increase dermal penetration and oral bioavailability. Niosomes have the property of reversibly reducing the barrier resistance of tissues, allowing the entrapped agents to reach target sites at a faster rate.

4. Components of niosomes are non-immunogenic, nontoxic, biocompatible and biodegradable making them relatively safe. Compared to ionic surfactants, nonionic surfactants have minimal interaction with biological systems, thus reducing their comparative toxicities.

5. Niosomes delay clearance and protect entrapped agents from biological environment thereby improving the therapeutic performance of their active ingredients.

6. Niosomes can act as depot systems that enable the release of entrapped agents over a long time period, achieving controlled release. This can also be altered accordingly to control the release rate.

7. They are less prone to chemical degradation than liposomes because they are majorly composed of more stable non-ionic surfactants.

Although niosomes hold great potential in improving active ingredient delivery in general, instability issues limit their use – physical stability is still a major concern during development [3]. Instabilities observed in niosomes results in fusion/aggregation of vesicles, leaching and hydrolysis of the entrapped agents, and the subsequent reduction of shelf-life [1]. A stable noisome is described as one which maintains the same vesicle dimensions upon storage, and recent research efforts are directed towards improving the stability of niosomes by altering their formulation parameters.

Cosmetic substances are intended for application on external parts of human body to clean, perfume, change appearance, protect or maintain good condition. Niosomes were introduced by researchers in the cosmetic industry in the 1970s. Generally, cosmetic care is concerned with equilibrating the moisture balance of the skin and niosomes contain lipids, which humidify the skin and replenish skin moisture and lipid structure, making them valuable cosmetic delivery systems [8]. Niosomes are widely reported in literature due to their advantages in topical and dermal delivery of cosmetic actives. These advantages include enhanced skin penetration, improved stability of entrapped active ingredients, improved surface adhesion, improved bioavailability and sustained release properties; when compared with conventional emulsion- based formulations [1]. This chapter will begin with a general discussion about niosomes as a delivery system, and thereafter focus on the application of niosomes as nanocarriers for cosmetics and cosmeceuticals.

8.2 TYPES OF NIOSOMES

Niosomes generally have a size range within the submicron group. However, they can be further classified based on their size into three groups:

- Small Unilamellar Vesicles (SUVs)
- Large Unilamellar Vesicles (LUVs)
- Multi-lamellar Vesicles (MLVs)

Major differences between these types of niosomes are discussed in Table 8.1.

In addition to being classified according to vesicle size, niosomes are also classified based on specialisation. These differences in specialisation highlight the effect of changing formulation parameters aimed at improving the stability of conventional niosomes. Some are discussed below:

8.2.1 PRONIOSOMES

Proniosomes are niosomes in a dry powder form. They are formulated from a suitable water-soluble carrier coated in non-ionic surfactant which upon adequate hydration just prior to use are converted into the niosomal suspension. Maltodextrin, sorbitol and mannitol are examples of water-soluble carriers commonly used because of their improved stability and are less likely to form aggregates, ultimately protecting against niosome leakage [9, 10].

TABLE 8.1

Classification of niosomes according to vesicle size and the methods used to fabricate each [2, 6, 9]

Type	Size (µm)	Features	Method of preparation
SUVs	0.025–0.05	– They have a single bilayer surrounding an aqueous core – Thermodynamically less stable than other types with a higher tendency to aggregate – Have poor loading capacity for hydrophilic agents	– Sonication of MLVs – Microfluidisation
LUVs	0.1–0.5	– They have a single bilayer surrounding an aqueous core – Entraps a larger volume of hydrophilic molecule and other bioactive agents as they have a larger aqueous/liquid compartment ratio	– Reverse phase Evaporation Technique (REV) – Ether injection method
MLVs	0.5–10	– They have a number of bilayers surrounding the aqueous lipid component separately – These niosomes are mechanically stable for longer periods – Very useful as carriers for lipophilic agents	– Hand shaking method (Thin Film Hydration TFH) – Trans-membrane pH gradient (inside acidic) drug uptake process (remote loading)

8.2.2 DEFORMABLE/ELASTIC NIOSOMES

These are specialised niosomes more commonly used in topical formulations due to their flexibility. They are composed of surfactants, ethanol and water and have the ability to pass through pores smaller than their size without destroying the vesicle [9]. Elastic niosomes can pass through the stratum corneum which is less than one-tenth its size, increasing their penetration efficiency. This flexible nature makes them superior to conventional niosomes. Singh *et al.* [10] reported a gel containing papain-loaded elastic niosomes that exhibited a higher reduction in hypertrophic scars due to their superior skin permeation.

8.2.3 DISCOSOMES

These thermoresponsive niosomes are shaped like discs with size 11–60 µm; their structure becoming less organised as temperature increases above 37 °C [9]. This feature enhances their effectiveness.

8.2.4 TRANSFERSOMES

These are niosomes composed of phospholipids which in aqueous environment self-assemble into a lipid bilayer and close to form vesicles. In addition to the phospholipids, transfersomes have an edge activator consisting of a single chain non-ionic surfactant which causes destabilisation of the lipid bilayer increasing its fluidity and elasticity. Transfersomes are suitable carriers for both low molecular weight and high molecular weight agents [9].

8.2.5 ASPASOMES

These are specialised niosomes whose vesicles are formed from a mixture of ascorbyl palmitate, cholesterol and a highly charged lipid. These niosomes are formed using the thin film hydration (TFH) technique followed by sonication and have been reported to enhance transdermal penetration across skin barriers [9]. Their penetration-enhancing ability is thought to be due to their lipophilicity which permits partitioning into the lipids of the skin as well as their amphiphilic character which alters the intracellular space. In addition to this property, aspasomes have intrinsic antioxidant activity due to the ascorbyl moiety and this indicates a promising future in cosmetics [9, 10].

8.3 FORMULATION OF NIOSOMES

The efficiency of cosmetic products depends on the active ingredient as well as the carrier employed [11]. Hence, understanding the physiochemical properties of each of the components of niosomes is key in the preparation of cosmetic niosomes with desired properties [9].

8.3.1 COMPOSITION OF NIOSOMES

Discussed below are the most common components of niosomes:

8.3.1.1 Non-ionic surfactants

Non-ionic surfactants (Figure 8.1) are amphiphilic molecules with a polar (hydrophilic) head group and a non-polar (hydrophobic) tail. They are the major components of conventional niosome structures and greatly determine the physiochemical property of niosomes. Non-ionic surfactants have no charge on their head groups and are more stable, biocompatible and less toxic than cationic, anionic and amphoteric surfactants – making them better preferred for the formation of stable niosomes [6, 9]. Generally, surfactants with a hydrophobic tail consisting of one or two alkyl groups or perfluoroalkyl groups or single steroidal groups are preferred [2, 9]. Although ester-type (-RCOOR') surfactants are less stable than ether-type (-ROR') surfactants, the former are preferred as they are less toxic and are biodegraded by esterases *in vivo* into triglycerides and fatty acids. These requirements are important as they influence vesicle properties such as size, entrapment efficiency, release and stability [2].

When considering non-ionic surfactants for niosome preparation, the hydrophilic-lipophilic balance (HLB) value, critical packing parameter (CPP) and the structure of the surfactant play vital roles in the way the vesicle's bilayer is formed [12]. *HLB* value range from 0 to 20 and express the relationship between the hydrophilic group and the hydrophobic group of compounds. Surfactants with more hydrophilic groups have a higher HLB value and are more water soluble while surfactants with more hydrophobic groups have a lower HLB value [9]. Therefore, surfactants with HLB values between 4 and 8 are used for the preparation of bilayered vesicles such as niosomes – surfactants with higher HLB values have difficulty forming bilayered membranes due to high aqueous solubility [3, 4]. The HLB value also influences the entrapment efficiency and vesicle size of the niosome formed [9]. Shahiwala and Misra [13] reported that entrapment efficiency decreased with a decrease in HLB value from 8 to 0; while Kauslya *et al.* [6] reported that the mean size of niosomes increased proportionally with increase in HLB value. This was due to the decrease in surface free energy recorded with increase in the hydrophobicity of the surfactant. The *CPP* of the surfactant (Eq 8.1) has been defined by Israelachvili [14] as the ratio of the volume of the hydrophobic group (v) to the length of the hydrophobic group (l_c) and the area of the hydrophilic head (a_0); and it influences the geometry of the vesicles formed [15]. Surfactants with CPP within the range 0.5 to 1 are considered optimal for the formation of bilayered vesicles such as niosomes. Surfactants with CPP less than 1 have larger head groups and form micelles while those with CPP greater than 1 form inverted micelles [9, 15].

$$CPP = \frac{v}{l_c a_0} \tag{8.1}$$

8.3.1.2 Cholesterol

Cholesterol is a lipid with a comparatively rigid structure that does not form bilayered vesicles on its own; however, when included in niosomal formulations it provides rigidity, appropriate shape and inflexibility to the preparation [12]. Cholesterol is used as an additive in niosomal formulations to reduce membrane fluidity, improve stability and also alter membrane permeability. This is a result of its molecular shape and solubility property which allows it to occupy empty spaces within the surfactant molecule, anchoring them to the bilayer structure [9, 16]. Abolition of the gel-to-liquid phase transition in niosomes by cholesterol also decreases drug leakage from niosomes and therefore influences the drug entrapment efficiency and the hydro-dynamic diameter of niosomes [17]. Cholesterol improves niosome stability by forming hydrogen bonds between its hydroxyl group and the surfactant hydrophilic head group, thus increasing the chain order of the liquid state bilayer [1, 18]. This is why increasing cholesterol concentration improves vesicle properties and loading of hydrophilic agents. However, increasing cholesterol concentration also decreases the release rate of the encapsulated agent [1]; and above a specific amount, increasing cholesterol inhibits the formation of the linear membrane structure as it competes with drug molecules for space, resulting in unstable niosomes [9]. Finally, the HLB value of the surfactant dictates the amount of cholesterol required; surfactants with HLB >6 require cholesterol to form bilayers, surfactants with HLB >8, such as Tweens, require more cholesterol to maintain the rigidity of the membrane while surfactants with HLB between 4 and 8 (Spans) require little or no cholesterol to form the bilayers [9].

8.3.1.3 Charge inducers

Stability problems of niosomes such as fusion and aggregation can be overcome by imparting a high surface charge [19]. Charge inducers are additives used in niosomal formulations to improve physical stability. They impart positive or negative charges on the surface of the vesicles, preventing aggregation by electrostatic stabilisation [9]. Charge inducers also enhance drug loading and improve entrapment efficacy. Dicetyl phosphate (DCP) and phosphatidic acid which impart negative charges, and stearyl amine or stearyl pyridinium chloride which impart positive charges are the commonly used charge inducers in niosomes [9, 20]. By imparting a zeta potential over ±30 mV, these molecules improve stability as electrical repulsion generated in particles with high zeta potential prevent aggregation [16]. Zeta potentials between |5| and |15| mV represent the region of limited flocculation (aggregation); while between |5| and |3| mV correspond to the region of maximum flocculation [20]. However, increasing the charge beyond a critical point can also prevent the formation of niosomes [3, 9]. In addition to improving stability, inclusion of charged molecules enhances active ingredient encapsulation efficiency, for skin permeation enhancement, and hybrid niosomal complex formation [20].

8.3.2 Factors Influencing Niosome Formation

8.3.2.1 Gel-liquid transition temperature (Tc) of the surfactant

The temperature at which the surfactant changes from gel to liquid state is an important factor affecting niosome formation and has a significant impact on membrane rigidity, membrane permeability, drug entrapment efficiency and overall stability of the niosomes. The T_c is determined by the presence of saturation within the alkyl chain of the surfactant; lack of which results in lower T_c. There is also a correlation between the chain length of the surfactant tail and the T_c: surfactants with short alkyl chains have low T_c leading to the formation of more flexible niosomes which could have poor properties such as leaky membranes and low stability. This is because the lower the T_c of the surfactant, the less likely for the surfactants to be in an ordered gel form that supports niosomal structure [9]. The high T_c of Span 60 has been found to account for its high ACI entrapment, making it one of the most commonly employed nonionic surfactants in niosomes [21, 22]. Cholesterol increases the transition temperature of vesicles by altering the fluidity of chains in bilayers; in sufficient concentrations, cholesterol abolishes the gel to liquid phase transition endotherm of surfactant bilayers; the more fluid the bilayer, the more leaky the niosomes formed and the lower the stability [9]. It should be noted that, during the manufacture of niosomes, the system must be maintained at temperatures slightly above the T_c.

8.3.2.2 Hydration temperature

The hydration temperature influences the shape and size of niosomes formed. Irrespective of the method used, the formation of niosomes involves hydration of the organic phase (which contains the membrane-forming ingredients) with the aqueous phase. This is a critical step in the process and must be carried out at a temperature above the T_c of the non-ionic surfactant/cholesterol [2]. The hydration temperature must be maintained using a water bath as temperature changes affect the assembly of surfactants into vesicles and could also result in vesicle shape transformation and instability. The duration of hydration and the volume of the hydrating medium are also important factors which if not carefully considered, result in the formation of fragile and/or leaky niosomes [2].

8.3.2.3   Nature of the encapsulated agent

The physiochemical properties of the agent encapsulated within the niosomes could affect the rigidity and charge of the bilayer, the vesicle size, the entrapment efficiency and/or the release characteristics. Alteration of the charge on the niosome formed could result in instability issues. The entrapped agent could interact with the surfactant head group to develop charges which increase the vesicle size by creating a mutual repulsion between the surfactant bilayers [2]. Alteration of charge by entrapped agent could also result in aggregation. Uchegbu and Vyas [17] reported that an amphiphilic agent entrapped in a negatively charged Span 60 niosome was found to form an aggregated dispersion requiring the addition of a stearic stabiliser. The hydrophobicity or hydrophilicity of entrapped agent could also affect the stability of the formed vesicles. Generally, hydrophobic agents could increase niosome stability and reduce chances of leakage, being entrapped within the lipid bilayer – while the reverse is seen for hydrophilic agents [2].

8.3.3   METHODS FOR MANUFACTURING NIOSOMES

There are multiple methods for preparing niosomes but the intended use of the niosome determines the choice of method to be used. This is because the method influences niosomal characteristics such as the number of bilayers, size, size distribution, entrapment efficiency within the aqueous phase, and the membrane permeability of the vesicles. The most common methods include:

- Thin film hydration (hand shaking) method
- Ether/ethanol injection method
- Reverse phase evaporation method
- Trans-membrane pH gradient active uptake process
- Emulsion method
- Bubble method
- Heating method
- Microfluidisation method
- Supercritical reverse phase evaporation method

The methods listed above can be grouped into either passive trapping, active trapping or miscellaneous methods. Active trapping describes methods in which active ingredient loading is performed after formation of the niosomes. Usually, ACI loading is performed maintaining a pH or ion gradient to facilitate uptake of active agent into niosomes. Advantages of active trapping methods include almost 100% entrapment, high ACI lipid ratios, absence of leakage, cost effectiveness and suitability for labile molecules [1]. Passive trapping describes methods in which ACI loading is performed during niosome formation i.e. the ACI is incorporated during the preparation of the niosome [1]. Table 8.2 provides details on these classifications as well as some advantages and disadvantages of the methods discussed below.

TABLE 8.2

Pros and cons of methods for manufacture of niosomes classified according to mode of entrapment of active ingredient within the vesicle

Technique	Classification based on active agent loading	Advantage	Disadvantage
Thin film hydration (hand shaking)	Passive Trapping	– Simplicity of the technique makes it suitable for laboratory research and the method can be scaled up easily – Produces niosomes with higher stability	– Involves the use of organic solvents which are toxic and can be difficult to remove – Produces multilamellar niosomes with a large size and low %EE – Niosomes formed are heterogeneous with great polydispersity index
Ether injection method	Passive Trapping	– Simplicity of the technique makes it suitable for laboratory research	– Involves the use of organic solvents which are toxic and can be difficult to remove – Unsuitable for heat labile active agents – Niosomes formed are heterogeneous with great polydispersity index – Produces multilamellar niosomes with low EE – Dissolving active agent in ether could be problematic
Reverse phase evaporation method	Passive Trapping	– Produces niosomes with high entrapment efficiency and more uniform particle size distribution – Suitable for encapsulation of small and large macromolecules	– Involves the use of organic solvents which are toxic and can be difficult to remove – Macromolecule denaturation due to exposure of macromolecules to organic solvent or sonication
Trans-membrane pH gradient active agent uptake process	Active Trapping	– Produces niosomes with high entrapment efficiency – Acidic nature of the interior of niosomes formed protects against leakage	– Involves the use of organic solvents which are toxic and can be difficult to remove – Produces heterogeneous niosomes with great polydispersity index – Need for additional steps such as sonication or extrusion – Low reproducibility and difficulty in standardisation of the technique
Emulsion method	Miscellaneous	– Simplicity of the technique makes it suitable for laboratory research	– Involves the use of organic solvents which are toxic and can be difficult to remove

(Continued)

TABLE 8.2 (Continued)

Technique	Classification based on drug loading	Advantage	Disadvantage
Micro fluidisation	Passive Trapping	– Non-toxic technique as it does not require the use of organic solvents – Produces unilamellar vesicles of defined size distribution	– Unsuitable for heat labile active agents – Susceptible to hydrolysis and/or oxidation – The tendency to aggregation and/or fusion – Leaking of encapsulated drug molecules
Bubble method	Passive Trapping	– Non-toxic technique as it does not require the use of organic solvents	– Unsuitable for heat labile active agents – Instability on prolonged storage
Heating Method	Miscellaneous	– Simple single step method that can be scaled up easily. – Non-toxic technique as it does not require the use of organic solvents	– Unsuitable for heat labile active agents
Supercritical reverse phase evaporation method	Passive trapping	– Non-toxic technique as it does not require the use of organic solvents – Technique can be easily scaled up	– Requires the use of special equipment

8.3.3.1 Thin film hydration (TFH) or hand shaking method

The simplicity of the TFH method makes it widely used in niosome formulation. THF had been previously employed in the fabrication of liposomes and involves the dissolution of membrane-forming materials (surfactant, cholesterol and other additives) in the organic phase and subsequent removal of the organic phase by rotary evaporation to produce a dry thin film. Hydration of this film with the aqueous phase with mechanical agitation (e.g. vortexing) results in the formation of niosomes [2, 9]. Agents for encapsulation within niosomes can either be incorporated into the thin film (lipophilic agents) or dissolved in the aqueous rehydration medium (hydrophilic agents). Beside its simplicity, this method produces highly stable niosomes and appreciable active ingredient entrapment compared to other methods [15]. TFH results in the formation of MLVs with size ranging from 0.5 to 10 μm [23]; however, the MLVs obtained can be converted to SUVs by further processes such as sonication [9]. Though a simple method, the success of the TFH method in formulating niosomes with good properties is dependent on the temperature at which the hydrations step is performed [2]. At temperatures above the Tc of the surfactant, the aqueous solution swells the thin film which is necessary for the formation of MLVs upon agitation [9].

8.3.3.2   Ether injection method

This method involves the slow introduction of a solution of surfactant/cholesterol dissolved in a volatile organic solvent (commonly diethyl ether or ethanol) into water maintained above the T_c of the surfactants. The mixture is injected using a 14-gauge needle. Subsequent evaporation of the organic solvent results in the formation of SUVs and LUVs with diameter ranging from 50 to 1000 nm [1, 6, 24]. To prevent inhalation of toxic fumes, this process must be carried out within a fume cupboard.

8.3.3.3   Reverse phase evaporation method

This method produces niosomes with high entrapment efficiency, large particle size and more uniform particle size distribution and is more suitable for encapsulating large hydrophilic macromolecules. In this method, the surfactant is dissolved in an organic solution of ether and chloroform and an aqueous phase containing the active agent is added to the mixture. The resulting immiscible phases are homogenised at 4–5 °C and the organic solvent evaporated at 40 °C under low pressure leading to the formation of a niosome suspension [7, 9].

8.3.3.4   Trans-membrane pH gradient active agent uptake process

This method is an adaptation of the TFH method and results in the formation of MLVs with a more acidic interior than the exterior. This acidic pH is advantageous in entrapping the active ingredient through ionisation and preventing leakage via the niosome membrane. It involves the formation of a thin lipid film achieved by dissolving surfactant and cholesterol in chloroform. This solution is then evaporated under reduced pressure to obtain a thin film on the wall of the round bottom flask. Citric acid (pH 4) is used to hydrate the film followed by three freeze-thaw cycles and sonication. The active agent is then added to the niosomal suspension as an aqueous solution with vortexing; the pH of this solution is adjusted to 7 and heated at 60 °C, producing niosomes [4, 9, 25].

8.3.3.5   Emulsion method

This is a simple method which involves addition of a mixture of surfactant and cholesterol in an organic solvent, to an aqueous phase containing the active agent to form an oil-in-water (o/w) emulsion. Subsequent evaporation of the organic phase results in the formation of a niosomal suspension in an aqueous medium.

8.3.3.6   Bubble method

The bubble method is a novel technique that is often described as a single-step technique that forms LUVs with average vesicle size range of 0.2–0.5 µm [24]. It involves the addition of surfactants, additives and buffer solution at 70 °C into a glass flask with three necks. Cooled water reflux is placed in the first neck, a thermometer is placed in second neck and nitrogen gas is passed through the third neck.

The dispersion is homogenised for 15 seconds and the flask immediately placed in a water bath and nitrogen gas passed through the solution [4, 6, 9].

8.3.3.7   Heating method

This is another method that does not require the use of organic solvents and is described as a one-step, eco-friendly, scalable and non-toxic technique. It involves hydration of surfactant, cholesterol and other additives in an aqueous phase such as buffer or distilled water. The hydration is usually carried out in presence of a polyol such as glycerol. Cholesterol is first hydrated in the aqueous phase at 120 °C to ensure proper dissolution before adding the surfactant and other additives and the resulting mixture is heated with continuous stirring (at low shear forces) until the vesicles are formed [1, 4].

8.3.3.8   Microfluidisation method

This is a more recently introduced method and it is used in the preparation of unilamellar vesicles of defined size distribution [7]. Generally, the method is rapid and involves the use of microchannels to provide a controlled mixing of surfactant-cholesterol solution and an aqueous solution [9]. The method is based on the submerged jet principle in which two fluidised streams interact at ultra-high velocities (100 mL/min), in precisely defined micro channels within the interaction chamber. The arrangement of impingement of thin liquid sheet along a common front is done such that the energy supplied to the system remains within the area where niosomes are formed. Advantages of this method includes greater uniformity within the niosomes, smaller size and better reproducibility [6].

8.3.3.9   Supercritical reverse phase evaporation method

This is another method that does not require the use of organic solvents and can be scaled up easily to form larger number of niosomes. It results in the formation of LUVs with sizes ranging from 100 to 500 nm. However, sonication or extrusion can be employed alongside to produce smaller niosomes [1]. Surfactants and other additives are added with buffer and ethanol into a view cell. CO_2 gas is then introduced into the cell and the mixture is magnetically stirred until equilibrium is reached after which the pressure is released and niosomal dispersion is obtained [4, 9].

8.3.4   CHARACTERISATION OF NIOSOMES

The quality of niosomes produced can be determined by evaluation of niosomal parameters such as size, size distribution, zeta potential, morphology, entrapment efficiency and release behaviour. Evaluation of these properties also help in examining the functionality of the niosomes as these parameters directly impact stability and *in vivo* performance. A summary of characterisation parameters for niosomes is provided in Table 8.3.

TABLE 8.3

Methods for characterising niosomal formulations to determine their functionality and efficacy [9, 15, 20]

Parameter	Function	Technique used
Particle size and size distribution	Provides information on physical properties and stability of the formulation	– Optical Microscopy (depending on size) – Dynamic light scattering (DLS) or photon correlation spectroscopy: The polydispersity index (PDI) obtained from the DLS is an indication of the distribution of niosome size, with PDI value of less than 0.5 indicating a monodispersed sample – Electron microscopy (for solid samples) – Transmission electronic microscopy (for liquid samples)
Morphology	Gives information on the shape, roundness, smoothness of the vesicles and formation of aggregates.	– Transmission electronic microscopy, – Negative-staining transmission electronic microscopy, – Freeze-fracture transmission electronic microscopy – Atomic force microscopy – Scanning electron microscopy – Cryo-scanning electron microscopy
Surface Charge (Zeta Potential)	Provides essential information in determining the physical stability of niosomes; the magnitude of zeta potential provides an indication of the degree of electrostatic repulsion between two adjacent particles. Niosome with a zeta potential higher than +30 mV or lower than −30 mV is considered to have enough repulsion to prevent particle aggregation.	– Laser doppler anemometry
Entrapment Efficiency (EE)	Indicates the percentage fraction of agent successfully entrapped within the niosomes. $$\%EE = \frac{Amount\ of\ agent\ entrapped}{Total\ agent\ added} \times 100$$ The free ACI is separated from the niosome via dialysis, filtration or centrifugation before the vesicles are burst and content evaluated using any of the techniques.	– UV-spectrometer – High performance liquid chromatography – Fluorescence
Release Properties	Reveals the *in vitro* release mechanism of the agent from the niosome.	– Dialysis – Franz diffusion cells

(Continued)

TABLE 8.3 (Continued)

Parameter	Function	Technique used
Bilayer characterisation	Shows if the niosome bilayer has either single layer (unilamellar) or multiple layers (multilamellar).	– Fluorescence polarisation and Fluorescence anisotropy: small angle X-ray scattering (SAXS), nuclear magnetic resonance spectroscopy (NMR) and AFM.
In vitro release/ permeation characteristics	Assesses the transport or diffusion of active ingredient out of the niosome and across relevant membranes	– Modified USP apparatus dissolution tests – Franz cell experiments using dialysis membranes or excised animal mucosa – Transport studies across cultured epithelial cells grown on transwell inserts
Biological toxicity	Profiles the toxicity of niosomes to relevant biological systems such as cultured cells or excised animal skin.	– 3-(4,5-dimethylthiazol-2-yl)-2,5-diphenyltetrazolium bromide) tetrazolium (MTT) assay – Egg yolk toxicity assay

8.4 APPLICATION OF NIOSOMES IN COSMETICS

In recent times, niosomes have gained research focus as alternatives to liposomes due to the ready availability of starting materials and a simpler production process [25]. Additionally, the higher stability of the surfactants resulting in good storage, reduced purity problems and low manufacturing cost, make them an attractive choice in cosmetics [11]. Considering topical cosmetics, the uniqueness of niosomes compared to other conventional delivery systems is in their similarity to the phospholipid bilayer of cell membranes which facilitates ACI transport to the site of action. Naturally, because of low penetration rate across the skin, conventional cosmetic forms like creams and ointments deliver substances in a concentration-dependent manner, however, niosomes have the ability to deliver substances at a higher concentration due to their unique structure [8]. Moisture content is another factor that varies across different skin layers. The stratum corneum consists of corneocytes embedded in a lipid matrix with moisture content 10–15%; while the viable epidermis has moisture content of 60%. This affects the delivery of substances across the skin resulting in the localisation of most products on the upper layer of the skin. However, the versatile nature of niosomes (hydrophilicity, hydrophobicity and amphiphilicity) is superior in driving cosmetic substances across skin layers when compared to conventional products [26]. Additionally, niosomes have the potential to improve the effectiveness of cosmetic products by improving the stability of the entrapped active ingredient, enhancing the bioavailability of poorly absorbed agents and increasing skin penetration. Generally, lipid-based delivery systems could significantly improve the effectiveness of cosmetics by improving the physiochemical stability of the skin system [11]. Compared to liposomes, another lipid-based system, niosomes are better preferred due to the higher chemical stability of non-ionic surfactants compared to liposome's phospholipids; low manufacturing cost and lesser purity problems [25]. Hence niosomes present a cheaper and simpler alternative for cosmetic delivery.

Other advantages of niosomes as cosmetic agents include:

1. Niosomes enhance substance accumulation at the site of administration and penetration into the lipid layers of the stratum corneum and epidermis facilitating dermal delivery [8].
2. Niosomes improve the stability and decrease the degradation of entrapped compounds including phytochemicals from environmental hazards [12].
3. Niosomes serve as solubilising agents (within the lipid bilayer) for the solubilisation of poorly water soluble substances like steroids resulting in increased ACI concentration [8].
4. Niosomes offer the possibility of surface functionalisation to allowing specific targeting into cellular and subcellular regions [11].
5. The versatile nature of the niosomes, having both hydrophilic and hydrophobic portions, is advantageous in delivering cosmetic substances across skin layers.
6. Niosomes improve the benefit/risk ratio of cosmetics due to their reduced toxicity [11].

8.4.1 POSSIBLE MECHANISMS FOR NIOSOMAL DERMAL/TRANSDERMAL DELIVERY

The skin which is the largest organ of the human body plays a vital role in the delivery of active cosmetic ingredients (ACIs) – the major problems with cosmetics being percutaneous penetration and permeation and also avoidance of absorption into the bloodstream [27]. Niosomes have enormous potential to achieve the major goal of an ideal ACI delivery system: improving delivery of ACI into the skin with greater stability, requiring lower quantities of the active agent and producing minor toxicological concerns [27, 28]. Niosomes thus provide an alternative way to improve dermatological therapy as their structure enables the transport of ACI through distinct skin layers. The surfactant molecules of niosomes greatly enhance skin permeation by modifying the stratum corneum lipid structure while the other components like cholesterol or PEG increase stability and mechanical properties [28].

Some mechanisms have been proposed to explain the penetration-enhancing effects of niosomes which are invaluable in the delivery of cosmetic actives [24]. These mechanisms are discussed below and depicted in Figure 8.2.

1. The adsorption and fusion of active-loaded niosomal vesicles onto the surface of the skin leading to a high thermodynamic activity gradient of the active ingredient at the surface of vesicles and stratum corneum, which acts as a driving force for noisome penetration across the stratum corneum.

2. A disruption of the densely packed lipids that fill extracellular spaces of the stratum corneum enhances the permeability of actives through the structural modification of the stratum corneum.

3. The non-ionic surfactants may play a crucial role in improving penetration by acting as penetration enhancers. Wherein vesicle bilayers enter the stratum corneum with subsequent modification of the intercellular lipids, which increases overall membrane fluidity.

4. Niosomes cause an alteration in the stratum corneum properties through a reduction in the trans-epidermal water loss, thus leading to an increase in subcutaneous hydration with the loosening of its closely packed cellular structure.

5. Transport of intact niosomes via the sebaceous glands or hair follicles of the skin.

8.4.2 NIOSOMAL DELIVERY FOR TREATMENT OF COMMON HAIR/SKIN CONDITIONS

Most cosmetic products commercially available are aimed at the treatment or the significant reduction of symptoms of skin conditions, howbeit transiently. Consequently, the popularity of beauty brands and their products are hinged on the efficacy of such products. Nanoparticulate delivery systems such as liposomes, nanovesicles, micelles etc. have played a major role in distinguishing the major players in the cosmetic industry from generic brands. This section will highlight some research which has investigated the use of niosomes in the treatment of common haircare and skincare concerns.

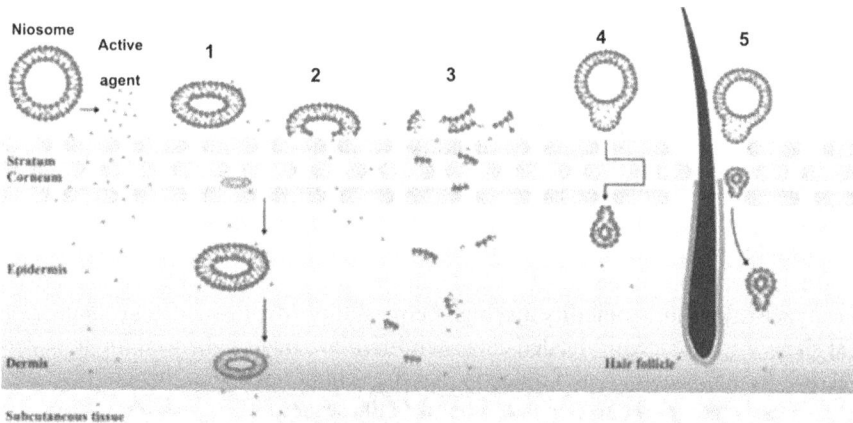

FIGURE 8.2 Possible mechanisms of action of niosomes for dermal and transdermal delivery. 1- active molecules released by a niosome; 2- adsorption of niosome and fusion with the stratum corneum; 3- intact niosome/elastic penetration through intact stratum corneum; 4- components of niosomes act as penetration enhancers to enhance active agent absorption; 5- niosome penetration through hair follicles or pilosebaceous units (adapted from Chen et al. [9])

8.4.2.1 Acne

Acne vulgaris is a chronic skin condition that affects the pilosebaceous unit – the hair follicle and sebaceous gland. It is caused by increased sebum production, inflammation, abnormalities in keratinisation, irregular microflora on the skin and a proliferation of the bacterium, *Propionibacterium acne*. Various cosmetic products formulated for acne-prone skin generally combat one or more of such symptoms. Antimicrobial agents remain a popular treatment for acne symptoms. Mohammadi *et al.* [29] reported that niosomal preparations of gold standard antimicrobials such as benzoyl peroxide and clindamycin showed higher efficacy in the treatment of acne lesions in comparison to conventional antimicrobial lotions; although, the results were statistically insignificant. However, dapsone-loaded niosomes showed noticeable clinical improvement after two weeks and remarkably significant improvement of acne lesions after eight weeks of treatment [30]. Novel niosomal formulations, loaded with rosmarinic acid, a naturally occurring ester of caffeic acid, have also shown statistically significant anti-microbial activity against *Propionibacterium acne* and *Staphylococcus aureus*, in comparison to conventional benzoyl peroxide preparations [31].

8.4.2.2 Psoriasis

Psoriasis refers to a chronic, immune-mediated inflammatory skin disorder, characterised by red coloured plaques on the skin, with a flaky, dry surface. An increased incidence of psoriasis presents as large plaque psoriasis. The plaques are characterised by a greatly thickened, horny layer (hyperkeratosis), and much of the pain, itching, and inflammatory change is due to the dryness and cracking of this layer. Psoriasis is considered an autoimmune condition, with a strong genetic background, and is dependent on environmental factors [23, 32]. Cosmetic products formulated for psoriatic symptoms, inflamed or extremely dry skin aim to offer soothing, anti-inflammatory or anti-irritant properties which may transiently alleviate such symptoms. Available treatments target certain pathways involved in the pathophysiology of the disorder. Diacerein, an anti-inflammatory, analgesic and antipyretic drug has been reported to inhibit the production of interleukin-1β by human monocytes *in vitro* and down regulates tumor necrosis factor-α and interleukin-12. Studies have demonstrated that interleukin-1, tumor necrosis factor-α and interleukin-12 are important mediators in the initiation and maintenance of psoriatic plaques. However, diacerein exhibits low solubility and high permeability. Moghassemi and Hadjizadeh [23] showed that diacerein-loaded cholesterol-rich niosomes enhanced the transdermal and dermal penetration of diacerin into the skin of *in vivo* rat models. Celastrol is a pentacyclic triterpene isolated from a Chinese herb and used mainly for the treatment of psoriasis due to its antioxidant, anti-inflammatory and antitumour activity. When formulated as niosomes, celastrol exhibited improved *in vitro* permeation compared to the neat (unencapsulated) agent. Meng *et al.* [32] reported that celastrol niosomes effectively alleviated erythema and scaling on the dorsal skin of psoriasis mouse models. Levels of cytokines, including IL-22, IL-23 and IL-17 decreased after the treatment, indicating the high therapeutic potential of this formulation for psoriasis. Similar results of enhanced permeation were observed when cyclosporin-loaded niosomes were investigated using *in vivo* and *ex vivo* rat models. In *ex vivo* permeability studies using rat skin, the niosomes showed high permeation (50.57% in 24 hours) in comparison to the cyclosporine suspension (10.13% in 24 hours) as reported by Pandey *et al.* [33].

8.4.2.3   Seborrheic dermatitis and dandruff

Seborrheic dermatitis (SD) and dandruff are a continuous spectrum of the same disease that affects the seborrheic areas of the body. Dandruff can be considered a mild form of SD, restricted to the scalp, and involves itchy, flaking skin without visible inflammation [34]. SD can affect the scalp as well as other seborrheic areas and involves itchy and flaking or scaling skin, with inflammation. An important pathogenic factor of SD is a *Malassezia* (fungal) infection. Its role in SD has been supported by evidence of positive correlation between yeast density on the skin and the severity of SD, as well as a high therapeutic efficacy of antifungal agents in SD. Consequently, anti-fungal and keratolytic agents such as ketonazole, bifonazole, miconazole, ciclopiroxolamine, hydrocortisone, metronidazole, etc., have proven to be effective treatments in the management of SD and dandruff. Borda and Wikramanayake [35] provide a comprehensive summary of their mechanisms of action and available formulations while Thomas and Khasraghi [36] have extensively explored and summarised studies on niosomal formulations incorporating the above antifungal agents, keratolytic agents, corticosteroids, etc. which offer enhanced permeation and bioavailability for the treatment of SD and dandruff.

8.4.3   COMMERCIAL AND PATENTED COSMETIC PRODUCTS CONTAINING NIOSOMES

The first patented cosmetic brand that employed niosomal technology was produced by L'Oréal. However, the first cosmetic product that launched into the market using niosomal technology was produced by Lancôme in 1987. The product carried the brand name NIOSOME®. An antiaging cream with a brand name 'Niosome Plus' was also developed and marketed by Lancôme, with a promise of better penetration of the skin and hence maximising the effectiveness of skin aging preparations [10, 37, 38]. Since then, various commercialised niosomal products have been developed for cosmetic use and are highlighted in Table 8.4; while Table 8.5 provides a summary of patents granted for cosmetic applications imbibing niosomal technology as a simple, safe and scalable technology.

8.4.4   RECENT ADVANCES IN THE APPLICATION OF NIOSOMES IN COSMETICS

8.4.4.1   Niosomal cosmeceuticals containing natural compounds

In recent times, the use of niosomes for formulating cosmeceuticals containing natural compounds has been explored to overcome their low stability to air and light which limit their cosmetic efficacy. Recent advances in cosmetic research have spotlighted bioactive ingredients extracted from plant materials in a bid to promote the trending "green beauty" movement. Such active ingredients exhibit proven efficacy (*in vitro/ in vivo*/Phase I clinicial trials) in combating a skin concern and have been further

TABLE 8.4

List of commercially available cosmetic products employing niosomal technology, classified by manufacturer and cosmetic application [9, 24]

Company/proprietor	Brand/product name	Application
Anne Moller (Barcelona, Spain)	Anne Moller anti-fatigue eye contour roll-on, with Niosome Elastic complex	Anti-puffiness and moisturising roll-on
Britney Spears	Britney Spears- Curious	Lip gloss
Britney Spears	Britney Spears- Curious	Body souffle
Elene	Elene- Eyecare	Day and Night eye programme
Estee Lauder	Beyond Paradise	Aftershave
Eusu (Bangkok, Thailand)	Eusu noisome makam pom whitening facial cream	Whitening cream
Gatineau	Gatineau-Moderactive Cleanser	Make-up remover
Givenchy	Blanc Parfait	Brightening spot corrector SP45
Givenchy	Givenchy Amariage	Eau de Toilette spray
Guinot	Guinot Night care	Whitening serum
Guinot	Guinot cleanser	Gentle face exfoliating cream
Hugo Boss	Hugo Boss- Boss Soul	Aftershave
Identik (Paris, France)	Identik masque floral repaire	Hair repair masque
Identik (Paris, France)	Identik Shampoo floral repair	Hair repair shampoo
Jean Paul Gaultier	Jean Paul Gaultier	Eau de Toilette spray
L'Oreal (Paris)	Prototype 37-C Lancome	Anti-aging cream
Lancaster	Suractif	Night Cream
Lancome	Niosome plus	Foundation, clear and balance skin tone
Lancome	Foundations and Complexions	Concealer
Lancome	Niosome plus perfected age treatment	Anti-wrinkle cream
Laon Cosmetics (Seoul, Korea)	Mayu niosome base cream	Whitening and hydrating cream
Liz Claiborne	Realities	Shower gel
Loriz Azzaro	Loriz Azzaro- Chrome	Eau de Toilette
Nina Ricchi	Love in Paris	Deodorant spray
Nouvelle-HAS Cosmetics (Italy)	Anti-age response cream	Anti-wrinkle cream
Orlane	Orlane- Lip colours and Lipsticks	Lip gloss
Shiseido	Shiseido bio-performance night care	Clarifying essence
Simply Man Match	Anti-age Response Cream	Slows down ageing process
White Shoulders	White Shoulders	Eau de Cologne spray

investigated in niosomal research to increase skin bioactive penetration, bioavailability in the stratum corneum and generally improve their effects on the skin. For instance, rice bran is a rich source for phytochemicals with antioxidant activity with

TABLE 8.5
List of patents granted for cosmetic products employing niosomal technology

No.	Patent publication title	Publication number & reference	Publication date	Applicant/country	Summary
1.	Stabilized niosomes of herbal extracts and color cosmetic materials	1020020065958 [47]	14.08.2002	Coreana Cosmetics Co. Ltd., Korea	Niosomes formulation that stabilize extracts of oriental herbs and the color cosmetic materials. It provides enhanced moisture maintenance and skin penetration. The formulation demonstrates excellent characteristics of adherence, and spreadability
2.	Methods of isoflavone niosome and cosmetic compositions comprising thereof	1020170080178 [48]	10.07.2017	Pulmuone Co. Ltd., Korea	Formulation produced stable, and effective whitening and anti-aging skin product.
3.	Cosmetic and pharmaceutical compositions containing niosomes and a water-soluble polyamide, and a process for preparing these compositions	4830857 [49] 1273870 [50] 0000003713492 [51]	16.05.1989 11.09.1991 29.10.1987	L'Oreal, USA L'Oreal, Canada L'Oreal, Germany	Formulations with niosomes for cosmetics applications.
4.	Cosmetic compositions containing niosomes of Graviola extract	KR1020180021503 [52]	05.03.2018	Cosmos Co. Ltd., Korea	A formulation that stabilizes the Graviola extract using sodium palmitoyl sarcosinate, cetyl phosphate and macadamia seed oil.
5.	Process for the preparation of foams which can be used in the cosmetics and pharmaceutical field and foams obtained by this process	5171577 [53]	15.12.1992	L'Oreal, USA	The preparation of cosmetics or pharmaceutical foam by foaming with the aid of a propellant. Niosomes is employed to stabilize the formulation which contain one or more non-ionic lipid layers encapsulating an aqueous phase.

(Continued)

TABLE 8.5 (Continued)

No.	Patent publication title	Publication number & reference	Publication date	Applicant/country	Summary
6.	Vectorisation of dsRNA by cationic particles and topical use	20070003609 [54]	04.01.2007	L'Oreal, USA	Niosomes used for topical application to skin, mucus membrane or eye.
7.	Double-stranded RNA oligonucleotide that inhibits tyrosinase expression	20132555511 [55]	26.12.2013	L'Oreal, Japan	A new double-stranded RNA oligonucleotides employing niosomes and is used for cosmetic and/or pharmaceutical applications to reduce tyrosinase expression.
8.	Use of hypoxia-inducible factor-1-alpha activator as an active agent in a topical cosmetic composition to enhance adipocyte mass and to combat or prevent aging including fine lines and wrinkles, preferably nasolabial fold	2988601 [56]	04.10.2013	Lucas Meyer Cosmetics, France	Niosomes containing formulation to enhance adipocyte mass via topical application.
9.	Cosmetic composition containing surfactant derived from olive and niosome with stabilized oat protein	1020140141331 [57]	10.12.2014	Cosmecca Korea Co. Ltd., Korea	A niosome formulation that maximizes the cosmetic effect of the active ingredients such as skin elasticity enhancement, continuous skin moisturising effect, and permeation effect.
10.	Cosmetic composition containing niosome including tranexamic acid and supercritical fluid extract of Phellinus Linteus to prevent skin wrinkles and whiten skin	1020040102766 [58]	08.12.2004	Nadri Cosmetics Co. Ltd., Korea	A niosome formulation that increases collagen synthesis and skin elasticity which prevents skin wrinkles, reduces melanogenesis, and whiten the skin. Formula composition contains up to 15 wt.% niosome that includes up to 20.0 wt.% a mixture of tranexamic acid and the supercritical fluid extract of Phellinus Linteus in a mixing ratio of 1:0.1–10.0.

(Continued)

TABLE 8.5 (Continued)

No.	Patent publication title	Publication number & reference	Publication date	Applicant/country	Summary
11.	Cosmetic composition containing niosome in which inner skin extract of Castanea crenata and Terminalia Chebula extract is stabilized	1020020001913 [59]	09.01.2002	Coreana Cosmetics Co. Ltd., Korea	The formula enhances skin elasticity.
12.	Cosmetic composition for skin care containing niosome stabilising extracts of Phellodendron Amurense Rupr., Cornus Officinalis Sieb, Rhei Undulati Rhizoma, Prunus Armeniaca L., Angelicae Gigantis Radix, and Eugenia Caryophyllata Thunb.	1020020027760 [60]	15.04.2002	Coreana Cosmetics Co. Ltd., Korea	The use of niosomes enhances the efficacy of the herbal components.
13.	Niosome containing Phellinus Linteus extract and cosmetic composition containing the same.	1020020068154 [61]	27.08.2002	Coreana Cosmetics Co. Ltd., Korea	The formula provides excellent skin moisturising capacity, skin elasticity, stability and skin whitening effect. Formulations include skin lotion, cream, cleansing foam, cleansing water, massage cream.
14.	KR1020010094549 – Cosmetic skin care composition containing Phytosphingosine and Morus Alba l. Extract	1020010094549 [62]	01.11.2001	Coreana Cosmetics Co. Ltd., Korea	The formula has improved anti-shadow effect, reduced formation of wrinkle and enhance elasticity.

(Continued)

TABLE 8.5 (Continued)

No.	Patent publication title	Publication number & reference	Publication date	Applicant/country	Summary
15.	Niosome composition comprising cultured wild ginseng root extract and personal cosmetic or sanitary composition comprising same	WO/2019/143041 [63]	25.07.2019	Hankook Cosmetics Manufacturing Co. Ltd., Korea	The formulation is stable and can effectively deliver a cultured wild ginseng root extract into the skin.
16.	Niosome typed multilamellar vesicle(mlv) containing vitamin A, vitamin C, vitamin E, linoleic acid, amaranth oil, and meadowfoam seed oil, and cosmetic compositions containing the same	1020040111227 [64]	31.12.2004	Able C & C Co. Ltd., Korea	The formulation improves skin condition by maximising moisturisation and antioxidation effects.
17.	Method for producing niosome	102177196 [65]	11.11.2020	Kim In Young, Korea	A multi-layered lamellar niosomes formula is formed in a supercritical state containing a cosmetic ingredient, for enhanced transdermal absorption.
18.	Method for producing niosome, niosome produced thereby, and use thereof	1020180038236 [66]	16.04.2018	Cha Biotech Co. Ltd., Korea	A simple scalable nontoxic method for manufacturing the niosome since no organic solvent is used.
19.	Cosmetic composition containing ursolic acid and synthetic palmitoyl pentapeptide	1020010094550 [67]	01.11.2001	Coreana Cosmetics Co. Ltd., Korea	The formulation provides stabilisation for ingredients and is used to reduce wrinkles and improve skin texture
20.	Cosmetic composition comprising niosome containing nitrofatty acid	1016648770000 [68]	12.10.2016	KIPO, Korea	The formulation has an anti-aging property as it imparts gloss and elasticity to the skin.

(Continued)

TABLE 8.5 (Continued)

No.	Patent publication title	Publication number & reference	Publication date	Applicant/country	Summary
21.	Low irritating cosmetic composition comprising extract of Persicaria Thunbergii H Gross, Lawsonia Inermis and Odenlandia Diffusa.	1020050088769 [69]	07.09.2005	OBS CO., LTD. Korea	The formulation enhances permeability through the skin and provides excellent stimulus alleviation effect. It is produced in lotions and creams.
22.	Method for producing niosome containing idebenone or coenzyme Q10, and cosmetic composition comprising same	102007469 [70]	05.08.2019	KIPO. Korea	Due to poor solubility of idebenone or coenzyme Q10 encapsulation within a niosomes enhanced this property. A stable transparent emulsion with excellent skin transdermal absorption, that promotes skin elasticity.
23.	Cosmetic composition for skin-whitening comprising Ramulus Mori extract and Linoleic Acid as active ingredients	1020040052418 [71]	23.06.2004	Coreana Cosmetics Co., Ltd. Korea	Effective skin whitening formulation using niosome technology.
24.	Cosmetic composition containing nanoliposome with stabilized corn bran using nonionic surfactant and vegetable phospholipid	1020100107856 [72]	06.10.2010	Saindang Cosmetics Co., Ltd. Korea	A cosmetic formulation containing stabilized corn bran extract to provides excellent skin penetration effect.
25.	A cosmetic composition	PI 2014000446 [73]	19.08.2015	Universiti Teknologi Malaysia	Formulation containing a mixture of virgin coconut oil and niosome
26.	Curcumin proniosomal/niosomal formulation, method for its preparation and use thereof	1288/DEL/2012 [74]	31.08.2016	Kiran Yadav/Deepak Yadav, India	Niosomes containing formulation that improves the lipophilic properties of curcumin

great cosmetic potential however, its low stability makes formulation into cosmetic products problematic. With the potential of niosomes to improve stability, rice bran niosomes have been formulated to have an improved stability with an increased potential to be used as UV filters, radical scavengers, skin elasticity enhancers and anti-wrinkle agent. These rice bran niosomes were reported to provide an increased clinical anti-aging activity – improving skin hydration, pigmentation and skin lightening [39]. Encapsulation of curcuminoids found in turmeric (with antioxidant, anti-inflammatory and anti-cancer properties) in niosomes was reported to enhance skin penetration. An *in vitro* penetration study showed that niosomes significantly enhanced permeation of curcuminoids when compared with the control solution [9]. From a commercial point of view, niosomes widen the prospects of agents formerly considered unstable for cosmetic use and increase their market potential [40]. Some more of these plant-derived bio-active cosmetic agents formulated as niosomes are highlighted in Table 8.6.

TABLE 8.6

Plant derived active ingredients employed for treatment of skin conditions and formulated as niosomes to overcome current delivery barriers

Bio-active/ active ingredient	Biological/cosmetic properties	Conventional delivery problems	Research focus	References
Curcumin	Antioxidant Anti-inflammatory Anti-cancer Anti-nociceptive	Poor solubility	Enhanced permeation Reduced skin irritation	[75–77]
Ellagaic acid	Potent Antioxidant Efficient skin whitener and pigmentation suppressor	Poor solubility Low permeability	Improved distribution in stratum corneum	[78]
Gallic acid	Astringent Anti-aging Antioxidant	Photolabile Oxidative instability	Improved transdermal bioavailability	[79]
Papain	Exfoliating enzyme	Limited transdermal permeation	Enhanced transdermal absorption Scar treatment	[80]
Caffeine	Antioxidan Astringent Anti-inflammatory Cellulite treatment	–	Increased dermal absorption and bioavailability	[79, 81, 82]
St. John's wort	Antimicrobial Anti-inflammatory Antioxidant effects Wound healing	–	Incorporation of hydrophilic and lipophilic components Improved skin bioavailability for wound healing	[83]

(Continued)

TABLE 8.6 (Continued)

Bio-active/ active ingredient	Biological/cosmetic properties	Conventional delivery problems	Research focus	References
Alpha tocopherol	Gold standard antioxidant in skincare	–	Optimisation of alpha tocopherol loaded niosomes	[84]
Resveratrol	Antioxidant Anti-inflammatory Chemo preventive	Photolabile Low oxidative stability Poor water solubility Short biological half-life Rapid metabolism and elimination	Enhanced transdermal delivery	[85]
Rutin	Antioxidant Anti-inflammatory Improves skin dermal density and elasticity	Low solubility Low bioavailability		[86, 87]
Rice bran extract	Antioxidant Anti-aging Skin lightening	Photolabile Oxidative instability	Evaluation of anti-aging effects of rice bran niosomes	[39]
Black tea extract (comprising caffeine and gallic acid)	Antioxidant	Lipid insoluble	As sunscreen agent	[88]

8.4.4.2 Formulation of specialised niosomes for cosmetics

The development of new generations of niosomes for formulation of cosmeceutical products in practice has been introduced with numerous outstanding features such as better stability, deformability, elasticity and enhanced skin permeation [9]. These specialised niosomes (discussed in Section 8.2) have further increased the potential for use of niosomes in cosmetics. For instance, transfersomes, because of their deformable nature, enhance skin penetration/permeation and ACI deposition through the influence of the water gradient between the skin surface and viable epidermis. Also, the simplicity of the manufacturing process makes it easily adaptable for commercialisation. Transferosomes loaded with hyaluronic acid and epigallocatechin gallate, in addition to enhancing skin penetration and ACI deposition, were reported to minimise collagen degradation and lipid peroxidation [28]. Elastic vesicles are also another form of specialised niosomes which differ from conventional niosomes by their characteristic fluid membrane with high elasticity. These features allow these vesicles to squeeze themselves through the intercellular regions of the skin under the influence of the transdermal water gradient. The deformability index values for elastic niosomes were reported to be significantly greater than that of conventional niosomes, indicating the higher flexibility of these elastic vesicles, and usefulness in achieving better skin permeation of ACIs [41]. A study on gallic acid isolated from *Terminalla chebula* galls revealed that elastic niosomes were able to enhance chemical stability and skin penetration of gallic acid when compared to non-elastic niosomes and could be useful carriers for skin anti-aging molecules [40].

8.4.4.3   Formulation strategies for sustained release

The liquid nature of niosomal formulations limits topical application because the product may slide off the application site before sufficient permeation is achieved. Introduction of gelling agents into niosome dispersions has been utilised to overcome this challenge and extend the application of niosomes. These niosomal gels can provide a reservoir of active agents in the subcutaneous layer for sustained release, which leads to high accumulation of these agents in the dermis and epidermis [9]. Such formulations also promote skin penetration due to the occlusion effects from the gel formation which can enhance skin hydration and subsequently increase the absorption [9]. These advances in the formulation of niosomes have improved niosome applicability in cosmetics. Furthermore, researchers have explored the incorporation of niosomes into topical films for the delivery of active pharmaceutical ingredients, to further control drug delivery [9]. This is an area requiring further research within the cosmetic industry.

8.4.5   SAFETY CONSIDERATIONS AND REGULATORY REQUIREMENTS FOR NIOSOME CONTAINING COSMETIC PRODUCTS

The delivery of ACIs require an effective, controlled and safe means of reaching the target site within the skin. Hence current studies on cosmetic niosomes are focused on developing more reliable and adequate safety risk assessments to ensure the safety of both the niosomes and nanosised active in humans. A point of concern in the production of niosomes lays in the use of toxic solvents in the formulation process and the sometimes expensive time-consuming techniques required [9]. Therefore, a critical step in the manufacturing process for niosomes involves the complete removal of residual solvent(s) for methods such as TFH and ether injection. Inert gases, like nitrogen, have been employed to facilitate residual solvent removal. Furthermore, newer techniques for production of niosomes such as the bubble method, supercritical reverse phase evaporation method, micro fluidisation and heating methods avoid or limit the use of toxic solvents and improve the toxicological profile of niosomes. These newer techniques also have the advantage of being easily scaled for large-scale commercial production of cosmetic niosomes with some of them being a single-step process. Formulation of active pharmaceutical ingredients (APIs) as niosomes has been shown to reduce the toxicity of the API, compared to the drug solution. Alyami *et al.* [15] reported that HCE-2 corneal epithelial cells treated with formulations of pilocarpine-loaded niosomes produced by TFH method showed significantly higher cell viability following the MTT toxicity assay, when compared to cells treated with the neat pilocarpine solution. Marianecci *et al.* [20] reported better safety profiles from niosomal formulations of lopinavir over ethosomal formulations, following histopathological studies in male Wistar rats. The safety of elastic niosomes and niosomal formulations of ciprofloxacine have also been established using the human lung carcinoma cell line [20]. It has been corroborated that the neutral nature of the non-ionic surfactants employed in niosomes ensure reduced interaction with biological systems, when compared to ionic/amphoteric counterparts; thus less toxicity is expected with these dosage forms [24]. Finally, although various studies have investigated the toxicity of niosomal formulations on mammalian cells or membranes, not much has been done to determine the toxicity of non-ionic surfactants, their degradation products and the metabolic pathways involved in their journey through the human body [24]. However, research into optimising cosmetic niosomal formulations and improving their safety is continuously ongoing for effective application of these delivery systems in the cosmetic industry.

Regulations pertinent to cosmetic preparations (EC) No. 1223/2009 is the main regulatory framework for cosmetic products to be placed on the EU market. The updated regulations focus on safety and contain new rules for the use of nanomaterials in cosmetic products [42]. Although there are no specific articles that relate to niosomes in particular; however, niosomes are covered under the nanotechnology sections. The term "nanomaterial" as per Article 2 point "k" is defined as "an insoluble or biopersistant and intentionally manufactured material with one or more external dimensions, or an internal structure, on the scale from 1 to 100 nm" [43]. According to EU regulations (Article 13), for a cosmetic product containing nanomaterials, the responsible person must make an electronic application six months before placing the product on the market. This electronic document should include the following [43]:

- The identification and description of the nanomaterial which needs to be aligned to Annexes II-VI.
- Specifications including particle size, as well as physical and chemical properties.
- Quantity within the formulation.
- The reasonably foreseeable exposure conditions.
- Toxicology profile and safety data.

The European commission may consult the Scientific Committee on Consumer Safety (SCCS) to assess the safety of the product and request their opinion within six months; and this information is made public [34]. The Committee published guidelines titled: "Guidance on the Safety Assessment of Nanomaterials in Cosmetics" that can help the manufacturers of cosmetic products with nanomaterials in safety assessment [44]. Based on the opinion of the SCCS, the commission may request amendments of Annex II or Annex III. Annex II is related to prohibited materials in cosmetics whereas Annex III is the list of restricted materials (restriction in concentration, purity or applications) [45, 46]. Furthermore, regulations mandate that manufacturers add clearly to the ingredient list, any ingredients present in the form of nanomaterials by placing the name of the material followed by the word "nano" in brackets [Chapter VI Consumer Information, Article 19 (Labelling) Point "g"] [43]. These regulations assure the safety of formulated cosmetic agents containing nanomaterials before they are released to the EU public.

8.5 CONCLUSION

Niosomes, as vesicular nanoparticulate delivery systems, have found various applications in the cosmetic industry for the delivery of bioactive and plant-derived cosmetic agents. Being a more "digital" version of the liposomal delivery system, niosomes can overcome issues such as high cost and low stability linked to liposome manufacture and handling. Current research into the use of niosomes as nanocarriers for cosmetics involves simplification and easy upscaling of the manufacturing process, the formulation of specialised niosomes for effective targeting and continuous research into reducing toxicity and documenting the safety and metabolic pathways for the degradation products of these vesicles. Future studies will be expected to consolidate on the aforementioned, as well as incorporation of niosomes into other structures such as films or inserts and evaluating their application in cosmetics.

REFERENCES

1. Kaur, D. and Kumar, S., Niosomes: Present scenario and future aspects. *Journal of Drug Delivery Therapeutics*, 2018. **8**(5): p. 35–43.
2. Kazi, K.M., Mandal, A.S., Biswas, N., Guha, A., Chatterjee, S., Behera, M., *et al.*, Niosome: A future of targeted drug delivery systems. *Journal of Advanced Pharmaceutical Technology Research*, 2010. **1**(4): p. 374.
3. Abdelkader, H., Alani, A.W., and Alany, R.G., Recent advances in non-ionic surfactant vesicles (niosomes): Self-assembly, fabrication, characterization, drug delivery applications and limitations. *Drug Delivery*, 2014. **21**(2): p. 87–100.
4. Ag Seleci, D., Seleci, M., Walter, J.-G., Stahl, F., and Scheper, T., Niosomes as nanoparticular drug carriers: Fundamentals and recent applications. *Journal of Nanomaterials*, 2016. **2016**: p. 1–13.
5. Kaul, S., Gulati, N., Verma, D., Mukherjee, S., and Nagaich, U., Role of nanotechnology in cosmeceuticals: A review of recent advances. *Journal of Pharmaceutics*, 2018. **2018**: p. 1–19.
6. Kauslya, A., Borawake, P.D., Shinde, J.V., and Chavan, R.S., Niosomes: A novel carrier drug delivery system. *Journal of Drug Delivery Therapeutics*, 2021. **11**(1): p. 162–170.
7. Madhav, N. and Saini, A., Niosomes: A novel drug delivery system. *International Journal of Research in Pharmacy Chemistry*, 2011. **1**(3): p. 498–511.
8. Betz, G., Aeppli, A., Menshutina, N., and Leuenberger, H., In vivo comparison of various liposome formulations for cosmetic application. *International Journal of Pharmaceutics*, 2005. **296**(1–2): p. 44–54.
9. Chen, S., Hanning, S., Falconer, J., Locke, M., and Wen, J., Recent advances in non-ionic surfactant vesicles (niosomes): Fabrication, characterization, pharmaceutical and cosmetic applications. *European Journal of Pharmaceutics Biopharmaceutics*, 2019. **144**: p. 18–39.
10. Singh, P., Ansari, H., and Dabre, S., Niosomes-a novel tool for anti-ageing cosmeceuticals. *Journal of Pharmaceutical Research*, 2016. **6**(10): p. 6691–6703.
11. Khezri, K., Saeedi, M., and Dizaj, S.M., Application of nanoparticles in percutaneous delivery of active ingredients in cosmetic preparations. *Biomedicine Pharmacotherapy*, 2018. **106**: p. 1499–1505.
12. Atif, M., Singh, S.P., and Kumar, A., Niosomes: A novel formulation for anti-ageing cosmeceuticals. *Journal of Drug Delivery Science Technology*, 2018. **5**(4): p. 370–375.
13. Shahiwala, A. and Misra, A., Studies in topical application of niosomally entrapped nimesulide. *Journal of Pharmacy and Pharmaceutical Sciences*, 2002. **5**(3): p. 220–225.

14. Israelachvili, J.N., *Intermolecular and surface forces*. 2011: Academic Press.
15. Alyami, H., Abdelaziz, K., Dahmash, E.Z., and Iyire, A., Nonionic surfactant vesicles (niosomes) for ocular drug delivery: Development, evaluation and toxicological profiling. *Journal of Drug Delivery Science Technology*, 2020. **60**: p. 102069.
16. Ge, X., Wei, M., He, S., and Yuan, W.-E., Advances of non-ionic surfactant vesicles (niosomes) and their application in drug delivery. *Pharmaceutics*, 2019. **11**(2): p. 55.
17. Uchegbu, I.F. and Vyas, S.P., Non-ionic surfactant based vesicles (niosomes) in drug delivery. *International Journal of Pharmaceutics*, 1998. **172**(1–2): p. 33–70.
18. Gharbavi, M., Amani, J., Kheiri-Manjili, H., Danafar, H., and Sharafi, A., Niosome: A promising nanocarrier for natural drug delivery through blood-brain barrier. *Advances in Pharmacological Sciences*, 2018. **2018**: p. 1–15.
19. Durak, S., Esmaeili Rad, M., Alp Yetisgin, A., Eda Sutova, H., Kutlu, O., Cetinel, S., *et al.*, Niosomal drug delivery systems for ocular disease—recent advances and future prospects. *Nanomaterials*, 2020. **10**(6): p. 1191.
20. Marianecci, C., Di Marzio, L., Rinaldi, F., Celia, C., Paolino, D., Alhaique, F., *et al.*, Niosomes from 80s to present: The state of the art. *Advances in Colloid Interface Science*, 2014. **205**: p. 187–206.
21. Shilakari Asthana, G., Sharma, P.K., and Asthana, A., In vitro and in vivo evaluation of niosomal formulation for controlled delivery of clarithromycin. *Scientifica*, 2016. **2016**: p. 1–10.
22. Kumar, G.P. and Rajeshwarrao, P., Nonionic surfactant vesicular systems for effective drug delivery—an overview. *Acta Pharmaceutica Sinica B*, 2011. **1**(4): p. 208–219.
23. Moghassemi, S. and Hadjizadeh, A., Nano-niosomes as nanoscale drug delivery systems: An illustrated review. *Journal of Controlled Release*, 2014. **185**: p. 22–36.
24. Masjedi, M. and Montahaei, T., An illustrated review on nonionic surfactant vesicles (niosomes) as an approach in modern drug delivery: Fabrication, characterization, pharmaceutical, and cosmetic applications. *Journal of Drug Delivery Science Technology*, 2021. **61**: p. 102234.
25. Yeo, P.L., Lim, C.L., Chye, S.M., Ling, A.K., and Koh, R.Y., Niosomes: A review of their structure, properties, methods of preparation, and medical applications. *Asian Biomedicine*, 2017. **11**(4): p. 301–314.
26. Schreier, H. and Bouwstra, J., Liposomes and niosomes as topical drug carriers: Dermal and transdermal drug delivery. *Journal of Controlled Release*, 1994. **30**(1): p. 1–15.
27. Hu, X. and He, H., A review of cosmetic skin delivery. *Journal of Cosmetic Dermatology*, 2021. **20**(7): p. 2020–2030.
28. Santos, A.C., Morais, F., Simões, A., Pereira, I., Sequeira, J.A., Pereira-Silva, M., *et al.*, Nanotechnology for the development of new cosmetic formulations. *Expert Opinion on Drug Delivery*, 2019. **16**(4): p. 313–330.
29. Mohammadi, S., Pardakhty, A., Khalili, M., Fathi, R., Rezaeizadeh, M., Farajzadeh, S., *et al.*, Niosomal benzoyl peroxide and clindamycin lotion versus niosomal clindamycin lotion in treatment of acne vulgaris: A randomized clinical trial. *Advanced Pharmaceutical Bulletin*, 2019. **9**(4): p. 578.
30. Al Sabaa, H., Mady, F.M., Hussein, A.K., Abdel-Wahab, H.M., and Ragaie, M.H., Dapsone in topical niosomes for treatment of acne vulgaris. *African Journal of Pharmacy Pharmacology*, 2018. **12**(18): p. 221–230.
31. Budhiraja, A. and Dhingra, G., Development and characterization of a novel antiacne niosomal gel of rosmarinic acid. *Drug Delivery*, 2015. **22**(6): p. 723–730.
32. Meng, S., Sun, L., Wang, L., Lin, Z., Liu, Z., Xi, L., *et al.*, Loading of water-insoluble celastrol into niosome hydrogels for improved topical permeation and anti-psoriasis activity. *Colloids Surfaces B: Biointerfaces*, 2019. **182**: p. 110352.

33. Pandey, S.S., Shah, K.M., Maulvi, F.A., Desai, D.T., Gupta, A.R., Joshi, S.V., *et al.*, Topical delivery of cyclosporine loaded tailored niosomal nanocarriers for improved skin penetration and deposition in psoriasis: Optimization, ex vivo and animal studies. *Journal of Drug Delivery Science Technology*, 2021. **63**: p. 102441.

34. Dessinioti, C. and Katsambas, A., Seborrheic dermatitis: Etiology, risk factors, and treatments: Facts and controversies. *Clinics in Dermatology*, 2013. **31**(4): p. 343–351.

35. Borda, L.J. and Wikramanayake, T.C., Seborrheic dermatitis and dandruff: A comprehensive review. *Journal of Clinical Investigative Dermatology*, 2015. **3**(2): p. 1–10.

36. Thomas, L.M. and Khasraghi, A.H., Nanotechnology-based topical drug delivery systems for management of dandruff and seborrheic dermatitis: An overview. *Iraqi Journal of Pharmaceutical Sciences*, 2020. **29**(1): p. 12–32.

37. Azeem, A., Anwer, M.K., and Talegaonkar, S., Niosomes in sustained and targeted drug delivery: Some recent advances. *Journal of Drug Targeting*, 2009. **17**(9): p. 671–689.

38. Ahmad, U., Ahmad, Z., Khan, A.A., Akhtar, J., Singh, S.P., and Ahmad, F.J., Strategies in development and delivery of nanotechnology based cosmetic products. *Drug Research*, 2018. **68**(10): p. 545–552.

39. Manosroi, A., Chutoprapat, R., Abe, M., Manosroi, W., and Manosroi, J., Anti-aging efficacy of topical formulations containing niosomes entrapped with rice bran bioactive compounds. *Pharmaceutical Biology*, 2012. **50**(2): p. 208–224.

40. Manosroi, A., Jantrawut, P., Akazawa, H., Akihisa, T., Manosroi, W., and Manosroi, J., Transdermal absorption enhancement of gel containing elastic niosomes loaded with gallic acid from terminalia chebula galls. *Pharmaceutical Biology*, 2011. **49**(6): p. 553–562.

41. Manosroi, A., Jantrawut, P., and Manosroi, J., Anti-inflammatory activity of gel containing novel elastic niosomes entrapped with diclofenac diethylammonium. *International Journal of Pharmaceutics*, 2008. **360**(1–2): p. 156–163.

42. European-Commission. Main legislation: Public consultation on the cosmetic products regulation. 2021; Accessed on: [23/03/2022]. Available from: https://ec.europa.eu/growth/sectors/cosmetics/legislation_en#:~:text=Regulation (EC) N° 1223,all operators in the sector.

43. European-Commission. Regulation (ec) no 1223/2009 of the european parliament and of the council on cosmetic products. 2019; Accessed on: [23/03/2022]. Available from: https://eur-lex.europa.eu/legal-content/EN/TXT/PDF/?uri=CELEX:02009R1223-20190813&from=EN.

44. Scientific-Committee-on-Consumer-Safety. Guidance on the safety assessment of nanomaterials in cosmetics. 2012; Accessed on: [23/03/2022]. Available from: https://ec.europa.eu/health/scientific_committees/consumer_safety/docs/sccs_s_005.pdf.

45. European-Commission. Annex ii: List of substances prohibited in cosmetic products. 2021; Accessed on: [23/03/2022]. Available from: https://ec.europa.eu/growth/tools-databases/cosing/pdf/COSING_Annex II_v2.pdf.

46. European-Commission. Annex iii: List of substances which cosmetic products must not contain except subject to the restrictions laid down. 2022; Accessed on: [24/03/2022]. Available from: https://ec.europa.eu/growth/tools-databases/cosing/pdf/COSING_Annex III_v2.pdf.

47. Cho, B.G., Kim, D.M., and Kim, S.R. Stabilized niosomes of herbal extracts and color cosmetic materials containing them. 2002; Database: [WIPO]. Accessed on: [22/03/2022]. Available from: https://patentscope.wipo.int/search/en/detail.jsf?docId=KR433712&_cid=P12-L15WFG-11367-1.

48. Kim, H.Y., Kim, T.S., Yeo, I.H., and Nam, S.W. Method for preparing isoflavone niosome and cosmetic composition comprising same. 2017; Database: [WIPO]. Accessed on: [23/3/2022]. Available from: https://patentscope.wipo.int/search/en/detail.jsf?docId=KR211049305&_cid=P12-L14GNO-40525-1.

49. Handjani, R.M., Ribier, A., Vanlerberghe, G., Zabotto, A., and Griat, J. Cosmetic and pharmaceutical compositions containing niosomes and a water-soluble polyamide, and a process for preparing these compositions. 1989; Database: [WIPO]. Accessed on: [23/3/2022]. Available from: https://patentscope.wipo.int/search/en/detail.jsf?docId=US37846979&_cid=P12-L14GNO-40525-1.

50. Handjani, R.-M., Ribier, A., and Vanlerberghe, G. Process for the preparation of a dispersion of niosomes in an aqueous phase and dispersion of niosomes produced in this way. 1987; Database: [WIPO]. Available from: https://patentscope.wipo.int/search/en/detail.jsf?docId=DE102657157&_cid=P12-L15WFG-11367-1.

51. Handjani, R.-M., Ribier, A., Vanlerberghe, G., Zabotto, A., and Griat, J. Composition with cosmetic or pharmaceutical use containing niosomes and at least one water-soluble polyamide and process for the preparation of the said composition. 1991; Database: [WIPO]. Available from: https://patentscope.wipo.int/search/en/detail.jsf?docId=CA93673640&_cid=P12-L15WFG-11367-1.

52. Lee, Y.J., Jang, K.H., Jeon, A.J., Lee, H.J., Kim, D.Y., Hong, I.K., *et al.* Cosmetic composition containing niosome stabilizing graviola extract. 2018; Database: [WIPO]. Accessed on: [23/03/2022]. Available from: https://patentscope.wipo.int/search/en/detail.jsf?docId=KR214752646&_cid=P12-L15WFG-11367-1.

53. Griat, J. and Ayache, L. Process for the preparation of foams which can be used in the cosmetics and pharmaceutical field and foams obtained by this process. 1992. Available from: https://patentscope.wipo.int/search/en/detail.jsf?docId=US38160768&_fid=CA94599647.

54. Collin-Djangone, C. and Simonnet, J.-T. Vectorization of dsrna by cationic particles and topical use. 2007; Database: [WIPO]. Accessed on: [23/03/2022]. Available from: https://patentscope.wipo.int/search/en/detail.jsf?docId=US41774137&_fid=EP14640627.

55. Collin-Djangone, C. and Simonnet, J.T. Double-stranded rna oligonucleotide that inhibits tyrosinase expression. 2013; Database: [WIPO]. Accessed on: [23/03/2022]. Available from: https://patentscope.wipo.int/search/en/detail.jsf?docId=JP273576615&_cid=P12-L15WFG-11367-1.

56. Bezivin, C. Use of hypoxia-inducible factor-1-alpha activator as an active agent in a topical cosmetic composition to enhance adipocyte mass and to combat or prevent aging including fine lines and wrinkles, preferably nasolabial fold. 2013; Database: [WIPO]. Accessed on: [23/03/2022]. Available from: https://patentscope.wipo.int/search/en/detail.jsf?docId=FR187794933&_cid=P12-L15WFG-11367-1.

57. Pack, S.K., An, H.C., Lee, J.J., Jeon, B.J., Kwon, Y.Y., and Jeon, S.Y. Cosmetic composition containing surfactant derived from olive and niosome with stabilized oat protein. 2014; Database: Accessed on: [23/03/2022]. Available from: https://patentscope.wipo.int/search/en/detail.jsf?docId=KR132663459&_cid=P12-L12MZJ-29239-1.

58. Hong, J.C. Cosmetic composition containing niosome including tranexamic acid and supercritical fluid extract of phellinus linteus to prevent skin wrinkles and whiten skin. 2004; Database: [WIPO]. Accessed on: [23/03/2022]. Available from: https://patentscope.wipo.int/search/en/detail.jsf?docId=KR660333&_cid=P12-L14GNO-40525-1.

59. Kim, B.J., Kim, G.J., Kim, J.H., and Lee, J.Y. Cosmetic composition containing niosome in which inner skin extract of castanea crenata and terminalia chebula extract is stabilized. 2002; Database: [WIPO]. Accessed on: [23/03/2022]. Available from: https://patentscope.wipo.int/search/en/detail.jsf?docId=KR369681&_cid=P12-L14GNO-40525-1.

60. Cho, B.G., Kim, G.J., Kim, J.H., Lee, J.Y., and Lee, S.H. Cosmetic composition for skin care containing niosome stabilizing extracts of phellodendron amurense rupr., cornus officinalis sieb, rhei undulati rhizoma, prunus armeniaca l., angelicae gigantis radix, and eugenia caryophyllata thunb. 2002; Database: [WIPO]. Accessed on: [23/03/2022]. Available from: https://patentscope.wipo.int/search/en/detail.jsf?docId=KR395525&_cid=P12-L14GNO-40525-1.

61. Kim, G.H. and Park, S.S. Niosome containing phellinus linteus extract and cosmetic composition containing the same. 2002; Database: [WIPO]. Accessed on: [23/03/2022]. Available from: https://patentscope.wipo.int/search/en/detail.jsf?docId=KR435908&_cid=P12-L14GNO-40525-1.

62. Cho, B.G., Kim, B.J., and Lee, G.S. Cosmetic skin care composition containing phytosphingosine and morus alba l. Extract 2001; Database: [WIPO]. Accessed on: [23/03/2022]. Available from: https://patentscope.wipo.int/search/en/detail.jsf?docId=KR360927&_cid=P12-L14GNO-40525-1.

63. Lee, G.Y., Lim, M.J., Park, S.J., and Kim, H.W. Niosome composition comprising cultured wild ginseng root extract and personal cosmetic or sanitary composition comprising same. 2019; Database: [WIPO]. Available from: https://patentscope.wipo.int/search/en/detail.jsf?docId=WO2019143041&_cid=P12-L14GNO-40525-1.

64. Kim, H.I., Lee, S.C., and Park, H.G. Niosome typed multilamellar vesicle(mlv) containing vitamin a, vitamin c, vitamin e, linoleic acid, amaranth oil, and meadowfoam seed oil, and cosmetic compositions containing the same. 2004; Database: [WIPO]. Accessed on: [23/03/2022]. Available from: https://patentscope.wipo.int/search/en/detail.jsf?docId=KR667639&_cid=P12-L14GNO-40525-1.

65. Young, K.I. Method for producing niosome. 2020; Database: [WIPO]. Accessed on: [23/03/2022]. Available from: https://patentscope.wipo.int/search/en/detail.jsf?docId=KR311156054&_cid=P12-L14GNO-40525-1.

66. Kim, A.R., Park, M.J., and Ban, E.M. Method for producing niosome, niosome produced thereby, and use thereof. 2018; Database: [WIPO]. Accessed on: [23/03/2022]. Available from: https://patentscope.wipo.int/search/en/detail.jsf?docId=KR217811090&_cid=P12-L14GNO-40525-1.

67. Cho, B.G., Kim, B.J., and Lee, J.Y. Cosmetic composition containing ursolic acid and synthetic palmitoyl pentapeptide. 2001; Database: [WIPO]. Accessed on: [23/03/2022]. Available from: https://patentscope.wipo.int/search/en/detail.jsf?docId=KR360928&_cid=P12-L14GNO-40525-1.

68. Kim, W.H., Kim, S.R., and Lee, K.K. Cosmetic composition comprising niosome containing nitrofatty acid. 2016; Database: [WIPO]. Accessed on: [23/03/2022]. Available from: https://patentscope.wipo.int/search/en/detail.jsf?docId=KR189707793&_cid=P12-L14GNO-40525-1.

69. Kim, J.H. Low irritating cosmetic composition comprising extract of persicaria thunbergii h gross, lawsonia inermis and odenlandia diffusa. 2005; Database: [WIPO]. Accessed on: [23/03/2022]. Available from: https://patentscope.wipo.int/search/en/detail.jsf?docId=KR756884&_cid=P12-L14GNO-40525-1.

70. Young, K.I. Method for producing niosome containing idebenone or coenzyme q10, and cosmetic composition comprising same. 2009; Database: [WIPO]. Available from: https://patentscope.wipo.int/search/en/detail.jsf?docId=KR250892853&_cid=P12-L14GNO-40525-1.

71. Cho, B.G., Jung, J.H., and Kim, S.U. Cosmetic composition for skin-whitening comprising ramulus mori extract and linoleic acid as active ingredients. 2004; Database: [WIPO]. Accessed on: [23/03/2022]. Available from: https://patentscope.wipo.int/search/en/detail.jsf?docId=KR610011&_cid=P12-L14GNO-40525-1.

72. Yoon, K.S., Jung, T.K., Im, K.R., and Choi, S.W. Cosmetic composition containing nanoliposome with stabilized corn bran using non-ionic surfactant and vegetable phospholipid. 2010; Database: [WIPO]. Accessed on: [23/03/2022]. Available from: https://patentscope.wipo.int/search/en/detail.jsf?docId=KR19741975&_cid=P12-L14GNO-40525-1.

73. Noor, N.M., Aziz, A.A., Sarmidi, M.R., Hamid, M.A., Aziz, R.A., Hisam, R.H., *et al.* A cosmetic composition. 2015; Database: [WIPO]. Accessed on: [23/03/2022]. Available from: https://patentscope.wipo.int/search/en/detail.jsf?docId=MY203063572&_cid=P12-L14GNO-40525-1.

74. Yadav, K., Yadav, D., Nanda, S., and Saroha, K. Curcumin proniosomal/niosomal formulation, method for its preparation and use thereof. 2016; Database: [WIPO]. Accessed on: [23/03/2022]. Available from: https://patentscope.wipo.int/search/en/ detail.jsf?docId=IN211774626&_cid=P12-L15WFG-11367-1.

75. Kumar, K. and Rai, A., Proniosomal formulation of curcumin having anti-inflammatory and anti-arthritic activity in different experimental animal models. *Die Pharmazie-An International Journal of Pharmaceutical Sciences*, 2012. **67**(10): p. 852–857.

76. Akbari, J., Saeedi, M., Enayatifard, R., Morteza-Semnani, K., Hashemi, S.M.H., Babaei, A., *et al.*, Curcumin niosomes (curcusomes) as an alternative to conventional vehicles: A potential for efficient dermal delivery. *Journal of Drug Delivery Science Technology*, 2020. **60**: p. 102035.

77. Sadeghi Ghadi, Z. and Ebrahimnejad, P., Curcumin entrapped hyaluronan containing niosomes: Preparation, characterisation and in vitro/in vivo evaluation. *Journal of Microencapsulation*, 2019. **36**(2): p. 169–179.

78. Junyaprasert, V.B., Singhsa, P., Suksiriworapong, J., and Chantasart, D., Physicochemical properties and skin permeation of span 60/tween 60 niosomes of ellagic acid. *International Journal of Pharmaceutics*, 2012. **423**(2): p. 303–311.

79. Teaima, M.H., Abdelhalim, S.A., El-Nabarawi, M.A., Attia, D.A., and Helal, D.A., Non-ionic surfactant based vesicular drug delivery system for topical delivery of caffeine for treatment of cellulite: Design, formulation, characterization, histological anti-cellulite activity, and pharmacokinetic evaluation. *Drug Development Industrial Pharmacy*, 2018. **44**(1): p. 158–171.

80. Manosroi, A., Chankhampan, C., Manosroi, W., and Manosroi, J., Transdermal absorption enhancement of papain loaded in elastic niosomes incorporated in gel for scar treatment. *European Journal of Pharmaceutical Sciences*, 2013. **48**(3): p. 474–483.

81. Khazaeli, P., Pardakhty, A., and Shoorabi, H.J.D.D., Caffeine-loaded niosomes: Characterization and in vitro release studies. *Drug Delivery*, 2007. **14**(7): p. 447–452.

82. Helal, D.A., Teaima, M.H., Abd El-Rhman, D., Abdel-Halim, S.A., and El-Nabaraw, M., Preparation and evaluation of niosomes containing an anticellulite drug. *Inventi Impact: Pharm Tech*, 2015. **2**: p. 95–101.

83. Ali, M., Abdel Motaal, A., Ahmed, M.A., Alsayari, A., and El-Gazayerly, O.N., An in vivo study of hypericum perforatum in a niosomal topical drug delivery system. *Drug Delivery*, 2018. **25**(1): p. 417–425.

84. Basiri, L., Rajabzadeh, G., and Bostan, A., A-*tocopherol-loaded niosome prepared by heating method and its release behavior*. *Food Chemistry*, 2017. **221**: p. 620–628.

85. Pando, D., Matos, M., Gutiérrez, G., and Pazos, C., Formulation of resveratrol entrapped niosomes for topical use. *Colloids Surfaces B: Biointerfaces*, 2015. **128**: p. 398–404.

86. Kamel, R., Basha, M., and Abd El-Alim, S.H., Development of a novel vesicular system using a binary mixture of sorbitan monostearate and polyethylene glycol fatty acid esters for rectal delivery of rutin. *Journal of Liposome Research*, 2013. **23**(1): p. 28–36.

87. Tran, T.H.Y., Hoang, T.H., and Vu, T.T.G., Preparation of nano niosomes loaded with rutin and aloe gel extract. *VNU Journal of Science: Medical Pharmaceutical Sciences*, 2020. **36**(1): p. 46–54.

88. Yeh, M.I., Huang, H.C., Liaw, J.H., Huang, M.C., Huang, K.F., and Hsu, F.L., Dermal delivery by niosomes of black tea extract as a sunscreen agent. *International Journal of Dermatology*, 2013. **52**(2): p. 239–245.

9 Nanoemulsions as drug delivery system in cosmetology
A recent update

*Kanika Verma, Akanksha Chaturvedi,
Swapnil Sharma and Sunil Kumar Dubey*

CONTENTS

DOI: 10.1201/9781003319146-9

9.1 INTRODUCTION

In recent decades, the cosmeceutical market has grown substantially following ameliorated consumer awareness for dermatologically beneficial products that promote disease prevention and good skin health. It is expected that the market of cosmetics could reach up to $463.5 billion by the year 2027, which suggests a compound annual growth rate (CAGR) of 5.3% from the year 2021 to 2027. The United States alone has the largest market for cosmetics with a revenue of 80.2 billion US dollars in the year 2021 followed by China with gross revenue of 51.7 billion US dollars and Japan (gross revenue of 37.8 billion US dollars) ("Cosmetics Industry – Statistics & Facts | Statista" 2022; "Cosmetics Market Size, Share, Industry Trends & Analysis 2021–2027" 2022). The application of cosmeceuticals has now expanded to anti-wrinkling, antiaging, tanning, and protection of skin, alongside other interesting uses such as in hair and nail care. Along with the increase in the market, the pressure on the manufacturers is also increasing to develop a glossy product, with less incorporation of surfactants, and smaller particle size. To enhance the properties of cosmetics, manufacturers are nowadays preferring nanotechnology. Nanotechnology is rapidly adopted by many manufacturers in the field of cosmetics, food, pharmaceutics, and agriculture. In cosmetics, nanotechnology is used to control the size and shape of particles at the scale of 1–100 nm (Dubey et al. 2022). The manufacturers use nanotechnology to reduce particle size, which enhances UV protection, and skin penetration, and improves the cosmetic texture. Nanoemulsion is a part of nanotechnology that provides ease in manufacturing cosmetics without the need for excess surfactants.

Nanoemulsions are lipid-based formulations that increase skin penetration of drugs containing amphiphilic, lipophilic, or hydrophilic components. Nanoemulsion is a colloidal dispersion of two immiscible liquids which are thermodynamically unstable. As nanoemulsions, reduce the droplet size of the formulation, the diameters of droplets in the formulations range from 10 to 200 nm. The nanoemulsion system comprises water, oil, and interfacial regions which stabilize the formulation. It provides better skin penetration and bioavailability by forming a thin lipid film on the skin (McClements 2012; Odile Sonneville-Aubrun, Yukuyama, and Pizzino 2018).

A nanoemulsion penetrates the skin through the stratum corneum and allows the drug to reach the dermis to exhibit its effect (Rapalli, Mahmood, et al. 2021). Stratum corneum is the major barrier in topical drug delivery, as compounds of size <500 Dalton can only pass the barrier. These compounds must also possess sufficient hydrophobicity to cross the membrane (Nastiti et al. 2017). Nanoemulsions make provision for penetration through the stratum corneum by reducing the size of compounds and by increasing the hydrophobicity. Skin provides three possible routes of penetration (Yukuyama et al. 2016; Shaker et al. 2019). This includes:

- Transfer of active constituents to cells via lipid bilayers and corneocytes – intracellular pathway
- Diffusion of active constituents through lipid layers around the corneocytes to cross Stratum corneum – intercellular pathway

- Penetration through hair follicles, sweat ducts, and sebaceous glands – follicular pathway

As nanoemulsions are lipid-based formulations, all the above-mentioned routes are favorable for skin penetration as shown in Figure 9.1. Nanoemulsions require less quantity of surfactants, which makes a non-irritant product. The formulation could be easily applied to the skin, and it penetrates even narrow gaps of skin such as hair scale and pilosebaceous follicles. It makes the formulation free from flocculation, creaming, coalescence, and sedimentation by reducing the droplet size. This small droplet size also provides hydration and moisture to the skin. Nanoemulsion retains its stability even when temperature varies or is exposed to water. The increased demand for skin care cosmetic-based nanoemulsions is because of their ability to mediate controlled delivery and also optimize the dispersion of the active ingredients into the desired layers of the skin (Sonneville-Aubrun, Simonnet, and L'Alloret 2004).

Nanoemulsions differ from emulsions in terms of size of the droplet, creaming rate, Brownian motion, interfacial area, and cost of manufacturing. Nanoemulsions reduce the creaming rate due to the small droplet size. Along with a low creaming rate, the associated Brownian motion of small droplets makes the emulsion homogenous and provides improved stability. Small droplets also expand the interfacial area, which makes possible faster release of the active substance from the oil droplets, increased penetration through the skin, and increased ease in spreading the formulation on the surface. Nanoemulsions also provide stability and therefore, it is preferred over emulsions to manufacture liquid or aerosol formulations. Natural origin compounds could also be incorporated into nanoemulsion to penetrate the skin with reduced droplet size (Sonneville-Aubrun, Simonnet, and L'Alloret 2004; Wiseva et al. 2021). Since the preparation of nanoemulsions holds significance to key industries, the present chapter aims to review the concept of nanoemulsion as a drug delivery system, its method of preparation and its associated applications in the cosmetic industry.

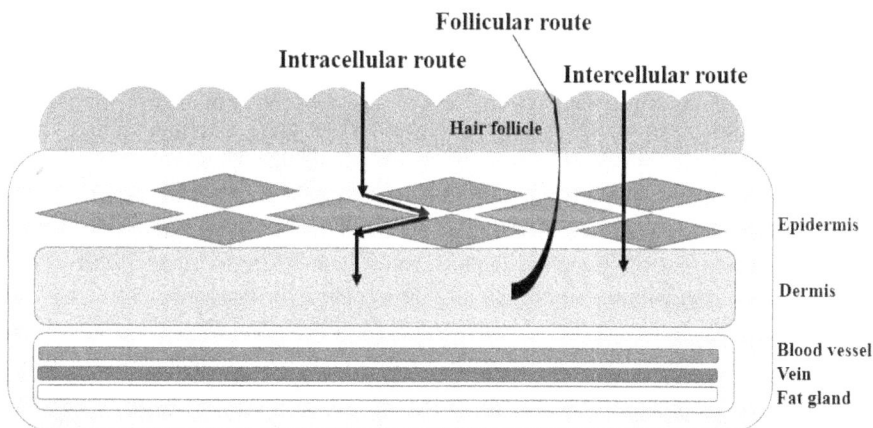

FIGURE 9.1 Transport pathways across human skin

9.2 METHODS OF PREPARATION FOR NANOEMULSIONS

The non-equilibrium nature of nanoemulsion depicts that the process requires an ample amount of energy from mechanical equipment and cannot be prepared spontaneously. For instance, an increased amount of mechanical energy is required for achieving droplet size to be small. The work required to modulate the interfacial area is represented by:

$$W: \text{Change in } A \times y$$

where W is work, A is the increase in the total interfacial area and y is the interfacial tension. This relationship advocates that a reduced amount of work is required when A is small (i.e. larger droplet size). Alternatively, nanoemulsions are also prepared by employing spontaneous techniques. The different techniques that are used to prepare nanoemulsions are suitably divided into high-energy and low-energy-based methods. High-energy techniques require much higher mechanical forces when compared with low-energy techniques. It is more prevalent in cosmeceutical industries to manufacture skin care products. It is generally used to manufacture o/w nanoemulsion. On the other hand, low-energy techniques are also getting attention while manufacturing cosmeceuticals to conserve energy. In general, the emulsification technique is preferred for the manufacturing of thermolabile active constituents. This technique includes phase inversion and stirring whereas the high-energy technique involves mechanical forces due to which droplets dissociate into nano-sized particles. Techniques for preparing nanoemulsions will be discussed in this section.

9.2.1 HIGH-ENERGY EMULSIFICATION METHODS

This technique involves the following methods that require high energy. It includes.

9.2.1.1 High-pressure homogenization

The method uses high-pressure homogenizers to break droplets into nano-sizes (up to 1 nm) using disruptive forces (Rai et al. 2018). The technique uses hydraulic shear, cavitation, and intense turbulence to reduce the size of coarse emulsion. It is important to maintain the time, temperature, and intensity of the homogenizer according to the sample (Qian and McClements 2011). Figure 9.2 shows that high intensity easily dissociates particles in small sizes. However, it should be noted that biopolymers must not be used as an emulsifier as it will make the particle bulky and therefore, small-sized molecules of surfactants should be selected. Coalescence is avoided by adding an excess of surfactants. The material must bear medium to low viscosity conditions to avoid the bulkiness of droplet size (more viscous compound, bulkier droplet). Nanoemulsion of omega-3 was formulated by Kabri et al. using a high-pressure homogenizer. It was prepared using oils such as miglyol, and salmon oil in oil phase, polyoxyethylene sorbitan monooleate (tween 80), and soy lecithin as surfactants, and deionized water (Kabri et al. 2011). Peppermint oil (Liang et al. 2012), quercetin (Karadag et al. 2013), and carotenoids (Salvia-Trujillo et al. 2014) were incorporated in the nanoemulsion prepared using a high-energy homogenization technique.

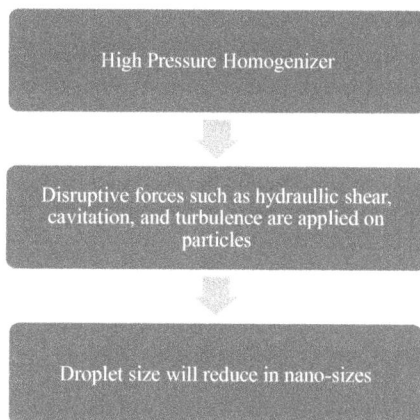

FIGURE 9.2 Flowchart of mechanism of high-pressure homogenizer

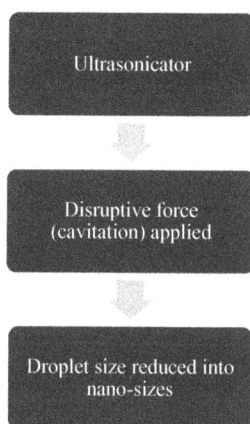

FIGURE 9.3 Flowchart of mechanism of ultrasonication

9.2.1.2 Ultrasonication

In this method, ultrasonicators are used to reduce droplet size by cavitation. Ultrasonic waves are applied to droplets and due to cavitation, particles are reduced into nanoparticles as shown in Figure 9.3. The particle size is decided by adjusting the intensity, time, and type of surfactant used in the ultrasound sonicator. Silva et al., designed a nanoemulsion of avocado oil incorporated with long chain of ethylene oxide as a surfactant. This nanoemulsion was prepared to act as a sun protectant. To the formulation, titanium dioxide and octyl methoxycinnamate were added to sunscreen to form SPF3 cream (Silva, Ricci-Júnior, and Mansur 2013). Eucalyptus oil, cinnamon oil, and capsaicin are also incorporated in nanoemulsion using ultrasonication.

9.2.1.3 Microfluidization

The microfluidizers use high pressure of around 500–20,000 psi. In this process, the fluids are allowed to pass through microchannels. These microchannels reduce the size of droplets by sending the emulsion to the interaction chamber. In this chamber, disruptive forces like cavitation, impact, and shear stress are applied to the molecules to reduce their sizes in nanometers as discussed in Figure 9.4. Microfluidizers reduce the particle size more efficiently than homogenizers. Fish oils, curcumin, D-limonene, essential oils, and beta-carotene are incorporated in nanoemulsions by using the microfluidization method (Kumar et al. 2019).

9.2.2 LOW-ENERGY EMULSIFICATION METHODS

The method employs the internal chemical energy of the system. The method requires gentle stirring to formulate the nanoemulsion. This technique involves the following methods for the preparation of nanoemulsion.

9.2.2.1 Phase inversion temperature (PIT) method

The PIT is a temperature-sensitive method that generally uses non-ionic surfactants such as polyethoxylated surfactants. Figures 9.5 and 9.6 depict that a mixture of oil, surfactant, and water all together at room temperature results in the formation of an oil-in-water (o/w) nanoemulsion. The temperature of the formulation is gradually increased to convert the surfactant into a lipophilic form. This helps the surfactant

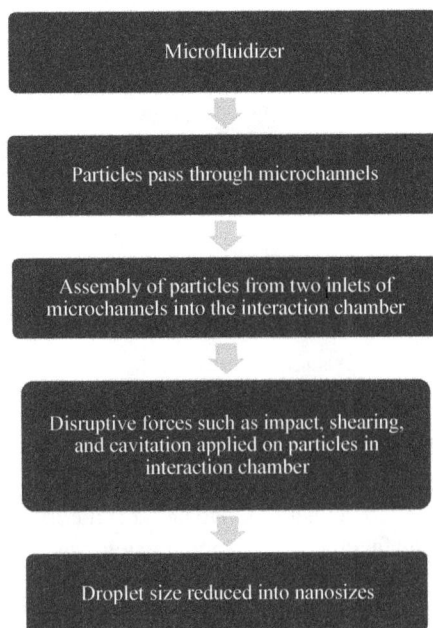

FIGURE 9.4 Flowchart of mechanism of micro fluidization

to easily mix in the oil phase. *Vellozia squamata* was formulated as hydroalcoholic extract and using the PIT technique, oil, and surfactant were heated together at 80 ± 2 °C, and water was also added at the same temperature to form a micelle (Quintão et al. 2013). Lemon oil, cinnamon oil, fisetin, mineral oil, and isohexadecane are formulated as nanoemulsions using PIT method (Kumar et al. 2019).

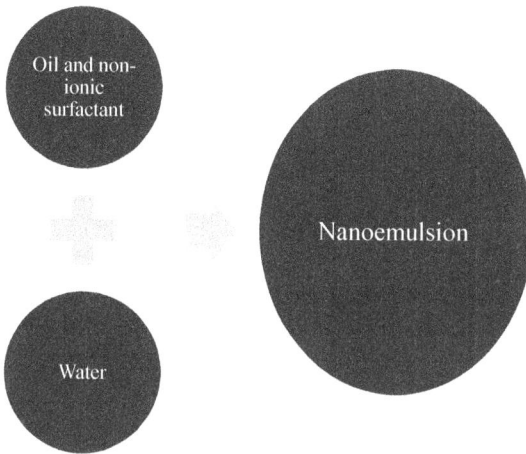

FIGURE 9.5   The PIT method for manufacturing nanoemulsion

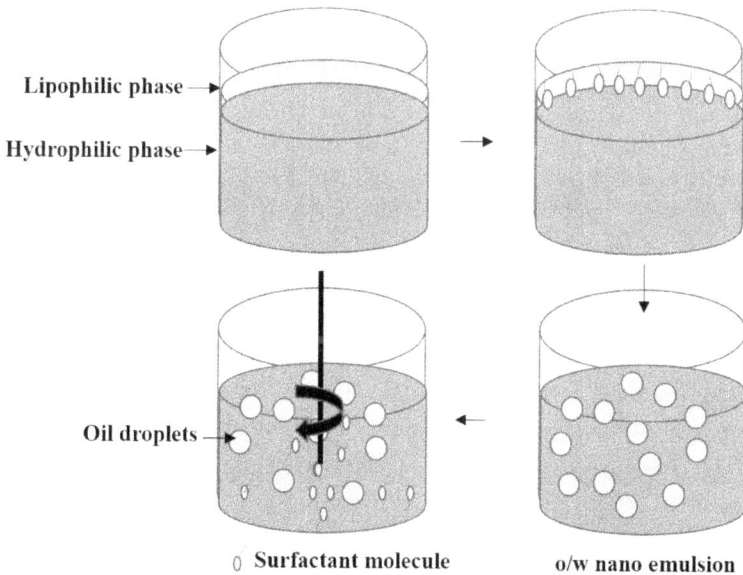

FIGURE 9.6   Production of an o/w nanoemulsion

9.2.2.2 Phase inversion composition (PIC) method

In the PIC method, phase inversion is changed by altering the composition of the formulation. For instance, in the oil phase, the incorporation of water causes phase inversion that results in the formation of nanoemulsion. The composition is manipulated until the affinity of surfactant towards oil and water reaches zero. These form a thermodynamically stable form of a microemulsion. Above the zero curvature, when the phase inversion is still performed, the emulsion becomes unstable and forms small particles which are kinetically stable. After achieving kinetic stability by the particles, any further change of composition does not affect the stable micelles. Sonneville-Aubrun et al., reported that a mixture of lidocaine and hydrogenated polyisobutene could be prepared using the PIC method (Sonneville-Aubrun et al. 2008). Hexadecane and oleic acid could also be mixed to form nanoemulsions using this method. In addition, paraffin o/w nanoemulsions are obtained using the PIC method. Palmitic ester of L-ascorbic acid and polysorbate 80 are also manufactured using this technique and size was reduced from 100 to 300 nm.

9.2.2.3 Spontaneous emulsification

This is also known as the self nano-emulsification method, another low-energy emulsification technique that requires a constant temperature to mix oil, surfactant, and water altogether to form an o/w nanoemulsion. In this process, two immiscible liquids are mixed without stirring, and heating to obtain a nanoemulsion. The solvent could be used with or without adding surfactants. The formulation prepared with the incorporation of oil, water, and water-soluble solvents and without adding surfactants, is known as the 'ouzo' effect. The oil part and hydrophilic surfactant are slowly added to the water phase to form a nanoemulsion (Solans, Morales, and Homs 2016). Vitamin E-incorporated nanoemulsions are prepared using spontaneous emulsification which showed a high impact on the droplet size of the oil phase.

Meanwhile, high-energy techniques are mostly used in cosmetology due to their capability to easily disrupt particles into small sizes, it also requires more energy. Because of high energy consumption, low-energy techniques are nowadays gathering more attention. The energy-saving nature of these techniques (except for PIT) is successfully processing micelle formations. This technique could also be employed on heat-sensitive compounds to which high-energy methods could not be applied. But high-energy techniques show faster emulsification than low-energy methods. It is also helpful in forming micelles of several kinds of compounds using high-energy processes. Therefore, both high-energy and low-energy techniques have their own advantages and drawbacks. Hence, it's the responsibility of a formulator to select the method, that would be optimum for the compound, and to obtain a final product.

9.3   HOW NANOEMULSIONS ARE DIFFERENT FROM EMULSIONS?

Although both emulsion and nanoemulsion are formulated by incorporating two immiscible liquids, they differ from each other in many aspects, whether in terms of appearance, efficacy, stability, methods of preparation, applications, etc. Emulsions are prepared by application of mechanical shear. The particle size of this formulation increases constantly in storage with time. This causes creaming in the emulsion which later results in cracking. This is the reason for this type of formulation to be considered thermodynamically unstable. On the other hand, nanoemulsions possess a small droplet size (in nanometers) due to which creaming of emulsion retards and the formulation retains stability (Sonneville-Aubrun, Simonnet, and L'Alloret 2004). These formulations also protect the compound against oxidation and hydrolysis.

Nanoemulsion differs from emulsion on the basis of advantages such as thermo-dynamic stability, reduction in the formation of flocculation, sedimentation, and phase inversion, and enhanced penetration and solubility of lipophilic drugs. The bioavailability and rate of absorption could also be increased using nanoemulsion and hence, this system could be used to formulate a controlled release and sustained drug delivery system. The nanoparticles of the formulation can even penetrate nar-row gaps in the skin which make it the best candidate to treat skin conditions such as xerosis and atopic dermatitis (Agero and Verallo-Rowell 2004; Evangelista, Abad-Casintahan, and Lopez-Villafuerte 2014; Karagounis et al. 2019; Chevalier and Bolzinger 2019).

The skin penetration by these droplets also helps to formulate a longer-acting skin moisturizer. The small size of droplets also develops a transparent and glossy formulation which attracts the buyers, whereas, an emulsion has a turbid appearance. Nanoemulsions could be easily scaled up due to spontaneous formation. In nano-emulsions, both hydrophilic and hydrophobic drugs could be easily encapsulated. It also has less cytotoxicity. Nanoemulsions are widely used in food, cosmetics, indus-tries, cancer treatment, diagnostic purposes, and therapeutic preparations (Gorain et al. 2020; Hejmady et al. 2020; Ashaolu 2021).

As every coin has two sides, nanoemulsion also comes with some disadvantages along with explicit advantages. One of the drawbacks is that its cost of manufacturing is quite expensive. As surfactant is a major component for the preparation of nano-emulsion, it must not be toxic. The formulation is influenced by temperature, pH, and other environmental parameters (Nimase, Patil, and Saudagar 2019).

Emulsion and nanoemulsion are also different based on their composition. Emulsions could be prepared by four ways; water-in-oil (w/o), oil-in-water (o/w), water-in-oil-in-water (w/o/w), and oil-in-water-in-oil (o/w/o). Contrarily, nanoemul-sions could be manufactured in the following ways, o/w, w/o, and bio-continuous methods (oil and water inter-dissociated in a system). The methods of preparation of these two systems are also unrelated as emulsions are manufactured using conven-tional procedures such as the dry gum method (emulsifier and oil are firstly triturated before adding water), wet gum method (oil and water triturated before adding an emulsifier), mechanical method (mechanically oil, water, and emulsifier are mixed all together), and *in situ* soap method (mixture of oil and calcium hydroxide solution (lime water) are stirred to form emulsion). Nanoemulsions are prepared using high-energy and low-energy processes. High-energy techniques include ultrasonication, high-pressure homogenization, and microfluidization, whereas, low energy tech-nique involves spontaneous emulsification methods, phase inversion composition, and temperature (Chevalier and Bolzinger 2019).

9.4   NANOEMULSION-BASED COSMETICS

Nanoemulsions are frequently used in cosmetology due to their ability to enhance stability, film formation, surface area, occlusiveness, uniform distribution, and the aesthetic look of cosmetics. Nanoemulsions have been reported to overcome the limitations such as sedimentation, flocculation, coalescence, and creaming associated with conventional cosmetics. Other than this, nanoemulsion makes the cosmetics glossier in appearance and provides better penetration and delivery of active constituents to the skin. Because of its multiple advantages, nanoemulsions are nowadays frequently opted for manufacturing of natural and synthetic cosmetic products (Wiseva et al. 2021).

9.4.1   Naturally Derived Nanoemulsion-Based Cosmetics

Due to the presence of side effects associated with synthetic cosmetics, many people around the globe are switching to natural products with minimal or no side effects. Plants contain compounds like flavonoids and polyphenols responsible for exhibiting antioxidant activity. Such plant-based nanoemulsions are used for the manufacturing of antiaging creams, hair products, etc. Nanoemulsions have reduced the size of droplets which increases their penetration across the skin (Aziz et al. 2016, 2019; Wiseva et al. 2021). In Tables 9.1 and 9.2 below, Hanifah M. et al. reported that plant extract of *Centella asiatica* when prepared as nanoemulsion lotion, helped to retard photoaging. It contains an active constituent asiaticoside, which is responsible for the synthesis of collagen type I in fibroblasts to heal stretch marks and allow improvement in wound healing in mice. Ethanolic extract of the plant is added in the oil phase (stearic acid, cetyl alcohol, olive oil, and tween-80), surfactant (propylene glycol and 96% ethyl alcohol), and water phase (isopropyl myristate, tetraethylammonium (TEA), methylparaben, propylene glycol, propylparaben, and aquademineralisata) to form nanoemulsion. In clinical trials (conducted on 20 female participants of age between 45 and 60 years) asiaticosides showed a protective effect on photoaged skin when 0.1% were applied topically along with 5% vitamin C (Haftek et al. 2008; Hanifah and Jufri 2018).

Similarly, a stable nanoemulsion of *Sesame indicum* and raspberry oils showed significant antioxidant activity. As red raspberry has a high content of carotenoids, tocopherols, and polyunsaturated fatty acids (PUFAs), it is widely used in cosmetics for skin protection. Nude mouse skin was treated with 750 mu/mL red raspberry oils for five days after exposure to UVB and it was observed that red raspberry reduced skin damage caused by UVB irradiation. Similarly, the hydroethanolic extract of *Rubus idaeus* showed inhibition of inflammatory factors IL-6, and IL-1β, and inhibited the release of matrix-metalloproteinase 1 and 3 induced by UVB. This extract showed photoaging effects by elevating the synthesis of procollagen. In Tables 9.1 and 9.2 below, another extract was prepared using red raspberry seed oil by incorporating the oils such as tocopheryl acetate, isostearyl isostearate, surfactants like polysorbate 80, and water phase containing *Rubus idaeus* (red raspberry) fruit extract, propylene glycol, and water. Along with this, extract of *Quercus petroea* (French oak fruit) and glycerine was also added to the preparation which manifests anticancer and antioxidant activity (Gao et al. 2018; Wang et al. 2019; Gledovic et al. 2020).

Mahdi E. S. et al. developed a nanoemulsion-based cream comprising of *Phyllanthus urinaria* which acted on skin aging. The nanoemulsion was prepared by incorporating palm kernel oil esters in 30% ethanolic extract. Surfactants like cetyl alcohol, span 80, tween 80, and glyceryl monosteareate were added to the formulation. The water phase included sodium benzoate and buffer phosphate. The extract showed good antioxidant activity and therefore, this could be used as an antiageing agent (Mahdi et al. 2011).

Oils like castor oil, virgin oil, argan oil, and coconut oils are also formulated as nanoemulsions in cosmetology. Hydroalcoholic extract of *Opuntia ficusindica* (L.) Mill was formulated as an antiaging cream. In this study, 12 volunteers were administered hydroalcoholic extract of the plant, and its efficacy was noted after five hours of application. Enhanced skin hydration and moisturizing were observed (De Azevedo Ribeiro et al. 2015; Damasceno et al. 2016). *Vellozia squamata* was also formulated as a nanoemulsion-based moisturizer and cream. The hydroalcoholic extract of the plant, distilled water, and surfactants like castor oil, babacu oil, and sorbitan monostearate was added to formulate nanoemulsion. It was noted that the formulation had good antioxidant properties. Another example of nanoemulsion is rice bran seed oil. It increased the hydration and oiliness of the skin. It also balanced skin pH; it could be used as an alternate approach for skin diseases, like psoriasis (Quintão et al. 2013; Rapalli et al. 2020). In Tables 9.1 and 9.2 below Wiseva et al. and Im et al. showed that *Garcinia mangostana* L. was formulated as nanoemulsion by adding surfactants like virgin coconut oil, tween 20 and span 20 and water and propylene glycol in the aqueous phase. Antioxidant and antimicrobial activity of the nanoemulsion was then reported in hairless mouse skin exposed with UVB radiations for 12 weeks due to the presence of Alpha-mangostin (Im et al. 2017; Wiseva et al. 2021).

Likewise, as mentioned in Tables 9.1 and 9.2 Marsup et al. (2020) prepared water extract of *Cordyceps militaris* to enhance its skin permeation and retention power using argan oil as a surfactant. It showed antioxidant activity due to the presence of phenolics, flavonoids, and cordycepin. It was prepared by adding plant extract, distilled water, and surfactants (jojoba, argan oil, sugar squalene, tween 20, tween 85 and tween 80). The effect of *Cordyceps militaris* was evaluated on D-galactose induced aging on mice for seven weeks. It was observed that plant extract was beneficial in inhibiting mitochondrial swelling, promoting antioxidant effect, enhanced dermal delivery and stability (Marsup et al. 2020).

Oil in water nanoemulsion formed by phytosphingosine (induced positive charge on emulsion) could be a promising carrier of low soluble drugs such as ceramides. Ceramides are lipids found in human skin. The formulation was prepared using a high-pressure homogenizer. It was noted that the nanoemulsion enhanced the elasticity and hydration of the skin (Yilmaz and Borchert 2005, 2006).

The stability of *Bixa orellana* nanoemulsion was enhanced by adding green coffee oil. The stable formulation was noted for seven days when kept at 85 °C without showing any notable variation. Another naturally derived example of nanoemulsion was passion fruit oil containing lavender essential oil. In Table 9.1 Lavender essential oil (2%) reduced the droplet size in less time when compared with the nanoemulsions prepared with passion fruit oil alone. In this nanoemulsion, natural extracts of areca, portulaca, and licorice enhanced its antioxidant activity and high stability for three months. Ethoxylated lanolin also enhanced the stability of nanoemulsion containing peach kernel oil, raspberry seed oil, and passion fruit oil (Rocha-Filho 2013).

TABLE 9.1

Natural and synthetically derived nanoemulsion-based cosmetics

Formulation type	Active constituent	Significance	In vivo findings	Reference
Natural				
Skin care	*Centella asiatica*, contains asiaticosides	Photoageing effect	0.2% asiaticoside when applied topically, showed an increase in non-enzymatic and enzymatic antioxidants in newly formed tissues	Hanifah and Jufri (2018), Kumar et al. (2019) and Haffek et al. (2008)
			Improvement in wound healing was observed in mice when 100 ng compound was applied on burn areas	
			In clinical trials (conducted on 20 female participants of age between 45 and 60 years) asiaticosides showed an effect on photoaged skin when 0.1% applied topically along with 5% vitamin C	
Skin care	Rubeus idaeus	Anti-inflammatory, antioxidant, antiphotoageing	Nude mouse skin was treated with 750 mu/mL red raspberry oils for five days after the exposure of UVB and it was observed that red raspberry reduced skin damage caused by UVB irradiation	Gledovic et al. (2020) and Wang et al. (2019)
Skin care	*Garcinia mangostana* L.	Antiwrinkle agent	Alpha-mangostin was applied on UVB exposed hairless mouse skin for 12 weeks. Reduction in skin damage and increase in SOD and CAT was noted	Wiseva et al. (2021) and Im et al. (2017)
Skin care	Opuntia ficusindica	Antiaging, skin hydration	Twelve volunteers were tested with hydroalcoholic extract of plant and its efficacy was noted after five hours of application. Enhanced skin hydration and moisturizing was noted in them	Damasceno et al. (2016) and De Azevedo Ribeiro et al. (2015)
Skin care	Cordyceps militaris	Antiwrinkle activity	Effect of *Cordyceps militaris* was evaluated on D-galactose induced aging on mice for seven weeks. It was observed that plant extract was beneficial in inhibiting mitochondrial swelling, and promoting antioxidant effect	Marsup et al. (2020)

(Continued)

TABLE 9.1 (Continued)

Hair care	Aloe barbedensis	Hair cleanser, sun protectant, hair color protectant	In a double-blind study, 44 people with seborrheic dermatitis were treated with topical aloe formulation twice a day for four to six weeks and noted reduced complaints of pruritis and scaliness	Umar and Noreen (2020)
Synthetic Skin care	Ceramides	Skin absorption, used to treat dry, and scaly skin	In vivo study was conducted on 14 healthy females Ceramides enhanced viscoelasticity and skin hydration	Yilmaz and Borchert (2006)
Hair care	Minoxidil-DPPC	Hair regrowth spray	2% minoxidil was applied for 30 days on denuded areas of albino rats and it was noted that after eight days of treatment, hair growth was initiated whereas after 20 days of treatment, complete hair growth was observed DPPC was tested on 39 patients for six months. The drug was when applied at the interval of 21 days, it showed better effect on hair growth	Rawat et al. (2010), Nowicka et al. (2018) and Aljuffali et al. (2014)
Skin care	Nile red dye	Hydrates skin and penetrates the skin fastly to treat psoriasis	Female Wistar rats' abdominal skin was shaved and Nile red dye incorporated lecithin nanoemulsion was applied on skin to understand skin permeation It was observed that nanoemulsion provided better skin hydration and permeation	S. Bhaskar et al. (2010), Hanifah and Jufri (2018) and Zhao et al. (2010)
Skin care	Tocopheryl acetate	Skin absorption, and nanoemulsion stability	Using hairless guinea pig skin, which was loaded in Franz type diffusion cell Absorption in skin increased with increase in lipid–polymer ratio	Nam et al. (2012)
Skin care	Flumethasone pivatate and fludrocortisone acetate	High skin permeation	Enhanced skin permeation of drugs was seen in porcine skin when applied in combination of phytosphingosine	Hoeller, Sperger, and Valenta (2009)

Many naturally derived nanoemulsion-based formulations are commercially available in the market as shown in Tables 9.1 and 9.2. Some of them are discussed as follows; Bayer HealthCare formulated a nanoemulsion named Bepanthol-Protect Facial Cream Ultra. It is widely used as an antiageing, antipollution, and moisturizing cream. Another example is of Vitacost Vita Nona-Vital skin toner manufactured by Vitacos cosmetics, South Korea. It is a skin moisturizer. The company also manufactured nanoemulsion-based skin moisturizer named Nanovital Vitanics Crystal Moisture cream (Dubey et al. 2022). It provides lightening and elastic effects along with moisturizing the skin. Korres developed a hair sun-protection spray named Korres red vine hair sun protection. It comprises active constituents of aloe. It prevents fading of hair color. Chanel manufactured a skin moisturizer called Precision solution Destressante solution and Nano Emulsion Peaux Sensitivity. It contains *Centella asiatica* which quickly penetrates and soothes the skin. Chanel also manufactured Coco Mademoiselle fresh moisture mist, which increased the fragrance effect. Coni Hyaluronic acid and Nanoemulsion Intensive Hydration Toner is also a skin moisturizer manufactured by Coni Beauty. It hydrates the skin and retains the skin moisture. Vital Nanoemulsions A-VC by Marie Louise and Phyto-Endorphin Hand Cream by Rhonda Allison is also nanoemulsions-manufactured to soothe and moisturize the skin (Kaul et al. 2018). NanoCacao by Mibelle Biochemistry manufactured a skin cream from a lipophilic fraction of cocoa beans. This cream showed antiaging activity. Other antiaging creams from Mibelle Biochemistry are Nano-Lipobelle DN CoQ10oA, NanoVit oA, and NanoMax. NanoVit oA shows photoprotection along with antiaging activity. NanoMax possesses antioxidant activity whereas Nano-Lipobelle DN CoQ10oA protects against photoaging by producing collagen. Nano-Lipobelle DN CoQ10oA is composed of coenzyme Q10, vitamin C derivatives, and tocopherol. NanoMax also contains coenzyme Q10, vitamin C and E. NanoVit oA is incorporated with omegas 3, 6, 7, and 9 (Souto et al. 2020). Sinerga marketed a nanoemulsion by the name Nanocream. It is formulated by adding palm glycerides, potassium lauroyl wheat amino acids, and caprloyl glycine. This O/W emulsion is used as wet wipes, sprayable and hyperfluid emulsions (Silva, Ricci-Júnior, and Mansur 2013; Faria-Silva et al. 2020).

TABLE 9.2
Examples of cosmeceutical products that are nanoemulsion-based

Brand name	Manufactured by	Formulation type	Application
Vitacost Vita Nona-Vital	Vitacos cosmetics, South Korea	Cream	Skin toner and moisturizer
Nanovital Vitanics Crystal Moisture	Vitacos cosmetics, South Korea	Cream	Lightening, and moisturizing effect on skin
NanoMax	Mibelle Biochemistry	Cream	Antioxidant and antiaging effect
Nano-Lipobelle DN CoQ10oA	Mibelle Biochemistry	Cream	Protection against photoaging and aging by producing collagen

(Continued)

TABLE 9.2 (Continued)

Brand name	Manufactured by	Formulation type	Application
NanoVitoA	Mibelle Biochemistry	Cream	Protection against photoaging and aging
NanoCacao	Mibelle Biochemistry	Cream	Antiaging activity
Phyto-Endorphin Hand Cream	Rhonda Allison	Cream	Moisturizing and skin-soothing properties
Kemira Nanogel	Tri-K industries	Gel	Moisturizes, protects against UV rays, and feels the skin fresh after application.
Naturel Kiss	SHE EmpireSdn. Bhd., Malaysia	Lip balm	Lip moisturizer with antiwrinkle effect. It also prevents dryness of lips.
Korres Red Vine Hair Sun Protection	Korres	Hair spray	Prevents fading of hair color and protects against UV rays
Coco Mademoiselle fresh moisture	Chanel	Spray	Prolongs fragrance duration
Bepanthol-Protect Facial Cream Ultra	Bayer HealthCare	Cream	Antiaging and antipollution effects along with moisturization
Nanocream	Sinerga	Wet wipes	Used to remove makeup from face or to cleanse face
Vital Nanoemulsions A-VC	Marie Louise	Cream	Soothes and moisturizes skin
Precision-Solution Déstressante Solution Nano Émulsion Peaux Sensitivity	Chanel	Lotion	Skin moisturizer
Coni Hyaluronic Acid & Nanoemulsion Intensive Hydration Toner	Coni Beauty	Face serum	Hydrates skin
Nanoemulsion Multi-Peptide Moisturizer	Hanacure	Balm	Skin moisturizer
Cosmeceutical Vitamin A, D, E, K	Vitalipid	Lotion	Cause skin hydration
Remains of the day	Croda	Lotion	Cleansing moisturizer and removes makeup from skin

9.4.2 SYNTHETICALLY DERIVED NANOEMULSION-BASED COSMETICS

Many synthetic-based drugs like anti-microbial and wound healing agents are incorporated in oil in water phase to allow easy penetration across the skin as given in Tables 9.1 and 9.2. Bifonazole, an azole is an antifungal drug, which acts against molds, dermatophytes, and other fungi. Yadav et al. prepared nanoemulsion of bifonazole by adding oil (caproyl oil), surfactant (tween 80), and cosurfactant (isopropyl alcohol) together. This nanoemulsion provides a sustained effect and overcame the drawbacks associated with bifonazole such as low permeability, and aqueous solubility (Lohani et al. 2014). Another nanoemulsion of synthetic derivatives is silicon oil. Silicone oil, a hydrophobic oil is essential for the lubrication of hair but due to its poor absorption in hair, its ability to enhance absorption is compromised. Span 80 and tween 80 were incorporated as surfactants in nanoemulsion to emulsify the formulation. This enhanced the absorption of silicone oil in hair. Park et al. also developed a method to enhance the stability of silicone oil by adding amphiphilic triblock copolymers. This makes it possible for the oil media to attach to the hair electrostatically (Park et al. 2018). Teo S.W. et al. designed a nanoemulsion for wound-healing. It was prepared using alkyd loaded with phenytoin. Phenytoin enhances keratinocyte growth which ultimately promotes wound healing (Teo et al. 2017). In another study, Aljuffali I.A. et al. formulated minoxidil-diphencyprone (DPCP) into a nanoemulsion to overcome all the challenges associated with these two drugs such as skin irritation and high chances of systemic absorption. Minoxidil is a vasodilator that is used in hair growth whereas DPCP is applied locally to induce hair regrowth. It treats alopecia when applied topically. The nanoemulsion retards the side effects caused by the drugs by controlling their release and selectively targeting on hair follicles. The selectivity for hair follicles is due to the fact that the formulation is composed of squalene, a sebum-derived lipid (Rawat et al. 2010; Aljuffali et al. 2014; Nowicka et al. 2018). In a double-blind study, reduced pruritis and scaliness were observed in 44 people with seborrheic dermatitis treated with topical aloe formulation twice a day for four to six weeks (Umar and Noreen 2020). Lu et al. prepared citral oil in nanoemulsion to overcome its instability and hydrophobicity. Citral oil was added in the water phase to make the nanoemulsion hydrophilic, and secure the antimicrobial activity of citral. Due to its antimicrobial effect, it is frequently utilized in cosmetics, food and drug industries (Lu et al. 2018). Koester L. et al., in a study, prepared a nanoemulsion to treat cutaneous leishmaniasis. Cutaneous leishmaniasis cause deformities in the skin by damaging the skin. Many antifungal drugs are prescribed to treat the deformity but the cure rate was found minimal. Then, the researchers formulated synthetic chalcone into a nanoemulsion. As this compound has low water solubility, this could be an ideal compound to penetrate the skin. Nanoemulsion reduced the size of droplets and enhanced the permeation power of the compound (de Mattos et al. 2015). Meyer and group prepared a nanoemulsion containing a very small particle size of polyglyceryl-4 laurate and dilauryl citrate. Surprisingly, this self-emulsifying nanoemulsion was manufactured using the phase inversion temperature (PIT) method. The nanoemulsion had low droplet size and low viscosity ("Preparing PIC Emulsions with Very Fine Particle Size | Cosmetics & Toiletries" 2022).

9.4.3 APPLICATIONS OF NANOEMULSIONS IN COSMECEUTICALS

Nanotechnology is not a new technology in the cosmetic industry, but it accelerated the growth of many cosmetic companies to showcase their beauty products to which all the qualities must exist. The nano-particle-based technology provides a variety of solutions in order to manufacture cosmetics that can provide deeper skin penetration, sustained effect, increased UV protection, and a better texture (Aziz et al. 2019). Nanotechnology is comprised of several types of nano-sized drug delivery system that provides simplified solutions over traditional therapies. This type of drug delivery system includes niosomes, nanoemulsions, dendrimers, liposomes, ethosomes, solid lipid nanoparticles (SLN), transferosomes, fullerenes, cubosomes, gold nanoparticles, and nanostructured lipid carriers (Bangale et al. 2012; Bapat et al. 2020; Waghule et al. 2020a, 2020b; Rapalli, Banerjee, et al. 2021; Dubey et al. 2022) as shown in Figure 9.7. The acceptance of this technology is due to its key characteristics, such as the ability to enhance penetration, better spreadability, quality, performance, protection, and entrapment.

Nanoemulsion is one of the nanotechnology-based drug delivery systems which is widely accepted by many cosmetic companies. These nanoemulsions form micellar nanoparticles that could easily penetrate the skin. Because of the formation of nanosized droplets, this technology is mostly used to design sunscreens, moisturizers, face cleansers, antiwrinkle creams, hair products, make-up removers, anti-acne lotions, etc. by preparing o/w nanoemulsions as shown in Figure 9.8. This delivery system also helps to prepare water-based cosmetics. About 1900 patents are granted in the last two decades for nanoemulsion-based formulations. These patents were mostly filed by major cosmetic industries that include L'Oreal, followed by Henkel, P&G, BASF, and AmorePacific (Sonneville-Aubrun, Yukuyama, and Pizzino 2018). Most of the patents were registered on skin care products and about one-third of patents were filed for hair formulations.

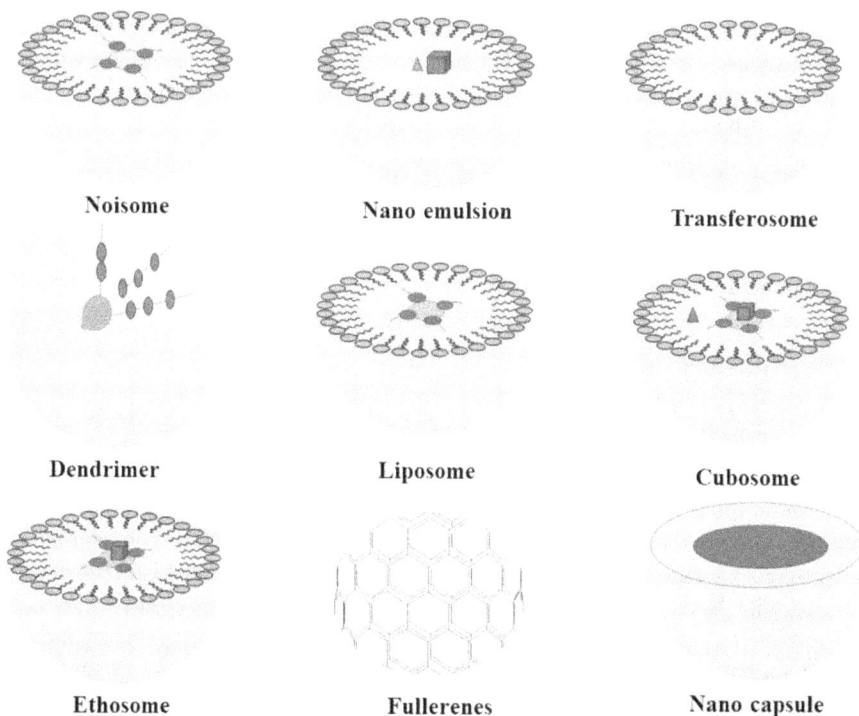

FIGURE 9.7 Types of nano-sized drug delivery system

Sunscreens

Make-up removers   Moisturizers

**Applications of nano
emulsion in cosmetics**

Hair care products   Face cleansers

Antiwrinkle creams

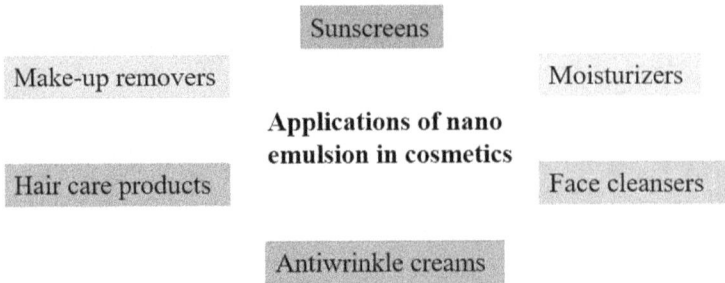

FIGURE 9.8   Application of nanoemulsion in cosmetics

Nanoemulsions are preferred because of the low incorporation of surfactants. This claims less or no chances of skin irritation. Nanoemulsion of emollient oils that soothes our skin could also be designed for better hydration. Coconut oil, an emollient, is incorporated into nanoemulsion to obtain better skin hydration (Karagounis et al. 2019). Lipids like olive oil (Katsouli, Giannou, and Tzia 2018; Katsouli and Tzia 2018), palm oil esters (Han et al. 2011), lanolin (Stone 2013), and vegetable oils (Bajerski et al. 2016) are added in various nanoemulsions to enhance occlusiveness and hydration of skin along with improving skin ailments. Nanoemulsions are also involved in designing antiaging creams. For instance, Nasir (2010) prepared an antiwrinkle cream by adding a neurotransmitter, gamma amino butyric acid (GABA), which has muscle relaxant activity. The nanoemulsion could hydrate the skin without changing the skin pH as noted by (Nasir 2010; Bernardi et al. 2011) while developing rice-bran oil-based and lanolin-based nanoemulsions.

There is a potential to formulate nanoemulsions in the form of sprays and liquids and design them to obtain a variety of skin and hair products. Due to its popularity, it is obvious that the demand for the nanoemulsion-based formulation is increasing and hence, many products are developed using this form of nanotechnology. Some of them will be discussed in this section.

9.4.4   Moisturizers

Moisturizers are used to protect the skin against dehydration caused due to evaporation of water from the stratum corneum. This makes the skin dry and prone to xerosis and eczema which forms cracks on the skin's surface. Moisturizers make a thin film on the skin that retards dehydration and retains moisture in the skin along with providing a better skin appearance (Lohani et al. 2014). Nanoemulsion-based moisturizers are readily available in the market as depicted in Table 9.2. For instance, Vitacost Vita Nona-Vital skin toner is manufactured by Vitacos cosmetics, a skin moisturizer that is commercially available in the market. Also, a company named Chanel manufactured a skin moisturizer called Precision solution, Destressante solution, Nano Emulsion Peaux Sensitivity, which quickly penetrates and soothes the skin. L'Oreal is a cosmetic brand that has mostly patented products based on nanoemulsion. These products are mostly based on skin, hair, and eye care. Many of them have the potential to hydrate skin for an adequate amount of time. Nile red dye incorporated with lecithin was formulated as nanoemulsions to treat dry skin and dry skin conditions such as psoriasis. The lipophilic formulation of flurbiprofen showed 4.4 times more effectiveness than the oral dose to treat skin dryness. Coni Hyaluronic acid and Nanoemulsion Intensive Hydration Toner, Vital Nanoemulsions A-VC, and Phyto-Endorphin Hand cream are also nanoemulsions manufactured to soothe and moisturize the skin (Bhaskar et al. 2009; Zhao et al. 2010; Hanifah and Jufri 2018; Kaul et al. 2018).

9.4.5 SUNSCREENS

Exposure to UV radiation causes high risks of skin cancer, sunburn, tanning, and other skin conditions (Watson, Holman, and Maguire-Eisen 2016). Overexposure to UV rays also causes photoaging of the skin. Therefore, it is essential to formulate a formulation that carries chemopreventive and photoprotective effects (Shi et al. 2012). Titanium oxide, an inorganic filter, acts against UV radiations by absorbing and scattering them. Nanoemulsions are preferred to formulate titanium oxide and zinc oxide into a sunscreen product, as nanoemulsions have better efficiency to protect against UV rays (Shi et al. 2012; Lu et al. 2015). The conventional technique to prepare a sunscreen leaves a white chalky layer on the skin surface after application, therefore, nanoemulsions are opted for by many cosmetic companies to develop a non-greasy, transparent, aesthetic texture, and odorless sunscreens (Lohani et al. 2014). Red raspberry oils (Gledovic et al. 2020) and asiaticosides (Hanifah and Jufri 2018) are designed to act against photoaging, as they have numerous antioxidants. In addition, other sunscreens like Nano-Lipobelle DN CoQ10oA, NanoMax, and NanoVitoA protect against photoaging by producing collagen (Souto et al. 2022) (Table 9.2).

9.4.6 ANTIAGING CREAMS

The aging of the skin is caused due to dehydration of skin, stress, skin contact with pathogens, chemicals, and skin exposure to UVB radiations. All these factors are responsible for skin aging, forming wrinkles, skin darkening, sagginess, and dryness (Tobin 2017). Therefore, it is important to design an antiwrinkle cream that could slow down the process of aging and rejuvenate the skin. Wrinkles could be removed from the skin by restoring collagen structure on the skin surface (Lephart 2016). This is possible by applying some antioxidants to the skin surface to avoid aging. This antioxidant includes vitamins, flavonoids, and polyphenols which extinguish free radicals and protect the skin from wrinkle formation. Many other compounds also enhance the production and metabolism of collagen. These compounds are collectively known as cell regulators which include, growth factors, retinol, and peptides (Boldyrev, Gallant, and Sukhich 1999). Vitamin A is a cell regulator which consists of tretinoin, retinol, and retinaldehyde (Sorg et al. 2006). They suppress Matrix Metalloproteinase-1 (MMP-1) expression which ultimately promotes the synthesis of collagen (Ganceviciene et al. 2012). Antiaging creams work by hydrating, moisturizing, toning, and whitening the skin. Some herbal plants such as *Garcinia mangostana* L., *Rubeus idaeus*, *Opuntia ficusindica*, *Centella asiatica*, and *Cordyceps militaris* have been used to protect the skin from aging and photoaging. Table 9.2 shows that Bayer HealthCare formulated a nanoemulsion named Bepanthol-Protect Facial Cream Ultra which is used as an antiaging, antipollution, and moisturizing cream (Kaul et al. 2018). Another creamlike Nano-Lipobelle DN CoQ10oA, NanoVit oA, and NanoMax. show photo-protection along with exhibiting an antiaging effect. NanoMax possesses antioxidant activity whereas Nano-Lipobelle DN CoQ10oA protects against photoaging by producing collagen (Souto et al. 2022).

9.4.7 FACE CLEANSER

A face cleanser is an essential need for our daily routine. It is essential to clear the clogged pores that may be filled with dirt, sweat, make-up, oil, sebum, or any other pollutants. These pollutants are responsible for making the skin itchy and dark. These particles also promote early skin aging. Face cleansers remove all the dirt by forming a micelle. Micelles are prepared by incorporating two immiscible liquids (oil phase and water phase) along with surfactants to stabilize the emulsion. Micelles are formed when interfacial tension between the oil phase and the water phase is reduced due to the presence of surfactants. The surfactant holds its hydrophilic portion in an aqueous phase, whereas, its hydrophobic part is attracted by an oily phase. After the addition of surfactants to a particular point, the formulation reaches its Critical Micelle Concentration or CMC. At CMC, surfactants trap all the oil globules and spread uniformly in the aqueous phase (Lee, Shenoy, and Sheel 2010). Nanoemulsions reduce the size of oil globules in nanometers and therefore, they more conveniently penetrate the skin to unclog the pores. Major cosmetic companies like L'Oreal, Bioderma, Avenue, La Roche-Posay, and other local or international companies are also claiming that the incorporation of nanotechnologies like nanoemulsions made the the product more effective along with its appearance. Many companies are marketing nanoemulsion-based face cleansers. For instance, Tri-K industries, New Jersey, has launched a nanogel available by the name of Kemira Nanogel, which acts as a sunscreen and a moisturizer and also claims that skin feels fresh after applying it. Croda developed an o/w nanoemulsion that moisturizes the skin and unclogs the pores of the skin. It is designed in a manner, which makes it compatible to permeate skin's lipid layers to clean the pores (Rigano and Lionetti 2016; Miastkowska et al. 2018).

9.4.8 HAIR PRODUCTS

Hair care products such as shampoos are quite essential for hair-cleansing. Shampoos help to reduce hair fall, make hair shinier and glossier, and make the hair surface smoother and silkier (Dubey et al. 2022). These products also cleanse hair scalps to avoid the accumulation of dandruff. Sericin is an active ingredient in cosmetology, which is mostly used in hair products to repair damage and provide shiny hair (Lohani et al. 2014).

As nanoemulsions are capable of penetrating even narrow gaps of sebaceous glands of hair follicles, they are nowadays preferred over conventional hair products.

Nanoemulsions retain the outer structure of hair fibers (cuticles) while penetrating the hair strands. Hence, this formulation does not harm the structure of hair fibers while treating the hair. Researchers have used nanoemulsions as a drug delivery system for developing therapies acting against alopecia for faster and longer action on the hair surface. For instance, silicone oil in nanoemulsion is used for enhancing the lubrication of hair follicles. The nanosize of the particles allows easy penetration through follicles and inhibits their accumulation on the scalp. Nanoemulsion utilizes non-ionic surfactants (cetyl alcohol, ceteareth-12, cetyl palmitate, glyceryl stearate, and ceteareth-20) for a stable and transparent appearance of hair care products. *Platycladus orientalis* contains cedrol as its active constituent which is responsible for hair growth (Park et al. 2018; Dhapte-Pawar et al. 2020; Deng et al. 2021). Minoxidil and DPCPare used to treat alopecia. In a study,) Aljuffali I. A. et al. (2014) developed a combination of these two drugs in the form of nanoemulsion which showed better efficacy in patients facing hair loss (Aljuffali et al. 2014). Korres developed a hair sun protection spray named Korres red vine hair sun protection. It comprises active constituents of aloe which prevents fading of hair color as mentioned in Table 9.2.

9.4.9 NANOEMULSION-BASED WET WIPES

Wet wipes are small to medium-sized moistened pieces of cloth that are designed for removing makeup and cleaning the face. Nowadays people are shifting toward products that are natural or free from harmful chemicals. Therefore, it is important to manufacture wet wipes that are PEG-free. Hence, to obtain PEG-free wet wipes with better permeation and efficacy, nanoemulsions are chosen as an alternative for their manufacturing. They are prepared by using the Phase Inversion method under low-energy methods. Firstly, a microemulsion is formed by adding a water phase to an oily phase, which then is converted into a less viscous nanoemulsion. This converts a w/o emulsion into an o/w emulsion. These wet wipes are used for sprays and lotions (Shah, Bhalodia, and Shelat 2010; Savardekar and Bajaj 2021). Some of the commercially available wet wipes are Nanocream by Sinerga, TEGO Wipe DE PF, and TEGO Wipe DE ("Sinerga | Cosmetic Contract Manufacturing | Cosmetic Ingredients and Formulations | Research and Development Formulation" 2022).

9.4.10 LIP CARE

Lip care includes products like lipsticks and lip balms that prevent the drying of lips. These products provide soft and glossy lips if applied regularly. Nanoemulsion-based lip care product exhibits wide spectra of colors using nanoparticles and provides a long-lasting effect. Moreover, nanoemulsions are made to homogenize all the ingredients uniformly (Dubey et al. 2022). SHE EmpireSdn. Bhd., Malaysia, launched a brand named Naturel Kiss or a natural nanoemulsion-based lip balm. It is made up of essential oils of peppermint, rose, rosehip, and lemon, which are effective in soothing lips, and have the potential to reduce wrinkles, scars, and pigmentation of the skin. This lip balm prevents drying and cracking of lips ("Naturel Kiss | ICC" 2022).

9.5 PATENTED NANOEMULSION FORMULATIONS IN COSMETICS

The use of nanoemulsions to formulate cosmetic products such as sunscreens, moisturizers, hair sprays, antiaging creams, etc. is very renowned. Many cosmetic brands have patented some of their nanoemulsion-based formulations to obtain their exclusive rights. However, some of them are not commercialized yet. Kemira patented a nanogel that claimed that the active ingredients of nanoemulsion enhance skin penetration, ameliorate dermal cells, and provide smooth skin texture. Another patent was filed by Natura Cosmeticos S.A. in the year 2005. The nanoemulsion had an average droplet size of 50 to 200 nm. The composition provides softness, smoothness, and moisture to the skin for around 24 hours. The opaque color nano-emulsion formulation possesses an antibacterial effect without adding preservatives (Google patent 2005 n.d.). Patent no. US6635676B2 patented a non-toxic antimicrobial composition in the year 2003 (expired in 2019). Along with patenting the nontoxic antimicrobial property of nanoemulsion, its methods to reduce morbidity, infection, and rate of mortality were also patented. The patent was filed by the University of Michigan. Another patent EP1411913A1 by the University of Michigan was also filed in the year 2003 (expired in 2019), it was about methods to prevent and treat microbial infections ("EP1411913A1 – Antimicrobial Nanoemulsion Compositions and Methods – Google Patents" 2022). Under the invention, an oil-in-water emulsion inactivated bacterial spores. The o/w nanoemulsion was comprised of oil, an organic phosphate-based solvent, and a surfactant (Google patents 2003c; 2004a). A patent number US6559183B1 describing nanoemulsion of 5-aminolevulinic acid was patented in the year 2003 (expired in the year 2019) by ASAT AG Applied Science ("US6559183B1 – Nano-Emulsion of 5-Aminolevulinic Acid – Google Patents" 2022). The invention claimed that 5-aminolevulinic acid detects proliferative cells and therefore, could be used in photodynamic therapy which is an emerging modality for treating cosmetic and skin-related concerns (Dey et al. 2022).

Among several patents filed by L'Oreal (Paris, France) for nanoemulsion, the transparent nanoemulsion (<100 nm) to use either in cosmetics or dermapharmaceuticals, was granted in 1998. Non-ionic amphiphilic lipids were incorporated into the formulation. The formulation consists of heat-sensitive active compounds (stable at a temperature less than 45 °C). The patent expired in the year 2016 (Google patents n.d.). Another patent by the brand was granted in the year 2014. Patent no. US20040076598A1 claims that nanoemulsions incorporated with either sugar fatty esters or ethers may be used in cosmetics to moisturize the skin and in hair care. In ophthalmology, it could be used as an eye lotion and is still active till 2025 ("US20040076598A1 – Nanoemulsion Based on Sugar Fatty Esters or on Sugar Fatty Ethers and Its Uses in the Cosmetics, Dermatological and/or Ophthalmological Fields – Google Patents" 2022). Nanoemulsion incorporated with ethylene oxide and propylene oxide block copolymers (Patent no. US20020015721A1) has better cosmetic properties and transparency. This formulation was patented by L'Oreal (Paris, France) in the year 2002 and it expired recently in the year 2020. The average size of droplets is less than 100 nm ("US20020015721A1 – Nanoemulsion Based on Ethylene Oxide and Propylene Oxide Block Copolymers and Its Uses in the Cosmetics, Dermatological and/or Ophthalmological Fields – Google Patents"

2022). Patent no. US6541018B1 claims that nanoemulsions are comprised of glycerol fatty esters which have better skin moisturizing effects and could also be used in hair treatment ("US6541018B1 – Nanoemulsion Based on Glycerol Fatty Esters, and Its Uses in the Cosmetics, Dermatological and/or Ophthalmological Fields – Google Patents" 2022). Patent no. JP3618612B2 was filed by L'Oreal in the year 2005. Oxyethylenated or non-oxyethyenated sorbitan fatty esters were formulated as nanoemulsions to use in cosmetology, dermatology, and ophthalmology. The formulation has good cosmetic properties as well as transparency. Another patented example of L'Oreal (Paris, France) is nanoemulsion-based phosphoric acid fatty acid esters (patent no. US6274150B1). The formulation is useful in cosmetics (skin, scalp, mucous membrane, and hair application) with excellent transparency, and is stable in storage ("US6274150B1 – Nanoemulsion Based on Phosphoric Acid Fatty Acid Esters and Its Uses in the Cosmetics, Dermatological, Pharmaceutical, and/or Ophthalmological Fields – Google Patents" 2022).

It has been anticipated that the sustainability of the future cosmeceutical industry will rely on the ability of manufacturers to embrace the advancements in the field of nanobiotechnology and nanotechnology, in line with increasing awareness of the versatile nature of technology and changes in consumer trends. Alternatively, we must seek novel ways to utilize nanobiotechnology and nanotechnology to develop nano products that could enhance the well-being of the general population. Moreover, nanotechnology has been progressively explored and proliferated in cosmetics, based on the increasing number of patents. It is supposed that consumers prefer nano-based cosmetics over conventional cosmetical products, as they become more informed of their benefits.

9.6 CONCLUSION

The aspect of reviewing the available methods and applications of nanoemulsions is discussed in this chapter, which provides the formulator with the knowledge that is according to specific needs. Nanoemulsions provide a common drug delivery system having several applications in the cosmeceutical and pharmaceutical industries. Nanoemulsions serve to be an efficient and versatile carrier system for delivering active ingredients to the targeted sites of interest. Thus, a firm knowledge of properly preparing nanoemulsions and their applications in cosmetic formulations is a prerequisite to attaining stable and sustained nanosized emulsions. The application of nanotechnology in the cosmetic industry is a favorable technology for future endeavors. It has attracted significant attention from related scientific researchers, something which is consequently reflected in the form of an increased number of publications in this area. This technology allows for the development of stable and sustainable cosmetics, delivering the consumers effective, cutting-edge cosmeceutical products.

REFERENCES

Agero, Anna Liza C., and Vermén M. Verallo-Rowell. 2004. "A Randomized Double-Blind Controlled Trial Comparing Extra Virgin Coconut Oil with Mineral Oil as a Moisturizer for Mild to Moderate Xerosis." *Dermatitis : Contact, Atopic, Occupational, Drug* 15 (3): 109–16. doi:10.2310/6620.2004.04006.

Aljuffali, Ibrahim A., Calvin T. Sung, Feng Ming Shen, Chi Ting Huang, and Jia You Fang. 2014. "Squarticles as a Lipid Nanocarrier for Delivering Diphencyprone and Minoxidil to Hair Follicles and Human Dermal Papilla Cells." *The AAPS Journal* 16 (1): 140–50. doi:10.1208/S12248-013-9550-Y.

Ashaolu, Tolulope Joshua. 2021. "Nanoemulsions for Health, Food, and Cosmetics: A Review." *Environmental Chemistry Letters* 19 (4): 3381–95. doi:10.1007/S10311-021-01216-9.

Azevedo Ribeiro, Renato Cesar De, Stella Maria Andrade Gomes Barreto, Elissa Arantes Ostrosky, Pedro Alves Da Rocha-Filho, Lourena Mafra Veríssimo, and Márcio Ferrari. 2015. "Production and Characterization of Cosmetic Nanoemulsions Containing Opuntia Ficus-Indica (L.) Mill Extract as Moisturizing Agent." *Molecules (Basel, Switzerland)* 20 (2): 2492–509. doi:10.3390/MOLECULES20022492.

Aziz, Zarith, Akil Ahmad, Siti Mohd-Setapar, Hashim Hassan, David Lokhat, Mohammad Kamal, and Ghulam Ashraf. 2016. "Recent Advances in Drug Delivery of Polymeric Nano-Micelles." *Current Drug Metabolism* 18 (1): 16–29. doi:10.2174/1389200217666 160921143616.

Aziz, Zarith Asyikin Abdul, Hasmida Mohd-Nasir, Akil Ahmad, Siti Hamidah Mohd Setapar, Wong Lee Peng, Sing Chuong Chuo, Asma Khatoon, Khalid Umar, Asim Ali Yaqoob, and Mohamad Nasir Mohamad Ibrahim. 2019. "Role of Nanotechnology for Design and Development of Cosmeceutical: Application in Makeup and Skin Care." *Frontiers in Chemistry* 7 (November): 739. doi:10.3389/fchem.2019.00739.

Bajerski, Lisiane, Luana Roberta Michels, Letícia Marques Colomé, Eduardo André Bender, Rodrigo José Freddo, Fernanda Bruxel, and Sandra Elisa Haas. 2016. "The Use of Brazilian Vegetable Oils in Nanoemulsions: An Update on Preparation and Biological Applications." *Brazilian Journal of Pharmaceutical Sciences* 52 (3): 347–63. doi:10.1590/S1984-82502016000300001.

Bangale, M. S., Sachin Mitkare, S. G. Gattani, and D. M. Sakarkar. 2012. "Recent Nanotechnological Aspects in Cosmetics and Dermatological Preparations." *International Journal of Pharmacy and Pharmaceutical Sciences* 4 (January): 88–97.

Bapat, Ranjeet A., Tanay V. Chaubal, Suyog Dharmadhikari, Anshad Mohamed Abdulla, Prachi Bapat, Amit Alexander, Sunil K. Dubey, and Prashant Kesharwani. 2020. "Recent Advances of Gold Nanoparticles as Biomaterial in Dentistry." *International Journal of Pharmaceutics* 586: 119596. doi:10.1016/j.ijpharm.2020.119596.

Bernardi, Daniela S., Tatiana A. Pereira, Naira R. Maciel, Josiane Bortoloto, Gisely S. Viera, Gustavo C. Oliveira, and Pedro A. Rocha-Filho. 2011. "Formation and Stability of Oil-in-Water Nanoemulsions Containing Rice Bran Oil: In Vitro and in Vivo Assessments." *Journal of Nanobiotechnology* 9 (1): 1–9. doi:10.1186/1477-3155-9-44/ FIGURES/8.

Bhaskar, Kesavan, Jayaraman Anbu, Velayutham Ravichandiran, Vobalaboina Venkateswarlu, and Yamsani Madhusudan Rao. 2009. "Lipid Nanoparticles for Transdermal Delivery of Flurbiprofen: Formulation, in Vitro, Ex Vivo and in Vivo Studies." *Lipids in Health and Disease* 8: 6. doi:10.1186/1476-511X-8-6.

Bhaskar, Sonu, Furong Tian, Tobias Stoeger, Wolfgang Kreyling, Jesús M. de la Fuente, Valeria Grazú, Paul Borm, Giovani Estrada, Vasilis Ntziachristos, and Daniel Razansky. 2010. "Multifunctional Nanocarriers for Diagnostics, Drug Delivery and Targeted Treatment across Blood-Brain Barrier: Perspectives on Tracking and Neuroimaging." *Particle and Fibre Toxicology* 7 (1): 3. doi:10.1186/1743-8977-7-3.

Boldyrev, Alexander A., Steven Ch. Gallant, and Gennady T. Sukhich. 1999. "Carnosine, the Protective, Anti-Aging Peptide." *Bioscience Reports* 19 (6): 581–7. doi:10.1023/A:1020271013277.

Chevalier, Yves, and Marie-Alexandrine Bolzinger. 2019. "Micelles and Nanoemulsions." In *Nanocosmetics*. Springer International Publishing, pp 47–72. doi:10.1007/978-3-030-16573-4_4.

"Cosmetics Industry – Statistics & Facts | Statista." 2022. Accessed September 7. https://www.statista.com/topics/3137/cosmetics-industry/.

"Cosmetics Market Size, Share, Industry Trends & Analysis 2021–2027." 2022. Accessed September 7. https://www.alliedmarketresearch.com/cosmetics-market.

Damasceno, Gabriel Azevedo de Brito, Rebeca Manuelle Alexandre da Costa Silva, Júlia Morais Fernandes, Elissa Arantes Ostrosky, Silvana Maria Zucolotto Langassner, and Márcio Ferrari. 2016. "Use of Opuntia Ficus-Indica (L.) Mill Extracts from Brazilian Caatinga as an Alternative of Natural Moisturizer in Cosmetic Formulations." *Brazilian Journal of Pharmaceutical Sciences* 52 (3): 459–70. doi:10.1590/S1984-82502016000300012.

Deng, Yaling, Feixue Huang, Jiewen Wang, Yumeng Zhang, Yan Zhang, Guangyue Su, and Yuqing Zhao. 2021. "Hair Growth Promoting Activity of Cedrol Nanoemulsion in C57BL/6 Mice and Its Bioavailability." *Molecules (Basel, Switzerland)* 26 (6). doi:10.3390/MOLECULES26061795.

Dey, Anuradha, Gautam Singhvi, Anu Puri, Prashant Kesharwani, and Sunil Kumar Dubey. 2022. "An Insight into Photodynamic Therapy towards Treating Major Dermatological Conditions." *Journal of Drug Delivery Science and Technology*, 103751. doi:10.1016/J.JDDST.2022.103751.

Dhapte-Pawar, Vividha, Shivajirao Kadam, Shai Saptarsi, and Prathmesh P. Kenjale. 2020. "Nanocosmeceuticals: Facets and Aspects." *Future Science OA* 6 (10). doi:10.2144/FSOA-2019-0109/ASSET/IMAGES/LARGE/FIGURE5.JPEG.

Dubey, Sunil Kumar, Anuradha Dey, Gautam Singhvi, Murali Manohar Pandey, Vanshikha Singh, and Prashant Kesharwani. 2022. "Emerging Trends of Nanotechnology In Advanced Cosmetics." *Colloids and Surfaces B: Biointerfaces*, 112440. doi:10.1016/j.colsurfb.2022.112440.

"EP1411913A1 – Antimicrobial Nanoemulsion Compositions and Methods – Google Patents." 2022. Accessed September 7. https://patents.google.com/patent/EP1411913A1/en.

Evangelista, Mara Therese Padilla, Flordeliz Abad-Casintahan, and Lillian Lopez-Villafuerte. 2014. "The Effect of Topical Virgin Coconut Oil on SCORAD Index, Transepidermal Water Loss, and Skin Capacitance in Mild to Moderate Pediatric Atopic Dermatitis: A Randomized, Double-Blind, Clinical Trial." *International Journal of Dermatology* 53 (1): 100–8. doi:10.1111/IJD.12339.

Faria-Silva, Catarina, Ana Costa, Andreia Ascenso, Helena Ribeiro, Joana Marto, Lidia Gonçalves, Manuela Carvalheiro, and Sandra Simões. 2020. "Nanoemulsions for Cosmetic Products." In *Nanocosmetics*, pp. 59–77. doi:10.1016/B978-0-12-822286-7.00004-8.

Ganceviciene, Ruta, Aikaterini I. Liakou, Athanasios Theodoridis, Evgenia Makrantonaki, and Christos C. Zouboulis. 2012. "Skin Anti-Aging Strategies." *Dermato-Endocrinology* 4 (3). doi:10.4161/DERM.22804.

Gao, Wei, Yu Shuai Wang, Eunson Hwang, Pei Lin, Jahyun Bae, Seul A. Seo, Zhengfei Yan, and Tae Hoo Yi. 2018. "Rubus Idaeus L. (Red Raspberry) Blocks UVB-Induced MMP Production and Promotes Type I Procollagen Synthesis via Inhibition of MAPK/AP-1, NF-Kβ and Stimulation of TGF-β/Smad, Nrf2 in Normal Human Dermal Fibroblasts." *Journal of Photochemistry and Photobiology. B, Biology* 185 (August): 241–53. doi:10.1016/J.JPHOTOBIOL.2018.06.007.

Gledovic, Ana, Aleksandra Janosevic Lezaic, Veljko Krstonosic, Jelena Djokovic, Ines Nikolic, Danica Bajuk-Bogdanovic, Jelena Antic Stankovic, et al. 2020. "Low-Energy Nanoemulsions as Carriers for Red Raspberry Seed Oil: Formulation Approach Based on Raman Spectroscopy and Textural Analysis, Physicochemical Properties, Stability and in Vitro Antioxidant/ Biological Activity." *PLoS One* 15 (4). doi:10.1371/JOURNAL.PONE.0230993.

Gorain, Bapi, Hira Choudhury, Anroop B. Nair, Sunil K. Dubey, and Prashant Kesharwani. 2020. "Theranostic Application of Nanoemulsions in Chemotherapy." *Drug Discovery Today* 25 (7): 1174–88. doi:10.1016/j.drudis.2020.04.013.

Haftek, Marek, Sophie Mac-Mary, Marie Aude Le Bitoux, Pierre Creidi, Sophie Seité, André Rougier, and Philippe Humbert. 2008. "Clinical, Biometric and Structural Evaluation of the Long-Term Effects of a Topical Treatment with Ascorbic Acid and Madecassoside in Photoaged Human Skin." *Experimental Dermatology* 17 (11): 946–52. doi:10.1111/J.1600-0625.2008.00732.X.

Han, Ng Sook, Mahiran Basri, Mohd Basyaruddin Abd Rahman, Raja Noor Zaliha Raja Abd Rahman, Abu Bakar Salleh, and Zahariah Ismail. 2011. "Phase Behavior and Formulation of Palm Oil Esters o/w Nanoemulsions Stabilized by Hydrocolloid Gums for Cosmeceuticals Application." *Journal of Dispersion Science and Technology* 32 (10): 1428–33. doi:10.1080/01932691.2010.513301.

Hanifah, Muthia, and Mahdi Jufri. 2018. "Formulation and Stability Testing of Nanoemulsion Lotion Containing Centella Asiatica Extract." *Journal of Young Pharmacists* 10 (4): 404–8. doi:10.5530/jyp.2018.10.89.

Hejmady, Siddhanth, Rajesh Pradhan, Amit Alexander, Mukta Agrawal, Gautam Singhvi, Bapi Gorain, Sanjay Tiwari, Prashant Kesharwani, and Sunil Kumar Dubey. 2020. "Recent Advances in Targeted Nanomedicine as Promising Antitumor Therapeutics." *Drug Discovery Today* 25 (12): 2227–44. doi: 10.1016/j.drudis.2020.09.031.

Hoeller, Sonja, Andrea Sperger, and Claudia Valenta. 2009. "Lecithin Based Nanoemulsions: A Comparative Study of the Influence of Non-Ionic Surfactants and the Cationic Phytosphingosine on Physicochemical Behaviour and Skin Permeation." *International Journal of Pharmaceutics* 370 (1–2): 181–86. doi:10.1016/J.IJPHARM.2008.11.014.

Im, A. Rang, Young Mi Kim, Young Won Chin, and Sungwook Chae. 2017. "Protective Effects of Compounds from Garcinia Mangostana L. (Mangosteen) against UVB Damage in HaCaT Cells and Hairless Mice." *International Journal of Molecular Medicine* 40 (6): 1941–9. doi:10.3892/IJMM.2017.3188.

Kabri, Tin Hinan, Elmira Arab-Tehrany, Nabila Belhaj, and Michel Linder. 2011. "Physico-Chemical Characterization of Nano-Emulsions in Cosmetic Matrix Enriched on Omega-3." *Journal of Nanobiotechnology* 9 (1): 1–8. doi:10.1186/1477-3155-9-41/FIGURES/3.

Karadag, Ayse, Xiaoqing Yang, Beraat Ozcelik, and Qingrong Huang. 2013. "Optimization of Preparation Conditions for Quercetin Nanoemulsions Using Response Surface Methodology." *Journal of Agricultural and Food Chemistry* 61 (9): 2130–39. doi:10.1021/JF3040463/ASSET/IMAGES/MEDIUM/JF-2012-040463_0006.GIF.

Karagounis, Theodora K., Julia K. Gittler, Veronica Rotemberg, and Kimberly D. Morel. 2019. "Use of 'Natural' Oils for Moisturization: Review of Olive, Coconut, and Sunflower Seed Oil." *Pediatric Dermatology* 36 (1): 9–15. doi:10.1111/PDE.13621.

Katsouli, Maria, Virginia Giannou, and Constantina Tzia. 2018. "A Comparative Study of O/W Nanoemulsions Using Extra Virgin Olive or Olive-Pomace Oil: Impacts on Formation and Stability." *Journal of the American Oil Chemists' Society* 95 (10): 1341–53. doi:10.1002/AOCS.12091.

Katsouli, Maria, and Constantina Tzia. 2018. "Development and Stability Assessment of Coenzyme Q10-Loaded Oil-in-Water Nanoemulsions Using as Carrier Oil: Extra Virgin Olive and Olive-Pomace Oil." *Food and Bioprocess Technology 2018 12:1* 12 (1). Springer: 54–76. doi:10.1007/S11947-018-2193-3.

Kaul, Shreya, Neha Gulati, Deepali Verma, Siddhartha Mukherjee, and Upendra Nagaich. 2018. "Role of Nanotechnology in Cosmeceuticals: A Review of Recent Advances." *Journal of Pharmaceutics* 2018 (March): 1–19. doi:10.1155/2018/3420204.

Kumar, Manish, Ram Singh Bishnoi, Ajay Kumar Shukla, and Chandra Prakash Jain. 2019. "Techniques for Formulation of Nanoemulsion Drug Delivery System: A Review." *Preventive Nutrition and Food Science* 24 (3): 225. doi:10.3746/PNF.2019.24.3.225.

Lee, Robert W., Dinesh B. Shenoy, and Rajiv Sheel. 2010. "Micellar Nanoparticles." *Handbook of Non-Invasive Drug Delivery Systems* 37–58. doi:10.1016/B978-0-8155-2025-2.10002-2.

Lephart, Edwin D. 2016. "Skin Aging and Oxidative Stress: Equol's Anti-Aging Effects via Biochemical and Molecular Mechanisms." *Ageing Research Reviews* 31 (November): 36–54. doi:10.1016/J.ARR.2016.08.001.

Liang, Rong, Shiqi Xu, Charles F. Shoemaker, Yue Li, Fang Zhong, and Qingrong Huang. 2012. "Physical and Antimicrobial Properties of Peppermint Oil Nanoemulsions." *Journal of Agricultural and Food Chemistry* 60 (30): 7548–55. doi:10.1021/JF301129K/SUPPL_FILE/JF301129K_SI_001.PDF.

Lohani, Alka, Anurag Verma, Himanshi Joshi, Niti Yadav, and Neha Karki. 2014. "Nanotechnology-Based Cosmeceuticals." In Edited by T. J. Ryan and T. Maisch. *ISRN Dermatology*. Hindawi Publishing Corporation, p. 843687. doi:10.1155/2014/843687.

Lu, Pei Jia, Shou Chieh Huang, Yu Pen Chen, Lih Ching Chiueh, and Daniel Yang Chih Shih. 2015. "Analysis of Titanium Dioxide and Zinc Oxide Nanoparticles in Cosmetics." *Journal of Food and Drug Analysis* 23 (3): 587–94. doi:10.1016/J.JFDA.2015.02.009.

Lu, Wen Chien, Da Wei Huang, Chiun C.R. Wang, Ching Hua Yeh, Jen Chieh Tsai, Yu Ting Huang, and Po Hsien Li. 2018. "Preparation, Characterization, and Antimicrobial Activity of Nanoemulsions Incorporating Citral Essential Oil." *Journal of Food and Drug Analysis* 26 (1): 82–89. doi:10.1016/J.JFDA.2016.12.018.

Mahdi, Elrashid Saleh, Azmin Mohd Noor, Mohamed Hameem Sakeena, Ghassan Z. Abdullah, Muthanna F. Abdulkarim, and Munavvar Abdul Sattar. 2011. "Formulation and in Vitro Release Evaluation of Newly Synthesized Palm Kernel Oil Esters-Based Nanoemulsion Delivery System for 30% Ethanolic Dried Extract Derived from Local Phyllanthus Urinaria for Skin Antiaging." *International Journal of Nanomedicine* 6: 2499–512. doi:10.2147/IJN.S22337.

Marsup, Pachabadee, Kankanit Yeerong, Waranya Neimkhum, Jakkapan Sirithunyalug, Songyot Anuchapreeda, Chaiwat To-Anun, and Wantida Chaiyana. 2020. "Enhancement of Chemical Stability and Dermal Delivery of Cordyceps Militaris Extracts by Nanoemulsion." *Nanomaterials (Basel, Switzerland)* 10 (8): 1–26. doi:10.3390/NANO10081565.

Mattos, Cristiane Bastos de, Débora Fretes Argenta, Gabriela de Lima Melchiades, Marlon Norberto Sechini Cordeiro, Maiko Luis Tonini, Milene Hoehr Moraes, Tanara Beatriz Weber, et al. 2015. "Nanoemulsions Containing a Synthetic Chalcone as an Alternative for Treating Cutaneous Leshmaniasis: Optimization Using a Full Factorial Design." *International Journal of Nanomedicine* 10 (1): 5529–42. doi:10.2147/IJN.S83929.

McClements, David Julian. 2012. "Nanoemulsions versus Microemulsions: Terminology, Differences, and Similarities." *Soft Matter* 8 (6): 1719–29. doi:10.1039/C2SM06903B.

Miastkowska, Małgorzata, Elwira Lasoń, Elżbieta Sikora, and Katarzyna Wolińska-Kennard. 2018. "Preparation and Characterization of Water-Based Nano-Perfumes." *Nanomaterials (Basel, Switzerland)* 8 (12). doi:10.3390/NANO8120981.

Nam, Yoon Sung, Jin Woong Kim, Jae Yoon Park, Jongwon Shim, Jong Suk Lee, and Sang Hoon Han. 2012. "Tocopheryl Acetate Nanoemulsions Stabilized with Lipid–Polymer Hybrid Emulsifiers for Effective Skin Delivery." *Colloids and Surfaces B: Biointerfaces* 94 (June): 51–57. doi:10.1016/J.COLSURFB.2012.01.016.

Nasir, Adnan. 2010. "Nanotechnology and Dermatology: Part II--Risks of Nanotechnology." *Clinics in Dermatology* 28 (5): 581–88. doi:10.1016/J.CLINDERMATOL.2009.06.006.

Nastiti, Christofori M. R. R., Thellie Ponto, Eman Abd, Jeffrey E. Grice, Heather A.E. Benson, and Michael S. Roberts. 2017. "Topical Nano and Microemulsions for Skin Delivery." *Pharmaceutics* 9 (4). doi:10.3390/PHARMACEUTICS9040037.

"Naturel Kiss | ICC." 2022. Accessed September 7. https://icc.utm.my/project/naturel-kiss/.

Nimase, S. A., P. B. Patil, and R. B. Saudagar. 2019. "A Recent Review on Nanoemulsion as a Topical Delivery System of Antipsoriatic Drugs." *Journal of Drug Delivery and Therapeutics* 9 (2s): 659–64. doi:10.22270/JDDT.V9I2-S.2545.

Nowicka, Danuta, Joanna Maj, Alina Jankowska-Konsur, and Anita Hryncewicz-Gwóźdź. 2018. "Efficacy of Diphenylcyclopropenone in Alopecia Areata: A Comparison of Two Treatment Regimens." *Postepy Dermatologii i Alergologii* 35 (6): 577–81. doi:10.5114/ADA.2018.77608.

Park, Hanhee, Kyounghee Shin, Jin Yong Lee, Ji Eun Kim, Hye Min Seo, and Jin Woong Kim. 2018. "Highly Stable, Electrostatically Attractive Silicone Nanoemulsions Produced by Interfacial Assembly of Amphiphilic Triblock Copolymers." *Soft Matter* 14 (27): 5581–87. doi:10.1039/C8SM00187A.

"Preparing PIC Emulsions with Very Fine Particle Size I Cosmetics & Toiletries." 2022. Accessed September 7. https://www.cosmeticsandtoiletries.com/research/methods-tools/article/21834006/preparing-pic-emulsions-with-very-fine-particle-size.

Qian, Cheng, and David Julian McClements. 2011. "Formation of Nanoemulsions Stabilized by Model Food-Grade Emulsifiers Using High-Pressure Homogenization: Factors Affecting Particle Size." *Food Hydrocolloids* 25 (5): 1000–8. doi:10.1016/J.FOODHYD.2010.09.017.

Quintão, Frederico J. O., Renata S. N. Tavares, Sidney A. Vieira-Filho, Gustavo H. B. Souza, and Orlando D. H. Santos. 2013. "Hydroalcoholic Extracts of Vellozia Squamata: Study of Its Nanoemulsions for Pharmaceutical or Cosmetic Applications." *Revista Brasileira de Farmacognosia* 23 (1): 101–7. doi:10.1590/S0102-695X2013005000001.

Rai, Vineet Kumar, Nidhi Mishra, Kuldeep Singh Yadav, and Narayan Prasad Yadav. 2018. "Nanoemulsion as Pharmaceutical Carrier for Dermal and Transdermal Drug Delivery: Formulation Development, Stability Issues, Basic Considerations and Applications." *Journal of Controlled Release : Official Journal of the Controlled Release Society* 270 (January): 203–25. doi:10.1016/J.JCONREL.2017.11.049.

Rapalli, Vamshi Krishna, Saswata Banerjee, Shahid Khan, Prabhat Nath Jha, Gaurav Gupta, Kamal Dua, Md Saquib Hasnain, Amit Kumar Nayak, Sunil Kumar Dubey, and Gautam Singhvi. 2021. "QbD-Driven Formulation Development and Evaluation of Topical Hydrogel Containing Ketoconazole Loaded Cubosomes." *Materials Science and Engineering: C* 119: 111548. doi:10.1016/j.msec.2020.111548.

Rapalli, Vamshi Krishna, Vedhant Kaul, Tejashree Waghule, Srividya Gorantla, Swati Sharma, Aniruddha Roy, Sunil Kumar Dubey, and Gautam Singhvi. 2020. "Curcumin Loaded Nanostructured Lipid Carriers for Enhanced Skin Retained Topical Delivery: Optimization, Scale-up, in-Vitro Characterization and Assessment of Ex-Vivo Skin Deposition." *European Journal of Pharmaceutical Sciences : Official Journal of the European Federation for Pharmaceutical Sciences* 152 (September): 105438. doi:10.1016/j.ejps.2020.105438.

Rapalli, Vamshi Krishna, Arisha Mahmood, Tejashree Waghule, Srividya Gorantla, Sunil Kumar Dubey, Amit Alexander, and Gautam Singhvi. 2021. "Revisiting Techniques to Evaluate Drug Permeation through Skin." *Expert Opinion on Drug Delivery* 18 (12): 1829–42. doi:10.1080/17425247.2021.2010702.

Rawat, M. s. M., Ajay Semalty, Geeta Joshi, and Mona Semalty. 2010. "In Vivo Hair Growth Activity of Herbal Formulations." *International Journal of Pharmacology* 6 (January): 53–57. doi:10.3923/ijp.2010.53.57.

Rigano, Luigi, and Nicola Lionetti. 2016. "Nanobiomaterials in Galenic Formulations and Cosmetics." *Nanobiomaterials in Galenic Formulations and Cosmetics: Applications of Nanobiomaterials*, 121–48. doi:10.1016/B978-0-323-42868-2.00006-1.

Rocha-Filho, P. A. 2013. "Nanoemulsions as a Vehicle for Drugs and Cosmetics." *Nanoscience & Technology: Open Access* 1 (1). doi:10.15226/2374-8141/1/1/00105.

Salvia-Trujillo, Laura, M. Alejandra Rojas-Graü, Robert Soliva-Fortuny, and Olga Martín-Belloso. 2014. "Impact of Microfluidization or Ultrasound Processing on the Antimicrobial Activity against Escherichia Coli of Lemongrass Oil-Loaded Nanoemulsions." *Food Control* 37 (1): 292–97. doi:10.1016/J.FOODCONT.2013.09.015.

Savardekar, Pranita, and Amrita Bajaj. 2021. "Nanoemulsions – A Review." *IJRPC.* Vol. 2016. Accessed March 8. www.ijrpc.com.

Shah, P., D. Bhalodia, and Pragna Shelat. 2010. "Nanoemulsion: A Pharmaceutical Review." *Systematic Reviews in Pharmacy* 1 (January). doi:10.4103/0975-8453.59509.

Shaker, Dalia S., Rania A. H. Ishak, Amira Ghoneim, and Muaeid A. Elhuoni. 2019. "Nanoemulsion: A Review on Mechanisms for the Transdermal Delivery of Hydrophobic and Hydrophilic Drugs." *Scientia Pharmaceutica* 87 (3): 17. doi:10.3390/SCIPHARM87030017.

Shi, Lei, Jingning Shan, Yiguang Ju, Patricia Aikens, and Robert K. Prud'homme. 2012. "Nanoparticles as Delivery Vehicles for Sunscreen Agents." *Colloids and Surfaces A: Physicochemical and Engineering Aspects* 396 (February): 122–29. doi:10.1016/J.COLSURFA.2011.12.053.

Silva, Flavia F. F., Eduardo Ricci-Júnior, and Claudia R. E. Mansur. 2013. "Nanoemulsions Containing Octyl Methoxycinnamate and Solid Particles of TiO2: Preparation, Characterization and in Vitro Evaluation of the Solar Protection Factor." *Drug Development and Industrial Pharmacy* 39 (9): 1378–88. doi:10.3109/03639045.2012.718787.

"Sinerga | Cosmetic Contract Manufacturing | Cosmetic Ingredients and Formulations | Research and Development Formulation." 2022. Accessed September 7. https://www.sinerga.it/en.

Solans, Conxita, Daniel Morales, and Maria Homs. 2016. "Spontaneous Emulsification." *Current Opinion in Colloid & Interface Science* 22 (April): 88–93. doi:10.1016/J.COCIS.2016.03.002.

Sonneville-Aubrun, O., D. Babayan, D. Bordeaux, P. Lindner, Gabriel Rata, and B. Cabane. 2008. "Phase Transition Pathways for the Production of 100 Nm Oil-in-Water Emulsions." *Physical Chemistry Chemical Physics* 11 (1): 101–10. doi:10.1039/B813502A.

Sonneville-Aubrun, O., J. T. Simonnet, and F. L'Alloret. 2004. "Nanoemulsions: A New Vehicle for Skincare Products." *Advances in Colloid and Interface Science* 108–109 (May): 145–49. doi:10.1016/J.CIS.2003.10.026.

Sonneville-Aubrun, Odile, Megumi N. Yukuyama, and Aldo Pizzino. 2018. "Application of Nanoemulsions in Cosmetics." In *Nanoemulsions: Formulation, Applications, and Characterization*, January. Academic Press, 435–75. doi:10.1016/B978-0-12-811838-2.00014-X.

Sorg, Olivier, Christophe Antille, Gürkan Kaya, and Jean Hilaire Saurat. 2006. "Retinoids in Cosmeceuticals." *Dermatologic Therapy* 19 (5): 289–96. doi:10.1111/J.1529-8019.2006.00086.X.

Souto, Eliana B., Amanda Cano, Carlos Martins-Gomes, Tiago E. Coutinho, Aleksandra Zielińska, and Amélia M. Silva. 2022. "Microemulsions and Nanoemulsions in Skin Drug Delivery." *Bioengineering (Basel, Switzerland)* 9 (4). doi:10.3390/BIOENGINEERING9040158.

Souto, Eliana B., Ana Rita Fernandes, Carlos Martins-Gomes, Tiago E. Coutinho, Alessandra Durazzo, Massimo Lucarini, Selma B. Souto, Amélia M. Silva, and Antonello Santini. 2020. "Nanomaterials for Skin Delivery of Cosmeceuticals and Pharmaceuticals." *Applied Sciences* 10 (5): 1594. doi:10.3390/APP10051594.

Stone, L. 2013. "Medilan: A Hypoallergenic Lanolin for Emollient Therapy." *British Journal of Nursing* 9 (1): 54–7. doi:10.12968/BJON.2000.9.1.6415.

Teo, Siew Yong, Mei Yeng Yew, Siang Yin Lee, Michael J. Rathbone, Seng Neon Gan, and Allan G.A. Coombes. 2017. "In Vitro Evaluation of Novel Phenytoin-Loaded Alkyd Nanoemulsions Designed for Application in Topical Wound Healing." *Journal of Pharmaceutical Sciences* 106 (1): 377–84. doi:10.1016/J.XPHS.2016.06.028.

Tobin, Desmond J. 2017. "Introduction to Skin Aging." *Journal of Tissue Viability* 26 (1): 37–46. doi:10.1016/J.JTV.2016.03.002.

Umar, Sanushi, and Sana Noreen. 2020. "A Close Look at Aloe Vera Barbadensis and It's Effect on Hair Health," July.

"US20020015721A1 – Nanoemulsion Based on Ethylene Oxide and Propylene Oxide Block Copolymers and Its Uses in the Cosmetics, Dermatological and/or Ophthalmological Fields – Google Patents." 2022. Accessed September 7. https://patents.google.com/patent/US20020015721A1/en.

"US20040076598A1 – Nanoemulsion Based on Sugar Fatty Esters or on Sugar Fatty Ethers and Its Uses in the Cosmetics, Dermatological and/or Ophthalmological Fields – Google Patents." 2022. Accessed September 7. https://patents.google.com/patent/US20040076598A1/en.

"US6274150B1 – Nanoemulsion Based on Phosphoric Acid Fatty Acid Esters and Its Uses in the Cosmetics, Dermatological, Pharmaceutical, and/or Ophthalmological Fields – Google Patents." 2022. Accessed September 7. https://patents.google.com/patent/US6274150B1/en.

"US6541018B1 – Nanoemulsion Based on Glycerol Fatty Esters, and Its Uses in the Cosmetics, Dermatological and/or Ophthalmological Fields – Google Patents." 2022. Accessed September 7. https://patents.google.com/patent/US6541018B1/en.

"US6559183B1 – Nano-Emulsion of 5-Aminolevulinic Acid – Google Patents." 2022. Accessed September 7. https://patents.google.com/patent/US6559183B1/en.

Waghule, Tejashree, Srividya Gorantla, Vamshi Krishna Rapalli, Pranav Shah, Sunil Kumar Dubey, Ranendra Narayan Saha, and Gautam Singhvi. 2020. "Emerging Trends in Topical Delivery of Curcumin Through Lipid Nanocarriers: Effectiveness in Skin Disorders." *AAPS PharmSciTech* 21 (7): 284. doi:10.1208/s12249-020-01831-9.

Waghule, Tejashree, Vamshi Krishna Rapalli, Srividya Gorantla, Ranendra Narayan Saha, Sunil Kumar Dubey, Anu Puri, and Gautam Singhvi. 2020. "Nanostructured Lipid Carriers as Potential Drug Delivery Systems for Skin Disorders." *Current Pharmaceutical Design* 26 (36): 4569–79. doi:10.2174/1381612826666200614175236.

Wang, Pei Wen, Yu Chen Cheng, Yu Chiang Hung, Chih Hung Lee, Jia You Fang, Wen Tai Li, Yun Ru Wu, and Tai Long Pan. 2019. "Red Raspberry Extract Protects the Skin against UVB-Induced Damage with Antioxidative and Anti-Inflammatory Properties." *Oxidative Medicine and Cellular Longevity* 2019. doi:10.1155/2019/9529676.

Watson, Meg, Dawn M. Holman, and Maryellen Maguire-Eisen. 2016. "Ultraviolet Radiation Exposure and Its Impact on Skin Cancer Risk." *Seminars in Oncology Nursing* 32 (3): 241–54. doi:10.1016/J.SONCN.2016.05.005.

Wiseva, Karina O., Frida Widyastuti, Hisa Faadhilah, and Nasrul Wathoni. 2021. "A Review on Various Formulation of Nanoemulsions in Cosmetics with Plant Extracts as the Active Ingredients." *International Journal of Applied Pharmaceutics* 13 (Special issue 4): 34–40. doi:10.22159/IJAP.2021.V13S4.43814.

Yilmaz, Erol, and Hans Hubert Borchert. 2005. "Design of a Phytosphingosine-Containing, Positively-Charged Nanoemulsion as a Colloidal Carrier System for Dermal Application of Ceramides." *European Journal of Pharmaceutics and Biopharmaceutics* 60 (1): 91–98. doi:10.1016/J.EJPB.2004.11.009.

——. 2006. "Effect of Lipid-Containing, Positively Charged Nanoemulsions on Skin Hydration, Elasticity and Erythema—An in Vivo Study." *International Journal of Pharmaceutics* 307 (2): 232–38. doi:10.1016/J.IJPHARM.2005.10.002.

Yukuyama, M. N., D. D.M. Ghisleni, T. J.A. Pinto, and N. A. Bou-Chacra. 2016. "Nanoemulsion: Process Selection and Application in Cosmetics – a Review." *International Journal of Cosmetic Science* 38 (1): 13–24. doi:10.1111/ICS.12260.

Zhao, Yi, Changguang Wang, Albert H. L. Chow, Ke Ren, Tao Gong, Zhirong Zhang, and Ying Zheng. 2010. "Self-Nanoemulsifying Drug Delivery System (SNEDDS) for Oral Delivery of Zedoary Essential Oil: Formulation and Bioavailability Studies." *International Journal of Pharmaceutics* 383 (1–2): 170–77. doi:10.1016/J.IJPHARM.2009.08.035.

10 Polymeric micelles and dendrimer drug delivery systems in cosmetics

Nitheesh Yanamandala, Pavan Kumar Achalla and Sunil Kumar Dubey

CONTENTS

10.1 INTRODUCTION

Nanotechnology is a major area of academic and industrial research in a variety of fields, including healthcare and cosmetics. Polymeric nanoparticles, Liposomes, Niosomes, and Solid lipid nanoparticles are some of the widely used nanomaterials for various cosmetic purposes including, anti-aging, wrinkles, hyperpigmentation, hair damage, etc. (Kaul et al., 2018a). Due to the unique physicochemical properties of various nanoparticles or nanocarriers, they are used as better alternatives to overcome various challenges in pharmaceutical and cosmetic formulations. In general, cosmetic formulations have problems with stability, solubility, and skin permeation. Many cosmetic active ingredients are often not able to penetrate the Stratum Corneum (SC) layer of the skin (Kandregula et al., 2021; Kim et al., 2004). In several studies, nanoparticle-based formulations using polymeric nano-carriers such as polymeric micelles and dendrimers have shown a greater ability to overcome these problems and deliver the active molecule at the targeted layer of skin (Chuo and Mohd Setapar, 2022; Tiwari et al., 2022).

Polymeric micelles are being studied for decades to deliver some of the key cosmetic ingredients including oleanolic acid, tretinoin, adapalene, etc. (An et al., 2020a; Makhmalzade and Chavoshy, 2018). In many studies, they have shown improved performance or efficacy with a better safety profile because of their stability, solubility, and loading capacity (Parra et al., 2021a). Dendrimer's application in cosmetics is also very significant. Some of the key players in the cosmetic market like L'Oreal, Unilever, and The Dow Chemical Company have patents for the use of dendrimers in the skin, hair, and nail care products. Revlon, another cosmetics company, has also investigated the use of poly amidoamine dendrimers in cosmetics and personal care (Bilal and Iqbal, 2020; Chauhan et al., 2019). Increased shelf life, controlled release, and enhanced solubility are some of the major benefits of dendrimers in cosmetics. Also, one of the critical properties of dendrimers is their lower intrinsic viscosity at higher molecular weights. This unique property of dendrimers makes them very suitable for skin applications (Gupta et al., 2018a).

Nanotechnology allows a product to be built from scratch and enables modification of its physicochemical properties to improve its performance. Theoretically, this means that using various nanoparticles researchers and manufacturers should be able to create new cosmetics with specific efficient uses, such as lipstick that lasts for days, a moisturizer that keeps the skin smooth for a week, and scents that last all day (Virmani and Pathak, 2021). However, there are various practical challenges involved in their application, the most critical of which is their potential in vivo toxicity. The fundamentals of polymeric micelles and dendrimers, as well as their various applications in cosmetics, were discussed in this chapter. This chapter also includes the potential for toxicity associated with the use of nano-cosmetic formulations, as well as their regulatory aspects.

10.2 FUNDAMENTALS OF POLYMERIC MICELLES AND DENDRIMERS

10.2.1 Polymeric micelles

Polymeric micelles are well-known nanocarriers formed by copolymers containing both hydrophilic (A) and hydrophobic (B) monomer blocks that self-assemble in aqueous solutions, resulting in a two-phase structure with a hydrophobic core and a hydrophilic surface (Figure 10.1) (Almeida et al., 2017). One of the special properties of this structure is solubilizing sparingly water-soluble molecules in the core whereas, with hydrophilic molecules in the outer shell (Hwang et al., 2020). The size of the micelles can range from 10 to 200 nm and the drug loading can increase the size further. Also, the length and molecular weight of the copolymer's hydrophilic and hydrophobic units can be used to determine their size (Guo et al., 2021). Micelles are typically spherical. However, depending on the copolymers used, worm-like, disc-like, hamburger-like, raspberry-like, and sheet-like shapes can be formed. The hydrophilic–lipophilic balance (HLB) of the copolymers and the solvent chosen for micelle formation influence the structure of the polymeric micelle (Güngör et al., n.d.; Middleton et al., 2021).

Generally, the molecule of interest is loaded into the core while the outer layer protects the hydrophobic section and therefore the loaded molecule from biological degradation. Additionally, the hydrophilic surface provides colloidal stability and decreases protein adsorption, resulting in longer blood circulation time and improved accumulation at the target site (Bose et al., 2021; Khan et al., 2018). It is

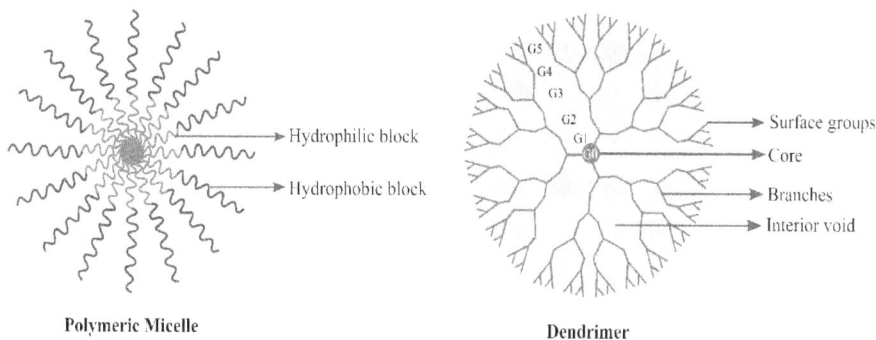

FIGURE 10.1 Schematic representation of polymeric micelles and dendrimers

well known that above the critical micelle concentration (CMC), surfactants aggregate to form micelles in aqueous media to reduce the surface free energy. Lower CMC values are often associated with improved system stability (Perinelli et al., 2020). As the CMC of polymeric micelles is several times lower than the classic micelles, they got special attention in the scientific community for drug delivery purposes (Ghezzi et al., 2021). Various properties of copolymers determine the CMC value. For instance, an increased length of the hydrophobic unit can lower the CMC value as the increased interactions between the hydrophobic fragments will lead to an equilibrium state (Kulthe et al., 2012; Wichit et al., 2012a). Micelle-forming copolymers are often made up of di-block (A–B) or tri-block (A–B–A) copolymers, such as poloxamines and poloxamers, graft copolymers with varying hydrophobicities (such as G-chitosan), or ionic copolymers. Furthermore, a reverse micelle is formed by the copolymers when synthesized in an organic solvent under appropriate conditions, resulting in a hydrophilic core and a hydrophobic surface. The self-aggregation of copolymers is a reversible process that depends on the CMC (Almeida et al., 2017; Braunová et al., 2019).

Direct dissolution, solvent evaporation, solid dispersion, oil-in-water emulsion, freeze-drying, and dialysis are the most commonly used methods for the preparation of polymeric micelles. The physicochemical properties of the active pharmaceutical ingredient and the copolymers (water solubility, molecular weight, HLB, etc.) play an important role in selecting the method of preparation (Miller et al., 2013).

10.2.2 DENDRIMERS

Dendrimers are highly branched synthetic polymeric macromolecules. High tree-like branches, monodispersed, circular, and radially symmetrical structures are the major distinguishable features of dendrimers (Abbasi et al., 2014; Malik et al., 2012; Nitheesh et al., 2021). As shown in Figure 10.1, the dendrimer structure starts with the core atom or molecule at the center. From the core, the branches, also called dendrons, are formed through a variety of chemical reactions (Figure 10.1) (Markelov et al., 2021). The exact structure of dendrimers is still debatable, particularly whether the end groups fold back into a dense core or are fully extended at the surface with high density (Abbasi et al., 2014). In both solid and liquid states, the surface functional groups are responsible for the physical properties of the dendrimers. The physicochemical properties of dendrimers can be changed by functionalizing the surface with selective compounds. The unique structure of the dendrimers with voids in between the branches and the modifiable surface provides a special opportunity to load and deliver the drug/cosmetic ingredients to the target site. The diameter of dendrimers can range from 1 to 15 nm (Mittal et al., 2021; Munavalli et al., 2019). A variety of dendrimers such as PEGylated, poly-amidoamine (PAMAM), poly-propylene imine (PPI), etc are available, and each has its own physicochemical and biological properties, including polyvalency, self-assembly, electrostatic interactions, solubility, low cytotoxicity, and chemical stability (Malik et al., 2012; Santos et al., 2019).

Dendrimers are synthetically prepared by a repeating step-by-step polymerization process. Theoretically, the dimensions of dendrimers can be regulated precisely using this repeating pattern of reaction steps (Malkoch and García-Gallego, 2020). Unlike classic linear polymers, which have polydisperse structures, the synthesis of dendrimers allows the formation of monodisperse, structure-controlled macromolecules similar to those found in biological systems (Gupta et al., 2018b). Dendrimers can be synthesized using either a divergent (core to the surface) or convergent (surface to the core) approach. Based on the number of monomer layers, dendrimers are classified into generations (G0.5, G1, G2, G3, G4, & G5), where each layer is referred to as a generation (Abbasi et al., 2014; Kesharwani et al., n.d.; Viswanath and Santhakumar, 2017). During the synthesis, dendrimers with generation four and above, change their conformation and becomes a densely packed globular structure, which exhibits different solution and bulk properties compared to the classic linear or branched polymers. The conformational change in the dendrimers with high generations will reduce their intrinsic viscosity, which is very useful in the formulation process. Additionally, the property of being soluble in several organic solvents grabs the attention of dendrimers to use in various drug delivery applications (Nilsson, 2008; Passeno et al., 2006).

The ability to entrap many components at once, including both hydrophilic and hydrophobic molecules, is one of the major advantages of dendrimers as a carrier of active ingredients of cosmetics. Dendrimers as drug delivery systems have some additional advantages such as controlled drug release, increased water solubility, prolonged stability, and protection of the encapsulated molecule. On the other side, the synthesis of dendrimers is time-consuming and complex, resulting in an expensive product with restricted manufacturing options (Aurelia Chis et al., 2020; Filipczak et al., 2021).

10.3 APPLICATIONS OF POLYMERIC MICELLES (PMS) AND DENDRIMERS IN COSMETICS

The application of nanoparticles for cosmetics has been the subject of research for their ease of drug penetration and controlled release. Cosmetic applications can be for skin, hair, lips, and mouth. An increase in life expectancy around the world has amplified interest in the aging process of the skin. Skin dryness, wrinkles, decreased collagen levels, epidermal atrophy, and loss of skin elasticity occurs as a result of the cumulative effects of the aging process, both preventable and unpreventable (Biniek et al., 2015). Also, acne vulgaris is the eighth most common skin disease in the world, affecting over 9.4% of the population around the world. Acne patients have increased sebum production, aberrant keratinocyte proliferation, differentiation, and Propionibacterium acnes hyperproliferation which causes inflammatory reactions due to the antigens and cytokines released during this process. In developing medications and cosmetics for use on the skin, the main problem is to overcome the physicochemical characteristics of the skin by improving penetration by transdermal route or using targeted delivery of the drug to hair follicles on the skin.

PMs have been explored for use in cosmetics for their ability to encapsulate the drug in their core and deliver the drug at the site at a controlled rate. Several investigations on the use of PM in skin care-based medicines and cosmetics have been undertaken in recent years. The majority of the application of PMs was observed in the anti-aging process, acne vulgaris, and other skin-based diseases (Parra et al., 2021b). One of the applications was found in the encapsulation of Oleanolic Acid (OA) in PM as it increases the production of ceramides, pro-collagen, and filaggrin while also inhibiting the MMP-1 enzyme which is an important molecule in collagen degradation. Since OA has poor aqueous solubility and penetration into the skin, PMs have been investigated as permeation enhancers for the skin. To see if nano-encapsulated OA was more effective at reducing wrinkles on the periocular skin, An et al. mixed OA with PM. The nano-encapsulated OA conjugated with PM was able to effectively reduce wrinkles with no signs of skin irritation (An et al., 2020b). In anti-aging formulations, the antioxidant activity of Coenzyme Q10 (CoQ10), which has high lipophilicity with low topical bioavailability, has been examined (Biniek et al., 2015). Since the amount of this coenzyme gets decreased in the body after 30 years of age, formulations that can help penetrate the exogenous CoQ10 to the epidermis have a lot of promise in anti-aging treatments. Šmejkalová et al. made CoQ10-loaded PMs for testing the effective decrease in oxidative stress for both in vitro testings using human fibroblasts and in vivo testing. The results were indicating the effective reduction of oxidative stress in human fibroblasts, also in comparison to others, skin hydration was two to three times higher, indicating that PM has the potential to boost payload bioactivity (Šmejkalová et al., 2017a). There are several anti-acne-based formulations of polymeric micelles used for effective treatment, which include conjugation with tretinoin (TRN), adapalene (ADA), and benzoyl peroxide (BPO). Wichit et al. used (polyethylene glycol)-conjugated with phosphatidylethanolamine (PEG-PE), and it was concluded that these polymeric micelles reduced the degradation of tretinoin when there is atmospheric oxygen (Wichit et al., 2012b). Using Pluronic F127, a non-ionic polymer authorized by FDA, Kahraman et al. created micellar nanocarriers of BPO, which were used to effectively deliver (Kahraman et al., 2016). The use of PEG and PE has been shown to improve skin medication absorption and, in some cases, reduce Stratum Corneum surface tension. Pluronic F127 was successful in delivering the medicine safely and efficiently by specifically acting on the PSU, making it a promising drug delivery agent in BPO formulation. However, more in vivo trials are needed to back up the in vitro and ex vivo findings. Kandekar et al. investigated PMs for the targeted delivery of ADA to the hair follicle under a defined dose with increased efficacy and fewer side effects. According to the findings, ADA-D-α-tocopheryl polyethylene glycol 1000 succinate (TPGS) was found to be efficient at PSU-targeted administration at a decreased ADA dosage than commercial formulations, thus resulting in less skin irritation and adverse effects, as well as improved patient compliance. The same research also shows that TPGS-based nanomicelles can be used as carriers for less soluble active pharmaceutical agents in PSU-related disease-based medications (Kandekar et al., 2018a).

Similarly, dendrimers are being employed as nanotechnology-based cosmeceuticals for a variety of applications, including hair, skin, and nail care. They have a network of extended symmetric branches with terminal ends containing functional groups. Because of their small size (2–20 nm), they have a high skin penetration and may easily deliver the cosmetic agent (Dubey et al., 2022). They can be found in a variety of cosmetic goods, including shampoos, sunscreens, hair styling gels, and anti-acne creams. For cosmetics intended for skin applications, L'Oreal has been utilizing terminal hydroxyl functionalized polyester dendrimers in combination with film-forming polymers. The L'Oreal patent explains the application of dendrimers in low-viscosity formulations which can provide excellent skin treatment. The dendrimers were found to boost the intensity of various L'Oreal cosmetics products. Revlon adopted the encapsulation technique for salicylic acid, which involves associating salicylic acid molecules to the surface amino groups of PAMAM dendrimer, resulting in salicylic acid stability. PAMAM dendrimers were investigated for their capacity to increase skin permeability. Vitamins play an important role by being the major component of skincare that can be supplied through cosmetic creams. Research shows how PAMAM dendrimer bioconjugates can improve the bioavailability of vitamin A and vitamin B6 by acting as a carrier for these nutrients. As seen in an in vitro model, these PAMAM dendrimer-vitamin formulations infiltrate the skin. Chauhan et al. published research on Dendrimer-trans-Resveratrol formulations which can increase the solubility and molecular stability in topical cream formulations. The compositions were found suitable to be used on the skin. Dendrimers have been found to improve the loading efficiency of resveratrol. The unusual design of a dendrimer can be effective for entrapping resveratrol, thus increasing the resveratrol's stability and aqueous solubility. Several patents for the use of dendrimers in cosmeceuticals have been issued to companies such as L'Oreal, Te Dow Company, Wella, and Unilever (Kaul et al., 2018b). They have the potential to improve medication solubility and bioavailability because of their network of branched structures, which can hold together molecules on the external layers and entrap them within the branched structure. The strong demand for dendrimers to develop various drug delivery mechanisms and products is due to these features. In Table 10.1. below, all the formulations of both polymeric micelles and dendrimers have been summarized with details about the polymer or drug used, its application, and the type of nanoparticle used. Figure 10.2 depicts the areas of applications of polymeric micelles and dendrimers used in cosmetics.

TABLE 10.1
Various nanoparticles for different cosmetic applications

S. no.	Type of nanoparticle	Polymer/drug used	Application	Reference
1	Polymeric micelles	Capryol 9o® and Poloxamer 407	Anti-aging (wrinkle reduction)	An et al. (2020b)
2	Polymeric micelles	Resveratrol	Anti-aging/anti psoriasis	Khurana et al. (2020)
3	Polymeric micelles	Adapalene (ADA)	Targeted delivery of ADA to the PSU	Kandekar et al. (2018a)
4	Polymeric micelles with biodegradable and biocompatible copolymer—diblock methoxy-poly(ethylene glycol)-poly(hexyl-substituted lactic acid) (MPEG-dihexPLA)	Tretinoin (TRA)	Targeted delivery to PSU	Kandekar et al. (2018b)
5	Polymeric micelles	Nile red/ Hyaluranon	Deeper penetration into the epidermis and dermis layers of the skin	Šmejkalová et al. (2017b)
6	Polymeric micelles	Benzoyl peroxide (BPO)	Better delivery of BPO to PSU	Kahraman et al. (2016)
7	Polymeric micelles	Coenzyme Q10	Anti-aging (protection of the skin against oxidative stress)	Lapteva et al. (2015)
8	Polymeric micelles	Tretinoin (TRA)	Prevention of TRA degradation	Wichit et al. (2012b)
9	G4-PAMAM dendrimer	Chlorhexidine gluconate	Anti-bacterial infection	Holmes et al. (2017)
10	Dendrimer	Resveratrol	Anti-aging	Pentek et al. (2017)
11	PAMAM dendrimer	Salicylic acid	Avoid the reaction of iron oxide pigments	Wolf et al. (n.d.)
12	Terminal hydroxyl functionalized polyester dendrimer	N/A	Increased intensity of skin care products (L'Oreal)	Lohani et al. (2014)
13	Dendrimer	Vitamins A and B6	Improve bioavailability of vitamins	Filipowicz and Wołowiec (2012)

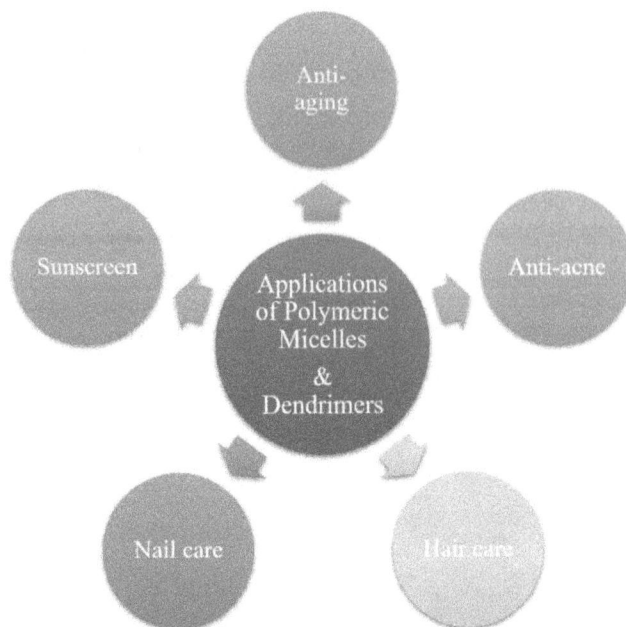

FIGURE 10.2   Applications of polymeric micelles and dendrimers in cosmetics

10.4   NANOTOXICITY

There are several nanoparticle-based cosmetic formulations available on the market. Nanoparticles are used in these products because they have several advantages, including effective and controlled delivery of active molecules, the photostability of potentially unbalanced cosmetic components, and skin tolerance to various UV filters. They also provide aesthetic texture to the final product (Kaul et al., 2018c; Raj et al., 2012; Salvioni et al., 2021). Although nanoparticles offer a variety of options and solutions to the challenges that are faced during the development of a cosmetic or pharmaceutical product, their use demands the evaluation and optimization of their physicochemical characteristics, such as shape, particle size, and surface charge, which affects their transport, reactivity, and toxicity (Ramanathan, 2019; Uchechi et al., 2014). The ability of nanoparticles to disrupt the normal physiology and damage the normal tissue and organs is the main reason for nanotoxicity. Recent studies suggest that oxidative stress and activation of the pro-inflammatory gene are some potential causes of toxic effects however, the exact mechanism is still not clear (Khanna et al., 2015; Sukhanova et al., 2018). Apart from particle-related aspects, loaded drug dose, administration route, and the tissue distribution extent are some crucial parameters in nano-cytotoxicity. One of the major challenges in nanoparticle use is extrapolating the in vitro concentrations to in vivo use, which, if not done properly, could result in nanoparticle overdosage in animals and humans. Therefore, the amount of administered dose reaching the target tissue and the nanoparticle-induced in vivo chemical changes that could go unnoticed are the two important points that need to be studied properly to reduce the nanoparticle cytotoxicity (de Jong and Borm, 2008; Hartung, 2018; Yildirimer et al., 2011).

Even though the penetration of nanoparticles in the cosmetic formulation through the dermis, epidermis, and systemic circulation is undesirable, small nanoparticles can penetrate the skin through hydrophilic channels. The problem is that if nanoparticles that enter the bloodstream are not biocompatible or biodegradable, they can accumulate and cause cytotoxicity in any organ. Therefore, biocompatible organic polymers are to be selected for the preparation of polymeric nanoparticles. Furthermore, by-products of the synthesis of nanoparticles can be harmful, thus it is critical to carefully consider the right production process of polymeric materials and related nano systems (Bahadar et al., 2016; Bolzinger et al., 2011; Kimura et al., 2012; Nohynek et al., 2008).

Nanoparticles can easily enter and deposit in the respiratory tract or lungs due to their small size. According to some animal studies, it is found that nanoparticles can enter systemic circulation through the lungs and distribute to other organs and cause cytotoxicity, including in the brain (Poh et al., 2018). The inhalation risk of nanoparticles depends on their tendency to become airborne, which is determined by their density, particle size, and inherent electrostatic forces. This risk will be higher in powder-based cosmetic products (Suriya Prabha et al., 2020). Therefore, the toxicity of nanoparticle-based cosmetics must be carefully studied in all aspects, both in vitro and in vivo, and regulatory bodies should provide further guidance and regulation of these studies.

10.5 REGULATORY ASPECTS

The lack of standard and uniform guidelines across the regulatory agencies in the world is the major challenge in regulating nanoparticle-based cosmetic products (Foulkes et al., 2020). Nanoparticles are difficult to define and classify. Given the number of characteristics that must be considered, as well as the wide range of nano-constituents, models, and approaches, this is a complex issue that has failed several attempts to define universal rules (Parra et al., 2021a). The Food and Drug Administration (FDA) and the European Medicines Union (EMU) have published various guidelines providing several standards for the preparation of a variety of nanoparticle-based products in the EU and the United States (Bowman et al., 2018). FDA has published guidance on the safety of nanoparticle-based cosmetics and identifying the nanoparticle-based products, detailing how the administration views the nanoparticles' use in the products that it regulates. According to FDA (2014a), the current safety assessment framework for nanoparticles is robust as well as flexible. In 2014b, the FDA recommended that product-specific features for nanoparticle-based cosmetics be thoroughly assessed. The FDA guidance (2014a) also suggests that the safety of nanoparticle-based cosmetics should be assessed thoroughly by clearly describing the nanoparticle used and evaluating its physicochemical properties ("Guidance for Industry: Safety of Nanomaterials in Cosmetic Products I FDA," n.d.). Because of the increased use of nanoparticles, FDA established a task force in 2007 to work on the FDA's future steps regarding nanoparticle-based products (Nitheesh et al., 2021). Furthermore, the European parliament and council decided to regulate nano-cosmetics as cosmetic products, but with higher regulations in place before they come to market. In EU Regulation 1223/2009, Chapter III of article 13 provides information regarding safety assessment and product information file (PIF), whereas chapters IV of 16, and VI of 19 have restrictions for specific substances and consumer information, respectively. Therefore, cosmetic products containing the nanoparticles should meet the requirements provided in these articles ("Regulation (EC) No. 1223/2009 of the European Parliament and of the Council of 30 November 2009 on cosmetic products (recast) (Text with EEA relevance)," n.d.).

In European Union (EU), six months before its commercialization, nanomaterials in the product should be notified in the EU Cosmetic Products Notification Portal (CPNP). The European Commission created a catalog of approved nanomaterials for cosmetic ("Cosmetic product notification portal," n.d.). Unlike the EU, the USA does not have any requirement for the approval of cosmetic ingredients except the colorant agents. Even though both the regulatory bodies are having a specific focus on nanoparticle-based products, the lack of standard guidelines and regulations regarding their safety assessment and the manufacturing processes is because of the wide and complex concept of nanotechnology. Thus, regulatory bodies should provide product-specific guidelines based on its physicochemical properties, composition, and route of administration. For example, since a small change in the copolymer can affect the efficacy and safety of the polymeric nanoparticle, a well-defined and controlled manufacturing process is necessary (Gupta et al., 2022; Parra et al., 2021a). Additionally, specific tests evaluating the impact of the selected copolymers on the bioactivity of polymeric nanoparticles should be provided. Furthermore, quality control should have product-specific standards for starting materials, important intermediates, and manufacturing contaminants, as well as process validation and evaluation.

10.6 CONCLUSION

Nanotechnology has become a prime interest for research in various fields including cosmetics. Several studies have shown that polymeric micelle/dendrimer-based products have various applications in the cosmetics industry, and it seems that there is a lot of potential in future. The major advantage of nanoparticle-based cosmetics is enhanced and long-lasting performance because of their controlled and target specific delivery of the active ingredient. The competition and evolution of the cosmetic industry are well known, and the modern world consumers are more aware of the ingredients and demanding galenic products. As a result, the industry is facing more pressure to produce novel products. In general, natural cosmetic ingredients have low hydrophilicity and solubility, which makes them difficult to formulate. Several studies have shown that formulations with polymeric micelle or dendrimer nanocarriers can increase the solubility and resolve the formulation issue by loading those ingredients in their hydrophobic core. From the review of some of the studies conducted in past few years, it is affirmed to say that both polymeric micelle and dendrimer-based formulations can be a better option for many cosmetic applications including anti-aging, anti-acne, hair care, and sunscreens. Although these nanoparticle-based products are gaining significant value in the cosmetics market, there is considerable argument about their safety and toxicity, demanding further research into their short- and long-term effects on the human body. Therefore, there should be well-defined regulations on both research and manufacturing of these nano-cosmetic products.

REFERENCES

Abbasi, E., Aval, S.F., Akbarzadeh, A., Milani, M., Nasrabadi, H.T., Joo, S.W., Hanifehpour, Y., Nejati-Koshki, K., Pashaei-Asl, R., 2014. Dendrimers: Synthesis, applications, and properties. *Nanoscale Research Letters* 9, 247. https://doi.org/10.1186/1556-276X-9-247

Almeida, M., Magalhães, M., Veiga, F., Figueiras, A., 2017. Poloxamers, poloxamines and polymeric micelles: Definition, structure and therapeutic applications in cancer. *Journal of Polymer Research* 25:1, 1–14. https://doi.org/10.1007/S10965-017-1426-X

An, J.Y., Yang, H.S., Park, N.R., Koo, T. Sung, Shin, B., Lee, E.H., Cho, S.H., 2020a. Development of polymeric micelles of oleanolic acid and evaluation of their clinical efficacy. *Nanoscale Research Letters* 15, 1–14. https://doi.org/10.1186/S11671-020-03348-3/FIGURES/13

An, J.Y., Yang, H.S., Park, N.R., Koo, T. Sung, Shin, B., Lee, E.H., Cho, S.H., 2020b. Development of polymeric micelles of oleanolic acid and evaluation of their clinical efficacy. *Nanoscale Research Letters* 15, 1–14. https://doi.org/10.1186/S11671-020-03348-3/FIGURES/13

Aurelia Chis, A., Dobrea, C., Morgovan, C., Arseniu, A.M., Rus, L.L., Butuca, A., Juncan, A.M., Totan, M., Vonica-Tincu, A.L., Cormos, G., Muntean, A.C., Muresan, M.L., Gligor, F.G., Frum, A., 2020. Applications and limitations of dendrimers in biomedicine. *Molecules* 25. https://doi.org/10.3390/MOLECULES25173982

Bahadar, H., Maqbool, F., Niaz, K., Abdollahi, M., 2016. Toxicity of nanoparticles and an overview of current experimental models. *Iranian Biomedical Journal* 20, 1. https://doi.org/10.7508/IBJ.2016.01.001

Bilal, M., Iqbal, H.M.N., 2020. New insights on unique features and role of nanostructured materials in cosmetics. *Cosmetics* 7, 24. https://doi.org/10.3390/COSMETICS7020024

Biniek, K., Kaczvinsky, J., Matts, P., Dauskardt, R.H., 2015. Understanding age-induced alterations to the biomechanical barrier function of human stratum corneum. *Journal of Dermatological Science* 80, 94–101. https://doi.org/10.1016/J.JDERMSCI.2015.07.016

Bolzinger, M.A., Briançon, S., Chevalier, Y., 2011. Nanoparticles through the skin: Managing conflicting results of inorganic and organic particles in cosmetics and pharmaceutics. *Wiley Interdisciplinary Reviews: Nanomedicine and Nanobiotechnology* 3, 463–78. https://doi.org/10.1002/WNAN.146

Bose, A., Burman, D.R., Sikdar, B., Patra, P., 2021. Nanomicelles: Types, properties and applications in drug delivery. *IET Nanobiotechnology* 15, 19–27. https://doi.org/10.1049/NBT2.12018

Bowman, D.M., May, N.D., Maynard, A.D., 2018. Nanomaterials in cosmetics: Regulatory aspects. In *Analysis of Cosmetic Products: Second Edition*, pp. 289–302. https://doi.org/10.1016/B978-0-444-63508-2.00012-6

Braunová, A., Kaňa, M., Kudláčová, J., Kostka, L., Bouček, J., Betka, J., Šírová, M., Etrych, T., 2019. Micelle-forming block copolymers tailored for inhibition of P-gp-mediated multidrug resistance: Structure to activity relationship. *Pharmaceutics* 11. https://doi.org/10.3390/PHARMACEUTICS11110579

Chauhan, A., Patil, C., Jain, P., Kulhari, H., 2019. Dendrimer-based marketed formulations and miscellaneous applications in cosmetics, veterinary, and agriculture. *Pharmaceutical Applications of Dendrimers* 325–34. https://doi.org/10.1016/B978-0-12-814527-2.00014-7

Chuo, S.C., Mohd Setapar, S.H., 2022. Application of nanotechnology for development of cosmetics. In *Nanotechnology for the Preparation of Cosmetics Using Plant-Based Extracts*, 327–44. https://doi.org/10.1016/B978-0-12-822967-5.00004-7

Cosmetic product notification portal [WWW Document], n.d. URL https://ec.europa.eu/growth/sectors/cosmetics/cosmetic-product-notification-portal_en (accessed 6.4.22).

de Jong, W.H., Borm, P.J.A., 2008. Drug delivery and nanoparticles: Applications and hazards. *International Journal of Nanomedicine* 3, 133. https://doi.org/10.2147/IJN.S596

Dubey, S.K., Dey, A., Singhvi, G., Pandey, M.M., Singh, V., Kesharwani, P., 2022. Emerging trends of nanotechnology in advanced cosmetics. *Colloids and Surfaces B: Biointerfaces* 214, 112440. https://doi.org/10.1016/J.COLSURFB.2022.112440

Filipczak, N., Yalamarty, S.S.K., Li, X., Parveen, F., Torchilin, V., 2021. Developments in treatment methodologies using dendrimers for infectious diseases. *Molecules* 26. https://doi.org/10.3390/MOLECULES26113304

Filipowicz, A., Wołowiec, S., 2012. Bioconjugates of PAMAM dendrimers with trans-retinal, pyridoxal, and pyridoxal phosphate. *International Journal of Nanomedicine* 7, 4819–28. https://doi.org/10.2147/IJN.S34175

Foulkes, R., Man, E., Thind, J., Yeung, S., Joy, A., Hoskins, C., 2020. The regulation of nanomaterials and nanomedicines for clinical application: Current and future perspectives. *Biomaterials Science* 8, 4653–64. https://doi.org/10.1039/D0BM00558D

Ghezzi, M., Pescina, S., Padula, C., Santi, P., del Favero, E., Cantù, L., Nicoli, S., 2021. Polymeric micelles in drug delivery: An insight of the techniques for their characterization and assessment in biorelevant conditions. *Journal of Controlled Release* 332, 312–36. https://doi.org/10.1016/J.JCONREL.2021.02.031

Guidance for Industry: Safety of Nanomaterials in Cosmetic Products | FDA [WWW Document], n.d. URL https://www.fda.gov/regulatory-information/search-fda- guidance-documents/guidance-industry-safety-nanomaterials-cosmetic-products (accessed 6.4.22).

Güngör, S., Kahraman, E., Özsoy, Y., 2015. Nano Based Drug Delivery Chapter: 14; IAPC Open Book and Monograph Platform (OBP) Editors: Jitendra Naik. Polymeric micelles for cutaneous drug delivery contents. 368–388. DOI: 10.5599/obp.8.14

Guo, Y., Gao, T., Fang, F., Sun, S., Yang, D., Li, Y., Lv, S., 2021. A novel polymer micelle as a targeted drug delivery system for 10-hydroxycamptothecin with high drug-loading properties and anti-tumor efficacy. *Biophysical Chemistry* 279, 106679. https://doi.org/10.1016/J.BPC.2021.106679

Gupta, A., Dubey, S., Mishra, M., 2018a. Unique structures, properties and applications of dendrimers. *Journal of Drug Delivery and Therapeutics* 8, 328–39. https://doi.org/10.22270/JDDT.V8I6-S.2083

Gupta, A., Dubey, S., Mishra, M., 2018b. Unique structures, properties and applications of dendrimers. *Journal of Drug Delivery and Therapeutics* 8, 328–39. https://doi.org/10.22270/JDDT.V8I6-S.2083

Gupta, V., Mohapatra, S., Mishra, H., Farooq, U., Kumar, K., Ansari, M.J., Aldawsari, M.F., Alalaiwe, A.S., Mirza, M.A., Iqbal, Z., 2022. Nanotechnology in cosmetics and cosmeceuticals—A review of latest advancements. *Gels* 8. https://doi.org/10.3390/GELS8030173

Hartung, T., 2018. Perspectives on in vitro to in vivo extrapolations. *Applied in Vitro Toxicology* 4, 305. https://doi.org/10.1089/AIVT.2016.0026

Holmes, A.M., Scurr, D.J., Heylings, J.R., Wan, K.W., Moss, G.P., 2017. Dendrimer pretreatment enhances the skin permeation of chlorhexidine digluconate: Characterisation by in vitro percutaneous absorption studies and Time-of-Flight Secondary Ion Mass Spectrometry. *European Journal of Pharmaceutical Sciences* 104, 90–101. https://doi.org/10.1016/J.EJPS.2017.03.034

Hwang, D., Ramsey, J.D., Kabanov, A. v., 2020. Polymeric micelles for the delivery of poorly soluble drugs: From nanoformulation to clinical approval. *Advanced Drug Delivery Reviews* 156, 80–118. https://doi.org/10.1016/J.ADDR.2020.09.009

Kahraman, E., Özhan, G., Özsoy, Y., Güngör, S., 2016. Polymeric micellar nanocarriers of benzoyl peroxide as potential follicular targeting approach for acne treatment. *Colloids and Surfaces B: Biointerfaces* 146, 692–99. https://doi.org/10.1016/J.COLSURFB.2016.07.029

Kandekar, S.G., Del Río-Sancho, S., Lapteva, M., Kalia, Y.N., 2018a. Selective delivery of adapalene to the human hair follicle under finite dose conditions using polymeric micelle nanocarriers. *Nanoscale* 10, 1099–110. https://doi.org/10.1039/C7NR07706H

Kandekar, S.G., Del Río-Sancho, S., Lapteva, M., Kalia, Y.N., 2018b. Selective delivery of adapalene to the human hair follicle under finite dose conditions using polymeric micelle nanocarriers. *Nanoscale* 10, 1099–110. https://doi.org/10.1039/C7NR07706H

Kandregula, B., Narisepalli, S., Chitkara, D., Mittal, A., 2021. Exploration of lipid-based nanocarriers as drug delivery systems in diabetic foot ulcer. *Molecular Pharmaceutics.* https://doi.org/10.1021/ACS.MOLPHARMACEUT.1C00970/ASSET/IMAGE S/MEDIUM/MP1C00970_0008.GIF

Kaul, S., Gulati, N., Verma, D., Mukherjee, S., Nagaich, U., 2018a. Role of nanotechnology in cosmeceuticals: A review of recent advances. *Journal of Pharmaceutics* 2018, 1–19. https://doi.org/10.1155/2018/3420204

Kaul, S., Gulati, N., Verma, D., Mukherjee, S., Nagaich, U., 2018b. Role of nanotechnology in cosmeceuticals: A review of recent advances. *Journal of Pharmaceutics* 2018, 3420204. https://doi.org/10.1155/2018/3420204

Kaul, S., Gulati, N., Verma, D., Mukherjee, S., Nagaich, U., 2018c. Role of nanotechnology in cosmeceuticals: A review of recent advances. *Journal of Pharmaceutics* 2018, 1–19. https://doi.org/10.1155/2018/3420204

Kesharwani, P., Tekade, R., Today, N.J.-D.Discov., 2015, undefined, n.d. Dendrimer generational nomenclature: The need to harmonize. *Drug Discovery Today* 20, 497–99.

Khan, I., Gothwal, A., Mishra, G., Gupta, U., 2018. *Polymeric Micelles*, pp 1–29. https://doi.org/10.1007/978-3-319-92066-5_11-1

Khanna, P., Ong, C., Bay, B.H., Baeg, G.H., 2015. Nanotoxicity: An interplay of oxidative stress, inflammation and cell death. *Nanomaterials* 5, 1163. https://doi.org/10.3390/NANO5031163

Khurana, B., Arora, D., Narang, R.K., 2020. QbD based exploration of resveratrol loaded polymeric micelles based carbomer gel for topical treatment of plaque psoriasis: In vitro, ex vivo and in vivo studies. *Journal of Drug Delivery Science and Technology* 59, 101901. https://doi.org/10.1016/J.JDDST.2020.101901

Kim, B., Cho, H.-E., Moon, S.H., Ahn, H.-J., Bae, S., Cho, H.-D., An, S., 2004. Penetration enhancers. *Advanced Drug Delivery Reviews* 56, 603–18. https://doi.org/10.1186/s41702-020-0058-7

Kimura, E., Kawano, Y., Todo, H., Ikarashi, Y., Sugibayashi, K., 2012. Measurement of skin permeation/penetration of nanoparticles for their safety evaluation. *Biological and Pharmaceutical Bulletin* 35, 1476–86. https://doi.org/10.1248/BPB.B12-00103

Kulthe, S.S., Choudhari, Y.M., Inamdar, N.N., Mourya, V., 2012. Polymeric micelles: Authoritative aspects for drug delivery. *Designed Monomers and Polymers* 15, 465–521. https://doi.org/10.1080/1385772X.2012.688328

Lapteva, M., Möller, M., Gurny, R., Kalia, Y.N., 2015. Self-assembled polymeric nanocarriers for the targeted delivery of retinoic acid to the hair follicle. *Nanoscale* 7, 18651–62. https://doi.org/10.1039/C5NR04770F

Lohani, A., Verma, A., Joshi, H., Yadav, N., Karki, N., 2014. Nanotechnology-based cosmeceuticals. *ISRN Dermatology* 2014, 1–14. https://doi.org/10.1155/2014/843687

Makhmalzade, B.S., Chavoshy, F., 2018. Polymeric micelles as cutaneous drug delivery system in normal skin and dermatological disorders. *Journal of Advanced Pharmaceutical Technology & Research* 9, 2. https://doi.org/10.4103/JAPTR.JAPTR_314_17

Malik, A., Chaudhary, S., Garg, G., Tomar, A., 2012. Dendrimers: A tool for drug delivery. *Advances in Biological Research* 6, 165–69. https://doi.org/10.5829/idosi.abr.2012.6.4.6254

Malkoch, M., García-Gallego, S., 2020. Chapter 1: Introduction to dendrimers and other dendritic polymers. In *Monographs in Supramolecular Chemistry 2020-January*, pp. 1–20. https://doi.org/10.1039/9781788012904-00001

Markelov, D.A., Semisalova, A.S., Mazo, M.A., 2021. Formation of a hollow core in dendrimers in solvents. *Macromolecular Chemistry and Physics* 222, 2100085. https://doi.org/10.1002/MACP.202100085

Middleton, J.M., Siefert, R.L., James, M.H., Schrand, A.M., Kolel-Veetil, M.K., Middleton, J.M., Siefert, R.L., James, M.H., Schrand, A.M., Kolel-Veetil, M.K., 2021. Micelle formation, structures, and metrology of functional metal nanoparticle compositions. *AIMS Materials Science* 4, 560–86. https://doi.org/10.3934/MATERSCI.2021035

Miller, T., van Colen, G., Sander, B., Golas, M.M., Uezguen, S., Weigandt, M., Goepferich, A., 2013. Drug loading of polymeric micelles. *Pharmaceutical Research* 30, 584–95. https://doi.org/10.1007/S11095-012-0903-5

Mittal, P., Saharan, A., Verma, R., Altalbawy, F.M.A., Alfaidi, M.A., Batiha, G.E.S., Akter, W., Gautam, R.K., Uddin, M.S., Rahman, M.S., 2021. Dendrimers: A new race of pharmaceutical nanocarriers. *BioMed Research International* 2021. https://doi.org/10.1155/2021/8844030

Munavalli, B.B., Naik, S.R., Torvi, A.I., Kariduraganavar, M.Y., 2019. Dendrimers. In *Functional Polymers*, pp. 289–345. https://doi.org/10.1007/978-3-319-95987-0_9

Nilsson, C., 2008. Dendrimers: Synthesis, characterization and use in thiol-ene networks. urn:nbn:se:kth:diva-9763; oai:DiVA.org:kth-9763.

Nitheesh, Y., Pradhan, R., Hejmady, S., Taliyan, R., Singhvi, G., Alexander, A., Kesharwani, P., Dubey, S.K., 2021. Surface engineered nanocarriers for the management of breast cancer. *Materials Science and Engineering: C* 130, 112441. https://doi.org/10.1016/J.MSEC.2021.112441

Nohynek, G.J., Dufour, E.K., Roberts, M.S., 2008. Nanotechnology, cosmetics and the skin: Is there a health risk? *Skin Pharmacology and Physiology* 21, 136–49. https://doi.org/10.1159/000131078

Parra, A., Jarak, I., Santos, A., Veiga, F., Figueiras, A., 2021a. Polymeric micelles: A promising pathway for dermal drug delivery. *Materials* 14. https://doi.org/10.3390/MA14237278

Parra, A., Jarak, I., Santos, A., Veiga, F., Figueiras, A., 2021b. Polymeric micelles: A promising pathway for dermal drug delivery. *Materials* 14. https://doi.org/10.3390/MA14237278

Passeno, L.M., Mackay, M.E., Baker, G.L., Vestberg, R., Hawker, C.J., 2006. Conformational changes of linear-dendrimer diblock copolymers in dilute solution. *Macromolecules* 39, 740–6. https://doi.org/10.1021/MA0517291/SUPPL_FILE/MA0517291SI20051105_094826.PDF

Pentek, T., Newenhouse, E., O'Brien, B., Singh Chauhan, A., 2017. Development of a topical resveratrol formulation for commercial applications using dendrimer nanotechnology. *Molecules* 2017, 22, 137. https://doi.org/10.3390/MOLECULES22010137

Perinelli, D.R., Cespi, M., Lorusso, N., Palmieri, G.F., Bonacucina, G., Blasi, P., 2020. Surfactant self-assembling and critical micelle concentration: One approach fits all? *Langmuir* 36, 5745–53. https://doi.org/10.1021/ACS.LANGMUIR.0C00420/ASSET/IMAGES/LARG E/LA0C00420_0006.JPEG

Poh, T.Y., Ali, N.A.T.B.M., mac Aogáin, M., Kathawala, M.H., Setyawati, M.I., Ng, K.W., Chotirmall, S.H., 2018. Inhaled nanomaterials and the respiratory microbiome: Clinical, immunological and toxicological perspectives. *Particle and Fibre Toxicology* 15, 1–16. https://doi.org/10.1186/S12989- 018-0282-0

Raj, S., Jose, S., Sumod, U.S., Sabitha, M., 2012. Nanotechnology in cosmetics: Opportunities and challenges. *Journal of Pharmacy & Bioallied Sciences* 4, 186. https://doi.org/10.4103/0975-7406.99016

Ramanathan, A., 2019. Toxicity of nanoparticles_challenges and opportunities. *Applied Microscopy* 49, 1–11. https://doi.org/10.1007/S42649-019-0004-6/TABLES/2

Regulation (EC) No 1223/2009 of the European Parliament and of the Council of 30 November 2009 on cosmetic products (recast) (Text with EEA relevance) [WWW Document], n.d. URL https://www.legislation.gov.uk/eur/2009/1223/contents (accessed 6.4.22).

Salvioni, L., Morelli, L., Ochoa, E., Labra, M., Fiandra, L., Palugan, L., Prosperi, D., Colombo, M., 2021. The emerging role of nanotechnology in skincare. *Advances in Colloid and Interface Science* 293, 102437. https://doi.org/10.1016/J.CIS.2021.102437

Santos, A., Veiga, F., Figueiras, A., 2019. Dendrimers as pharmaceutical excipients: Synthesis, properties, toxicity and biomedical applications. *Materials (Basel)* 13, 65. https://doi.org/10.3390/MA13010065

Šmejkalová, D., Muthný, T., Nešporová, K., Hermannová, M., Achbergerová, E., Huerta-Angeles, G., Svoboda, M., Čepa, M., Machalová, V., Luptáková, D., Velebný, V., 2017a. Hyaluronan polymeric micelles for topical drug delivery. *Carbohydrate Polymers* 156, 86–96. https://doi.org/10.1016/J.CARBPOL.2016.09.013

Šmejkalová, D., Muthný, T., Nešporová, K., Hermannová, M., Achbergerová, E., Huerta-Angeles, G., Svoboda, M., Čepa, M., Machalová, V., Luptáková, D., Velebný, V., 2017b. Hyaluronan polymeric micelles for topical drug delivery. *Carbohydrate Polymers* 156, 86–96. https://doi.org/10.1016/J.CARBPOL.2016.09.013

Sukhanova, A., Bozrova, S., Sokolov, P., Berestovoy, M., Karaulov, A., Nabiev, I., 2018. Dependence of nanoparticle toxicity on their physical and chemical properties. *Nanoscale Research Letters* 13, 1–21. https://doi.org/10.1186/S11671-018-2457-X

Suriya Prabha, A., Dorothy, R., Jancirani, S., Rajendran, S., Singh, G., Senthil Kumaran, S., 2020. Recent advances in the study of toxicity of polymer-based nanomaterials. *Nanotoxicity: Prevention and Antibacterial Applications of Nanomaterials* 143–65. https://doi.org/10.1016/B978-0-12-819943-5.00007-5

Tiwari, N., Osorio-Blanco, E.R., Sonzogni, A., Esporrín-Ubieto, D., Wang, H., Calderón, M., 2022. Nanocarriers for skin applications: Where do we stand? *Angewandte Chemie International Edition* 61, e202107960. https://doi.org/10.1002/ANIE.202107960

Uchechi, O., Ogbonna, J.D.N., Attama, A.A., 2014. Nanoparticles for dermal and trans-dermal drug delivery. *Application of Nanotechnology in Drug Delivery*. https://doi.org/10.5772/58672

Virmani, R., Pathak, K., 2021. Consumer nanoproducts for cosmetics. In *Handbook of Consumer Nanoproducts*, pp. 1–32. https://doi.org/10.1007/978-981-15-6453-6_58-1

Viswanath, V., Santhakumar, K., 2017. Perspectives on dendritic architectures and their biological applications: From core to cell. *Journal of Photochemistry and Photobiology B: Biology* 173, 61–83. https://doi.org/10.1016/J.JPHOTOBIOL.2017.05.023

Wichit, A., Tangsumranjit, A., Pitaksuteepong, T., Waranuch, N., 2012a. Polymeric micelles of PEG-PE as carriers of all-trans retinoic acid for stability improvement. *AAPS PharmSciTech* 13, 336–43. https://doi.org/10.1208/S12249-011-9749-0

Wichit, A., Tangsumranjit, A., Pitaksuteepong, T., Waranuch, N., 2012b. Polymeric micelles of PEG-PE as carriers of all-trans retinoic acid for stability improvement. *AAPS PharmSciTech* 13, 336–43. https://doi.org/10.1208/S12249-011-9749-0

Wolf, B., F Snyder — US Patent 5, 449,519, 1995, undefined, n.d. Cosmetic compositions having keratolytic and anti-acne activity. Google Patents.

Yildirimer, L., Thanh, N.T.K., Loizidou, M., Seifalian, A.M., 2011. Toxicology and clinical potential of nanoparticles. *Nano Today* 6, 585–607. https://doi.org/10.1016/J.NANTOD.2011.10.001

11 Nanoparticles mediated drug-delivery system in cosmetics

Sristi, Afsana Sheikh, Amirhossein Sahebkar,
Mohammad Sarwar Alam, Waleed H.
Almalki and Prashant Kesharwani

CONTENTS

11.1 INTRODUCTION

Materials that fall within the nanoscale, or under 100 nm, are known as nanoparticles (NMs). The relevance of these materials in modern cosmetics has increased recently. Cosmetics are "particles intended to be applied onto human bodies or any part thereof for cleansing, beautifying, promoting attractiveness, or altering the appearance" according to the Food and Drug Administration (FDA) ("Importing Cosmetics I FDA," n.d.). The term "nanocosmeceutical" was first used to describe nanotechnology when it was first introduced to the cosmetics sector in the early 1980s. The cosmetic industry was one of the first to adopt nanotechnology because of its considerable

DOI: 10.1201/9781003319146-11

contribution to better skin penetration, stability, and component release through the skin barrier, which results in greater aesthetic effects. Titanium oxide and zinc oxide are the two most prevalent varieties of manmade nanomaterials. However, additional nanosized materials have started to be employed in cosmetic applications, including metals like gold and silver, metal oxides, liposomes, nanocapsules, cubosomes, dendrimers, niosomes, and solid lipid nanoparticles. Numerous cosmeceuticals, including sunscreens, moisturizing and whitening creams, face cleansers, shampoos for hair repair, and anti-aging products, incorporate nanotechnology. Nanoparticles can be created using a variety of techniques, including chemical, physical, and biological ones. Even though the chemical technique of synthesis produces a huge number of nanoparticles in a short amount of time, this method needs a variety of capping agents to stabilize the size of the nanoparticles. Chemicals employed in the synthesis and stabilization of nanoparticles may be hazardous and produce byproducts that affect the environment. The growing interest in biological methods that do not use harmful chemicals as byproducts is a result of the demand for environmentally safe synthetic methods for the manufacture of nanoparticles (Ghoshal and Singh, 2022). The cosmetic industry uses NMs to achieve stable products with long-lasting effects. Nanomaterials' large surface areas enable more efficient delivery of the chemicals through the skin. One of the major purposes of using nanomaterials in cosmetics might be the efficient penetration into the skin for improved delivery of the product's ingredients, new color components (e.g., in lipsticks and nail polishes), transparency (e.g., in sunscreens), and long-lasting effects (e.g., in makeup). The use of nanomaterials in the field of cosmetic industry prompted the development of transdermal delivery system. The breakthrough in the mechanism of such nanoparticle related cosmetic system, it could be expected the employment of novel innovations in the coming future. As cosmetics products do not call for clinical assessment, the production of cosmetic products is uncomplicated in terms of financial assistance and time. However, the current regulatory guidelines are necessary to be followed to adhere to the safety guidelines laid in the interest with cosmetic field.

11.2   TOPICAL ADMINISTRATION THROUGH THE SKIN

One of the primary roles of the skin is to serve as a barrier and protection from external substances. However, it is also well known that many substances have the ability to penetrate and permeate this barrier and, in some cases, even cause local effects, such as irritability or sensitization events, or even translocate into blood vessels and affect systemic circulation. Since it has been shown that some NPs can pass through the stratum corneum layer of the skin's surface and into the deeper dermal layer, where they can enter the systemic circulation, the interaction of NPs with the skin is still a subject of research. The skin, one of the larger organs in the human body, has a variety of purposes, including preventing the advent of Langerhans cells, which may digest antigens and initiate the inflammatory response in exposure to irritants from the outside world, are responsible for this latter function. In addition, skin contributes to the creation of essential chemicals such as keratin, collagen, melanin, lipids, and carbohydrates. Resident heat, touch, and pain receptors also enable the skin to serve as a neurosensory organ. The dermis and epidermis make up the two layers that constitute the skin barrier. A stratified squamous keratinized epithelium, which includes both keratinocytes and non-keratinocytes as primary cells, constitutes the epidermis. The former cells are organized into layers called the stratum basale, stratum spinosum, stratum granulosum, stratum lucidum, and stratum corneum, whereas the latter cells have immune and UV-light protection roles (melanocytes, Merkel cells, and Langerhans cells) (Larese Filon et al., 2015). The stratum corneum, the skin's top layer, serves as the body's initial line of defense against outside influences and controls how quickly substances like chemicals, medicines, and particles can penetrate deeper skin layers. The literature on permeation of nanosystem through the skin shows dependency on particle size, permeation mechanism considering both trans-appendageal and transcellular in addition to depth of permeation. There is no size limit for trans-appendageal route, means the size of the particles should be within nanometric range, however, their entry is limited in small parts of human system. On contrary, the paracellular or transcellular pathway, due to the compact structure of stratum corneum limit the entry of molecules which are lipophilic in nature considering the molecular weight below 200 Da. The design of skin is such that the stratum corneum comprising of the corneocytes are designed as clusters which are surrounded by canyons. Having a size range of 20–30 nm which is larger than 0.4 nm sized inter-corneocytes spaces, only inter-cluster pathway is allowed for permeation. Owing to the chemical composition, the canyons act as reservoir for lipophilic constituents.

11.3 NANOMATERIALS USED IN THE COSMETICS DOMAIN

11.3.1 INORGANIC NANOPARTICLES

Numerous inorganic nanoparticles have the potential to be employed as carriers for the cellular delivery of different drugs. To meet the strict requirements for cellular delivery, such as good biocompatibility, the strong affinity between the carriers and biomolecules, the high charge density of the nanohybrids, site-specificity, etc., the majority of inorganic nanoparticles should be subjected to chemical and/or biological modification (Xu et al., 2006). Compared to organic nanoparticles, inorganic nanoparticles are non-toxic, hydrophilic, biocompatible, and highly stable (Fytianos et al., 2020). UV protection and antibacterial activity are the two main uses for inorganic nanoparticles. The most frequently used types of inorganic materials are metal and metal oxide nanoparticles. Based on their ability to enhance skin appearance, protect the skin from external causes, and perform cleaning duties, inorganic nanoparticles are utilized as active ingredients in cosmetic compositions. Inorganic nanoparticles' distinctive physicochemical characteristics are used by many products to attain desired functions (Vinod and Jelinek, 2019). The section below describes the use of various inorganic nanoparticles in the delivery of cosmetic products.

11.3.2 TITANIUM OXIDE AND ZINC OXIDE BASED NANOPARTICLES

UV filters made of titanium oxide (TiO_2) and zinc oxide (ZnO) nanoparticles have been successfully incorporated into a variety of cosmetic items. In actuality, ZnO is capable of reflecting UVA radiation while TiO2 is responsible for reflecting UVB rays. When these two oxides are used together, it provides effective sun protection as well as desirable qualities like transparency, spreadability, and superior texture without causing skin irritation, which is typically a side effect of chemical UV filters. Due to the deposition of TiO_2 and ZnO NPs on the stratum corneum's external surface, the UV protection offered by these systems is widely utilized. ZnO has also shown to be a desirable solution for the cosmetic and pharmaceutical industries because of its antibacterial property. Its antibacterial properties are a result of the generation of ROS and subsequent release of Zn^{2+}, which is toxic to bacteria and causes the breakdown of their cell walls. Consequently, a synergistic effect can be seen in the combination of ZnO and $TiO_2.$ The chemicals employed in topical formulations have a significant role in the ZnO NP's efficiency since they can interact and reduce the ZnO antibacterial activity, as is the case of antioxidants and EDTA [5]. ZnO NPs were made using a *P. austroarabica* (Yemeni mistletoe) extract, which were studied using a variety of techniques, including FTIR, PXRD, UV-Vis, SEM, EDS, TEM, and FL spectroscopy. The Zn^{2+} was reduced to ZnO NPs according to the FTIR analysis results, and the phytochemicals in the *P. austroarabica* extract served as stabilizing and capping agents as well as agents that reduced Zn^{2+} to ZnO NPs. They worked well as a nanocatalyst for the breakdown of harmful organic dyes. In comparison to *P. austroarabica* extract and ascorbic acid, they also demonstrated high antioxidant activity. This suggests that the environmentally friendly, non-toxic, and inexpensive green synthesized ZnO NPs from *P. austroarabica* extract could be a promising material as a bio-environmental probe to identify and eliminate harmful pollutants in a variety of fields. They could also be used as antioxidants (Alahdal et al., 2022). Zinc oxide (ZnO) nanoparticles are utilized in cosmetics, sun protection, coatings, and antibacterial products. It is primarily found in ointments, powders, and other cosmetic products.

Numerous cosmetic products use titanium oxide (TiO_2) nanoparticles as UV filters. The UVB radiation is reflected by titanium oxide. Because of this, using TiO_2 and ZnO NPs together offers good sun protection without causing skin discomfort, which is generally a sign of chemical UV filters (Abu Hajleh et al., 2021). The TiO_2 and ZnO NP's separate use could result in undesired opaqueness, however, their separate nanonization and micronization could help to overcome such problems (Smijs and Pavel, 2011).

11.3.3 SILVER-BASED NANOPARTICLES

Silver nanoparticles (AgNPs) are one of the most extensively explored metallic nanoparticles due to their potential for numerous uses. One of AgNPs' most admired characteristics is the potent inhibitory effect on a wide range of microbial species. Additionally, AgNPs exhibit remarkable anti-inflammatory properties that are essential for wound healing and medicinal applications. AgNPs are used to create skin cleansers, lotions, creams, deodorants, shampoos, and toothpaste because of their much-recognized medicinal properties (Ong and Nyam, 2022). AgNPs work by adhering to bacterial cell membranes promoting their effective internalization. The respiratory chain, cell division, and ultimately cell necrosis are the areas that nanoparticles tend to target. Silver ions are released into the bacterial cells, increasing their bactericidal activity (Gajbhiye et al., 2016). Similar to other nanoparticles, AgNPs can be made in a variety of ways, although the traditional method is expensive and uses a lot of energy because the heat treatment is also required. As a result, the green production of silver nanoparticles uses a variety of biological resources, including plants and microbes (yeast, fungi, and bacteria). The presence of biomolecules such as amino acids, proteins, vitamins, enzymes, and polysaccharides in these sources' extracts, together with secondary metabolites, aids in the reduction reaction (Ong and Nyam, 2022).

Fenugreek leaf extract, which contains the essential bioactive for reducing silver ions to silver nanoparticles, was used in the green synthesis process to create silver nanoparticles. The synthesized particles had a spherical form and had an average size of about 25 nm. The optimization process took into account the concentration, absorbance, and average particle size. With an average size of 30 nm and a spherical form, AgNPs with a size range between 4 and 30 nm were visible in FESEM images taken at an accelerating voltage of 20 KV. In comparison to other bacteria, *P. aeruginosa* was found to be more susceptible to the inhibitory effects of the produced AgNPs. Due to their small surface area, bigger nanoparticles are less active than smaller ones (Ghoshal and Singh, 2022).

According to a previous study, the tea includes phenolics, terpenoids, polysaccharides, flavonoids, and other phytochemicals that can be used to produce nanosilver by reducing silver nitrate. Additionally, these phytochemicals can interact with the surface of nanoparticles to stabilize them. According to X-ray diffraction (XRD) and the Scherrer equation, the average size of silver nanoparticles was 15.41 nm; the morphology is relatively uniform and evenly disseminated; the zeta potential was 13.91 0.89 mV, and the aggregate particle size distribution in water was around 773.49 nm. Moreover, *Pseudopestalotiopsis theae (P. theae) and Escherichia coli* (*E. coli*) were effectively inhibited by the produced nano silver, with inhibition ratios of 96.05% and 61.87%, respectively. Additionally, intermolecular forces may be responsible for the reactivity of silver nanoparticles with bacteria and fungi rather than electrostatic interactions (Fang et al., 2021).

Tyrosinase catalyzes tyrosine oxidation to dopaquinone, which is subsequently further oxidized to generate melanin. As a result, tyrosinase activity can be inhibited to restrict the development of melanin and therefore achieve a whitening effect. AgNPs are said to be able to prevent mushroom tyrosinase from activation. When it comes to apoptosis inhibition and wrinkle prevention, AgNPs can be very beneficial. Based on the measurement of kojic acid equivalents (KAE/g), it was discovered that AgNPs produced using the leaf extract of *E. ulmoides* were able to inhibit tyrosinase activity and behaved better than AgNPs produced using plant extracts from other sources *(Sideritis brevidens, S. lycia, and Bidens Frondosa)*. In both in vitro and in-cell tests, the AgNPs produced by *E. ulmoides* leaf extract showed potent anti-TYR activity and antioxidant characteristics without impairing cell viability. AgNPs can therefore be utilized in the domain of cosmetics as a whitening agent (Xi et al., 2022).

The creation of silver-doped ZnO (ZnOS) in SPA improved the performance of the liquid cleaning solution by establishing germicidal efficacies. In comparison to AR grade and nano ZnO, the morphological, optical, and germicidal characteristics of silver-doped zinc oxide (ZnOS) nanoparticles were examined. The outcomes demonstrated the successful synthesis of zinc oxide (ZnOS) doped with silver, which had an average crystalline size of 15.14 nm and a blue shift in absorbance peaks between 349 nm and 340 nm that may have been caused by the Burstein-Moss effect in the metal-doped system. XRD diffractogram analysis has been used to investigate crystallite size, dislocation density, specific surface area, and crystallinity characteristics. The SEM result reveals that the nano ZnOS surface shape is composed of densely packed, spherical particles. Researchers have looked at the anisotropic crystal structure of nano ZnOS and probable dopants of silver at the Zn site, dopants of silver at the O site, and Ag at the interstitial sites. The germicidal behavior of AR grade ZnO, nano ZnO, and ZnOS was also examined, and the results showed that nano ZnOS had higher levels of germicidal activity than either the AR grade or nano ZnO. The reduction in particle diameter has improved the germicidal activity of nano ZnO and nano ZnOS against *E. coli and S. aureus* when compared to AR-grade ZnO. As previously noted, as doping was increased, the band gap energy of nano ZnOS increases, and the dimensions of the nanomaterials shrink. Their stability and germicidal effectiveness were significantly impacted by this. The surface area to volume ratio rises with decrease in size. This suggests that the greater the number of silver atoms on the surface in contact with bacterial cells, the smaller the size portion. The surface of the nano ZnOS has a lot of edges, which makes the surface site potentially active. These nanoparticles' interactions with cell membranes and the resulting bacterial surface damage have been linked to germicidal activity. Due to the corners and edges on the abrasive surface of the nanomaterial, when these materials adhere to cell membranes, significant damage results. Nanoparticle microbial activity is significantly influenced by their surface shape (Bhalla et al., 2022).

11.3.4 GOLD-BASED NANOPARTICLES

In cosmetics, gold nanoparticles have shown their potential as anti-inflammatory, anti-aging, and lotions to treat skin wounds. Pollution, exposure to UV radiation from the sun, cigarette smoke, and other variables harm the skin cells causing the generation of reactive oxygen species (ROS). The expression of collagen- and elastin-degrading matrix metalloproteinases (MMP) is increased when there is an excess of reactive oxygen species, which in turn causes oxidative stress that destroys cells, DNA, and protein and accelerates the aging process of the skin. Giving the skin an additional supply of antioxidants is crucial to assisting it in protecting itself. Through the process of "green synthesis," bioactive components in plant extracts can transform metal ions into nanoparticles. *Panax ginseng* is an adaptogen and an oriental medicinal herb that has traditionally been used to treat a wide range of illnesses. The *P. ginseng* (Pg) leaves-mediated gold nanoparticles are also novel materials with added value. Pg-Au-NPs demonstrated the ability to retain moisture and efficiently inhibited mushroom tyrosinase. Pg-Au-NPs were not toxic to human dermal fibroblast and B16 cells, according to 3-(4,5-dimethyl-thiazol-2yl)-2,5-diphenyl tetrazolium bromide results. They also significantly decreased the amount of melanin, tyrosinase activity, and the mRNA expression of tyrosinase and melanogenesis-associated transcription factor in B16 cells (Jiménez-Pérez et al., 2018).

An indigenous plant from Reunion Island in the Indian Ocean called Hubertia ambavilla has been used both internally and externally for healing and as an anti-inflammatory. Metal salts can be turned into nanoparticles by polyphenolic compounds obtained through aqueous phase extractions. Gold nanoparticles produced by *Hubertia ambavilla* have the potential to scavenge free radicals, are non-toxic to human dermal fibroblasts, and can guard against UVA radiation's harmful effects on fibroblasts and dermal cells (Sakulwech et al., 2018). Various shapes of gold nanoparticles have been demonstrated in number of studies that range from nanorods, nanoshells, nanotriangles, nanocubes, nanostars to nanospheres. The driving force for optical behavior and cellular uptake is the state of such particles. Moreover, their anti-bacterial and anti-fungal properties enable their use further use in cosmetics industry. Soothing effect of gold establishes their use in the treatment of sunburn, inflammation, and hypersensitivity while improving the skin texture, flexibility, and grace (Akturk et al., 2016; Irshad et al., 2019).

Snail slime was used to create gold nanoparticles, giving them intriguing features, as an alternative and creative usage. The major components of the slime were used to decorate the inorganic metallic core of the 14 ± 6 nm wide hybrid gold nanoparticles, which were created using an eco-friendly, one-pot method (Rizzi et al., 2021). Interestingly, the photostability of gold nanoparticles, which was studied using a solar simulator lamp, points to their possible usage as a substitute for the inorganic sunscreen components that are often found in commercial cosmetic sunscreen solutions. The theoretical Sun Protection Factor was assessed, and values between 0 and 12 were obtained.

11.4 POLYMERIC NANOPARTICLES

The application of polymeric nanoparticles in skincare opens up many new possibilities for maintaining the physiological functions and health of the skin as well as improving cosmetics. Additionally, it enhances the physicochemical characteristics of conventional cream emulsions and improves the bioactive ingredients' ability to penetrate the skin's deep layers. The key benefits of employing polymeric nanoparticles in dermo-cosmetics and drug delivery entail building robust platforms that allow for cytotoxicity reduction, overcoming many biological barriers, and giving a targeted, sustained release of bioactive agents (Cantin et al., 2015).

11.4.1 CHITOSAN-BASED NANOPARTICLES

The special features of chitosan nanoparticles (CT NPs) make them suitable for biomedical applications. Sodium tripolyphosphate (TPP) and Acacia were used as crosslinkers to create CT NPs by the ionic gelation process. The molecular weight of chitosan and the concentration of both chitosan and crosslinkers seemed to have an impact on the particle sizes of NPs. By looking at their zeta potential values, CT NPs was positively charged. SEM and TEM micrographs demonstrates the structure of CT-Acacia and CT: TPP NPs, were both smooth, spherical in shape, and uniformly dispersed with a size range of 200–300 nm. Over 48 hours, the CT: TPP NPs were able maintain 98% of the added water on average. This indicate their capacity to entrap polar actives in cosmetics and release the encapsulated actives in low polarity skin circumstances. CT-Acacia NPs demonstrated significant moisture absorption but decreased moisture retention capacity. Confocal laser scanning microscopy demonstrated that CT NPs with a particle size of 530 nm and a marker of fluorescein sodium salt could pass through the skin of pigs and gather in the dermis (Ta et al., 2021). NPs can be employed as cosmetics and dermal medicine delivery systems since they have the ability to transfer active ingredients and cosmetic components via the skin.

The use of Hyaluronic acid in cosmetics is expanded by nanoparticles made from quaternized cyclodextrin-grafted chitosan. Hyaluronic acid (HA, 20–50 kDa) is a hydrophilic macromolecule that has moisturizing and anti-wrinkle actions. The benefits of this substance are, however, restricted to topical effects due to its high molecular weight, which prevents it from permeating into the deeper layers of the skin. To facilitate skin penetration and give anti-wrinkle and moisturizing properties, HA can be hydrolyzed to have a molecular weight of 20–50 kDa. As a result, HA is a top contender to be included with QCD-g-CS as a negatively charged macromolecule with a variety of uses. Fourier-transform infrared spectroscopy was used to confirm that the carboxylic moieties of HA and the amides of QCD-g-CS were conjugated. To produce nanoparticles with a nano-size (235.63 nm), a narrow polydispersity index (0.13 0.02), and a zeta potential of 16.07 0.65 mV, the system was tuned. Using ultra-performance liquid chromatography, the association efficiency and loading efficiency were found to be, respectively, 86.77 0.69% and 10.85 0.09%. The resulting nanoparticles' spherical shape was verified by transmission electron microscopy. Additionally, the in-vitro hydrating capacity was noticeably greater ($P \sim 0.001$) than that of bulk HA (3.29 ± 0.41 and 1.71 ± 0.05 g water/g sample, respectively) (Sakulwech et al., 2018).

By using an ionic gelation technique modified by sodium tripolyphosphate and genipin, chitosan (CS) based nanoparticles were created that were simultaneously loaded with (−)-epigallocatechin gallate (EGCG) and ferulic acid (FA) (G-CS-EGCG-FA NPs). In H_2O_2-induced cells, G-CS-EGCG-FA NPs (200 g/mL) had an oxidative repaired capacity that was over 100% higher than that of free EGCG or FA at the same dose. Additionally, the tyrosinase inhibitory activity of G-CS-EGCG-FA NPs (25 g/mL) (84.6%) was stronger than that of free EGCG (55.3%), free FA (47.1%), and kojic acid, demonstrating the G-CS-EGCG-FA NPs' potent skin-whitening and skin-repairing properties (Li et al., 2022).

Pickering emulsions were stabilized using chitosan/collagen peptide nanoparticles for potential cosmetic applications. The hydrophilic properties of the nanoparticles were evident from their contact angle, which was 78.02° 2.04°. Their typical dimensions were 32.27 nm, and their average zeta potential was +59.7 mV. The nanoparticles are well-adsorbed at the oil-water interface, according to confocal laser scanning microscopy (CLSM). The emulsions had a viscosity that reduced under shear and a texture similar to gel. Their typical droplet size varied from 7.63 to 15.72 micrometers. Following ex vivo skin application of the pickering emulsions, CLSM skin tracking of the nanoparticles showed that the emulsion droplets could penetrate the stratum corneum and deposit in deeper skin layers. The concentration of nanoparticles in the emulsion and the length of skin contact determine the degree of penetration (Sharkawy et al., 2021).

11.4.2 STARCH-BASED NANOPARTICLES

Starch nanoparticles (SNPs) are defined as particles larger than a single molecule yet with at least one dimension smaller than 1000 nm. Since, they are organic, regenerative, and biodegradable, starch nanoparticles (SNPs) are gaining popularity as a sustainable alternative to other types of nanomaterials. Starch granules can be broken down into SNPs in a variety of ways, including approaches that use both physical and chemical means. The operational conditions and the synthesis method employed both have a significant impact on the final properties of the SNPs. For several bio-applications, controlled and monodispersed size is critical. Due to their great biocompatibility, possible functionalization, and high surface/volume ratio, SNPs are regarded as a promising carrier to enhance the controlled release of several bioactive chemicals in various fields of research (Caldonazo et al., 2021; Campelo et al., 2020; Morán et al., 2021). Titania is a white pigment extensively employed in cosmetic industry due to excellent whitening properties, high refractive index and covering potential. However, the electrostatic interaction and high density, a successful dispersion of titania is hard to achieve in water-based systems. To tackle this issue, very recently, research was published wherein they encapsulated titania in Octenyl succinic anhydride (OSA) starch nanoparticle using spray drying method followed by dissolving the NP in κ-carrageenan solution to improve the biocompatibility. The process turned to be effective in successful dispersion of titania pigment, which could further be employed for easy dispersion of other pigments as well (Chen et al., 2022).

11.4.3 POLY-LACTIC ACID (PLA)-BASED NANOPARTICLES

The creation of pharmaceutical and cosmetic products based on PLA polymers led to the creation of a unique delivery system with multifunctional properties for possible use in the biomedical and cosmetics industries. PLA polymers have been used in cosmeceuticals and have shown to be effective for facial, and soft tissue fillers, body deodorants, and cleaning care products. These PLA polymers have distinct physical, chemical, biodegradable, and biocompatible properties that distinguish them from other polymers and can be thought of as the basis of a unique methodology in healthcare applications (Abu Hajleh et al., 2020).

Hollow polymeric nanoparticles have drawn a significant amount of attention as UV-shielding materials in personal care products due to their excellent light-scattering properties and low density. The migration of poly (vinyl alcohol) (PVA) (stabilizing agent) from the oil droplet to the oil/water interface while entangled with cross-linked P(LA-co-GMA) chains and the rapid evaporation rate of the chloroform solvent plays an essential role in the hollow structure formation. Monodispersed hollow nanoparticles with an ideal size range of 500–700 nm and good colloidal stability can be produced. The hollow nanoparticles consisting of biocompatible/degradable poly (lactic acid-*co*-glycidyl methacrylate) are highly UV-shielding and have low toxicity (Thananukul et al., 2022).

The loading of olive leaf extract into PLA NPs produces nanoparticles with adequate properties that can be successfully incorporated into cosmetic emulsions without degrading their stability or appearance. To create cosmetic compositions with favorable properties, sensitive polyphenolic extracts are enclosed in biodegradable PLA nanoparticles (Kesente et al., 2017).

11.4.4 POLYGLYCOLIC ACID (PGA) -BASED NANOPARTICLES

It is generally known that polymeric nanoparticles, particularly poly (D, L-lactic-co-glycolic acid) (PLGA), are the most frequently employed hydrophobic drug delivery systems. The US Food and Drug Administration has approved the use of PLGA in humans since it can be hydrolyzed into non-toxic, biodegradable metabolite monomers such as lactic acid and glycolic acid. Food and drug administration (FDA) has approved and marked PLGA safe for dermal delivery which is due to its biocompatibility and bioavailability profile. One of the major application of PLGA NP is their ability to entrap volatile oil at room temperature. Thymol loaded PLGA nanoparticles were developed having zeta potential of -28 mV and size below 200 nm for the treatment of acne. The anti-microbial activity demonstrated strong inhibition against *Cutibacterium acnes* while minimal effect was noticed on *Staphylococcus epidermi (an important* healthy skin microbiota resident).The nanoparticle here played an important role showing deeper dermal penetration, high retention and slow release (Folle et al., 2021). Thus, PLGA nanoparticles could be explored in near future for the treatment and increasing the aesthetics.

Astaxanthin (xanthophyll carotenoid) involves in highly beneficial biological processes like antioxidant activity and scavenging oxygen free radicals, although its use is restricted due to limited bioavailability and poor water solubility. Compared to pure astaxanthin, PLGA NP demonstrated superior antioxidant activity in the UVB radiation photodamage model. The AST-PLGA NP has the potential to be used in the area of cosmetics and could be considered as a potential treatment element for skin disorders (Kong et al., 2019; Pandita et al., 2015).

11.4.5 CELLULOSE-BASED NANOPARTICLES

Due to its morphology, mechanical strength, high purity, high water uptake, non-toxicity, chemical stability, and biocompatibility, cellulose is a potential novel material for a variety of biomedical and cosmetic applications (Mbituyimana et al., 2021). A wide variety of plants and animals can be used to extract cellulose. Cellulose nanocrystals grafted with diethyl sinapate to create a powerful UV-blocking nanomaterial (CNC-DES). Due to the remarkable water-dispersibility of CNC nanoparticles, the water-insoluble DES can be grafted onto them to make them soluble in these water-based formulations. Importantly, grafting DES gave CNCs excellent anti-UV characteristics. Strong anti-UV characteristics of CNC-DES (≥ 0.5 wt %) dispersed in aqueous glycerol are demonstrated, blocking the transmission of light in all UVA, UVB, and UVC regions. When added to an oil-free moisturizing cream, CNC-DES has a greater SPF than commercial sunscreen with the same number of active components. By absorbing the harmful energy from UV rays and distributing it into the environment in the form of heat, the CNC-DES is successful in this regard (Mendoza et al., 2021).

Phloretin (Phl) was enclosed in gliadin/sodium carboxymethyl cellulose nanoparticles (G/CMC) for improved stability and bioaccessibility through simple antisolvent precipitation. G/CMC nanoparticle production was primarily driven by electrostatic interaction and hydrogen bonds. Phloretin-loaded G/CMC nanoparticles (G/CMC-Phl) demonstrated strong pH shift, heat treatment, and UV irradiation resistance. Due to their high loading capacity, protection, and controlled release behavior, G/CMC nanoparticles are anticipated to be used as carriers for liposoluble bioactive chemicals in the commercial industry (He et al., 2022).

A polymer of great interest to scientists is bacterial cellulose. Due to its great biocompatibility with human tissues. Because of its high-water retention capacity and permeability, bacterial cellulose can be employed to permeate the skin with hydrophilic actives. In addition to creating a mask with a whitening effect, it is utilized as a moisturizer and anti-aging product. Since it maintains a moist environment, acts as an efficient physical barrier against pathogens, fosters tissue regeneration, and is recommended for temporary skin wound coverage, bacterial cellulose has been utilized in wound care, including burns. Because collagen and other elements of the cell matrix are related, there is biocompatibility (Bianchet et al., 2020).

11.4.6 NANOCAPSULES

Nanocapsules are polymeric nanomaterial capsules that are encased in an oily or liquid phase. In cosmetics, nanocapsules are used to protect chemicals, lessen chemical odors, and fix compatibility problems between formulation elements. Polymeric nanocapsule suspensions can be used as a finished product to apply directly to the skin or as an ingredient in semisolid formulations. The polymer and surfactant used as basic ingredients can influence how deeply a substance penetrates the skin. Through the nanoprecipitation process, stabilized poly-l-lactic acid nanocapsules with a diameter of around 115 nm were created. By encasing fragrance molecules in a polymeric nano-carrier, a prolonged release of fragrance was made possible. The future of deodorant products may be significantly impacted by this form of encapsulation of chemicals in biocompatible nanocapsules (Fytianos et al., 2020).

An emulsion-diffusion method was used to create nanocapsules that contained hinokitiol (HKL). Cetyltrimethylamonium Chloride (CTAC) was used as a cationic emulsifier, Poly (caprolactone) (PCL) as a wall material, and HKL dissolved in octylsalicylate (OS) as a core material in the emulsification process of creating nanocapsules. TEM images demonstrated the submicron-sized nanoparticle. The mean diameters were 223 nm, and the sizes ranged from 55 to 234 nm, according to a dynamic light scattering method. Shampoo and hair tonic were two different types of preparations that contained hinokitiol (HKL) nanocapsules. The in vivo hair growth-promoting effects were comparable to those of a solution of minoxidil (Hwang and Kim, 2008).

The coenzyme Q10 (CoQ10), also known as ubiquinone or 2,3-dimethoxy-5-methyl-6-decaprenil-benzoquinone, is a cellular endogenous antioxidant. When exposed to light, it becomes photo-unstable. Their inclusion in a cosmetic formulation appropriate for topical application is hampered by the lack of chemical stability. The inability to regulate how much coenzyme Q10 permeates the skin is another restriction. The interfacial deposition was used to create coenzyme Q10-containing nanocapsule solutions. The nanocapsules exhibited qualities suitable for cutaneous application. With low polydispersity and negative zeta potential, the particles have a size between 213 and 248 nm. Coenzyme Q10 (CoQ10) can become more permeable, more stable, and more resistant to photodegradation because of nano encapsulation. Additionally, antioxidant activity was elevated by the nanoencapsulation of Q10 (CoQ10) (Pohlmann and Am, n.d.).

11.4.7 SOLID LIPID NANOPARTICLES

Solid lipid nanoparticles (SLNs), commonly referred to as lipid carriers, have undergone extensive research on a global scale and have developed into potential nanotechnology drug delivery systems (Sharma et al., 2022). Beginning in the 1990s, solid lipid nanoparticles (SLN) were created as an alternative carrier system to emulsions, liposomes, and polymeric nanoparticles. The liquid lipid (oil) of an o/w emulsion is replaced with a solid lipid or a combination of solid lipids to create SLN, where the lipid particle matrix is solid at both room temperature and body temperature. Beginning in the 1990s, solid lipid nanoparticles (SLN) were created as an alternative carrier system to emulsions, liposomes, and polymeric nanoparticles. The liquid lipid (oil) of an o/w emulsion is replaced with a solid lipid or a combination of solid lipids to create SLN, where the lipid particle matrix is solid at both room temperature and body temperature. The typical SLN particle size is in the submicron range, falling between 40 and 1000 nm. In the lipid nanoparticle technology's second generation, the particles were made utilizing mixtures of liquid and solid lipids (oils). Nanostructured lipid carriers are the second generation of nanoparticles (NLC). These second-generation submicron particles can include both pharmaceutical and cosmetic active ingredients. To get over some SLN-related possible drawbacks, NLCs were created. In comparison to SLN, NLC has a greater loading capacity for a variety of active compounds, the lower water content of the particle suspension, and the ability to prevent or decrease potential active chemical expulsion during storage. Numerous characteristics of NLC and SLN are beneficial for dermal application. They are colloidal carriers that offer numerous compounds'-controlled release profiles. They have a very high level of tolerance because they are made of lipids that are physiological and biodegradable and have little toxicity and cytotoxicity. A tight contact with the stratum corneum is ensured by the small size, which may increase the amount of drug that is absorbed via the skin (Müller et al., 2000; Pardeike et al., 2009). An increased skin hydration impact is seen as a result of the occlusive features of lipid nanoparticles. Additionally, lipid nanoparticles can improve the chemical stability of substances susceptible to light, oxidation, hydrolysis, and other environmental stresses.

Tretinoin, a vitamin A metabolite, is used topically to treat a variety of inflammatory and proliferative skin conditions, including psoriasis, acne, photoaging, and epithelial skin cancer. Localized skin irritation, including erythema, peeling, and burning, as well as increased photosensitivity, are some of the main drawbacks of using tretinoin topically. Tretinoin was added to SLN to solve these issues. The penetration profile of SLN-based tretinoin gel is similar to that of commercial tretinoin cream.

SLN are physical sunscreens with more UV protection than comparable reference emulsions. Molecular sunscreens added to SLN have synergistic UV-blocking effects. To retain the level of protection, a formulation's molecular sunscreen concentration might be reduced (Wissing and Müller, 2003). Sarhadi et al., formulated and compared moisturization potential of SLN and NLC using various solid lipids such as Precirol®, Tripalmitin, and Compritol® in deionized and magnetized water. Both the formulations were developed using ultrasound and high shear homogenization techniques. The result of the finding revealed that 5% SLN with Compritol® showed maximum moisturization potential in vitro while 5% SLN with Precirol® had maximum moisturization potential in vivo. Thus, the use of magnetized water could accelerate the stability and effectiveness of moisturizing agents (Sarhadi et al., 2020).

11.5   SIDE EFFECTS

These readily available nanoparticles have certain short- and long-term negative effects that may restrict their use. The side effects are influenced by concentration, time, and the chosen administration route. Cosmetics provide a danger of systemic toxicities after accidental consumption because they are designed to be applied to the lips or contouring areas. Sticks, glosses, and coloring agents fall under this category. The negative effects are dependent on several variables; some are related to the active components' formulation characteristics, such as their surface area, coating materials, particle size, and tendency to aggregate, while others are connected to the cosmeceutical agent's chemical makeup. Cosmetics that are meant to be applied to the lips or other contouring areas run the risk of causing systemic toxicity if accidentally consumed. Lipsticks, glosses, and coloring agents are some of these. The harmful effects are dependent on some variables; some of them are related to the active ingredient's formulation properties, such as their surface area, coating materials, particle size, and tendency to aggregate, while others are related to the chemical makeup of the cosmeceutical agent. After pulmonary access, ultrafine titanium dioxide particles may cause inflammatory reactions and damage to the lung tissue. Even though metallic nanoparticles offer antibacterial properties, research on animals has shown that using them reduces RBC count, weight, and spleen index. Due to their ability to cross the placenta, they should not be used during pregnancy. Zinc and titanium oxide nanoparticle-containing substances that, particularly after exposure to UV light, damage membranes by producing free radicals Conversely, dendrimers damage cell membranes by interacting with the phospholipid bilayers. Unfortunately, because of their relative instability, lipid-based NP carriers for cosmeceuticals don't exhibit a safer profile. The risk could be greater if the organ's surface matches that of the nanostructure. In general, an increase in surface area to volume reflects an increase in both chemical and biological reactivity. As a result, toxicities in nanoparticles with low solubilities are evident and somehow connected to breast cancer. In conclusion, the use of nanomaterials is linked to systemic and localized health risks, and further research into their safety profile is required (Abu Hajleh et al., 2021).

11.6 FUTURE POSSIBILITIES FOR NANO-COSMECEUTICAL

Around 40 years ago, liposome-based moisturizing creams introduced nanotechnology to the cosmetics and healthcare industries. However, the earliest evidence of its use dates back more than 4000 years, when scholars in ancient Egypt, Greece, and Rome used nanotechnology in the preparation of hair dyes (Singh and Sharma, 2016). According to a 2006 European Commission investigation, nanoparticles are included in at least 5% of cosmetic formulations. The use of nanoparticles in cosmetic products has been a controversial subject in the media, among scientists, and among policymakers for the past few years. Due to studies being published on the safety of nanoparticles and the scientists' lack of understanding, there have been some toxicity issues concerning the dermal application of nanoparticles. Various kinds of nanoparticles have been either proposed or used in cosmetics. Nanoparticles' enhanced self-cleaning or self-adhesive characteristics are made possible by their reduced size. In addition, a small size enhances a unique material's toughness, provides greater friction resistance, etc. Biodegradable nanoparticles (such as liposomes, polycyanoacrylate, chitosan, and poly lactic-co-glycolic acid) and nonbiodegradable nanoparticles (such as polystyrene, ZnO, silica-based nanoparticles, and quantum dots) are the two types of nanoparticles that will be employed in the cosmetics business. The kind and physicochemical characteristics of nanoparticles and carriers, the composition of the substance, and the state of the skin significantly affect how well nanoparticles are absorbed and penetrated. Concerns over how these nanomaterials affect the environment and how they might affect people's health have grown as well. The growing use of nanoparticles in the cosmetics industry is an obvious sign of the bright future that nanotechnology holds for the industry. An increasing number of products are being created and produced that use nanotechnology. Currently, the fields of cosmetics and consumer skin care account for a good proportion of patents relating to discoveries based on nanotechnology. In terms of the total number of nanoparticle patents, they hold the top spot. They improve the effectiveness of deodorants, sunscreens, eye shadows, lipsticks, moisturizers, after-shave products, shampoos, and conditioners. The development of nanotechnology is still in its early stages. It has excellent opportunities for both research and business, and it embodies the key technologies of the 21st century. The involvement of major and minor manufacturers, as well as local businesses worldwide, has contributed to the cosmetics industry's rapid rise. Risks to consumer health and safety have also increased with the development and commercialization of nanotechnology. Therefore, nanotechnology in the cosmetics sector should be applied in a method that is the safest for both the environment and the health of customers.

REFERENCES

Abu Hajleh, M.N., Abu-Huwaij, R., AL-Samydai, A., Al-Halaseh, L.K., Al-Dujaili, E.A., 2021. The revolution of cosmeceuticals delivery by using nanotechnology: A narrative review of advantages and side effects. *J. Cosmet. Dermatol.* 20, 3818–28. https://doi.org/10.1111/JOCD.14441

Abu Hajleh, M.N., AL-Samydai, A., Al-Dujaili, E.A.S., 2020. Nano, micro particulate and cosmetic delivery systems of polylactic acid: A mini review. *J. Cosmet. Dermatol.* 19, 2805–11. https://doi.org/10.1111/JOCD.13696

Akturk, O., Kismet, K., Yasti, A.C., Kuru, S., Duymus, M.E., Kaya, F., Caydere, M., Hucumenoglu, S., Keskin, D., 2016. Collagen/gold nanoparticle nanocomposites: A potential skin wound healing biomaterial. *J. Biomater. Appl.* 31, 283–301. https://doi.org/10.1177/0885328216644536

Alahdal, F.A.M., Qashqoosh, M.T.A., Manea, Y.K., Salem, M.A.S., Khan, A.M.T., Naqvi, S., 2022. Eco-friendly synthesis of zinc oxide nanoparticles as nanosensor, nanocatalyst and antioxidant agent using leaf extract of P. austroarabica. *OpenNano* 8, 100067. https://doi.org/10.1016/J.ONANO.2022.100067

Bhalla, N., Jayaprakash, A., Ingle, N., Patel, H., Patri, S.V., Haranath, D., 2022. Fabrication and infusion of potent silver doped nano ZnO aimed to advance germicidal efficacy of health and hygiene products. *J. Sci. Adv. Mater. Devices* 7, 100487. https://doi.org/10.1016/J.JSAMD.2022.100487

Bianchet, R.T., Vieira Cubas, A.L., Machado, M.M., Siegel Moecke, E.H., 2020. Applicability of bacterial cellulose in cosmetics – Bibliometric review. *Biotechnol. Reports* 27. https://doi.org/10.1016/J.BTRE.2020.E00502

Caldonazo, A., Almeida, S.L., Bonetti, A.F., Lazo, R.E.L., Mengarda, M., Murakami, F.S., 2021. Pharmaceutical applications of starch nanoparticles: A scoping review. *Int. J. Biol. Macromol.* 181, 697–704. https://doi.org/10.1016/J.IJBIOMAC.2021.03.061

Campelo, P.H., Sant'Ana, A.S., Pedrosa Silva Clerici, M.T., 2020. Starch nanoparticles: Production methods, structure, and properties for food applications. *Curr. Opin. Food Sci.* 33, 136–40. https://doi.org/10.1016/J.COFS.2020.04.007

Cantin, M., Zepeda, K., Vilos, C., 2015. Polymeric nanoparticles in dermocosmetic nanopartículas poliméricas en dermocosmética. *Int. J. Morphol.* 33, 1563–68.

Chen, X., Wu, K., Zeng, S., Chen, D., Yao, L., Song, S., Wang, H., Sun, M., Feng, T., 2022. Stabilization and dispersion of OSA starch-coated titania nanoparticles in kappa-carrageenan-based solution. *Nanomaterials* 12. https://doi.org/10.3390/NANO12091519

Fang, Y., Hong, C.Q., Chen, F.R., Gui, F.Z., You, Y.X., Guan, X., Pan, X. hong, 2021. Green synthesis of nano silver by tea extract with high antimicrobial activity. *Inorg. Chem. Commun.* 132. https://doi.org/10.1016/J.INOCHE.2021.108808

Folle, C., Marqués, A.M., Díaz-Garrido, N., Espina, M., Sánchez-López, E., Badia, J., Baldoma, L., Calpena, A.C., García, M.L., 2021. Thymol-loaded PLGA nanoparticles: An efficient approach for acne treatment. *J. Nanobiotechnol.* 19, 359. https://doi.org/10.1186/S12951-021-01092-Z

Fytianos, G., Rahdar, A., Kyzas, G.Z., 2020. Nanomaterials in cosmetics: Recent updates. nanomater. https://doi.org/10.3390/nano10050979

Gajbhiye, S., Sakharwade, S., Gajbhiye, S., Sakharwade, S., 2016. Silver nanoparticles in cosmetics. *J. Cosmet. Dermatol. Sci. Appl.* 6, 48–53. https://doi.org/10.4236/JCDSA.2016.61007

Ghoshal, G., Singh, M., 2022. Characterization of silver nano-particles synthesized using fenugreek leave extract and its antibacterial activity. *Mater. Sci. Energy Technol.* 5, 22–29. https://doi.org/10.1016/J.MSET.2021.10.001

He, J.R., Zhu, J.J., Yin, S.W., Yang, X.Q., 2022. Bioaccessibility and intracellular antioxidant activity of phloretin embodied by gliadin/sodium carboxymethyl cellulose nanoparticles. *Food Hydrocoll.* 122, 107076. https://doi.org/10.1016/J.FOODHYD.2021.107076

Hwang, S.L., Kim, J.C., 2008. In vivo hair growth promotion effects of cosmetic preparations containing hinokitiol-loaded poly(ε-caprolacton) nanocapsules. 25, 351–56. http://dx.doi.org/10.1080/02652040802000557 Importing Cosmetics | FDA [WWW Document], n.d.

Irshad, A., Zahid, M., Husnain, T., Rao, A.Q., Sarwar, N., Hussain, I., 2019. A proactive model on innovative biomedical applications of gold nanoparticles. *Appl. Nanosci.* 10, 2453–65. https://doi.org/10.1007/S13204-019-01165-4

Jiménez-Pérez, Z.E., Singh, P., Kim, Y.J., Mathiyalagan, R., Kim, D.H., Lee, M.H., Yang, D.C., 2018. Applications of Panax ginseng leaves-mediated gold nanoparticles in cosmetics relation to antioxidant, moisture retention, and whitening effect on B16BL6 cells. *J. Ginseng Res.* 42, 327–33. https://doi.org/10.1016/J.JGR.2017.04.003

Kesente, M., Kavetsou, E., Roussaki, M., Blidi, S., Loupassaki, S., Chanioti, S., Siamandoura, P., Stamatogianni, C., Philippou, E., Papaspyrides, C., Vouyiouka, S., Detsi, A., 2017. Encapsulation of olive leaves extracts in biodegradable PLA nanoparticles for use in cosmetic formulation. *Bioengineering* 4, 75. https://doi.org/10.3390/BIOENGINEERING4030075

Kong, F., Wei, K., Hu, F., Liu, W., Yan, L., 2019. Optimization and characterization of poly(lactic-co-glycolic acid) nanoparticles loaded with astaxanthin and evaluation of anti-photodamage effect in vitro. https://doi.org/10.1098/rsos.191184

Larese Filon, F., Mauro, M., Adami, G., Bovenzi, M., Crosera, M., 2015. Nanoparticles skin absorption: New aspects for a safety profile evaluation. *Regul. Toxicol. Pharmacol.* 72, 310–22. https://doi.org/10.1016/J.YRTPH.2015.05.005

Li, G., Lee, Y.Y., Lu, X., Chen, J., Liu, N., Qiu, C., Wang, Y., 2022. Simultaneous loading of (−)-epigallocatechin gallate and ferulic acid in chitosan-based nanoparticles as effective antioxidant and potential skin-whitening agents. *Int. J. Biol. Macromol.* 219, 333–45. https://doi.org/10.1016/J.IJBIOMAC.2022.07.242

Mbituyimana, B., Liu, L., Ye, W., Ode Boni, B.O., Zhang, K., Chen, J., Thomas, S., Vasilievich, R.V., Shi, Z., Yang, G., 2021. Bacterial cellulose-based composites for biomedical and cosmetic applications: Research progress and existing products. *Carbohydr. Polym.* 273, 118565. https://doi.org/10.1016/J.CARBPOL.2021.118565

Mendoza, D.J., Maliha, M., Raghuwanshi, V.S., Browne, C., Mouterde, L.M.M., Simon, G.P., Allais, F., Garnier, G., 2021. Diethyl sinapate-grafted cellulose nanocrystals as nature-inspired UV filters in cosmetic formulations. *Mater. Today Bio* 12. https://doi.org/10.1016/J.MTBIO.2021.100126

Morán, D., Gutiérrez, G., Blanco-López, M.C., Marefati, A., Rayner, M., Matos, M., 2021. Synthesis of starch nanoparticles and their applications for bioactive compound encapsulation. *Appl. Sci.* 11, 4547. https://doi.org/10.3390/APP11104547

Müller, R.H., Mäder, K., Gohla, S., 2000. Solid lipid nanoparticles (SLN) for controlled drug delivery – A review of the state of the art. *Eur. J. Pharm. Biopharm.* https://doi.org/10.1016/S0939-6411(00)00087-4

Ong, W.T.J., Nyam, K.L., 2022. Evaluation of silver nanoparticles in cosmeceutical and potential biosafety complications. *Saudi J. Biol. Sci.* 29, 2085–94. https://doi.org/10.1016/J.SJBS.2022.01.035

Pandita, D., Kumar, S., Lather, V., 2015. Hybrid poly(lactic-co-glycolic acid) nanoparticles: Design and delivery prospectives. *Drug Discov. Today*. https://doi.org/10.1016/j.drudis.2014.09.018

Pardeike, J., Hommoss, A., Müller, R.H., 2009. Lipid nanoparticles (SLN, NLC) in cosmetic and pharmaceutical dermal products. *Int. J. Pharm.* 366, 170–84. https://doi.org/10.1016/j.ijpharm.2008.10.003

Pohlmann, A., Am, L.J.,…, n.d. Development of semi-solid cosmetic formulations containing coenzyme Q10-loaded nanocapsules *Lat. Am. J. Pharm.* 28(6), 819–826.

Rizzi, V., Gubitosa, J., Fini, P., Nuzzo, S., Agostiano, A., Cosma, P., 2021. Snail slime-based gold nanoparticles: An interesting potential ingredient in cosmetics as an antioxidant, sunscreen, and tyrosinase inhibitor. *J. Photochem. Photobiol. B Biol.* 224, 112309. https://doi.org/10.1016/J.JPHOTOBIOL.2021.112309

Sakulwech, S., Lourith, N., Ruktanonchai, U., Kanlayavattanakul, M., 2018. Preparation and characterization of nanoparticles from quaternized cyclodextrin-grafted chitosan associated with hyaluronic acid for cosmetics. *Asian J. Pharm. Sci.* 13, 498–504. https://doi.org/10.1016/J.AJPS.2018.05.006

Sarhadi, S., Gholizadeh, M., Moghadasian, T., Golmohammadzadeh, S., 2020. Moisturizing effects of solid lipid nanoparticles (SLN) and nanostructured lipid carriers (NLC) using deionized and magnetized water by in vivo and in vitro methods. *Iran. J. Basic Med. Sci.* 23, 337. https://doi.org/10.22038/IJBMS.2020.39587.9397

Sharkawy, A., Barreiro, M.F., Rodrigues, A.E., 2021. New Pickering emulsions stabilized with chitosan/collagen peptides nanoparticles: Synthesis, characterization and tracking of the nanoparticles after skin application. *Colloids Surfaces A Physicochem. Eng. Asp.* 616, 126327. https://doi.org/10.1016/J.COLSURFA.2021.126327

Sharma, G., Khanna, G., Gupta, S., Ramzan, M., Singh, J., Singh, M., Mudgill, U., Gulati, J.S., Kaur, I.P., 2022. Scope of solid lipid nanoparticles per se as all-purpose moisturising sunscreens. *J. Drug Deliv. Sci. Technol.* 75, 103687. https://doi.org/10.1016/J.JDDST.2022.103687

Singh, T.G., Sharma, N., 2016. Nanobiomaterials in cosmetics: Current status and future prospects. *Nanobiomater. Galen. Formul. Cosmet. Appl. Nanobiomater.* 149–74. https://doi.org/10.1016/B978-0-323-42868-2.00007-3

Smijs, T.G., Pavel, S., 2011. Titanium dioxide and zinc oxide nanoparticles in sunscreens: Focus on their safety and effectiveness. *Nanotechnol. Sci. Appl.* 4, 95. https://doi.org/10.2147/NSA.S19419

Ta, Q., Ting, J., Harwood, S., Browning, N., Simm, A., Ross, K., Olier, I., Al-Kassas, R., 2021. Chitosan nanoparticles for enhancing drugs and cosmetic components penetration through the skin. *Eur. J. Pharm. Sci.* 160. https://doi.org/10.1016/J.EJPS.2021.105765

Thananukul, K., Kaewsaneha, C., Sreearunothai, P., Petchsuk, A., Buchatip, S., Supmak, W., Nim, B., Okubo, M., Opaprakasit, P., 2022. Biocompatible degradable hollow nanoparticles from curable copolymers of polylactic acid for UV-shielding cosmetics. *ACS Appl. Nano Mater.* 5, 4473–83. https://doi.org/10.1021/ACSANM.2C00606/SUPPL_FILE/AN2C00606_SI_001.PDF

Vinod, T.P., Jelinek, R., 2019. Inorganic nanoparticles in cosmetics. *Nanocosmetics* 29–46. https://doi.org/10.1007/978-3-030-16573-4_3

Wissing, S.A., Müller, R.H., 2003. Cosmetic applications for solid lipid nanoparticles (SLN). *Int. J. Pharm.* 254, 65–68. https://doi.org/10.1016/S0378-5173(02)00684-1

Xi, J., Kan, W., Zhu, Y., Huang, S., Wu, L., Wang, J., 2022. Synthesis of silver nanoparticles using Eucommia ulmoides extract and their potential biological function in cosmetics. *Heliyon* 8, e10021. https://doi.org/10.1016/J.HELIYON.2022.E10021

Xu, Z.P., Zeng, Q.H., Lu, G.Q., Yu, A.B., 2006. Inorganic nanoparticles as carriers for efficient cellular delivery. *Chem. Eng. Sci.* 61, 1027–40. https://doi.org/10.1016/J.CES.2005.06.019

12 Nanosilver and nanogold delivery system in nanocosmetics
A recent update

Akansha Bisht, Shruti Richa, Shivangi Jaiswal, Jaya Dwivedi and Swapnil Sharma

CONTENTS

DOI: 10.1201/9781003319146-12

12.1   INTRODUCTION

The inclusion of nanomaterials utilizing nanotechnology has shown enormous potential in the field of cosmeceuticals, thus providing new approaches for the development of nanotechnology-based products in the cosmetic industries. Nanocosmetic products offer various benefits over traditionally utilized cosmetics such as enhanced bioavailability, controlled release and prolonged action (Nanda et al. 2016; Zed et al. 2019). Nanotechnology being an amalgamation of both scientific and technological approaches have led to advancements in cosmetics, thereby enhancing consumer demand globally. Presently, nanocosmetics is one of the fastest emerging sectors that have massively expanded their market and research in recent years (Souto et al. 2020). Attesting the same with statistics, the global market of nanocosmetics, particularly gold and silver-based nanocosmetics, was raised by 8.5 billion in 2019 and is expected to rise further by 12.1% in the upcoming years. The characteristic features of gold and silver nanomaterials offer a wide range of benefits over traditional cosmetics including preservation of active compounds, higher efficiency, less reactivity, enhanced hydration power, excellent stability, and higher efficiency that ameliorate their interaction with the microenvironment (Barbosa et al. 2016; Matsuura-Sawada et al. 2022). Nanogold and nano silver cosmetics are successfully marketed as nail care products, skincare products, hair care products, and many more (Srinivas 2016). It is a well-known fact that dosage form affects the bioavailability of an active drug; hence, the application of these nanoparticles does not alter the properties of cosmetics but enhances their therapeutic efficacy, skin adherence, skin penetration, ultraviolet protection, fragrance, and anti-aging effects (Nanda et al. 2016). Some other benefits of nanogold and nanosilver-based nanocosmetics have been reported in the literature viz. glossy, healthy and nourishing skin, whitening and rejuvenating skin, detanned skin, smoothening and enhanced hair texture, reducing lip and foot cracks, eye glittering, nail shimmering, and imparting fragrance etc. (Dubey, Dey, et al. 2022). Although gold and silver nanocosmetics possess a wide array of benefits, they exhibit certain drawbacks related to toxicity, stability, cost, scalability, etc. Toxicity and safety profiles of gold and silver nanomaterials are a matter of concern as they impart toxicity by increasing the concentration of active molecule reaching the bloodstream. Some of the merits and demerits of gold and silver nanoparticles are highlighted in Figure 12.1.

Therefore, this chapter covers the concept of nanogold and nanosilver delivery systems in cosmeceutical industries, their synthesis, their penetration, and various

ADVANTAGES	DISADVANTAGES
Gold nanoparticles	**Gold nanoparticles**
• High stability due to gold-sulphur bond • Large surface area • Chemically inert • No photo bleaching or blinking • Biocompatibility	• High cost for large scale production • Lack of standard protocol for translation into clinics • Non-biodegradable • Acute toxicity
Silver nanoparticles	**Silver nanoparticles**
• Inexpensive • Eco-friendly • Use of non-toxic chemicals • Less energy is used to synthesize NPs	• Short half-life • Formation of by products • Release of sweet smelling amines • High power cost • Requires long maintenance • Non operative for all dyes

FIGURE 12.1 Merits and demerits of gold and silver nanoparticles.

commercially available nanocosmetics. This chapter also addresses nanotoxicity, safety concerns, and regulation of the nano cosmeceuticals which will surely help regulators and consumers to gain awareness pertaining to benefits as well as toxicities associated with the long-term application of these nanocosmetics, thereby encouraging their judicious use.

12.2 METAL-BASED NANOCOSMETICS: NANOGOLD (AUNPS) AND NANOSILVER (AGNPS) NANOPARTICLES AS COSMECEUTICALS

Interestingly, gold and silver nanoparticles have been utilized as active ingredients in various cosmetic preparations such as shower gels, face creams, masks, soaps, toothpaste, nail enamel etc. due to their antimicrobial property (Bapat et al. 2020). There are a huge variety of nanoparticles for the cosmeceutical industries to choose from as an option but amongst them, metal nanoparticles are the ones globally trending for beautifying purposes. They exhibit a higher distribution of active components and have been in use for several years, but the optimum usage of several available commercially cosmetic formulations is still the focus of the investigation. The metal-based nanoparticles consist only of nanoparticles obtained from metallic sources such as titanium, platinum, silver, gold, and aluminum, among others. The size range of gold nanoparticles is 5–400 nm, whereas silver nanoparticles vary from 5 nm to 200 nm (Dubey, Dey, et al. 2022). Nanogold and nanosilver materials are widely utilized nano-materials due to their expansive applications. The metal particles used in cosmetics must be chosen with care to ensure their suitability for usage and avoidance of metal-based toxicity (Bocca et al. 2014). The broad application of these nanomaterials is because of their excellent physical and chemical properties, small size, stability, optical flexibility, cytoprotective, and high surface area (Alaqad and Saleh 2016).

12.2.1 Nanogold (AuNPs) derived cosmetic products

Nanogold materials are widely employed in several cosmetic formulations, including face masks, packs, scrubs, and creams. Due to their excellent antiseptic, antibacte-rial, and antifungal properties, they help in skin rejuvenation, enhance skin texture and elasticity, prevent skin aging, and enhance blood circulation. Furthermore, it is demonstrated that gold is inert, nonreactive, noncytotoxic, stable, and highly pen-etrable, making it suitable for aesthetic applications (Herizchi et al. 2016; Khan et al. 2014; Kokura et al. 2010; Nasir, Sl, and Kaur 2012; Singh 2021; Varun, Sonia, and Patil 2022). It also exhibits anti-wrinkle and anti-redness properties. Nanogold can occur in conjugated or unconjugated forms. However, it is often more stable in liquid and dry states. They are utilized to load a substantial amount of actives and to target cells hidden deep inside the skin. Their absorption into the stratum corneum heavily depends on their chemical surface, shape, and compatibility with the lipid domains of the skin, along with their size, as nanoparticles (NPs) have a higher tissue pen-etration. Numerous qualities of nanogold are helpful for aesthetic purposes such as their anti-inflammatory potential, antibacterial potential, promote elasticity of the skin, and enhance blood circulation and anti-aging potential. Because of their nano size, which can enhance collagen production by 20–200 times (Singh 2021), they are utilized in the fabrication of masks and lotions for anti-aging purposes. The color of gold nanoparticles varies from violet to red, then red to blue, and sometimes the color appears black (Khan et al. 2014). Nanogold properties of gold are in fashion for vital-izing skin metabolism (Tangau, Chong, and Yeong 2022). Large cosmetic firms like L'Oreal employ gold nanoparticles to boost the effectiveness of their products and when they are coupled with silk, essential oils, and herbal extracts, they possess anti-oxidant effects, smoothen the skin, and impart clarity and radiance (Dubey, Parab, et al. 2022). Another example includes skin care products by Chantecaille like Nano Gold Illuminating Eye Serum and Nano Gold Stimulating Cream. In these products, the gold powder is conjugated with coffee seeds, pumpkin seeds, hyaluronic acid, and vitamins A, C, and E. Nano gold anti-aging lifting serum is another nanogold-based skin care product with anti-aging properties (Souto et al. 2020).

12.2.2 Nanosilver (AgNPs) derived cosmetic products

Nanosilver particles are of greater use in different sectors of cosmetic indus-tries. They are also used as antibacterial/antifungal/antiviral and could be used in wound healing and dental care products and thus, are utilized on a versatile level. A proceeding for further discoveries shows that silver nanomaterials exhibit antimicrobial effects by inhibiting the growth of infectious microbial strains such as *Candida albicans* and *Candida glabrata*. Silver nanoparticles have been reported in the treatment of skin issues, acne, microbial infections and photo-damage. They have also been employed as a preservative in the cosmeceutical products such as body foams, toothpastes, lip products, shampoos, wet wipes, creams, soaps, and many more due to their broad-spectrum antibacterial effects. Nanosilver-containing cleansers and soaps are well known to exhibit fungicidal and bactericidal effects and are used to prevent or treat sunburned skin and acne. Dental products and dentifrices also contain nanosilver for whitening of the teeth. Nail paints are also in this category as they look good on the nails and added beauty plus, and above all they effectively mitigate fungal infections of the toe, hands, and nails (Oberdörster, Oberdörster, and Oberdörster 2005). The studies have also reported higher stability of silver nanomaterials for more than one year without sedimentation (Souto et al. 2020).

Pulit-Prociak et al. (2019) studied the versatile applications of nanogold and nanosilver materials in cosmetic formulations. They observed a noticeable difference between both the nanomaterials with respect to the structure of a cream. They reported agglomeration of silver nanoparticles after incorporation into the cream whereas gold nanomaterials did not exhibit agglomeration upon introduction into the cream. This study signifies the higher electrokinetic potential of gold nanoparticles in comparison with silver nanoparticles. The penetration of gold and silver nanoparticles of 100–200 mg/kg concentration into the skin is of a matter concern as evidenced by a dermal membrane study model (Pulit-Prociak et al. 2019). The characterization of gold and silver nanomaterials is very difficult due to the subtle composition of cosmetics (Cao et al. 2016).

12.3 CHARACTERIZATION OF GOLD AND SILVER NANOPARTICLES

The distinctive way nanomaterials act, their clear definition, and detailed portrayal based on their distinct properties are essential for a safety evaluation. Characterization should be done with the raw resources during the process of developing the cosmetic, along with carrying out an evaluation when the people are using it so that a toxicological evaluation can be completed in its entirety. Measurements must also be completed using general and comprehensive documentation, which should also be given. Even though the size of the particles is the essential element, it must be accurate in more than a way to be sure. Electron microscopy is among the most accurate ways to accomplish this. Below some characterization techniques are listed for gold and silver nanoparticles.

12.3.1 ELEMENTAL ANALYSIS

Mass spectroscopy is extensively used to determine the nature and content of undesirable and toxic chemical elements.

12.3.2 X-ray diffraction (XRD)

XRD provides data on the size and state of the units comprising the nanoparticles. XRD aids in analyzing the crystalline structure, and the lattice-related parameters (Kim et al. 2014). The structure and shape of the crystalline moieties are also determining factors. This system also demonstrates how it could be employed to look at nanostructures. The width and the state of reflective surfaces reveal information about the material requirements, and structural components, including the sizes of micro crystallites, small-scale breaks in a grid, and separation structures (Dorofeev et al. 2012).

12.3.3 Fourier transform infrared (FTIR) spectroscopy

This type of spectroscopy is used to discover whether the nanoparticle does have a functional group. It can be discovered by figuring out how long the electromagnetic radiation with frequency range in the mid-infrared area stays in the area (4000–400 cm–¹). A recorded range provides an understanding of the existing groups and how powerful and stable the bonds are. The above clarifies how well the atoms are placed and work together (Mourdikoudis, Pallares, and Thanh 2018).

12.3.4 Mass spectra

Mass spectroscopy can provide a familiar means for determining the core's size and any ligands that may have been added to the surface for functionalization. It also measures the size dispersity, the average mass and full mass distribution.

12.3.5 NMR spectra (nuclear magnetic resonance)

This spectroscopy aids in investigating the surface, chemistry, and physical properties. For instance, Thiol protected gold nanostructures. Thus, it plays its role in understanding the molecular structure of molecules bound to the surfaces of these nanoparticles as well as a method to detect the size of nanoparticles in solution.

12.3.6 Microscopic techniques

Some different microscopic techniques such as scanning electron microscopy (SEM), transmission electron microscopy (TEM), etc. are used to examine in depth the gold and silver nanoparticles.

12.3.7 Scanning electron microscopy (SEM)

SEM is used to determine the measurement, morphology, and shape by providing high-resolution images. The working principle of SEM is the same as that of the optical microscope, but the only difference is that it measures the scattering of electrons from the test samples.

12.3.8 TRANSMISSION MICROSCOPY (TEM)

In TEM, a light emission passes through a fragile sample and appears to work together during the same. While the electrons move through the sample, they work together to make an image. The image is enlarged, put in the center point of an imaging system, and named (Kim et al. 2014).

12.4 VARIOUS METHODS TO SYNTHESIZE GOLD AND SILVER NANOPARTICLES

Various techniques have evolved to develop eco-friendly approaches to synthesize the environmentally safe and less-expensive methods involving physical mode, chemical mode or via green synthesis mode using plants and microorganisms (Table 12.1).

TABLE 12.1
Different methods to synthesize gold and silver nanoparticles

S. no.	Nanoparticle	Synthesis	Explanation	Merits	Demerits	Reference
1.	Silver nanoparticles	Chemical reduction synthesis	Silver ions undergoing reduction reaction to form metallic silver, being carried out in either aqueous/ non-aqueous solutions with reducing agents of either organic/ inorganic nature in the presence of OH groups	Higher yield Lesser cost No aggregation of particles	Higher reaction rate leads to very small particle sizes Lower reaction rate leads to formation of agglomerates	Kim et al. (2014)
		Physical synthesis	Laser-ablation Evaporation-condensation UV radiation Synthesis reaction is carried out in ceramic heaters. Nucleation is the principle for its growth	Thermal stability Uniformly sized smaller particles High yield	Higher energy requirement Costly methods More time required for obtaining thermal stability Space required is high for evaporation-condensation tube furnace	Jung et al. (2006)

(Continued)

TABLE 12.1   (Continued)

S. no.	Nanoparticle	Synthesis	Explanation	Merits	Demerits	Reference
		Green synthesis	Biological resources example: plants, microbes etc. are used. Several biomolecules (amino acids/ proteins/ vitamins/ enzymes/secondary metabolites etc.) present in the extracts mediate reduction reaction.	Ecofriendly, lowers the risk of chemical usage, less usage of physical instruments	Needs special precaution while handling the microbe	Arya, Mishra, and Chundawat (2019), Siddiqi, Husen, and Rao (2018)
			Example: Green algae (*Botryococcus braunii*) was used to synthesize cubical/ spherical/ triangle-shaped silver nanoparticles having average size of 88.8 nm Spherical nanoparticles (14 nm size) from *Withania coagulans*			
2.	Gold nanoparticles	Green synthesis (using plants and microbes)	Different parts of the plant are washed, dried, grinded, boiled, then filtered/ centrifuged for purification. Gold salt solution undergoes reduction to yield AuNPs	Cost effective Circumvent exposure to chemicals and harmful by-products No requirement of external stabilizer Ease of	–	Santhosh, Genova, and Chamati (2022)
			Different bacteria, for example, *B. subtilis, E. coli, P. aeruginosa* and *B. megaterium* are well reported to take part in the green synthesis of gold nanoparticles	availability of raw materials Reproducible Energy-efficient Ease of scale-up		

12.5 PENETRATION AND ABSORPTION OF GOLD AND SILVER NANOPARTICLES

Nanoparticles can enter through the skin, which is dependent on their chemical and physical properties, like their form, size, and surface charge (Kaul et al. 2018). Nanomaterials always permeate across the skin in either of three ways: between the cells, within the cells, or through appendages (Rapalli et al. 2021). In the inter-cellular route, nanoparticles disperse across bilayers as well as the corneocytes matrix. In the intracellular route, aqueous regions encircled by polar lipids create microchannels, and the trans-appendageal routes use sweat glands or hair follicles (Katz, Dewan, and Bronaugh 2015). Nanomaterials can be inactively moved through the stratum corneum by bilayers in the spaces between the cells, corneocytes matrix, and higher concentration of proteins below the stratum corneum. The penetration phenomena of gold and silver nanoparticles are depicted in Figure 12.2.

12.5.1 PENETRATION/ABSORPTION OF GOLD NANOPARTICLES

A study on Sprague-Dawley rats was conducted where the penetration ability of gold nanoparticles was determined and the higher penetration of nanoparticles was observed in all the layers of the skin viz. epidermis, dermis, and stratum corneum (Fernandes et al. 2015; Raju et al. 2018).

12.5.2 PENETRATION/ABSORPTION OF SILVER NANOPARTICLES

In a study, microscopic evaluation of skin isolated from Sprague-Dawley rats treated with colloidal silver nanoparticles (10–20 nm size) demonstrated no abnormality in the epidermis layer of rats as compared to control group rats (Maneewattanapinyo et al. 2011). In another study, a sleeve incorporating silver nanoparticles at 3.6 mm concentration was tied to the rat's forearm for five days. At the end of the study, the presence of silver nanoparticle aggregates was observed on the surface of the skin

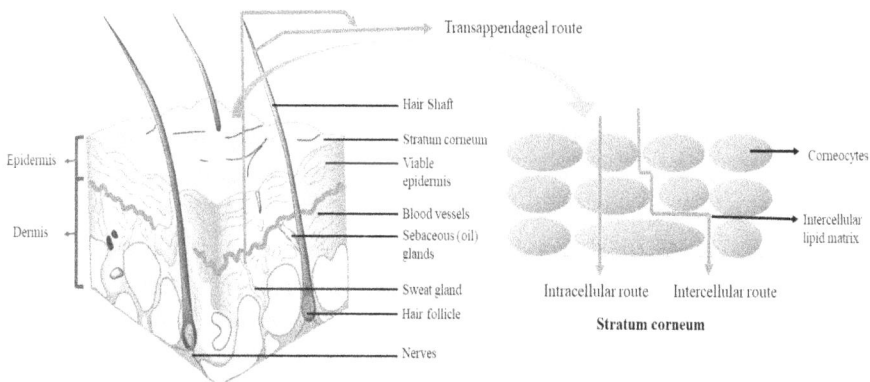

FIGURE 12.2 Penetration of gold and silver nanoparticles into the skin

(Bianco et al. 2016). The possibility of silver nanoparticles penetrating into the skin was confirmed by George et al. (2014); their study revealed the presence of silver nanoparticle clusters in the deeper layer of the epidermis and as well as the stratum corneum (George et al. 2014). Despite being deposited in the skin layers, the silver nanoparticles did not produce systemic responses as they did not reach the systemic circulation (George et al. 2014).

Nanoparticle studies on the skin were studied utilizing laser scanning confocal microscopy to determine the penetration efficiency of fluorescently tagged particles of 20–200 nm size. The findings demonstrated that the nanoparticles which were partially damaged or tightly intact do not reach the epidermis or go any beyond; rather, they do not even penetrate the skin barrier, hence their delivery has no harmful effects and is safe for cosmeceuticals (Campbell et al. 2012). The silver nanoparticles with sizes 5µm -8µm reach only the stratum corneum while the nanoparticles with sizes less than 10 nm reach the deeper layers of the stratum corneum (Malakar et al. 2021). However, in-depth studies are required to determine the behavior of silver and gold nanoparticles, and studies should be conducted to evaluate whether the particles get absorbed into the bloodstream and reach the different organs or they get absorbed at the skin surface or remain in the stratum corneum. Further research and studies are still going on for confirming the penetration of these nanoparticles via BBB (blood-brain barrier).

12.6 TYPES OF COMMERCIALLY AVAILABLE COSMETIC PRODUCTS CONTAINING GOLD AND SILVER NANOPARTICLES

Gold and silver nanosized particles are utilized in the formulation of various cosmetics products including skin care products, hair care products, lip care products, eye care products, nail products, etc. as enlisted in Table 12.2.

12.6.1 SKINCARE PRODUCTS

When moisturizers or creams are applied to the skin, their primary goal is to prevent the skin from drying out, protecting it, and making it appear smoother. Skin is the soft tissue on the outside of the body. It is protected by a hydrolipid film, a natural shield of protection against illness (Srinivas 2016). AuNPs and AgNPs are increasingly being incorporated into skincare products to enhance their appearance.

12.6.2 SUNSCREENS

Sunscreens with gold and silver nanoparticles are better, clear, and less oily. They also impart a much better appearance. The sunscreen gives a more even coating to the skin. Eg: NanoGold BB Cream having sun-protection factor (SPF) 50 (Dhawan, Sharma, and Nanda 2020) is an example of gold nanoparticle sunscreens. It gives protection against sunburn and is hence is popular in the market.

TABLE 12.2
Commercially available cosmetics products containing gold and silver nanoparticles

S. no.	Type of cosmeceuticals	Type of formulation/ cosmeceutical	Company	Commercially available products name	Claims for their application
Gold nanoparticle-based cosmeceuticals					
1.	Skincare	Face elixir	RAS Luxury Oils	Gold Radiance Beauty Boosting Face Elixir	Brightens the skin
					Reduces wrinkle, fine line and dryness
		Day cream	Ayouthveda	Sparkling Gold day cream	Works against uneven skin tone, dullness & acne, reduces blemishes Promotes youthfulness in aging and damaging skin by stimulating skin cells while reducing blemishes and restoring skin elasticity
		Face serum	Craggy, Pilgrim, Eveline, Belora Paris	Craggy Face serum, 24k Gold Serum Advanced++	Whitens the skin, cure dull skin, wide pores, damaged skin activity, increases elasticity and prevents sagging, eliminates wrinkles and fine lines, gives smooth, shiny and silky skin, removes patches, gives anti-aging effect and works as anti-wrinkle product
		Cellular facial glow kit	Lotus Herbals	Radiant Gold	Reduces fine lines, wrinkles and sagging skin, fights sun damage, acts as anti-aging, deeply nourishes the skin, polishes skin
		Mineral ampoule	Eu.mei	Mineral Gold Ampoule	For whitening and antiwrinkle effect
		Face paint	Graftobian, Diamond FX	Graftobian ProPaint Metallic Face Paints	Gives nongreasy and even coverage and shimmer effects to face
		Moisturizer	Belora Paris	Belora Paris Illuminating Moisturizer	Nourishes and keeps the skin moisten
		Derma roller	Gubb	Gubb Women Derma Roller	Rejuvenates skin, tightens the skin, gives anti-aging effect and is scar treatment, promotes glowing skin, makes skin plumper & supple, banishes wrinkles, fine lines and dark circles

(Continued)

TABLE 12.2 (Continued)

S. no.	Type of cosmeceuticals	Type of formulation/ cosmeceutical	Company	Commercially available products name	Claims for their application
2.	Haircare	Shampoo and conditioner	Shine Hair	Banho Ouro Gold bath Nanoparticles Softness Nourishing Treatment Kit	Protects & nourishes hair, makes them soft & shiny
		Shampoo and conditioner	Absolute Color	Banho Ouro Golden Bath Repair Hydrating Luminous Treatment Kit	Cleans and repairs hair gently, gives intense shining, strengthen the hair, promotes revitalization and hydration to promote healthy hair
3.	Eyecare	Eye pencil	Nyka, Seven Seas, Kajal and Kohls	Kay Beauty Gold Gel Kajal	Not indicated
		Eyeshadow	e.I.f.	Liquid glitter eyeshadow	For shining of eyeshadows
		Eyeliner	Ronzille, Incolor, ONE on ONE, Recode Sudios, Shopaarel	Metallic Eyeliner	High precision mark on eyes with endless longevity
		Contact lens	DivaLens, Bella	Gold Cat Eye Contact, Bella Elite Collection Color Contact Lens	Keeps eye it vulnerable to dryness from the environment and blinking. For protecting eyes from accidental exposure to a green laser
		Browzer	Hollywood Browzer	Hollywood Browzer Derma planting Kit	Removes eye and facial hair and exfoliates
4.	Lipcare	Lip mask	Quinn, Knesko	24K Gold Overnight Lip Mask	Soften and smoothens the lip texture, brightens and blushes the lips like baby lips
		Lip fonduo	Rohto (Japan)	Mentholatum Lip Fonduo	As a base for lip
5.	Nailcare	Nail polish	CENIDE- UDE	Goldfinger	Adds shining
	Silver nanoparticle-based cosmeceuticals				

(Continued)

TABLE 12.2 (Continued)

S. no.	Type of cosmeceuticals	Type of formulation/ cosmeceutical	Company	Commercially available products name	Claims for their application
6.	Skincare	Cream	SilverPure	Silver Pure Skin Cream- 35 K	Soothens and hydrates the skin
		Facial	VLCC	VLCC Natural Sciences Silver Facial Kit	Restores pH balance and evens out uneven skin tone, controls oil secretion, gives a matt and shiny skin, removes dead skin cell, accumulated impurities and clears the skin pores, purifies and detoxifies the skin, provides intense nourishment and making the skin supple and youthful
7.	Hair care	Shampoo	Biotop Professional19,	Biotop Professional 19 Pro Silver Shampoo,	Restores radiant, neutralizes unwanted yellow and brassy tones, moisturizes hair and controls frizz
			GK Hair Silver Bombshell	Silver Bombshell Shampoo	Enhances hair color, smoothens and shines the hair, removes yellow and brassy tones
		Hair mask	Biotop Professional 19	19 Silver Hair Mask	Banishes brassiness and brightens blondes, deeply nourishes hair, moisturizes and adds shining, creates healthy looking blonde hair, fights frizz and smooths damaged hair, locks in moisture and enhances softness
		Oil	Biotop Professional 19	19 Pro- Silver Oil	Nourishes hair from inside to out, smooths the cuticle & reduces frizz, hydrates and softens, adds shining, provides heat and UV protection, reduces the appearance of split ends and breakage
8.	Eyecare	Eyeliner	Aya, Faces Canada	Aya Waterproof Eyeliner	Gives a glossy look, keeps the intense color locked down for 24 hours
9.	Lipcare	Lip balm	Silverpure	Nanosilver Lip Balm	Keeps lip hydrated all day, Reduces chafing and cracking of lips
10.	Nailcare	Nail polish	Beauty People- Wow, Half N Half Mirror, Faces Canada, Bella Voste	Range Nail Polish Beauty People- Wow, Half N Half Mirror, Faces Canada, Bella Voste Metallic Nail Paint, Bella Voste Sugar Baby Nail Paint	Provides silver shining to the nails, shimmers them

12.6.3 Moisturizers

Moisturizing ingredients are used in the cosmetic products industry to maintain the skin to be moist, soft, and smooth, and also to stop it from cracking. Most skincare products have components that keep the skin from becoming drying out, which is the reason they are termed "moisturizers." Despite the fact that layers of skin already contain some moisture, moisturizing creams are still applied to keep the skin from becoming too dry and withstand sudden changes in environmental conditions.

12.6.4 Skin cleansers

Skin cleansers are one of the vital personal care products that are applied to remove dirt, makeup, oil, pollutants, and dead skin cells from the skin. Skin cleansers containing gold or silver nanoparticles enhance the efficacy of active molecules and skin penetration, thereby providing protection to the skin. Nano Cyclic Inc. had produced a pink cleansing bar and cyclic nanosilver cleanser containing natural ingredients and silver nanoparticles that effectively prevent acne, inhibit bacterial and fungal growth, recuperate sun-damaged skin, exfoliate dead skin, diminish age spots, and remove makeup gently (Lohani et al. 2014).

12.6.5 Anti-aging products

Almost 90% of skin aging is caused by ultraviolet radiation exposure or influenced by lifestyle-related choices (e.g., stress, smoking, drinking, and not getting enough sleep). Both of these can speed up skin aging, along with other factors such as environmental changes (e.g., environmental damage) and poor health-related conditions. The indications of skin aging include loss of skin elasticity, wrinkles, and dryness (Souto et al. 2020). Anti-aging creams also neutralize the harm that has been caused to the sun-damaged, along with recuperating the same (Dhawan, Sharma, and Nanda 2020). Nano anti-aging serums are also available with good results in the market.

12.6.6 Gold-loaded facial masks

Gold face masks are commonly used in beauty spas and salons these days. It works by improving the flow of blood and improving skin elasticity. Normal face masks on the market are made of cotton and are already moistened with skin nutrients. The aqueous phase of a mask that has already been moistened can speed up the rate at which unstable ingredients like ascorbic acid break down, thus forfeiting their purpose of use. To solve this problem, scientists have made a new synthetic polymer face mask that can be loaded with a wide range of nutrients, such as ascorbic acid, retinoic acid, gold, and collagen, ensuring their stability. Skin permeation research shows that spherical gold nanoparticles are not intrinsically harmful to the cells of the human tissue (Kaufman and Alexis 2020).

12.6.7 HAIRCARE PRODUCTS

This category has several items, like shampoos, conditioners, hair growth stimulants, hair dye, hair masks, hair colorants, hair oil, etc. that maintain the smoothness, silkiness, shine and health of the hair. R & D facilities are paying attention to nanotechnology incorporation in hair care items in order to develop methods for understanding the mechanisms of using gold and silver nanoparticles as a remedy to treat male pattern baldness, plushness, and strengthening of hair (Nanda et al. 2016). Hair preparations with silver nanoparticles are easily available for treating hair depilation (Lohani et al. 2014). These nanoparticles do not cause any damage to the cuticles of the hair and impart shine, smoothness, and softness, and enhance disentangling of hair. An increase in the quantity of the active ingredients delivered to the hair follicles can be achieved owing to the intrinsic properties and the unique sizes of nanoparticles to improve hair care products (Gavazzoni Dias 2015). This nanoparticle moisturizes hair cuticles by developing the formation of a protective layer as well as increasing their holding time in hair follicles. Sericin adheres to the hair surface easily combined with these nanoparticles for hair cosmeceuticals and can treat the damaged hair cuticles. Another advantage of gold and silver nanoparticles in hair care product category is the alteration of the hair color.

12.6.8 EYECARE PRODUCTS

Gold and silver nanocosmeceuticals of this section impart high precision while applying on eyes with endless longevity and gives a gorgeous shine to the eyes. The available products are: Eyelashes, eyeliners, and eyeshadows. E.g.: Kay Beauty Gold Gel Kajal, an eye pencil.

12.6.9 LIPCARE PRODUCTS

Golden and silver nanoparticles are generally utilized in lipsticks as color pigments and to maintain color for longer durations. Lipcare products include lip gloss, lipsticks, lip masks and lip balms. Different shades in lipsticks are added by the inclusion of pigments and dyes. However, lead toxicities are associated with these pigments. The nanosilver materials exhibit yellow color while nanogold materials demonstrate red color. These nanomaterials are incorporated into the lipsticks to provide desired shades and colored pigments which are harmless, safe, and show excellent dispersion of pigments (Nanda et al. 2016). The addition of nanogold and nanosilver materials into the lip gloss and lipsticks provides softness and texture to the lips by preventing migration of pigments and transdermal water loss from the lips, thus maintaining lip color for a longer time period. An example of nanogold lip care product is 24K Gold Overnight Lip Mask, which is also available online.

12.6.10 BREAST CARE CREAM

Breast cream is also one of the categories to be considered in beauty products that claim to increase the size. E.g.: St Herb Nano Breast Cream, that "expands the cellular substructure and facilitates development of the lobules and alveoli of the breasts" (Mamillapalli 2016).

12.6.11 Fragrance related products

Fragrances are aroma-imparting products. There are multiple kinds of perfumes and colognes prepared from gold and silver nanoparticles available in the market; every perfume has its own unique scent as signature (Cayuela, Soriano, and Valcárcel 2015). Example: Bogart One Man Show Gold Deo.

12.6.12 Nailcare products

Nanosilver and nanogold materials are highly utilized nanoparticles in nail care products such as nail polishes and nail paints to improve toughness and firmness and to retain durability. Silver and gold nanoparticles are used to take care of the nails and perform to enhance beauty and aesthetics and also disinfectant as well as manage fungal infections of nails. Nail care prevents nail problems by giving protection against fungi and ingrown toe nails. Cosmeceuticals containing gold and silver nanomaterials are of major advantage since they improve the toughness (Katz, Dewan, and Bronaugh 2015). Silver nanoparticles are specially designed and appreciated for their antifungal effects (Katz, Dewan, and Bronaugh 2015). One of the marketed preparations in this category is Range Nail Polish by Beauty People-Wow.

12.7 NANOTOXICITY OF NANOCOSMECEUTICALS: AN EMERGING PROBLEM

Although gold and silver nanoparticles have a great deal of potential, they are harmful once they get into the bloodstream (Jatana et al. 2016). Nanoparticles have the potential to precipitate harmful reactions because they lead to the generation of free radicals, which induce oxidative stress and consequently lead to cellular death. Nanoparticles might gain entry into our systems by either being breathed in, absorbed through the skin, or consumed unknowingly. They have the potency to elicit skin irritation and harm the lungs, brain, and other organs after being transported via the blood (Jia et al. 2017; Kakoty et al. 2022). Celular toxicity to keratinocytes and fibroblasts occurs when the concentration of silver in a product is high.

12.7.1 Safety of nanocosmeceuticals

Concerns have been expressed about the safety and toxic effects of these nanomaterials when they are applied onto the skin or even other outer parts and the skin is exposed to metallic nanoparticles (Wani, Ara, and Usmani 2015). Changes in the physical and chemical properties of gold and silver nanomaterials in the final product as well as impurities in them need to be taken into consideration (Gupta et al. 2013). The characteristics of these nanomaterials may change as they are managed and stored, making it hard to predict how they will react and what risks they pose (Dréno et al. 2019). During the process of developing and formulating new cosmeceuticals, one should evaluate the safety, asses their activity, stability, and measure how satisfactory are the results brought forth by them (Kaul et al. 2018). The European Commission's Joint Research Centre (JRC) was initiated as the world's first reference repository for nanomaterials so that national and international standardization bodies could evaluate the safety of their products/formulations. The reference nanomaterials available here may be used as a reference when developing new platforms, protocols, and experimental studies (Nanda et al. 2016).

12.7.2 MAJOR CONSIDERATIONS TO ASSESS THE SAFETY OF NANOMATERIALS

Based on the physical interconnections amongst the various constituents of a product there may be safety concerns about nanoparticles having gold and/or silver in them. Consequently, it should be made mandatory that changes in the characteristics and biological effects be reported during the monitoring of gold and silver nanoparticles (Lu et al. 2015). It is vital to provide exhaustive details on these nanoparticles. Analyzing these nanoparticles coated with metal is necessary to identify their precise physicochemical characteristics (Aziz et al. 2019; Fytianos, Rahdar, and Kyzas 2020). The pharmacodynamic evaluations should be undertaken and consolidated. The toxicity profiles of the gold and silver nanoparticles must be created and readily accessible. If any distinguishing attribute or biological activity of a cosmetic product containing gold and silver nanomaterials is discovered, the same should be documented. Conventional testing methods must be used to evaluate the product's toxicity. Gaining an understanding of how the physicochemical properties of nanomaterials affect the toxic effects and how these properties impact the formulation development of cosmetic products is essential to improve the existing practices and technologies related to the same. Every segment of the scientific community, researchers, regulators, and industrialists should have easy access to an in-depth collection of details and data regarding the use safely of cosmetic products (Dhawan, Sharma, and Nanda 2020; Dubey, Dey, et al. 2022).

12.8 CONCLUSIONS AND FUTURE PERSPECTIVES

Presently, nanogold and nanosilver materials are considered revolutionizing and promising in the field of nanotechnology and are widely employed in dermatological, biomedical and cosmeceutical applications. The inclusion of gold and silver nanomaterials in nanocosmetics have increased their demand among customers and opened up new avenues for the cosmetic industries to keep foraying. Various gold and silver-based nanocarriers such as nanoemulsions, cubosomes, liposomes, solid lipid nanoparticles, ethosomes, niosomes, cubosomes, etc. are being explored for the fabrication and development of different cosmetics such that they provide higher efficiency and efficacy over traditional cosmetics. Today these nanocarriers-based cosmetic formulations are an essential part of everyone's daily routine as they deliver the formulations into the skin through multiple mechanisms, thus imparting several properties including glossy and hydrating skin, acne-free skin, sun protection, prevent wrinkles, moisturization and many more. Even though the market value of gold and silver-based nanocosmetics has expanded to a greater extent, there is a debate ongoing concerning their toxicity and safety in humans, hence demanding further investigations. Therefore, it is essential to provide a reference list of ingredients by the cosmetic legislation encompassing all ingredients which produce undesired effects. Also, carcinogenic studies and long-term toxicity studies of nanocosmetics should be carried out before their commercialization. In addition to this, their manufacturing should be done in such a way that they provide benefits to the consumer's health. Preclinical and clinical trials of nanogold and nanosilver-based cosmetic products should be conducted and attested along with the study summaries so as to ensure their safety in animals as well as in humans. Clinical development of gold- and silver-incorporated nanoformulations employing well-characterized delivery systems will be effective, provided the development strategy is meticulously established based on the efficacy criteria of the gold and silver nanoparticles. Strict rules and regulations should be imposed on the import, manufacturing, marketing, and storage of cosmetics, as well as the incorporated gold and silver nanoparticles. Moreover, global regulatory agencies in collaboration with researchers should form standard regulations pertaining to careful use of these nanocarriers and help to address the existing gaps in the cosmetic industry. The governmental bodies and non-governmental organizations should conduct seminars through video and multimedia with the aim to generate awareness regarding the judicious use of nanocosmetics amongst consumers. Other than this, it is essential to harmonize the regulations in order to establish effective regulatory structure for marketing, efficacy, and safety of products which will definitely help cosmeceutical industries to protect customers from the deleterious health hazards.

REFERENCES

Alaqad, Khalid, and Tawfik A Saleh. 2016. "Gold and Silver Nanoparticles: Synthesis Methods, Characterization Routes and Applications towards Drugs." *Undefined* 6 (4): 384. doi:10.4172/2161-0525.1000384.

Arya, Anju, Vaibhav Mishra, and Tejpal Singh Chundawat. 2019. "Green Synthesis of Silver Nanoparticles from Green Algae (Botryococcus Braunii) and Its Catalytic Behavior for the Synthesis of Benzimidazoles." *Chemical Data Collections* 20 (February): 100190. doi:10.1016/j.cdc.2019.100190.

Aziz, Zarith Asyikin Abdul, Hasmida Mohd-Nasir, Akil Ahmad, Siti Hamidah Mohd Setapar, Wong Lee Peng, Sing Chuong Chuo, Asma Khatoon, Khalid Umar, Asim Ali Yaqoob, and Mohamad Nasir Mohamad Ibrahim. 2019. "Role of Nanotechnology for Design and Development of Cosmeceutical: Application in Makeup and Skin Care." *Frontiers in Chemistry* 7 (November): 739. doi:10.3389/fchem.2019.00739.

Bapat, Ranjeet A, Tanay V Chaubal, Suyog Dharmadhikari, Anshad Mohamed Abdulla, Prachi Bapat, Amit Alexander, Sunil K Dubey, and Prashant Kesharwani. 2020. "Recent Advances of Gold Nanoparticles as Biomaterial in Dentistry." *International Journal of Pharmaceutics* 586: 119596. doi:10.1016/j.ijpharm.2020.119596.

Barbosa, Gustavo P, Henrique S Debone, Patrícia Severino, Eliana B Souto, and Classius F da Silva. 2016. "Design and Characterization of Chitosan/Zeolite Composite Films — Effect of Zeolite Type and Zeolite Dose on the Film Properties." *Materials Science and Engineering: C* 60: 246–54. doi:10.1016/j.msec.2015.11.034.

Bianco, Carlotta, Maaike J Visser, Olivier A Pluut, Vesna Svetličić, Galja Pletikapić, Ivone Jakasa, Christoph Riethmuller, et al. 2016. "Characterization of Silver Particles in the Stratum Corneum of Healthy Subjects and Atopic Dermatitis Patients Dermally Exposed to a Silver-Containing Garment." *Nanotoxicology* 10 (10): 1480–91. doi:10.1080/1743 5390.2016.1235739.

Bocca, Beatrice, Anna Pino, Alessandro Alimonti, and Giovanni Forte. 2014. "Toxic Metals Contained in Cosmetics: A Status Report." *Regulatory Toxicology and Pharmacology: RTP* 68 (3): 447–67. doi:10.1016/j.yrtph.2014.02.003.

Campbell, Christopher S J, L Rodrigo Contreras-Rojas, M Begoña Delgado-Charro, and Richard H Guy. 2012. "Objective Assessment of Nanoparticle Disposition in Mammalian Skin after Topical Exposure." *Journal of Controlled Release: Official Journal of the Controlled Release Society* 162 (1): 201–7. doi:10.1016/j.jconrel.2012.06.024.

Cao, Mingjing, Jiayang Li, Jinglong Tang, Chunying Chen, and Yuliang Zhao. 2016. "Gold Nanomaterials in Consumer Cosmetics Nanoproducts: Analyses, Characterization, and Dermal Safety Assessment." *Small (Weinheim an Der Bergstrasse, Germany)* 12 (39): 5488–96. doi:10.1002/smll.201601574.

Cayuela, Angelina, M Laura Soriano, and Miguel Valcárcel. 2015. "Reusable Sensor Based on Functionalized Carbon Dots for the Detection of Silver Nanoparticles in Cosmetics via Inner Filter Effect." *Analytica Chimica Acta* 872 (May): 70–76. doi:10.1016/j. aca.2015.02.052.

Dhawan, Surbhi, Pragya Sharma, and Sanju Nanda. 2020. "Cosmetic Nanoformulations and Their Intended Use." In, 141–69. doi:10.1016/B978-0-12-822286-7.00017-6.

Dorofeev G, A Streletskii, Ivan Povstugar, Andrey Protasov, and E Elsukov. 2012. "Determination of Nanoparticle Sizes by X-Ray Diffraction." *Colloid Journal* 74 (November). doi:10.1134/S1061933X12060051.

Dréno B, A Alexis, B Chuberre, and M Marinovich. 2019. "Safety of Titanium Dioxide Nanoparticles in Cosmetics." *Journal of the European Academy of Dermatology and Venereology : JEADV* 33 Suppl 7 (November): 34–46. doi:10.1111/jdv.15943.

Dubey, Sunil Kumar, Anuradha Dey, Gautam Singhvi, Murali Manohar Pandey, Vanshikha Singh, and Prashant Kesharwani. 2022. "Emerging Trends of Nanotechnology In Advanced Cosmetics." *Colloids and Surfaces B: Biointerfaces* 112440. doi:10.1016/j. colsurfb.2022.112440.

Dubey, Sunil Kumar, Shraddha Parab, Vaishnav Pavan Kumarr Achalla, Avinash Narwaria, Swapnil Sharma, B H Jaswanth Gowda, and Prashant Kesharwani. 2022. "Microparticulate and Nanotechnology Mediated Drug Delivery System for the Delivery of Herbal Extracts." *Journal of Biomaterials Science, Polymer Edition* April: 1–24. doi: 10.1080/09205063.2022.2065408.

Fernandes, Rute, Neil R Smyth, Otto L Muskens, Simone Nitti, Amelie Heuer-Jungemann, Michael R Ardern-Jones, and Antonios G Kanaras. 2015. "Interactions of Skin with Gold Nanoparticles of Different Surface Charge, Shape, and Functionality." *Small (Weinheim an Der Bergstrasse, Germany)* 11 (6): 713–21. doi:10.1002/smll.201401912.

Fytianos, Georgios, Abbas Rahdar, and George Z Kyzas. 2020. "Nanomaterials in Cosmetics: Recent Updates." *Nanomaterials.* doi:10.3390/nano10050979.

Gavazzoni Dias, Maria Fernanda Reis. 2015. "Hair Cosmetics: An Overview." *International Journal of Trichology* 7 (1): 2–15. doi:10.4103/0974-7753.153450.

George, Robert, Steve Merten, Tim T Wang, Peter Kennedy, and Peter Maitz. 2014. "In Vivo Analysis of Dermal and Systemic Absorption of Silver Nanoparticles through Healthy Human Skin." *The Australasian Journal of Dermatology* 55 (3): 185–90. doi:10.1111/ ajd.12101.

Gupta, Sanjeev, Radhika Bansal, Sunita Gupta, Nidhi Jindal, and Abhinav Jindal. 2012. "Nanocarriers and Nanoparticles for Skin Care and Dermatological Treatments." *Indian Dermatology Online Journal* 4 (4): 267–72. doi:10.4103/2229-5178.120635.

Herizchi, Roya, Elham Abbasi, Morteza Milani, and Abolfazl Akbarzadeh. 2016. "Current Methods for Synthesis of Gold Nanoparticles." *Artificial Cells, Nanomedicine, and Biotechnology* 44 (2): 596–602. doi:10.3109/21691401.2014.971807.

Jatana, Samreen, Linda M Callahan, Alice P Pentland, and Lisa A DeLouise. 2016. "Impact of Cosmetic Lotions on Nanoparticle Penetration Through Ex Vivo C57BL/6 Hairless Mouse and Human Skin: A Comparison Study." *Cosmetics* 3 (1). doi:10.3390/cosmetics3010006.

Jia, Xiaochuan, Shuo Wang, Lei Zhou, and Li Sun. 2017. "The Potential Liver, Brain, and Embryo Toxicity of Titanium Dioxide Nanoparticles on Mice." *Nanoscale Research Letters* 12 (1): 478. doi:10.1186/s11671-017-2242-2.

Jung, Jae Hee, Hyun Cheol Oh, Hyung Soo Noh, Jun Ho Ji, and Sang Soo Kim. 2006. "Metal Nanoparticle Generation Using a Small Ceramic Heater with a Local Heating Area." *Journal of Aerosol Science* 37 (12): 1662–70. doi:10.1016/j.jaerosci.2006.09.002.

Kakoty, Violina, KC Sarathlal, Meghna Pandey, Sunil Kumar Dubey, Prashant Kesharwani, and Rajeev Taliyan. 2022. "Chapter 18- Biological Toxicity of Nanoparticles." In, edited by Prashant Kesharwani and K B T Kamalinder, *Nanoparticle Therapeutics*, 603–28. Academic Press. doi:10.1016/B978-0-12-820757-4.00016-8.

Katz, Linda M, Kapal Dewan, and Robert L Bronaugh. 2015. "Nanotechnology in Cosmetics." *Food and Chemical Toxicology : An International Journal Published for the British Industrial Biological Research Association* 85 (November): 127–37. doi:10.1016/j.fct.2015.06.020.

Kaufman, Bridget P, and Andrew F Alexis. 2020. "Randomized, Double-Blinded, Split-Face Study Comparing the Efficacy and Tolerability of Two Topical Products for Melasma." *Journal of Drugs in Dermatology : JDD* 19 (9): 822–27. doi:10.36849/JDD.2020.10.36849/JDD.2020.5353.

Kaul, Shreya, Neha Gulati, Deepali Verma, Siddhartha Mukherjee, and Upendra Nagaich. 2018. "Role of Nanotechnology in Cosmeceuticals: A Review of Recent Advances." *Journal of Pharmaceutics* 2018 (March): 1–19. doi:10.1155/2018/3420204.

Khan, A K., R Rashid, G Murtaza, and A Zahra. 2014. "Gold Nanoparticles: Synthesis and Applications in Drug Delivery." *Tropical Journal of Pharmaceutical Research* 13 (7): 1169–77. doi:10.4314/tjpr.v13i7.23.

Kim, Byung Hyo, Michael J Hackett, Jongnam Park, and Taeghwan Hyeon. 2014. "Synthesis, Characterization, and Application of Ultrasmall Nanoparticles." *Chemistry of Materials* 26 (1): 59–71. doi:10.1021/cm402225z.

Kokura, Satoshi, Osamu Handa, Tomohisa Takagi, Takeshi Ishikawa, Yuji Naito, and Toshikazu Yoshikawa. 2010. "Silver Nanoparticles as a Safe Preservative for Use in Cosmetics." *Nanomedicine: Nanotechnology, Biology, and Medicine* 6 (4): 570–74. doi:10.1016/j.nano.2009.12.002.

Lohani, Alka, Anurag Verma, Himanshi Joshi, Niti Yadav, and Neha Karki. 2014. "Nanotechnology-Based Cosmeceuticals." In edited by T J Ryan and T Maisch. *ISRN Dermatology* 2014. Hindawi Publishing Corporation: 843687. doi:10.1155/2014/843687.

Lu, Pei Jia, Shou Chieh Huang, Yu Pen Chen, Lih Ching Chiueh, and Daniel Yang Chih Shih. 2015. "Analysis of Titanium Dioxide and Zinc Oxide Nanoparticles in Cosmetics." *Journal of Food and Drug Analysis* 23 (3): 587–94. doi:10.1016/J.JFDA.2015.02.009.

Malakar, Arindam, Sushil R Kanel, Chittaranjan Ray, Daniel D Snow, and Mallikarjuna N Nadagouda. 2021. "Nanomaterials in the Environment, Human Exposure Pathway, and Health Effects: A Review." *The Science of the Total Environment* 759. Nebraska Water Center, part of the Robert B. Daugherty Water for Food Global Institute 2021 Transformation Drive, University of Nebraska, Lincoln, NE 68588-0844: 143470. doi:10.1016/j.scitotenv.2020.143470.

Mamillapalli, Vani. 2016. "Nanoparticles for Herbal Extracts." *Asian Journal of Pharmaceutics (AJP)* 10 (2 Se-Review Articles). doi:10.22377/ajp.v10i2.623.

Maneewattanapinyo, Pattwat, Wijit Banlunara, Chuchaat Thammacharoen, Sanong Ekgasit, and Theerayuth Kaewamatawong. 2011. "An Evaluation of Acute Toxicity of Colloidal Silver Nanoparticles." *The Journal of Veterinary Medical Science* 73 (11): 1417–23. doi:10.1292/jvms.11-0038.

Manikanika, Jagdeep Kumar, and S Jaswal. 2021. "Role of Nanotechnology in the World of Cosmetology: A Review." *Materials Today: Proceedings* 45 (February). doi:10.1016/j. matpr.2020.12.638.

Matsuura-Sawada, Yuka, Masatoshi Maeki, Takaaki Nishioka, Ayuka Niwa, Jun Yamauchi, Masashi Mizoguchi, Koichi Wada, and Manabu Tokeshi. 2022. "Microfluidic Device-Enabled Mass Production of Lipid-Based Nanoparticles for Applications in Nanomedicine and Cosmsetics." *ACS Applied Nano Materials* 5 (6): 7867–76. doi:10.1021/acsanm.2c00886.

Mourdikoudis, Stefanos, Roger M Pallares, and Nguyen T K Thanh. 2018. "Characterization Techniques for Nanoparticles: Comparison and Complementarity upon Studying Nanoparticle Properties." *Nanoscale* 10 (27): 12871–934. doi:10.1039/C8NR02278J.

Nanda, Sanju, Arun Nanda, Shikha Lohan, Ranjot Kaur, and Bhupinder Singh. 2016. "Nanocosmetics: Performance Enhancement and Safety Assurance." In, 47–67. doi:10.1016/B978-0-323-42868-2.00003-6.

Nasir, Ali, Harikumar Sl, and Amanpreet Kaur. 2012. "Niosomes: An Excellent Tool for Drug Delivery." *International Journal of Research In Pharmacy and Chemistry* 2: 479–487.

Oberdörster, Günter, Eva Oberdörster, and Jan Oberdörster. 2005. "Nanotoxicology: An Emerging Discipline Evolving from Studies of Ultrafine Particles." *Environmental Health Perspectives* 113 (7): 823–39. doi:10.1289/ehp.7339.

Pulit-Prociak, Jolanta, Aleksandra Grabowska, Jarosław Chwastowski, Tomasz M Majka, and Marcin Banach. 2019. "Safety of the Application of Nanosilver and Nanogold in Topical Cosmetic Preparations." *Colloids and Surfaces B: Biointerfaces* 183: 110416. doi:10.1016/j.colsurfb.2019.110416.

Raju, Gayathri, Neeraj Katiyar, Sajini Vadukumpully, and Sahadev A Shankarappa. 2018. "Penetration of Gold Nanoparticles across the Stratum Corneum Layer of Thick-Skin." *Journal of Dermatological Science* 89 (2): 146–54. doi:10.1016/j. jdermsci.2017.11.001.

Rapalli, Vamshi Krishna, Arisha Mahmood, Tejashree Waghule, Srividya Gorantla, Sunil Kumar Dubey, Amit Alexander, and Gautam Singhvi. 2021. "Revisiting Techniques to Evaluate Drug Permeation through Skin." *Expert Opinion on Drug Delivery* 18 (12): 1829–42. doi:10.1080/17425247.2021.2010702.

Santhosh, Poornima B, Julia Genova, and Hassan Chamati. 2022. "Green Synthesis of Gold Nanoparticles: An Eco-Friendly Approach." *Chemistry*. doi:10.3390/chemistry4020026.

Siddiqi, Khwaja Salahuddin, Azamal Husen, and Rifaqat A K Rao. 2018. "A Review on Biosynthesis of Silver Nanoparticles and Their Biocidal Properties." *Journal of Nanobiotechnology* 16 (1): 14. doi:10.1186/s12951-018-0334-5.

Singh, Archana. 2021. "Carbon Nanofiber in Cosmetics." In, 341–63. doi:10.1002/ 9781119769149.ch14.

Souto, Eliana B., Ana Rita Fernandes, Carlos Martins-Gomes, Tiago E. Coutinho, Alessandra Durazzo, Massimo Lucarini, Selma B. Souto, Amélia M. Silva, and Antonello Santini. 2020. "Nanomaterials for Skin Delivery of Cosmeceuticals and Pharmaceuticals." *Applied Sciences 2020* 10 (5): 1594. doi:10.3390/APP10051594.

Srinivas, Kurapati. 2016. "The Current Role of Nanomaterials in Cosmetics." *Journal of Chemical and Pharmaceutical Research* 8 (5): 906–14. https://www.jocpr.com/abstract/ the-current-role-of-nanomaterials-in-cosmetics-5888.html.

Tangau, Mason Jarius, Yie Kie Chong, and Keng Yoon Yeong. 2022. "Advances in Cosmeceutical Nanotechnology for Hyperpigmentation Treatment." *Journal of Nanoparticle Research* 24 (8): 155. doi:10.1007/s11051-022-05534-z.

Varun, Thakur, Arora Sonia, and Vishal Patil. 2022. "Niosomes and Liposomes-Vesicular Approach Towards Transdermal Drug Delivery." *International Journal of Pharmaceutical and Chemical Sciences ISSN: 2277-5005* 1 (3): 29. www.ijpcsonline.com.

Wani, Ab Latif, Anjum Ara, and Jawed Ahmad Usmani. 2015. "Lead Toxicity: A Review." *Interdisciplinary Toxicology* 8 (2): 55–64. doi:10.1515/intox-2015-0009.

Zed, Salman, Irshad Ullah, Ali Karim, Wali Muhammad, Naimat Ullah, Mehmand Khan, and Warda Komal. 2019. "A Review on Nanotechnology Applications in Electric Components." *Nanoscale Reports* 2 (2 SE-Review): 32–38. doi:10.26524/nr1924.

13 Using nanostructured materials to increase safety and efficacy of organic UV filters

André Luis Máximo Daneluti, André Rolim Baby and Yogeshvar N. Kalia

CONTENTS

13.1 INTRODUCTION

The purpose of sunscreens is to protect the skin from damage caused by ultraviolet (UV) radiation. Sunscreen formulations can be composed of organic filters and inorganic filters or a combination of both (Kockler *et al.*, 2012; Abid *et al.*, 2017).

DOI: 10.1201/9781003319146-13

Organic UV filters are aromatic compounds conjugated with carboxylic groups, capable of absorbing UVA and UVB radiation (Velasco *et al.*, 2008; Shaath, 2010).

Organic UV filters must remain on the skin surface and penetrate as little as possible into the surface layers of the skin tissue (stratum corneum or SC) to unfold their photoprotective properties and to avoid systemic toxicity (Benech-Kieffer *et al.*, 2000; Potard *et al.*, 2000; Klimová, Hojerová and Beránková, 2015). However, many organic UV filters can permeate through the deeper layers of the skin, reaching the blood circulation and consequently causing systemic effects on the endocrine system. In addition, many of these compounds are photo-unstable, impairing their photoprotective efficacy as well as causing photo sensibilization and cutaneous phototoxicity for the user (Krause *et al.*, 2012; Klimová, Hojerová and Pažoureková, 2013; Klimová, Hojerová and Beránková, 2015).

Given these challenges, scientific researchers in the field of organic UV filters seek to innovate new technologies to increase the safety and efficacy of these products. New nanoparticles or nanostructured carrier materials such as cyclodextrins, polymers, gelatine nanoparticles, nanostructured lipid carriers, liposomes and mesoporous materials are the focus of new studies aimed at preventing skin permeation while increasing the photostability of organic UV filters (Puglia *et al.*, 2012; Ambrogi, Latterini, Marmottini, Pagano, *et al.*, 2013; Ambrogi, Latterini, Marmottini, Tiralti, *et al.*, 2013; de Oliveira *et al.*, 2016a,b; Gilbert *et al.*, 2016; Zhou *et al.*, 2018; Daneluti *et al.*, 2019).

In this chapter, the use of nanostructure materials to increase the safety and efficacy of organic UV filters will be discussed. In addition, the regulatory and toxicology aspects of encapsulated UV filters will be described.

13.2   UV RADIATION AND DELETERIOUS EFFECTS

The spectral distribution of UV radiation reaching the surface of the Earth is composed of UVC (100–280 nm), UVB (290–320 nm) and UVA. UVA can be divided into UVA II (320–340 nm) and UVA I (340–400 nm) (Diffey *et al.*, 2000; Diffey, 2015). Both UVA and UVB radiation are responsible for degenerative processes of the skin such as aging and cancer (Wang, Balagula and Osterwalder, 2010; Levi, 2013). UVA is considered long-wave radiation and interacts both with the epidermis and dermis; it causes oxidative stress on the skin via reactive oxygen (ROS) and nitrogen species, which can contribute to skin cancer. In addition, UVA induces pigmentation and skin photoaging. UVB, on the other hand, mainly damages the epidermis, might cause serious erythema in a short time period and has deleterious effects on the epidermal DNA, thereby creating a predisposition for skin cancer if exposed for a long term (Levi, 2013; Rai, Deep and Tasduq, 2022). The third type of UV radiation, UVC, has no real impact on the skin, as it is effectively absorbed by the ozone layer. In addition to UV, there is increasing evidence that infrared radiation (IR) might also release free radicals, accelerate aging and play a role as well in skin damage (Levi, 2013; Gubitosa *et al.*, 2020).

Clinically, UV rays can cause skin changes such as wrinkles, dryness, telangiectasia, aging and pigmentation. Histologically, skin aging is reflected in the disorganization of collagen fibers, loss of elastin fibers, flattening of the dermo-epidermal junction and dilation of blood vessels. Furthermore, morphological and quantitative changes occur in keratinocytes, melanocytes, fibroblasts and Langerhans cells. Chronic exposure to UV rays has been associated with the development of actinic keratosis, squamous cell carcinoma and basal cell cancer, while intermittent exposure is associated with the development of melanoma, the most dangerous type of skin cancer (Djavaheri-Mergny *et al.*, 1996; Hanson *et al.*, 2015; Gilbert *et al.*, 2016; Abid *et al.*, 2017). To reduce these damages, the skin possesses self-defence mechanisms that can be distinguished as endogen and exogen processes. The tanning process during exposure to UV rays is considered as the main endogenous defence mechanism and is induced through melanin formation by UVB stimulation. The secondary endogenous defensive mechanisms include SC thickening, excretion of urocanic acid (sweating process), deposition of β-carotene in the hypodermis, and activation of DNA repair mechanisms (Gubitosa *et al.*, 2020). However, these endogenous defensive mechanisms may be insufficient to avoid skin tissue damage caused by excessive exposure to UV radiation. Additional measures should be adopted, such as the use of protective clothing and accessories (hats, sunglasses and umbrellas) to physically block UV radiation. Moreover, the correct and frequent application of sunscreen formulations is recognized as being the most efficient against sunburn and sun damage to the skin (Cestari, Oliveira and Boza, 2012; Miksa *et al.*, 2016; Lim, Arellano-Mendoza and Stengel, 2017).

13.3   SUNSCREEN FORMULATIONS

Sunscreen formulations can be classified into different groups, such as: emulsions (creams and lotions), oils, alcoholic solutions, aerosols, moulded forms (sticks and lip balms), gels and powders. The most common formulation on the market are oil-in-water (O/W) emulsions. Due to its versatility in conveying compounds of different polarities, it is possible to select the most suitable system of filters and excipients to obtain formulations with better sensory, efficacy and safety (Tanner, 2006). However, the most important point is how consumers use these products. In general, consumers commit three types of mistakes when applying sunscreens: non-uniform application; absence of reapplication; and application in reduced amounts, all of which are responsible for the decrease in sunscreen efficacy (Tanner, 2006).

Ou-Yang et al. (2012) evaluated the relationship between the amount of sunscreen applied and the Sun Protection Factor (SPF) values obtained during in vivo efficacy studies, obtaining a linear relationship between the two variables. Knowing that the tests validated for the evaluation of sunscreen efficacy use controlled amounts of formulation (2.0 mg cm^{-2} of skin), it was proven that the sun protection obtained by consumers was lower than what was indicated on the labels of products, because epidemiological studies have shown that the population applies an average of 0.5–1.0 mg cm^{-2} (Ou-yang *et al.*, 2012).

Sunscreen formulations contain UV filters that act on the skin surface, involving two distinct mechanisms, in general: absorption (organic or chemical filters) or reflection (inorganic or physical filters) of UV radiation (Cestari, Oliveira and Boza, 2012). Currently, sunscreen formulations contain a combination of organic and inorganic UV filters that provide a broad spectrum of protection. Organic UV-filters can be further subdivided into UVA-filters which only absorb UVA-light, UVB-filters (only absorbing UVB-light) and broad-spectrum filters which absorb both UVA- and UVB-light (Maier *et al.*, 2005; Kockler *et al.*, 2012; Gilbert *et al.*, 2013; Miksa *et al.*, 2016).

13.4 INORGANIC UV FILTERS

Inorganic filters are metallic oxides that act by reflection or scattering of UV radiation and are represented by two types: ZnO (zinc oxide) offers better UVA protection whereas TiO2 (titanium dioxide) provides superior UVB protection. It is noteworthy that the phenomena of reflection and scattering depend on the particle size of these chemical compounds, among other factors (DIFFEY and GRICE, 1997; Leong *et al.*, 2016). These compounds are dispersed in the formulations and are mainly responsible for the opacity and the white colour of the sunscreen formulations. Inorganic filters are effective and safe with rare cases of photosensitization or photoallergies, however, their cosmetic acceptability represents an obstacle to their wide use. (Wang, Balagula and Osterwalder, 2010; Sambandan and Ratner, 2011; Morlando *et al.*, 2016).

13.5 ORGANIC UV FILTERS

Chemical UV filters are essentially aromatic compounds conjugated with carboxylic groups and usually have a donor group electron such as an amine or methoxyl in the ortho or para position of the aromatic ring absorbing radiation in different spectral regions (UVA or UVB). The main organic UVA filters present in sunscreen formulations include benzophenones (mainly oxybenzone), avobenzone (Figure 13.1), terephthalidenedicamphor sulfonic acid and drometrizoltrisiloxane. Mentyl anthranilate is classified as a UVA filter, but it is rarely used (González, Fernández-Lorente and Gilaberte-Calzada, 2008a). Among the main UVB filters are cinnamates, salicylates and camphor derivatives (Figure 13.2), while the main broad spectrum organic UV filters are Bisoctrizole (methylene-bis-benzotriazolyltetramethylbutylphenol),

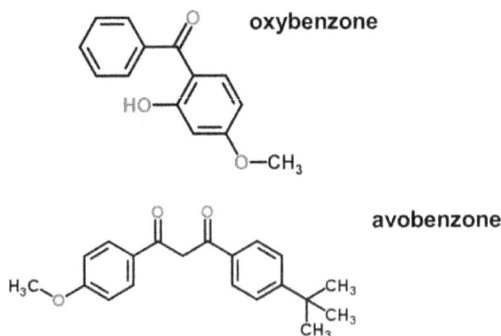

FIGURE 13.1   Molecular structures of oxybenzone UVA and avobenzone

Octyl triazone and Bemotrizinol (bis-ethylhexyloxyphenol methoxy phenyltriazine) (Figure 13.3). For a sunscreen to be effective, organic UV filters must be used in combination with other compounds. At least two types of organic UV filters are commonly associated. These UV filters might contain other compounds such as antioxidants, which are also important to protect the skin from the effects of exposure to UV light and to photostabilize some organic UV filters (Kockler *et al.*, 2012; de Oliveira *et al.*, 2015; Peres *et al.*, 2018; Tomazelli *et al.*, 2018).

Oxybenzone is an aromatic ketone absorbed in the UVAII/UVB regions, with absorption peaks between 290 and 360 nm. Studies demonstrate that oxybenzone has high photostability, even after being irradiated for a long period of time (Palm and

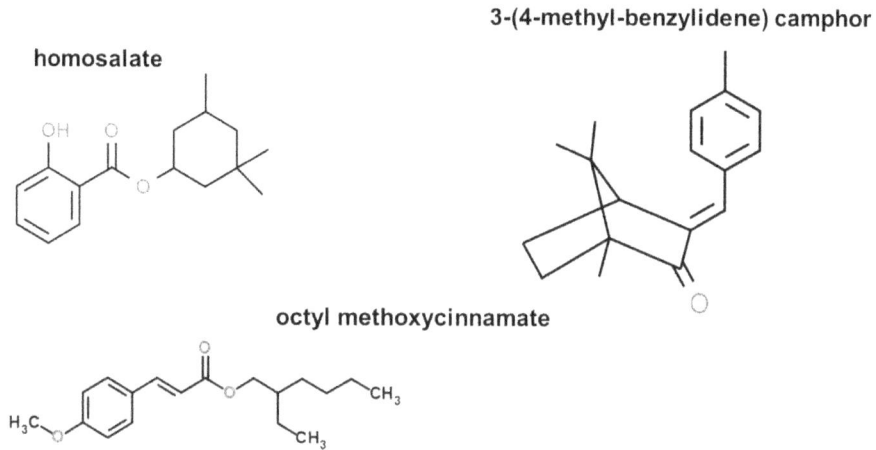

FIGURE 13.2   Molecular structure of UVB filters

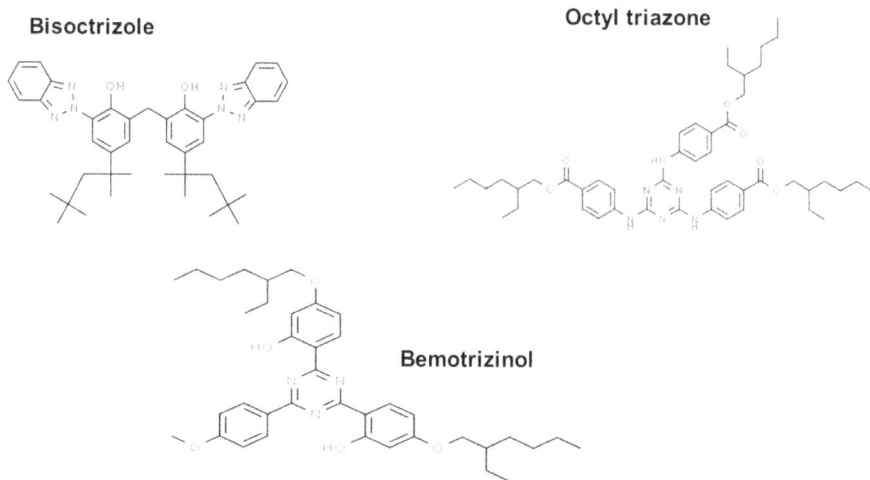

FIGURE 13.3   Molecular structures of mainly broad-spectrum organic UV filters

O'Donoghue, 2007; González, Fernández-Lorente and Gilaberte-Calzada, 2008b; Abid *et al.*, 2017).

However, benzophenones are considered by the FDA as one of the most allergenic organic UV filter compounds, which has a higher incidence of photo allergenicity, photo contact when compared to other compounds (Palm and O'Donoghue, 2007; FDA, 2011; Ambrogi, Latterini, Marmottini, Pagano, *et al.*, 2013). In addition, Oxybenzone has the capacity to permeate the skin, and consequently systemic absorption, causing endocrine problems, especially estrogenic effects. (Ambrogi, Latterini, Marmottini, Pagano, *et al.*, 2013; Kim and Choi, 2014; Klimová, Hojerová and Beránková, 2015).

Avobenzone is one of the most effective UVA filters used in photoprotective formulations. This compound was the first to feature UVA-I protection, covering the wavelength range from 310 to 400 nm. However, avobenzone undergoes significant degradation upon exposure to light (Afonso *et al.*, 2014).

Cinnamates have similar structures where the aromatic molecule is disubstituted with both an electron-releasing group (OCH3) and an electron-accepting group (the ester group that it is further conjugated with a double bond). This permits the extended delocalization of electrons enabling these molecules to absorb in the region of 310 nm as illustrated in Figure 13.4 (Palm and O'Donoghue, 2007; Shaath, 2010).

Octyl p-methoxycinnamate (OMC) is the most used cinnamate worldwide. OMC is the most potent UVB absorber widely in use. It is approximately one order of magnitude weaker than Padimate O and absorbs wavelengths 270–328 nm. In addition, this sunscreen has desirable physical properties for photoprotective formulations, providing water resistance to the preparations (Palm and O'Donoghue, 2007).

Oxybenzone, avobenzone and octyl methoxycinnamate are organic UV filters that are widely used in commercial sunscreen products (Janjua *et al.*, 2008; Klinubol, Asawanonda and Wanichwecharungruang, 2008; de Oliveira *et al.*, 2015). However, avobenzone and octyl methoxycinnamate are photo-chemically unstable and exposure to sunlight reduces their absorbance (de Oliveira *et al.*, 2015; Shetty *et al.*, 2015; Leong *et al.*, 2016; Daneluti *et al.*, 2019). Furthermore, AVO degrades significantly when exposed to UV, creating photodegradation products that may interact with the skin causing phototoxicity or photoallergic contact dermatitis (Gaspar *et al.*, 2012; Afonso *et al.*, 2014).

Several studies have shown that some organic UV filters have the capacity to permeate through the skin, undergo metabolism in the body and be excreted. This phenomenon may be explained by the low molecular weight and the lipophilic character of these compounds, facilitating their partitioning into the SC. This may cause local side effects (i.e. allergic contact and dermatitis), and/or systemic side effects, e.g.

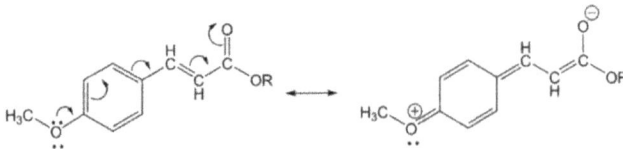

FIGURE 13.4   The electron delocalization in a cinnamate molecule (Shaath, 2010)

mutagenic and estrogenic activity (Janjua *et al.*, 2008; Krause *et al.*, 2012; Klimová, Hojerová and Beránková, 2015). For instance, oxybenzone and avobenzone can cause significant allergic reactions, while oxybenzone, octyl methoxycinnamate, 3-benzylidene camphor and 3-(4-methyl-benzylidene) camphor have been detected in systemic circulation, such as in human breast milk or urine, and can also act as endocrine disruptors (estrogenic and progestogenic activity) (Janjua *et al.*, 2008; Krause *et al.*, 2012; Kim and Choi, 2014).

Scientific research in the area of photoprotection has been dedicated to the development of new technologies or new carriers to increase the safety of these compounds. Therefore, new carrier particles such as cyclodextrins, polymers, lipid microparticles and mesoporous materials have been studied to prevent skin permeation and increase the photostability of organic UV filters (Blasi *et al.*, 2011; Wu *et al.*, 2014; Daneluti *et al.*, 2019, 2021).

13.6   NATURAL BIOACTIVE COMPOUNDS

There is a growing interest in natural bioactive compounds with both photoprotector and antioxidant potential. Carotenoids, flavonoids, polyphenols, walnut oil, rice germ oil, helichrysum oil, algal extracts and Punica granatum extracts are natural compounds used as antioxidants in cosmetic formulations, preventing or reducing potentially photogenerated reactive species (de Oliveira *et al.*, 2015, 2016a,b; Peres *et al.*, 2018; Tomazelli *et al.*, 2018; Road, 2021; Sauce *et al.*, 2021). Moreover, these compounds could reduce the concentration of synthetic UV filters and add multifunctional characteristics to the formulations (Velasco *et al.*, 2008).

One of many natural compounds used in cosmetic formulations, Rutin (3,3′, 4′, 5,7-pentahydroxyflavone-3-rhamnoglucoside; quercetin-3-rutinoside) (Figure 13.5) is a flavonoid with characteristics that could benefit sunscreens. The UV absorption

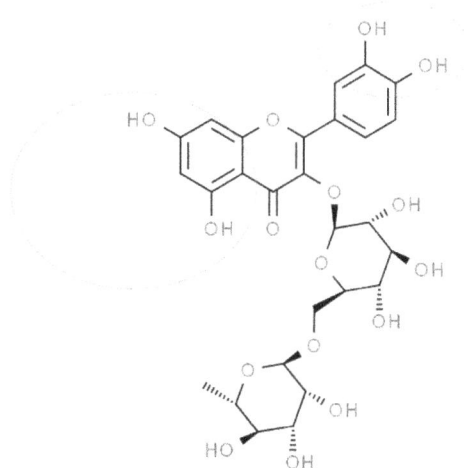

FIGURE 13.5   Chemical structure of rutin. Free radical scavenging capacity is attributed to the high reactivity of hydroxyl groups highlighted

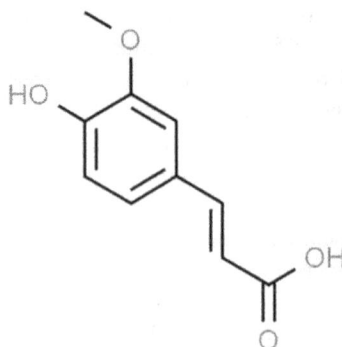

FIGURE 13.6   Structure of ferulic acid

spectrum of this compound has a peak between 320 and 385 nm, covering a considerable part of the UVA spectrum and therefore marking its potential as a photoprotective molecule (de Oliveira *et al.*, 2015). Rutin has been studied as a photostabilizer, antioxidant and sun protection factor (SPF) enhancer (Velasco *et al.*, 2008; de Oliveira *et al.*, 2015; de Oliveira *et al.*, 2016; Peres *et al.*, 2016; Tomazelli *et al.*, 2018).

Another antioxidant used in cosmetic formulations is ferulic acid (Figure 13.6) (FA) (4-hydroxy-3-methoxy cinnamic acid), a natural phenolic compound, which belongs to the class of hydroxycinnamic acids and is found mainly in rice, citrus fruits, wheat, corn, roasted coffee and several other vegetables (Srinivasan *et al.*, 2005; Gerin *et al.*, 2016). The association of ferulic acid with two organic UV filters (ethylhexyl triazone and bis-ethylhexyloxyphenol methoxyphenyl triazine) showed an increase of in vivo SPF by 37% and 26% of the UVA protection factor (UVA-PF) (Peres *et al.*, 2018).

13.7   NANOSTRUCTURE AND NANOPARTICLES APPLIED IN SUNSCREEN FORMULATIONS

Nanotechnology (NT) is a generic term for techniques, materials and devices that operate at the nanometer scale. It has been defined as the design, characterization, production and application of structures, devices and systems, which maintain shape and size at the nanometer scale. NT represents one of the most promising technologies of the 21st century and has been considered a new industrial revolution (Nohynek and Roberts, 2007; Papakostas and Rancan, 2011).

Nanotechnology in cosmetic formulations has been used for over 40 years and has been part of countless publications. There have been patent applications since 1983 by the Shiseido Company for a skin cosmetic formulation containing hydrophobic titanium dioxide particles with a maximum size of less than 0.1 μm and an average size of 30–40 nm (Gubitosa *et al.*, 2020). The increased application of nanomaterials in cosmetic products is indicative of the great potential that nanotechnology represents for the cosmetics industry and its consumers. Nanomaterials are used in make-up, skin care and sunscreens product to make fragrances last longer, to enhance SPF efficacy of organic UV filters or to claim face creams as being effective for antiaging

(Chaudhri, Soni and Prajapati, 2015). Other benefits of nanomaterials in cosmetic products are increased efficiency, transparency, unique texture, protection of active ingredient and, overall, higher consumer compliance (Gubitosa *et al.*, 2020).

Nanoemulsions are emulsions having small droplet size (20–300 nm). Nanoemulsions containing droplets above 100 nm appear white, whereas dispersions around 70–100 nm appear opaque and below that become transparent (Chanchal and Swarnlata, 2008). The droplets' small size gives them inherent stability against creaming, sedimentation, flocculation and coalescence, and allows the effective transport of active ingredients to the skin (Guglielmini, 2008).

The term "particle" derives from the Latin "particle", which means a small part. Nano-scale particles are described as nanoparticles (NP). Currently, there is no globally accepted definition of its dimensions, although the British Standards Institution (BSI) and FDA establish 100.0 nm as the upper limit and 1.0 nm as the lower limit (BSI, 2007; FDA, 2014). The Europe regulation defines a nanomaterial as "an insoluble or biopersistent and intentionally manufactured material with one or more external dimensions, or an internal structure, on the scale from 1 to 100 nm" (EC, 2009). However, there are several records in the literature of particles with biological applications labelling a diameter between 200.0 and 400.0 nm as nanoparticles (BSI, 2007; Nahar *et al.*, 2008; Almeida *et al.*, 2010).

As it was mentioned before, inorganic UV filters such as TiO2 and ZnO have the ability to reflect UV light due to a high refractive index. Nanoparticle grades of these materials are often incorporated into sunscreen formulations. The nano sized ZnO and TiO2 provide superior UV protection than the larger pigments (Nohynek and Roberts, 2007; Wang, Balagula and Osterwalder, 2010; Morabito *et al.*, 2011; Wang and Tooley, 2011; Gilbert *et al.*, 2013).

13.8 ENCAPSULATION AND INCORPORATION OF ORGANIC UV FILTERS INTO NANOMATERIALS/NANOPARTICLES

With the progress of research in the area of nanotechnology, several nanomaterials and nanoparticles have been proposed for entrapping/encapsulating organic UV filters to improve their photostability, to increase the sun protection factor (SPF) and to decrease skin penetration (Gilbert *et al.*, 2013; Deng *et al.*, 2015; Daneluti *et al.*, 2019; Souto *et al.*, 2022).

13.8.1 LIPOSOMES

Liposomes are spherical vesicular structures composed of a hydrophilic core and one or more phospholipid bilayers, with sizes ranging from 20 to 1000 nm. These materials can encapsulate hydrophilic substances in their aqueous core, whereas lipophilic molecules can be incorporated into the acyl chains of the lipid bilayer fatty acids (Figure 13.7). According to their size and number of bilayers (lamellarity), liposomes can be grouped into three classes: small unilamellar vesicles, large unilamellar vesicles and multilamellar vesicles (Santos *et al.*, 2022).

The phospholipid bilayer is composed of natural and synthetic phospholipid molecules that are relatively biocompatible, biodegradable and non-immunogenic

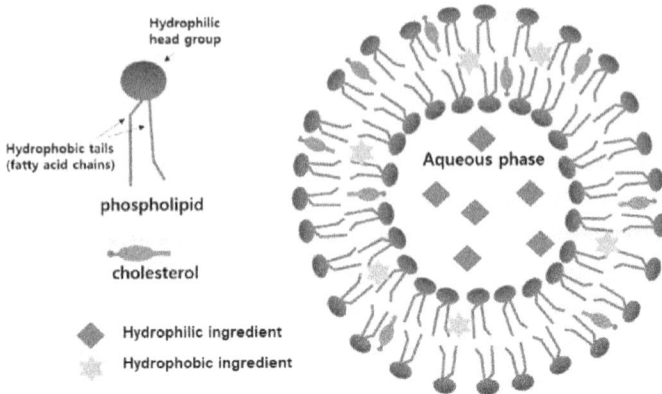

FIGURE 13.7   Structural representation of liposomes encapsulating hydrophilic and hydrophobic ingredients (Lee, 2020)

material (Laouini *et al.*, 2012). Furthermore, the liposome membrane may also contain cholesterol, which increases the stability and fluidity of the bilayer as well as its hydrophobicity, thus preventing the passage of water-soluble molecules (Santos *et al.*, 2022). Once liposomes are applied on the surface of the skin, they remain in the upper layer of the SC, acting as an ingredient reservoir. When in contact with SC cells, the liposomal bilayer fuses with the phospholipid bilayer of the cell membrane, releasing the encapsulated content (Ascenso *et al.*, 2015; Santos *et al.*, 2022).

The use of liposomes in sunscreen formulations enables the incorporation of lipophilic organic UV filters in the phospholipid bilayers (Santos *et al.*, 2022). Mota et al. performed in vivo and in vitro studies using encapsulated octyl methoxycinnamate (OMC) into liposomes (Mota *et al.*, 2013). HET-CAM results demonstrated that the liposome/OMC had no potential to cause ocular irritation. The formulation containing OMC encapsulated in liposome demonstrated higher SPF in vivo (11.5 ± 2.7) compared with the non-encapsulated OMC formulation (7.0 ± 1.6). Moreover, the stripping method using volunteers showed increased uptake of OMC in the SC, yielding an amount of 22.64 ± 7.55 µg cm^{-2} of OMC, which was higher than the amount found for the conventional formulation (14.57 ± 2.30 µg cm^{-2}). The results demonstrated that the inclusion of OMC in liposomes was able to reduce the skin permeation of this filter (Mota *et al.*, 2013).

13.8.2   SOLID LIPID NANOPARTICLES AND NANOSTRUCTURED LIPID CARRIERS

Solid lipid nanoparticles (SLN) and nanostructured lipid carriers (NLC) can be considered as most interesting carrier systems used in topical formulations. Given their lipid composition, SLN and NLC are suitable for the loading of lipophilic compounds such as UV filters within their matrices (Figure 13.8) (Scioli Montoto, Muraca and Ruiz, 2020; Souto *et al.*, 2022). SLNs are composed solely of a solid lipid, whereas NLCs are lipid nanoparticles composed of a mixture of solid lipids and liquids; both SLN and NLC are stabilized by surfactant molecules. SLNs possess characteristics of inorganic UV filters on their own. In addition, they have a high affinity for the SC allowing the organic UV filters to adhere to the skin as a protective film, improving by both means the efficacy of the photoprotective system (Carlotti *et al.*, 2005; Scioli Montoto, Muraca and Ruiz, 2020; Souto *et al.*, 2022)

SLNs have demonstrated a synergistic effect when combined with organic UV filters such as oxybenzone and octyl p-methoxycinnamate. This synergic effect can be explained by the ability of SLN to reflect and disperse incident UV radiation. The reflection and dispersion properties of SLNs are related to the crystallinity of the lipid matrix (Wissing and Müller, 2002; Carlotti *et al.*, 2005).

Following the trend of the new generation of lipid carriers, namely NLC, several studies with organic UV filters have been performed that demonstrate their considerable ability to encapsulate and absorb UV radiation (Xia, Saupe and Mu, 2007), their physicochemical stability (Lacerda, Cerize and Ré, 2011) and, more specifically, their synergistic effect (Nikolić *et al.*, 2011). Nikolíc and collaborators (2011) evaluated the encapsulation of octyl methoxycinnamate, ethylhexyl triazone and Tinosorb® S (bis-ethylhexyloxyphenol methoxyphenyl triazine) in carnauba wax-based NLCs. The SPF value increased by 45.0% compared to a conventional nano-emulsion. The authors suggested that carnauba wax interacted with the lipid matrix of the carrier and this interaction created particles in the form of crystals with UV radiation reflective properties, consequently enhancing the SPF of the formulation (Nikolić *et al.*, 2011).

Puglia et al evaluated the encapsulation of different and widespread UVA or UVB organic UV filters (ethyl hexyltriazone (EHT), diethylamino hydroxybenzoyl hexyl benzoate (DHHB), bemotrizinol (Tinosorb S), octyl methoxycinnamate and avobenzone) in nanostructured lipid carriers (NLC) and nanoemulsions (NE). Studies using excised human skin showed that the permeation of incorporated UV filters in NLC was drastically reduced, as the filters remained mainly on the surface of the skin. The photostability studies demonstrated that EHT, DHHB and Tinosorb S still maintain their photostability when incorporated in these carriers, whereas octyl methoxycinnamate and avobenzone were not photostable. Finally, the authors concluded that the photo-protective efficacy of incorporated UV filters was increased (Puglia *et al.*, 2014).

Gilbert et al. compared the percutaneous absorption and cutaneous bioavailability of oxybenzone loaded SLN, NLC, nanostructured polymeric lipid carriers (NPLC) and nanocapsules (NC). The results demonstrated that polymeric lipid carriers (NPLC and NC), significantly reduced oxybenzone skin permeation while exhibiting the highest in vitro SPF (Gilbert *et al.*, 2016).

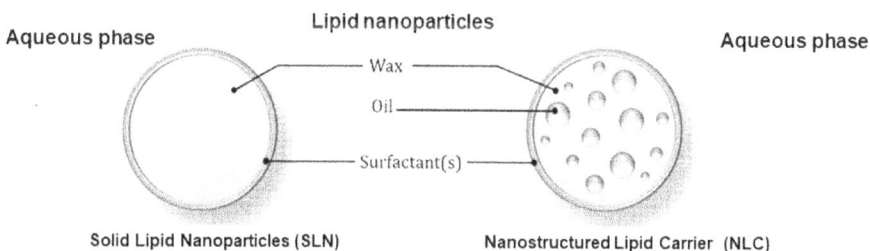

FIGURE 13.8 Schematic representation of the morphology of solid lipid nanoparticles (SLN) and nanostructured lipid carriers (NLC) (Gilbert *et al.*, 2016)

13.8.3 Polymeric nanoencapsulation

Polymeric nanoparticles (Figure 13.9) have been studied by many research groups and applied in various applications in the field of cosmetics (Nastiti *et al.*, 2017). These nanoparticle materials can be prepared from pre-formed polymer (nanoprecipitation or mini/emulsification and solvent evaporation) and in situ polymerization (miniemulsion polymerization). The miniemulsion polymerization technique is frequently deployed due to easier encapsulation of hydrophobic actives in a single reaction step, producing more homogeneous and stable nanoparticles (Nastiti *et al.*, 2017; Frizzo *et al.*, 2019). The application of biodegradable polymers is preferred because they are naturally eliminated from the organisms, thus, preventing their accumulation and cytotoxicity. For instance, poly-lactic acid, poly(glycolic-co-lactic acid) (PLGA), and poly(ε-caprolactone) are biodegradable polymers widely used in cosmetic formulations (Vettor *et al.*, 2010; Marcato *et al.*, 2011; Papakostas and Rancan, 2011).

Polymeric nanoparticles have been used to encapsulate organic UV filters, reducing transdermal absorption, increasing the sun protection factor (SPF), and improving photostability and safety. Nanoparticles have shown significant advantages compared with conventional delivery systems, creams, sprays and gels (Wu *et al.*, 2014; Nastiti *et al.*, 2017; Frizzo *et al.*, 2019).

A study conducted by Perugini et al. (2002) showed that nanoencapsulation of octyl p-methoxycinnamate in poly(D, L-lactide-co-glycolide) increased the photostability of this compound (Perugini *et al.*, 2002). Similarly, Lee et al. encapsulated avobenzone into poly methyl methacrylate) (PMMA) demonstrating that the photostability of avobenzone was improved. Wu *et al.* (2014) demonstrated that the photoprotective ability and photostability of encapsulated organic UV filters, such oxybenzone, avobenzone, octyl methoxycinnamate and diethylamino hydroxybenzoyl hexyl benzoate in PMMA increased (Wu *et al.*, 2014).

Marcato et al. (2014) carried out a study with poly(ε- caprolactone) nanoparticles and solid lipid nanoparticles of oxybenzone, aiming at improving safety of sunscreen

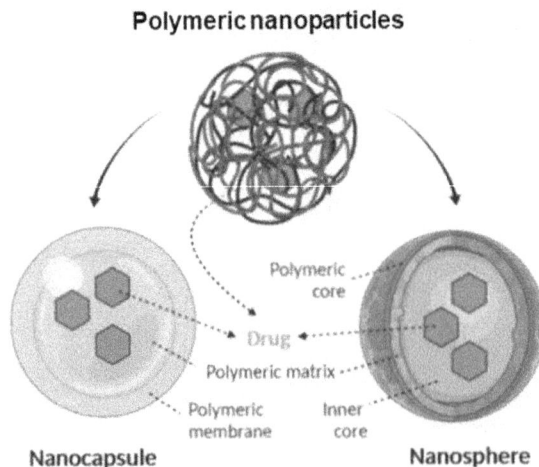

Polymeric nanoparticles

Polymeric core

Drug

Polymeric matrix

Polymeric membrane

Inner core

Nanocapsule

Nanosphere

FIGURE 13.9 Schematic representation of polymeric nanoparticles (Zielińska *et al.*, 2020)

formulations by increasing the sun protection factor (SPF), reducing oxybenzone skin penetration and decreasing the oxybenzone concentration in the final formulation. The SPF increased when oxybenzone was encapsulated in both nanostructures. However, oxybenzone encapsulated in poly(ε- caprolactone) nanoparticles decreased its skin permeation more than solid lipid nanoparticles of oxybenzone. After 24 hours of exposure the authors observed an epidermal and dermal retention profile of 3.4 and 4.4 times lower than that of the non-encapsulated compound. The cytotoxicity of the in vitro assay demonstrated that oybenzone encapsulated in SLN did not exhibit cytotoxic or phototoxic effects in human keratinocytes (HaCaT cells) and BABL/c 3T3 fibroblasts, whereas polymeric nanoparticles with oxybenzone showed phototoxic potential in HaCaT cells (Marcato et al., 2011).

Other investigations conducted by Frizzo et al. (2019) showed that the simultaneous encapsulation of zinc oxide nanoparticles (ZnO) and octocrylene in poly-styrene-co-methyl methacrylate (PMMA/PS) nanoparticles resulted in a sunscreen with SPF > 30 (moderate photoprotection) and classified as a good UVA sunscreen. Moreover, it presented a low cytotoxicity on human dermal fibroblasts and was therefore released for use in humans (Frizzo et al., 2019).

Another type of polymeric nanoparticles is bioadhesive nanoparticles (BNPs). BNPs represent a technological evolution of polymeric nanoparticles as carriers of organic UV filters. The bioadhesion of these PNs guarantees that the UV filters remain on the surface of the skin without penetrating the deeper layers. For organic UV filters this quality is essential because it prevents their penetration through the skin and through hair follicles and prevents access to the bloodstream. Moreover, it also decreases the generation of ROS responsible for cell and DNA damage (Deng et al., 2015). The advantages mentioned previously were emphasized in a study by Deng et al. (2015), which used BNPs to encapsulate padimate-O and compare their characteristics with padimate-O in non-encapsulate and in non-bioadhesive nanoparticles. It was demonstrated that only the BNPs-encapsulated padimate-O remained on the surface of the skin. Regarding oxidative stress, the results showed a significant reduction of DNA double-strand breaks when the BNPs were used. In addition, it was observed that for a given anti-UV effect, the amount of UV filters required when incorporated into BNPs was significantly lower than the amount of UV filters required in commercial formulations (Deng et al., 2015).

13.8.4 CYCLODEXTRIN COMPLEXATION

Cyclodextrins are chemical compounds obtained from the enzymatic degradation of one of the most essential polysaccharides, starch. They are cyclic oligiosaccharides composed of five or more a-d-glucopyranoside units. Depending on the number of α-glucose units cyclodextrins can be divided into three main types: α-Cyclodextrins, β-Cyclodextrins and γ-Cyclodextrins, consisting of six, seven and eight glucopyranose units linked by one to four bonds, respectively (Figure 13.10) (Morabito et al., 2011; Crini, 2014; Dahabra et al., 2021). Cyclodextrins have a truncated-cone-shaped structure with hydrophilic surfaces and hydrophobic cavities. This structure enables them to entrap poorly soluble drugs moieties, such as natural antioxidants and anticancer drugs, in their hydrophobic cavities to form inclusion complexes (Morabito et al., 2011; Dahabra et al., 2021).

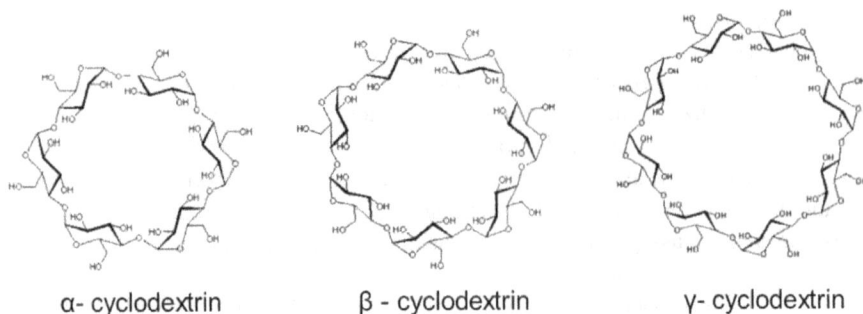

α- cyclodextrin   β - cyclodextrin   γ- cyclodextrin

FIGURE 13.10 Chemical structure of cyclodextrin (Li *et al.*, 2007)

Cyclodextrins have been applied to entrap organic UV filters to prevent photodegradation and to reduce their skin permeation (Dahabra *et al.*, 2021). For instance, organic UV filters such oxybenzone, octocrylene and ethylhexyl-methoxycinnamate entrapped in β-Cyclodextrin showed a significant photostability enhancement compared with non-entrapped UV filters (Al-Rawashdeh, Al-Sadeh and Al-Bitar, 2013). Scalia *et al.* reported that the light-induced decomposition of 4-methylbenzylidene camphor (4-MBC) in emulsion vehicles was markedly decreased by complexation with methyl-β-Cyclodextrin. The extent of degradation determined by HPLC, was 7.1% for the complex compared to 21.1% for free 4-MBC (Scalia, Tursilli and Iannuccelli, 2007).

Another study conducted by Scalia *et al.*, showed that the amount of avobenzone penetrating the SC was significantly decreased after encapsulation in hydroxypropyl-β-cyclodextrin (Scalia, Coppi and Iannuccelli, 2011). In addition, the entrapment of oxybenzone in sulfobutylether-β-Cyclodextrin increased the aqueous solubility of this compound, while significantly limiting its percutaneous absorption (Simeoni *et al.*, 2006).

The effect of employing CD-UV filters inclusion complexes on the overall SPF value of the formulation was investigated. Some researchers reported negligible to minor increases in the formulation SPF when CD-UV filters inclusion complexes were used (Sarveiya, Templeton and Benson, 2004; Monteiro *et al.*, 2012). Others reported more significant effects. For instance, Felton et al. investigated the influence of cyclodextrin complexation (HP-β-CD) on the in vivo photoprotective effects of oxybenzone, and to compare these novel sunscreen products to a commercial SPF 30 sunscreen product. It was reported that a 5% HP-β-CD inclusion complex formulation provided sun protection equivalent to a commercial sunscreen with SPF 30 (Felton, Wiley and Godwin, 2004; Dahabra *et al.*, 2021).

13.8.5 MESOPOROUS MATERIALS

Mesoporous materials are inorganic polymers, constituted of siloxane groups (Si-O-Si) in its interior and by silanol groups (Si-OH) in the pore surface. The latter are responsible for the reactivity of the material, giving the silica polar structures, which are considered efficient adsorption sites and can be easily hydrated.

The electronic density distribution in the silanol groups causes them to have an acidic behaviour, which confers reactivity to the material (Kresge *et al.*, 1992; Airoldi and Farias, 2000; Matos *et al.*, 2001; da Silva *et al.*, 2015). According to IUPAC, porous materials are divided into three classes: microporous (<2 nm), mesoporous (2–50 nm) and macroporous (>50 nm). The zeolites are well known, classified as microporous and have excellent catalytic properties due to the crystalline aluminosilicate network (Ciesla and Schüth, 1999).

The synthesis of mesoporous materials are usually prepared using tetraethyl orthosilicate (TEOS) as a silica precursor and supramolecular non-ionic self-assembly between ethylene oxide/propylene oxide copolymer-template (Pluronic P123), which act as structure-directing agents, in range of temperature from 30°C to 120°C, followed by calcination at high temperature in order to degrade the surfactant (Figure 13.11) (Crucianelli, Bizzarri and Saladino, 2019). Depending on the type of mesoporous materials, the synthesis can be performed in acid medium (SBA-15) or in alkaline medium (MCM-41) (Kresge *et al.*, 1992, 1998; Matos *et al.*, 2001; de Ávila, Silva and Matos, 2016).

The amorphous mesoporous silica (MCM-41, MCM-48, SBA-15, FDU-1 etc.) have been proposed as drug and UV filters carriers due to their nontoxic nature, their ordered mesoporous structure, their adjustable diameter and pore volume and their great superficial area with many silanol groups on the surface of the pore (Ambrogi, Latterini, Marmottini, Pagano, *et al.*, 2013; Ambrogi, Latterini, Marmottini, Tiralti, *et al.*, 2013; Daneluti *et al.*, 2018, 2019). Moreover, these materials possess high thermal, hydrothermal, chemical and mechanical stability. The storage properties and drug release in mesoporous silica ordered as MCM-41 and SBA-15 demonstrated that the size and proper pore volume of these materials make them viable media for encapsulation, allowing subsequent release of a wide variety of molecules with therapeutic activity (Daneluti *et al.*, 2018). SBA-15 shows structural similarity with MCM-41, since both have hexagonal structures (Figure 13.11). However, SBA-15 has a higher pore size and thicker pore walls. In addition, its pore interconnectivity confers higher hydrothermal stability as well as better thermal and mechanical properties (Kruk *et al.*, 2000; Matos *et al.*, 2001; Mariano-Neto *et al.*, 2014).

In this context, mesoporous materials have been used to encapsulate organic UV filters to increase photostability, photoprotection and to reduce the skin permeation of these compounds (Ambrogi *et al.*, 2007; Ambrogi, Latterini, Marmottini, Pagano, *et al.*, 2013; Daneluti *et al.*, 2019, 2021).

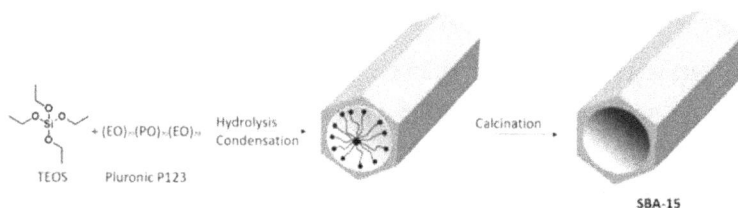

FIGURE 13.11 Schematic representation of the synthesis for SBA-15 (Crucianelli, Bizzarri and Saladino, 2019)

Ambrogie *et al.* (2013) investigated the use of mesoporous silica MCM-41 to increase the photostability and safety of octyl methoxycinnamate. The results showed that the entrapment of octyl methoxycinnamate inside the pores of the silicate MCM-41 allows a broader photoprotection range, enhances its photostability and moreover reduces sunscreen release in comparison with formulation content of a non-entrapped UV filter (Ambrogi, Latterini, Marmottini, Pagano, *et al.*, 2013).

A recent study aimed at investigating the cutaneous deposition and permeation of incorporated organic UV filters (avobenzone, oxybenzone and octyl methoxycinnamate) in mesoporous silica (SBA-15). The application of stick formulations containing incorporated UV filters to porcine skin significantly demonstrated the reduction of oxybenzone and avobenzone skin deposition in comparison with stick formulations containing these UV filters in free form. For instance, the oxybezone permeation across the skin was 30-, 12- and 1.5-fold lower after 6, 12 and 24 hours, respectively, following application of stick formulations containing incorporated UV filters (Daneluti *et al.*, 2019). Given the known thickness of the different porcine skin layers (~20–40 µm for SC, ~70–160 µm for viable epidermis and approximately 1.86 mm for the dermis) (Jacobi *et al.*, 2007) and knowing the number of cryotome slices removed (21) in the present study, the investigation into the biodistribution profile of UV filters enabled the quantification of filters from the SC down to the upper/mid dermis (Lapteva *et al.*, 2014). Therefore, the cutaneous biodistribution profiles of avobenzone and oxybenzone to 800 µm evidenced a significant decrease in the amounts in the viable epidermis and dermis. The authors have also performed in vitro photoprotective efficacy studies. The results showed that adsorption/entrapment of UV filters enhanced the sun protection factor by 94% (Daneluti *et al.*, 2019).

The same authors conducted preclinical and clinical studies to evaluate cutaneous biodistribution, safety and efficacy of avobenzone, oxybenzone, octyl methoxycinnamate encapsulated in mesoporous silica SBA-15. Cutaneous deposition and transdermal permeation of OXY in and across human skin were 3.8- and 13.4-fold lower, respectively, after application of stick-entrapped filters. Biodistribution results showed that encapsulation in SBA-15 decreased AVO and OXY penetration reaching porcine and human dermis. HET-CAM results demonstrated that the incorporated UV filters had no potential to cause ocular irritation. In addition, clinical (irritation and dermal sensitivity assays as well as the assessment of the photoirritation and photosensitivity potential) showed that the stick-incorporated filters had good biocompatibility in vivo and safety profiles, even under sun-exposed conditions. Moreover, the entrapment of UV filters improved the SPF in vivo by 26% and produced the same SPF profile as a marketed stick (Daneluti *et al.*, 2021).

13.9 REGULATORY ASPECTS OF SUNSCREEN PRODUCTS AND NANOSTRUCTURED MATERIALS: EFFICACY

13.9.1 Efficacy of Sunscreen Formulations

Nowadays, there are in vitro and in vivo techniques for evaluating SPF of sunscreen formulations. The most recognized and relevant measure of sunscreen formulation efficacy is the sun protection factor (SPF) test. The SPF test is the ratio of the time for artificial solar UV radiation to produce a barely perceptible, uniform erythema in protected and unprotected skin 16–24 hours after exposure. Therefore, if it takes ten seconds to produce a barely perceptible uniform erythema in unprotected skin and 100 seconds to produce an identical response in skin where 2 mg cm^{-2} of sunscreen had been applied, the SPF is 10 (100/10) (Nash, 2006; Food and Drug Administration, 2012). The SPF is a parameter referring only to the UVB radiation that induces the erythema, while the UVA Protection Factor (UVA-PF) indicates the level of protection against UVA radiation; and the critical wavelength (λ crit), which provides the amplitude of protection considering the full spectrum of UV radiation (FDA, 2011).

The SPF assay is performed in vivo with volunteers since no in vitro methodology has been validated. The SPF is defined as the UV energy required to produce a minimum erythemal dose (MED) on protected skin (MEDp), divided by the UV energy required to produce a minimum erythemal dose on unprotected skin (MEDu) (Eq. 13.1) (COLIPA, 2006).

$$SPF = \frac{MED \text{ of protected skin}(MEDp)}{MED \text{ of unprotected skin}(MEDu)} \tag{13.1}$$

The spectrum generated by the UV solar simulator is filtered and provides emission in the range of 290–400 nm, with a limit of 1500 W m^{-2} in the total irradiation. The variable irradiance should be used to expose several small sites on volunteers' skin (phototypes I, II, or III) between the waist and shoulder line in incremental erythemal doses. Areas exposed include an area of unprotected skin, area(s) of skin protected by the sunscreen(s) under test and an area of skin protected by an SPF reference formulation. By incrementally increasing the UV dose, varying degrees of skin erythema are generated. The delayed erythemal responses are visually assessed for redness 16–24 hours after UV radiation, by the judgement of a trained evaluator to determine the MED for protected skin (MEDp) and unprotected skin (MEDu) (COLIPA, 2006; FDA, 2014).

In addition, the SPF evaluation must consider that a sunscreen product should be used during bathing, with the risk of a drastic reduction in the effective protection following contact with water (50% of its effectiveness) Indeed, the effectiveness of sunscreen is dose-dependent, and the commonly applied dose, required to perform the labelled SPF value, is much lower than that used in the tests. To address this issue, the SPF is measured before (static SPF) and after (SPF wet) the interested area is immersed in water. The products claimed as "water resistant" on the label must guarantee that the wet SPF is not lower than the static SPF by more than a defined % (COLIPA, 2006; FDA, 2011; Gubitosa et al., 2020).

The UVA-protection factor (UVA-PF) can be measured in vivo or in vitro. The in vivo test-method, also called the Persistent-Pigment Darkening (PPD) measures the minimal darkening effect of UVA-radiation on the skin before and after exposure to UVA-light. The UVA-PF is defined as minimal pigmenting dose (MPD = UVA-dose when darkening of the skin is visible) on protected skin divided by the MPD on unprotected skin (Eq. 13.2) (COLIPA, 2011; Kockler *et al.*, 2012).

$$\text{UVA-PF} = \frac{\text{MPD of protected skin (MPDp)}}{\text{MPD of unprotected skin (MPDu)}} \tag{13.2}$$

The PPD response in unprotected skin areas has been measured to be at a relatively high dose of irradiance of 10.4 and 8.3 J cm^{-2} in volunteers with Fitzpatrick skin types III and IV (Hwang *et al.*, 2011).

Although in vivo methodologies remain approved for the evaluation of UVA photoprotection, their use is not recommended, since there are validated and regulated in vitro methodologies, in contrast to what occurs with the SPF assay (COLIPA, 2006; FDA, 2011). The reflectance spectrophotometry with an integrated sphere is used to determine the UVA-PF and the critical wavelength. The spectrophotometer presents a xenon arc lamp, with a spectral range from 250 to 450 nm, capable of processing transmittance data at each wavelength and transforming them into parameters of photoprotective efficacy (Springsteen *et al.*, 1999).

In addition to providing parameters of photoprotective efficacy quickly, economically and safely (since it saves volunteers from exposure to UV radiation used in in vivo analyses), in vitro assays also find an application in the research and development of photoprotective formulations, providing, including estimated FPS values for screening and improvement of formulations prior to their submission to in vivo clinical tests (Tanner, 2006; Andreassi, 2011).

13.9.2 LABELLING OF SUNSCREEN PRODUCTS AND NANOSTRUCTURED MATERIALS

The European Commission publishes a Commission Recommendation on the efficacy of sunscreen products, in which they provide advice regarding the hazards of UV radiation, the need of sunscreens and recommendations about labelling and testing. An industry association, Colipa (European Cosmetic, Toiletry and Perfumery Association) was created to develop the industry standards on testing, labelling and consumer education (COLIPA, 2006; Kockler *et al.*, 2012).

In 2006, following extensive cooperation from industry, the European Commission published the Recommendation (2006/647/EC) "on the efficacy of sunscreen products and the claims made relating thereto" (OJ L265, 26/9/06). This initiative aimed to standardize and simplify as much as possible the way products are tested and labelled throughout Europe (EC, 2006). The subsequent recommendation Colipa n°. 23 of 2009 contains important information relating to sunscreen products to be reported on the label as already indicated in the European Commission document (COLIPA, 2009). The following set of important information is described in this guideline:

- Sunscreen products should protect against both UVA and UVB radiation
- The minimum SPF value required for sunscreen products is 6 (Products with an SPF below 6 should not be labelled as sunscreen product)

- The SPF numbers should be accompanied with an indication of the type of protection that product provides i.e., low, medium, high or very high
- The protection category should be labelled at least as prominently as the SPF number
- For the determination of UVB protection, the International SPF Test Method CTFA-SA/ Colipa/JCIA 2006 (now ISO 24444:2010 method) should be indicated
- The SPF values to be reported on the label are 8

The protection factor obtained with the test is approximated by default to the nearest number among the eight values reported in the SPF classification table. The minimum SPF value is 6; the maximum value is 50 (also according to the FDA method), corresponding to the highest protection category. To report SPF 50 on the label, the value obtained with the test must be equal to or greater than 60 (FDA, 2011).

- As recommended also by FDA, sunscreen products should have UVA protection of 1/3 of the labelled sun protection factor (i.e., an SPF 15 product should have a UVA protection factor of at least 5) and should have a critical wavelength of 370 nm.
- Compliance with both requirements is indicated on the label with a logo that represents a small circle enclosing the initials UVA, having the same dimensions as the character used to indicate the value of the SPF.

Moreover, since 2002, Colipa has recommended that no sunscreen product can provide 100% protection, and this should not be claimed or implied. It means that the term 'sunblock' should not be used (COLIPA, 2009).

Under the new FDA regulations, sunscreen products are required to designate UVA protection properties based on a 4-star rating system, which will be derived from in vivo and in vitro tests (FDA, 2011; Gubitosa *et al.*, 2020). The UVA logo and SPF value must be reported on the front of the primary and secondary packaging. Harmonized labelling for UVA protection claim is important in the way to make easier for consumers to choose a sunscreen product with a "broad spectrum" that protects against both UVA and UVB rays (FDA, 2011; Gubitosa *et al.*, 2020).

In addition, the following important instructions and warnings should be labelled:

- Apply sunscreen product before sun exposure
- Reapply the product frequently, especially after bathing, sweating and after drying
- Apply a correct amount of product generously (The Commission recommends a quantity similar to the one used for testing (i.e. 2 mg cm^{-2}) which is approximately 36 grams for the body of an average adult, which is equal to six teaspoons of product
- Do not stay too long in the sun, even while using a sunscreen product
- Keep children away from direct sunlight (COLIPA, 2009)

13.10 SAFETY

13.10.1 SAFETY OF SUNSCREEN PRODUCTS

Sunscreen products are required to be regulated because they are expected to be sufficiently effective in protecting the skin from the harmful effects of UV rays. The regulations of sunscreens differ, depending on the country and the requirements of the relevant regulatory agency. In general, a list of approved UV filters as well as the appropriate labelling of sunscreen formulations and requirements for the measuring of SPF, UVA-PF and water resistance can be included in these regulations (Kockler et al., 2012; Santos et al., 2022).

In Europe, sunscreens are considered cosmetic products and are regulated under "Regulation (EC) N° 1223/2009 of the European Parliament and of the Council" (EC, 2009). The safety of UV filters is evaluated at the European Commission level by a Scientific Committee on Consumer Safety (SCCS). For the safety evaluation of cosmetic substances including sunscreen products all available scientific data is considered, including the physical and chemical properties of the compounds under investigation, results obtained from (Q)SAR {(quantitative/qualitative) structure activity relationship}, chemical categories, grouping, read across, physiologically based pharmacokinetics (PBPK)/toxicokinetic (PBTK) modelling, in vitro experiments and data obtained from animal studies (in vivo). Also, clinical data, epidemiological studies, information derived from misuse of incidents and any other human data are taken into consideration. Safety of cosmetic products in the EU are based on the safety of the ingredients which are evaluated by toxicological testing (Sargent and Travers, 2016; SCCS, 2019). Phototoxicity in vitro tests such as 3T3 Neutral Red Uptake Phototoxicity Test (3T3 NRU PT) is mandatory in the EU as well as recommended under ICH Guidelines. This test is a validated in vitro method based on a comparison of the cytotoxicity of chemicals when tested in the presence and in the absence of exposure to a non-cytotoxicity dose of UV light (Sargent and Travers, 2016). In vitro skin permeation studies are required for cosmetics products and should be performed on all UV filter compounds (ICH, 2013; Sargent and Travers, 2016).

In Australia, most sunscreens are 'listed' medicines under the Therapeutic Goods Act 1989, some can be 'exempt' and others are required to be 'registered (Kockler et al., 2012; TGA, 2016). In the United States, the cosmetic products labelled as sunscreens are considered over the counter (OTC) drugs under the supervision of the FDA (Federal Drug Administration) The Federal Register contains the final monograph with regulations for sunscreens (list of allowed UV filters, their required maximum concentration and the allowed combinations of these compounds). The same monograph contains requirements for labelling and testing of the SPF and water resistance, but not for UVA protection (FDA, 1999, 2011; Kockler et al., 2012).

Concerning the safety evaluation in the United States and Australia, the following safety studies are usually required for a new sunscreen ingredient or new excipient:

1. Photostability
   - UV absorption spectra Acute
2. Acute toxicity (oral and dermal; animal data)

3. Local tolerance and Allergenicity
- skin irritation and sensitization (animal data and/or human repeat insult patch test - HRIPT
- phototoxicity (animal and/or human data)
- eye irritation (animal in vivo or in vitro test)

4. Toxicokinetics:
- oral and dermal bioavailability
- ADME (absorption, distribution, metabolism and excretion) studies

5. Repeat dose toxicity

6. Genotoxicity
- In vitro bacterial (Ames) assay
- In vitro mammalian cell line assay
- In vitro and in vivo chromosome aberration assay

7. Reproductive toxicity
- For assessment of developmental and fertility effects
- Endocrine disruption potential needs to be addressed. This could be examined during the repeat-dose toxicity and/or reproductive toxicity studies

8. Carcinogenicity
- In vivo carcinogenicity and photocarcinogenicity bioassays or a justification for not providing these studies

Human safety testing of both prescription and OTC sunscreens is required by the FDA. Unlike animal dermal safety studies conducted on individual sunscreen ingredients, human dermal safety studies are conducted on formulated product(s) (FDA, 2019). Relevant human studies are acceptable in the assessment of potential skin irritation and sensitization using the repeat 'insult patch test' or other relevant validated tests. For instance, the human repeated insult patch test (HRIPT) is a clinical study involving human volunteers. This study is widely used as a test to confirm safety of skin allergic sensitization reactions (late hypersensitivity immune reactions), after application of sunscreen products and organic encapsulated UV filters (Basketter, 2009; de Oliveira et al., 2016a,b; Daneluti et al., 2021). Typically, 9 × 24-hour or 48-hour exposures are delivered over a three-week period to each of approximately 100 volunteers. The gender, age and skin phototype may be specified. After a two-week break, a challenge exposure is made on both the induction site and a naïve site, again using a 24/48-hour patch. The skin reactions are scored over the subsequent few days. and new material is applied for two more weeks. The next two weeks are normally characterized as a rest period, no samples are applied. After this period, fresh patches with the samples should be applied for one last week, the challenge phase. The scores normally used are 0 for no erythema, 1 for well-defined erythema, 2 for erythema and induration, and 3 for vesiculation and bullous reaction (Basketter, 2009; de Oliveira et al., 2016a,b; Tomazelli et al., 2018; Daneluti et al., 2021). In addition, photoirritation and photosensitivity potential are important tests to assure the safety of sunscreen formulations. The purpose of the photoirritation test is to verify the absence of any irritant potential of a product applied to the skin when exposed to UVA radiation, while the photosensitivity test aims to prove the absence of allergenic potential (late hypersensitivity immune reactions) of a product applied to the skin after exposure to UVA radiation. Both assays are performed on human volunteers (de Oliveira *et al.*, 2016a,b; Tomazelli *et al.*, 2018; Daneluti *et al.*, 2021).

It is very important to mention that, given the many differences in regulatory requirements for sunscreen products safety assessment between the United States and EU, a scientifically acceptable testing scheme should be harmonized (Gubitosa *et al.*, 2020). In the EU, human safety testing of sunscreen products or ingredients is considered unethical. The SCCS further envisaged a skin compatibility test with human volunteers confirming that there are no harmful effects when applying a cosmetic product for the first time. Such testing according to SCCS is only scientifically and ethically appropriate after animal and or in vitro alternative tests have identified no toxicological concerns (SCCS, 2012). In addition, the EU does not require human dermal absorption data but rather uses in vitro data exclusively to assess dermal absorption/bioavailability for purposes of human safety assessment (SCCS, 2012).

13.10.2 SAFETY CONCERNS OF NANOSTRUCTURED MATERIALS

In recent years, a number of nanostructured materials have been widely studied and used in topical formulations, personal care and sunscreen products (de Oliveira *et al.*, 2016a,b; Kandekar *et al.*, 2019; Lapteva *et al.*, 2019; Daneluti *et al.*, 2021). However, topically applied nanostructured materials have been increasing concerns because of their potential toxicity for the consumer (Katz, Dewan and Bronaugh, 2015; Santos *et al.*, 2022). Normally, most of the published studies available do not highlight the adverse health effects of nanostructure materials, alarming the consumers. Therefore, Considerable effort has been spent at the international level to establish programs to access the safety and hazard potential for nanostructured materials (Gubitosa *et al.*, 2020; Santos *et al.*, 2022).

The Scientific Committee on Consumer Safety (SCCS) has published the Guidance on safety assessment of nanomaterials in cosmetics (SCCS/1484/12), which is currently being revised, a memorandum on the relevance, adequacy and quality of the expected data in nanomaterial safety dossiers (SCCS/1524/13, revision of 27 March 2014) and a checklist for manufacturers that submit dossiers on nanomaterials as cosmetic ingredients (SCCS/1588/17) (EC, 2009; Gubitosa *et al.*, 2020). In addition, the SCCS has considered information available in published literature as well as other relevant documents (SCCS, 2019; Gubitosa *et al.*, 2020).

The SCC reported that nanostructured materials might exhibit certain physico-chemical properties, biokinetic behaviour, biological interactions and/or toxicological effects that are different from conventional or bulk form of the same ingredients. Thus, topically applied nanocarriers should not permeate across the SC. However, there are still concerns about the ability of nanostructure materials to penetrate the surface of the skin and reach viable tissues, such as blood vessels for systemic distribution, where they might interfere with biological processes (Katz, Dewan and Bronaugh, 2015; de Oliveira *et al.*, 2016a,b; Daneluti *et al.*, 2021).

It is well known that larger particles tend to agglomerate (such as TiO2 and ZnO) and then cannot permeate across the intact skin. However, it was demonstrated that particle sizes smaller than 100 nm may nevertheless reach the viable skin layers (Shegokar and Müller, 2010; Katz, Dewan and Bronaugh, 2015; Santos *et al.*, 2022).

Nanostructured materials containing UV filters should cause skin irritation or other health issues. However, when sunscreen formulations that contain not only organic UV filters but also TiO_2 nanoparticles penetrate living skin layers or are sprayed with consequent inhalation, they may interact with living cells, causing adverse effects. Moreover, some authors have proposed that these components may also reach the deepest layers of the skin if their integrity is compromised or can reach systemic circulation when they penetrate through the hair follicles (Gulson *et al.*, 2015; Klimová, Hojerová and Beránková, 2015; Lademann *et al.*, 2015). Others have argued that the penetration of TiO_2 nanoparticles depend on the skin conditions. It was concluded that damaged skin (e.g. scarring, sunburn and depilated skin) is more susceptible to penetration of TiO_2 nanoparticles than healthy skin (Lin *et al.*, 2011; Touloumes *et al.*, 2020). Titanium levels were found in healthy epidermis and dermis of patients after long-term application of sunscreen formulations containing TiO_2 nanoparticles (Gulson *et al.*, 2015; Shakeel *et al.*, 2016; Santos *et al.*, 2022).

The permeation mechanisms of nanostructured materials have not been clarified yet. Nevertheless, some authors have suggested that nanostructured materials may permeate through different pathways, such as transcellular and paracellular routes, hair follicles (trans appendageal), sweet glands, skin porous and skin fold (Filipe *et al.*, 2009; Wu *et al.*, 2009).

Regarding the use of ZnO nanoparticles in sunscreen formulations and their dermal or oral absorption, no available results were observed. The SCC concluded that ZnO nanoparticles do not demonstrate hazardous effects after their application to the skin. It was assumed that these ZnO nanoparticles completely dissolve and Zn^{2+} ions are released. However, it was shown that the inhalation of No nanoparticles caused serious local effects in the lungs after sprayable products were applied (Committee and Sccs, 2013; Santos *et al.*, 2022).

REFERENCES

Abid, A. R. *et al.* (2017) 'Photo-stability and photo-sensitizing characterization of selected sunscreens' ingredients', *Journal of Photochemistry and Photobiology A: Chemistry*, 332, pp. 241–50. doi: 10.1016/j.jphotochem.2016.08.036.

Afonso, S. *et al.* (2014) 'Photodegradation of avobenzone: Stabilization effect of antioxidants', *Journal of Photochemistry and Photobiology B: Biology*, 140, pp. 36–40. doi: 10.1016/j.jphotobiol.2014.07.004.

Airoldi, C. and Farias, R. F. (2000) 'The use of organofuntionalized silica gel as sequestrating agent for metals', *Química Nova*, 23(4), pp. 496–503. doi: 10.1590/S0100-40422000000400012.

Al-Rawashdeh, N. A. F., Al-Sadeh, K. S. and Al-Bitar, M. B. (2013) 'Inclusion complexes of sunscreen agents with β-cyclodextrin: Spectroscopic and molecular modeling studies', *Journal of Spectroscopy*, 1(1). doi: 10.1155/2013/841409.

Almeida, J. S. *et al.* (2010) 'Nanostructured systems containing rutin: In vitro antioxidant activity and photostability studies', *Nanoscale Research Letters*, 5(10), pp. 1603–10. doi: 10.1007/s11671-010-9683-1.

Ambrogi, V. *et al.* (2007) 'Mesoporous silicate MCM-41 containing organic ultraviolet ray absorbents: Preparation, photostability and in vitro release', *Journal of Physics and Chemistry of Solids*, 68(5–6), pp. 1173–77. doi: 10.1016/j.jpcs.2007.02.033.

Ambrogi, V., Latterini, L., Marmottini, F., Pagano, C., *et al.* (2013) 'Mesoporous silicate MCM-41 as a particulate carrier for Octyl methoxycinnamate: Sunscreen release and photostability', *Journal of Pharmaceutical Sciences*, 102(5), pp. 1468–75. doi: 10.1002/jps.23478.

Ambrogi, V., Latterini, L., Marmottini, F., Tiralti, M. C., *et al.* (2013) 'Oxybenzone entrapped in mesoporous silicate MCM-41', *Journal of Pharmaceutical Innovation*, 8(4), pp. 212–17. doi: 10.1007/s12247-013-9161-2.

Andreassi, M. (2011) 'Sunscreens and photoprotection', *Expert Review of Dermatology*, 6(5), pp. 433–35. doi: 10.1586/edm.11.59.

Ascenso, A. *et al.* (2015) 'Development, characterization, and skin delivery studies of related ultradeformable vesicles: Transfersomes, ethosomes, and transethosomes', *International Journal of Nanomedicine*, 10, pp. 5837–51. doi: 10.2147/IJN.S86186.

Basketter, D. A. (2009) 'The human repeated insult patch test in the 21st century: A commentary', *Cutaneous and Ocular Toxicology*, 28(2), pp. 49–53. doi: 10.1080/15569520902938032.

Benech-Kieffer, F. *et al.* (2000) 'Percutaneous absorption of sunscreens in vitro: Interspecies comparison, skin models and reproducibility aspects', *Skin Pharmacology and Physiology*, 13(6), pp. 324–35. doi: 10.1159/000029940.

Blasi, P. *et al.* (2011) 'The real value of novel particulate carriers for sunscreen formulation', *Expert Review of Dermatology*, 6(5), pp. 509–17. doi: 10.1586/edm.11.57.

BSI (2007) 'Terminology for medical, nanotechnology applications of health and personal care', pp. 1–14.

Carlotti, M. E. *et al.* (2005) 'Study on the photostability of Octyl-p-methoxy cinnamate in SLN', *Journal of Dispersion Science and Technology*, 26(6), pp. 809–16. doi: 10.1081/DIS-200063141.

Cestari, T. F., Oliveira, F. B. de and Boza, J. C. (2012) 'Considerations on photoprotection and skin disorders.', *Annales de dermatologie et de vénéréologie*, 139(Suppl), pp. 135–43. doi: 10.1016/S0151-9638(12)70125-4.

Chanchal, D. and Swarnlata, S. (2008) 'Novel approaches in herbal cosmetics', *Journal of Cosmetic Dermatology*, 7(2), pp. 89–95. doi: 10.1111/j.1473-2165.2008.00369.x.

Chaudhri, N., Soni, G. C. and Prajapati, S. K. (2015) 'Nanotechnology: An advance tool for nano-cosmetics preparation', *International Journal of Pharma Research & Review*, 4(44), pp. 28–40.

Ciesla, U. and Schüth, F. (1999) 'Ordered mesoporous materials', *Microporous and Mesoporous Materials*, 27, pp. 131–49. doi: 10.1016/S1387-1811(98)00249-2.

COLIPA (2006) 'International sun protection factor (SPF) test method', *Cosmetic Europe*, pp. 1–78.

COLIPA (2009) 'Important usage and labelling instructions for sun protection products', (February), pp. 1–14.

COLIPA (2011) *COLIPA Guidelines. In vitro method for the determination of the UVA protection factor and '"critical wavelength"' values of sunscreen products.*

Committee, S. and Sccs, C. S. (2013) *Scientific committee on consumer opinion on zinc oxide (nano form).*

Crini, G. (2014) 'Review: A history of cyclodextrins', *Chemical Reviews*, 114(21), pp. 10940–75. doi: 10.1021/cr500081p.

Crucianelli, M., Bizzarri, B. M. and Saladino, R. (2019) 'SBA-15 anchored metal containing catalysts in the oxidative desulfurization process', *Catalysts*, 9(12), p. 984. doi: 10.3390/catal9120984.

da Silva, L. C. C. *et al.* (2015) 'Adsorption/desorption of Hg(II) on FDU-1 silica and FDU-1 silica modified with humic acid', *Separation Science and Technology*, 50(7), pp. 984–92. doi: 10.1080/01496395.2014.983246.

Dahabra, L. *et al.* (2021) 'Sunscreens containing cyclodextrin inclusion complexes for enhanced efficiency: A strategy for skin cancer prevention', *Molecules*, 26(6), p. 1698. doi: 10.3390/molecules26061698.

Daneluti, A. L. M. *et al.* (2018) 'Evaluation and characterization of the encapsulation/entrapping process of octyl methoxycinnamate in ordered mesoporous silica type SBA-15', *Journal of Thermal Analysis and Calorimetry*, 131(1), pp. 789–98. doi: 10.1007/s10973-017-6265-9.

Daneluti, A. L. M. *et al.* (2019) 'Using ordered mesoporous silica SBA-15 to limit cutaneous penetration and transdermal permeation of organic UV filters', *International Journal of Pharmaceutics*, 570, p. 118633. doi: 10.1016/j.ijpharm.2019.118633.

Daneluti, A. L. M. *et al.* (2021) 'Preclinical and clinical studies to evaluate cutaneous biodistribution, safety and efficacy of UV filters encapsulated in mesoporous silica SBA-15', *European Journal of Pharmaceutics and Biopharmaceutics*, 169, pp. 113–24. doi: 10.1016/j.ejpb.2021.10.002.

de Ávila, S. G., Silva, L. C. C. and Matos, J. R. (2016) 'Optimisation of SBA-15 properties using Soxhlet solvent extraction for template removal', *Microporous and Mesoporous Materials*, 234, pp. 277–86. doi: 10.1016/j.micromeso.2016.07.027.

de Oliveira, C. A. *et al.* (2015) 'Functional photostability and cutaneous compatibility of bioactive UVA sun care products', *Journal of Photochemistry and Photobiology B: Biology*, 148, pp. 154–59. doi: 10.1016/j.jphotobiol.2015.04.007.

de Oliveira, C. A. *et al.* (2016a) 'Safety and efficacy evaluation of gelatin-based nanoparticles associated with UV filters', *Colloids and Surfaces B: Biointerfaces*, 140, pp. 531–37. doi: 10.1016/j.colsurfb.2015.11.031.

de Oliveira, C. A. *et al.* (2016b) 'Cutaneous biocompatible rutin-loaded gelatin-based nanoparticles increase the SPF of the association of UVA and UVB filters', *European Journal of Pharmaceutical Sciences*, 81, pp. 1–9. doi: 10.1016/j.ejps.2015.09.016.

Deng, Y. *et al.* (2015) 'A sunblock based on bioadhesive nanoparticles.', *Nature Materials*, 14(12), pp. 1278–85. doi: 10.1038/nmat4422.

Diffey, B. L. (2015) 'Solar spectral irradiance and summary outputs using excel', *Photochemistry and Photobiology*, 91(3), pp. 553–57. doi: 10.1111/php.12422.

Diffey, B. L. and Grice, J. (1997) 'The influence of sunscreen type on photoprotection', *British Journal of Dermatology*, 137(1), pp. 103–05. doi: 10.1046/j.1365-2133.1997.17761863.x.

Diffey, B. L. *et al.* (2000) 'In vitro assessment of the broad-spectrum ultraviolet protection of sunscreen products', *Journal of the American Academy of Dermatology*, 43(6), pp. 1024–35. doi: 10.1067/mjd.2000.109291.

Djavaheri-Mergny, M. *et al.* (1996) 'Ultraviolet-A induces activation of AP-1 in cultured human keratinocytes', *FEBS Letters*, 384(1), pp. 92–96. doi: 10.1016/0014-5793(96)00294-3.

EC (2009) *Regulation (EC) No 1223/2009 of the european parliament and of the council of 30 November 2009 on cosmetic products*.

FDA (1999) 'Sunscreen drug products for over-the counter human use; Final Monograph Federal Register', 64(98), pp. 27666–93. Available at: www.fda.gov.

FDA (2011) 'Labeling and effectiveness testing;Sunscreen drug products for over-the-counter human use', 76(117), pp. 1–34. Available at: http://www.fda.gov/Drugs/.

FDA (2014) 'Guidance documents – Guidance for industry: Safety of nanomaterials in cosmetic products', pp. 1–37.

FDA (2019) 'Sunscreen drug products for over-the-counter human use: Proposed rule.', *Federal Register*, 84(38), pp. 6204–75. Available at: https://www.federalregister.gov/documents/2019/02/26/2019-03019/sunscreen-drug-products-for-over-the-counter-human-use.

Felton, L. A., Wiley, C. J. and Godwin, D. A. (2004) 'Influence of cyclodextrin complexation on the in vivo photoprotective effects of oxybenzone', *Drug Development and Industrial Pharmacy*, 30(1), pp. 95–102. doi: 10.1081/DDC-120027516.

Filipe, P. *et al.* (2009) 'Stratum corneum is an effective barrier to TiO_2 and ZnO nanoparticle percutaneous absorption', *Skin Pharmacology and Physiology*, 22(5), pp. 266–75. doi: 10.1159/000235554.

Food and Drug Administration (2012) 'Guidance for industry labeling and effectiveness testing : Sunscreen drug products for over- guidance for industry labeling and effectiveness testing '.

Frizzo, M. S. *et al.* (2019) 'Simultaneous encapsulation of zinc oxide and octocrylene in poly (methyl methacrylate-co-styrene) nanoparticles obtained by miniemulsion polymerization for use in sunscreen formulations', *Colloids and Surfaces A: Physicochemical and Engineering Aspects*, 561(2018), pp. 39–46. doi: 10.1016/j.colsurfa.2018.10.062.

Gaspar, L. R. *et al.* (2012) 'Skin phototoxicity of cosmetic formulations containing photounstable and photostable UV-filters and vitamin A palmitate', *Toxicology in Vitro*, 27(1), pp. 418–25. doi: 10.1016/j.tiv.2012.08.006.

Gerin, F. *et al.* (2016) 'The effects of ferulic acid against oxidative stress and inflammation in formaldehyde-induced hepatotoxicity'. doi: 10.1007/s10753-016-0369-4.

Gilbert, E. *et al.* (2013) 'Commonly used UV filter toxicity on biological functions: Review of last decade studies', *International Journal of Cosmetic Science*, 35(3), pp. 208–19. doi: 10.1111/ics.12030.

Gilbert, E. *et al.* (2016) 'Percutaneous absorption of benzophenone-3 loaded lipid nanoparticles and polymeric nanocapsules: A comparative study', *International Journal of Pharmaceutics*, 504(1–2), pp. 48–58. doi: 10.1016/j.ijpharm.2016.03.018.

González, S., Fernández-Lorente, M. and Gilaberte-Calzada, Y. (2008a) 'The latest on skin photoprotection', *Clinics in Dermatology*, 26(6), pp. 614–26. doi: 10.1016/j.clindermatol.2007.09.010.

González, S., Fernández-Lorente, M. and Gilaberte-Calzada, Y. (2008b) 'The latest on skin photoprotection', *Clinics in Dermatology*, 26(6), pp. 614–26. doi: 10.1016/j.clindermatol.2007.09.010.

Gubitosa, J. *et al.* (2020) 'Nanomaterials in sun-care products', in *Nanocosmetics*. Elsevier, pp. 349–73. doi: 10.1016/B978-0-12-822286-7.00022-x.

Guglielmini, G. (2008) 'Nanostructured novel carrier for topical application', *Clinics in Dermatology*, 26(4), pp. 341–46. doi: 10.1016/j.clindermatol.2008.05.004.

Gulson, B. *et al.* (2015) 'A review of critical factors for assessing the dermal absorption of metal oxide nanoparticles from sunscreens applied to humans, and a research strategy to address current deficiencies', *Archives of Toxicology*, 89(11), pp. 1909–30. doi: 10.1007/s00204-015-1564-z.

Hanson, K. M. *et al.* (2015) 'Photochemical degradation of the UV filter octyl methoxycinnamate in solution and in aggregates.', *Photochemical & Photobiological Sciences*, 14(9), pp. 1607–16. doi: 10.1039/c5pp00074b.

Hwang, Y. J. *et al.* (2011) 'Immediate pigment darkening and persistent pigment darkening as means of measuring the ultraviolet A protection factor in vivo: A comparative study', pp. 1356–61. doi: 10.1111/j.1365-2133.2011.10225.x.

ICH (2013) *ICH guideline S10 Photosafety evaluation of pharmaceuticals, ICH Guidelines*. Available at: EMA/CHMP/ICH/752211/2012.

Jacobi, U. *et al.* (2007) 'Porcine ear skin: An in vitro model for human skin', *Skin Research and Technology*, 13(1), pp. 19–24. doi: 10.1111/j.1600-0846.2006.00179.x.

Janjua, N. R. *et al.* (2008) 'Sunscreens in human plasma and urine after repeated whole-body topical application', *Journal of the European Academy of Dermatology and Venereology*, 22(4), pp. 456–61. doi: 10.1111/j.1468-3083.2007.02492.x.

Kandekar, S. G. *et al.* (2019) 'Polymeric micelle nanocarriers for targeted epidermal delivery of the hedgehog pathway inhibitor vismodegib: Formulation development and cutaneous biodistribution in human skin', *Expert Opinion on Drug Delivery*, 5247. doi: 10.1080/17425247.2019.1609449.

Katz, L. M., Dewan, K. and Bronaugh, R. L. (2015) 'Nanotechnology in cosmetics', *Food and Chemical Toxicology* 85, pp. 127–37. doi: 10.1016/j.fct.2015.06.020.

Kim, S. and Choi, K. (2014) 'Occurrences, toxicities, and ecological risks of benzophenone-3, a common component of organic sunscreen products: A mini-review', *Environment International*, 70, pp. 143–57. doi: 10.1016/j.envint.2014.05.015.

Klimová, Z., Hojerová, J. and Beránková, M. (2015) 'Skin absorption and human exposure estimation of three widely discussed UV filters in sunscreens – In vitro study mimicking real-life consumer habits', *Food and Chemical Toxicology*, 83, pp. 237–50. doi: 10.1016/j.fct.2015.06.025.

Klimová, Z., Hojerová, J. and Pažoureková, S. (2013) 'Current problems in the use of organic UV filters to protect skin from excessive sun exposure', *Acta Chimica Slovaca*, 6(1), pp. 82–88. doi: 10.2478/acs-2013-0014.

Klinubol, P., Asawanonda, P. and Wanichwecharungruang, S. P. (2008) 'Transdermal penetration of UV filters', *Skin Pharmacology and Physiology*, 21(1), pp. 23–29. doi: 10.1159/000109085.

Kockler, J. *et al.* (2012) 'Photostability of sunscreens', *Journal of Photochemistry and Photobiology C: Photochemistry Reviews*, 13(1), pp. 91–110. doi: 10.1016/j.jphotochemrev.2011.12.001.

Krause, M. *et al.* (2012) 'Sunscreens: Are they beneficial for health? An overview of endocrine disrupting properties of UV-filters', *International Journal of Andrology*, 35(3), pp. 424–36. doi: 10.1111/j.1365-2605.2012.01280.x.

Kresge, C. T. *et al.* (1992) 'Ordered mesoporous molecular sieves synthesized by a liquid-crystal template mechanism', *Nature*, 359(6397), pp. 710–12. doi: 10.1038/359710a0.

Kresge, C.T., J. C. V. W. J. R. J. S. B. S. B. M. Vartuli, J.C., Roth, W.J., Beck, J.S., McCullen, S.B., Kresge, C.T. (1998). The Synthesis and Properties of M41S and Related Mesoporous Materials. In: Synthesis. Molecular Sieves, vol 1. Springer, Berlin, Heidelberg. https://doi.org/10.1007/3-540-69615-6_4.

Kruk, M. *et al.* (2000) 'Characterization of the porous structure of SBA-15', *Chemistry of Materials*, 12(7), pp. 1961–68. doi: 10.1021/cm000164e.

Lacerda, S. P., Cerize, N. N. P. and Ré, M. I. (2011) 'Preparation and characterization of carnauba wax nanostructured lipid carriers containing benzophenone-3', *International Journal of Cosmetic Science*, 33(4), pp. 312–21. doi: 10.1111/j.1468-2494.2010.00626.x.

Lademann, J. *et al.* (2015) 'Hair follicles as a target structure for nanoparticles', *Journal of Innovative Optical Health Sciences*, 08(04), 1530004. doi: 10.1142/S1793545815300049.

Laouini, A. *et al.* (2012) 'Preparation, characterization and applications of liposomes: State of the art', *Journal of Colloid Science and Biotechnology*, 1(2), pp. 147–68. doi: 10.1166/jcsb.2012.1020.

Lapteva, M. *et al.* (2014) 'Polymeric micelle nanocarriers for the cutaneous delivery of tacrolimus: A targeted approach for the treatment of psoriasis', *Molecular Pharmaceutics*, 11(9), pp. 2989–3001. doi: 10.1021/mp400639e.

Lapteva, M. *et al.* (2019) 'Self-assembled mPEG-hexPLA polymeric nanocarriers for the targeted cutaneous delivery of imiquimod', *European Journal of Pharmaceutics and Biopharmaceutics*, 142(January), pp. 553–62. doi: 10.1016/j.ejpb.2019.01.008.

Lee, M. (2020) 'Liposomes for enhanced bioavailability of water-insoluble drugs: In vivo evidence and recent approaches'. doi: 10.3390/pharmaceutics12030264.

Leong, H. J. *et al.* (2016) 'Preparation of alpha-bisabolol and phenylethyl resorcinol/TiO2 hybrid composites for potential applications in cosmetics', *International Journal of Cosmetic Science*, pp. 524–34. doi: 10.1111/ics.12339.

Levi, K. (2013) 'UV damage and sun care: Deciphering mechanics of skin to develop next generation therapies', *Journal of the Mechanical Behavior of Biomedical Materials*, 28, pp. 471–73. doi: 10.1016/j.jmbbm.2013.02.008.

Li, Z. *et al.* (2007) 'γ-Cyclodextrin: A review on enzymatic production and applications', *Applied Microbiology and Biotechnology*, 77(2), pp. 245–55. doi: 10.1007/s00253-007-1166-7.

Lim, H. W., Arellano-Mendoza, M.-I. and Stengel, F. (2017) 'Current challenges in photoprotection', *Journal of the American Academy of Dermatology*, 76(3), pp. S91–99. doi: 10.1016/j.jaad.2016.09.040.

Lin, L. L. *et al.* (2011) 'Time-correlated single photon counting for simultaneous monitoring of zinc oxide nanoparticles and NAD(P)H in intact and barrier-disrupted volunteer skin', *Pharmaceutical Research*, 28(11), pp. 2920–30. doi: 10.1007/s11095-011-0515-5.

Maier, H. *et al.* (2005) 'Ultraviolet protective performance of photoprotective lipsticks: Change of spectral transmittance because of ultraviolet exposure', *Photodermatology Photoimmunology and Photomedicine*, 21(2), pp. 84–92. doi: 10.1111/j.1600-0781.2005.00143.x.

Marcato, P. D. *et al.* (2011) 'Nanostructured polymer and lipid carriers for sunscreen. Biological effects and skin permeation.', *Journal of Nanoscience and Nanotechnology*, 11, pp. 1880–86. doi: 10.1166/jnn.2010.3135.

Mariano-Neto, F. *et al.* (2014) 'Physical properties of ordered mesoporous SBA-15 silica as immunological adjuvant', *Journal of Physics D: Applied Physics*, 47(42), 425402. doi: 10.1088/0022-3727/47/42/425402.

Matos, J. R. *et al.* (2001) 'Toward the synthesis of extra-large-pore MCM-41 analogues', *Chemistry of Materials*, 13(5), pp. 1726–31. doi: 10.1021/cm000964p.

Miksa, S. *et al.* (2016) 'New approach for a reliable in vitro sun protection factor method – Part II: Practical aspects and implementations', *International Journal of Cosmetic Science*, 38(5), pp. 504–11. doi: 10.1111/ics.12327.

Monteiro, M. S. de S. de B. *et al.* (2012) 'Evaluation of octyl p-methoxycinnamate included in liposomes and cyclodextrins in anti-solar preparations: Preparations, characterizations and in vitro penetration studies', *International Journal of Nanomedicine*, 7, pp. 3045–58. doi: 10.2147/IJN.S28550.

Morabito, K. *et al.* (2011) 'Nanoparticles and their applications in ultraviolet protection: A review', *Analytical Chemistry*, (1), pp. 1–10.

Morlando, A. *et al.* (2016) 'Titanium doped tin dioxide as potential UV filter with low photocatalytic activity for sunscreen products', *Materials Letters*, 171, pp. 289–92. doi: 10.1016/j.matlet.2016.02.094.

Mota, A. de C. V. *et al.* (2013) 'In vivo and in vitro evaluation of octyl methoxycinnamate liposomes', *International Journal of Nanomedicine*, 8, pp. 4689–4700. doi: 10.2147/IJN.S51383.

Nahar, M. *et al.* (2008) 'Development, characterization, and toxicity evaluation of amphotericin B – loaded gelatin nanoparticles', 4, pp. 252–61. doi: 10.1016/j.nano.2008.03.007.

Nash, J. F. (2006) 'Human safety and efficacy of ultraviolet filters and sunscreen products', *Dermatologic Clinics*, 24(1), pp. 35–51. doi: 10.1016/j.det.2005.09.006.

Nastiti, C. *et al.* (2017) 'Topical nano and microemulsions for skin delivery', *Pharmaceutics*, 9(4), p. 37. doi: 10.3390/pharmaceutics9040037.

Nikolić, S. *et al.* (2011) 'Skin photoprotection improvement: Synergistic interaction between lipid nanoparticles and organic UV filters', *International Journal of Pharmaceutics*, 414(1–2), pp. 276–84. doi: 10.1016/j.ijpharm.2011.05.010.

Nohynek, G. J. and Roberts, M. S. (2007) 'Grey Goo on the skin ? Nanotechnology, cosmetic and sunscreen safety J urgen', pp. 251–77. doi: 10.1080/10408440601177780.

Ou-yang, H. *et al.* (2012) 'High-SPF sunscreens (SPF ≥ 70) may provide ultraviolet protection above minimal recommended levels by adequately compensating for lower sunscreen user application amounts', *Journal of American Dermatology*, 67(6), pp. 1220–27. doi: 10.1016/j.jaad.2012.02.029.

Palm, M. D. and O'Donoghue, M. N. (2007) 'Update on photoprotection', *Dermatologic Therapy*, pp. 360–76. doi: 10.1111/j.1529-8019.2007.00150.x.

Papakostas, D. and Rancan, F. (2011) 'Nanoparticles in dermatology', pp. 533–50. doi: 10.1007/s00403-011-1163-7.

Peres, D. D. A. *et al.* (2016) 'Rutin increases critical wavelength of systems containing a single UV filter and with good skin compatibility', *Skin Research and Technology*, pp. 325–33. doi: 10.1111/srt.12265.

Peres, D. D. A. *et al.* (2018) 'Ferulic acid photoprotective properties in association with UV filters: multifunctional sunscreen with improved SPF and UVA-PF', *Journal of Photochemistry and Photobiology B: Biology*, 185, pp. 46–49. doi: 10.1016/j.jphotobiol.2018.05.026.

Perugini, P. *et al.* (2002) 'Effect of nanoparticle encapsulation on the photostability of the sunscreen agent, 2-ethylhexyl-p-methoxycinnamate', *International Journal of Pharmaceutics*, 246(1–2), pp. 37–45. doi: 10.1016/S0378-5173(02)00356-3.

Potard, G. *et al.* (2000) 'The stripping technique: In vitro absorption and penetration of five UV filters on excised fresh human skin', *Skin Pharmacology and Physiology*, 13(6), pp. 336–44. doi: 10.1159/000029941.

Puglia, C. *et al.* (2012) 'Lipid nanoparticles as carrier for octyl-methoxycinnamate: In vitro percutaneous absorption and photostability studies', 101(1), pp. 301–11. doi: 10.1002/jps.

Puglia, C. *et al.* (2014) 'Evaluation of nanostructured lipid carriers (NLC) and nanoemulsions as carriers for UV-filters: Characterization, in vitro penetration and photostability studies', *European Journal of Pharmaceutical Sciences*, 51(1), pp. 211–17. doi: 10.1016/j.ejps.2013.09.023.

Rai, R., Deep, A. and Tasduq, S. (2022) 'Journal of ayurveda and integrative medicine herbal products as skincare therapeutic agents against ultraviolet radiation-induced skin disorders', *Journal of Ayurveda and Integrative Medicine*, 13(1), p. 100500.

Road, G. X. (2021) 'Natural components in sunscreens: Topical formulations with sun protection factor (SPF)', *Biomedicine & Pharmacotherapy*, 134(2020), p. 111161.

Sambandan, D. R. and Ratner, D. (2011) 'Sunscreens: An overview and update', *Journal of the American Academy of Dermatology*, 64(4), pp. 748–58. doi: 10.1016/j.jaad.2010.01.005.

Santos, A. C. *et al.* (2022) 'Nanotechnology-based sunscreens—A review', *Materials Today Chemistry*, 23. doi: 10.1016/j.mtchem.2021.100709.

Sargent, E. V and Travers, J. B. (2016) 'Examining the differences in current regulatory processes for sunscreens and proposed safety assessment paradigm', *Regulatory Toxicology and Pharmacology*, 79, pp. 125–41. doi: 10.1016/j.yrtph.2016.03.008.

Sarveiya, V., Templeton, J. F. and Benson, H. A. E. (2004) 'Inclusion complexation of the sunscreen 2-hydroxy-4-methoxy benzophenone (oxybenzone) with hydroxypropyl-β-cyclodextrin: Effect on membrane diffusion', *Journal of Inclusion Phenomena*, 49(3–4), pp. 275–81. doi: 10.1007/s10847-004-6098-6.

Sauce, R. *et al.* (2021) 'Ex vivo penetration analysis and anti-inflammatory efficacy of the association of ferulic acid and UV filters', *European Journal of Pharmaceutical Sciences*, 156(May 2020), 105578. doi: 10.1016/j.ejps.2020.105578.

Scalia, S., Coppi, G. and Iannuccelli, V. (2011) 'Microencapsulation of a cyclodextrin complex of the UV filter, butyl methoxydibenzoylmethane: In vivo skin penetration studies', *Journal of Pharmaceutical and Biomedical Analysis*, 54(2), pp. 345–50. doi: 10.1016/j.jpba.2010.09.018.

Scalia, S., Tursilli, R. and Iannuccelli, V. (2007) 'Complexation of the sunscreen agent, 4-methylbenzylidene camphor with cyclodextrins: Effect on photostability and human stratum corneum penetration', *Journal of Pharmaceutical and Biomedical Analysis*, 44(1), pp. 29–34. doi: 10.1016/j.jpba.2007.01.016.

SCCS (2012) 'The SCCS'S notes of guidance for the testing of cosmetic substances and their safety evaluation', 8th(December), pp. 1–133.

SCCS (2019) *Guidance on the safety assessment of nanomaterials in cosmetics*. doi: 10.2875/40446.

Scioli Montoto, S., Muraca, G. and Ruiz, M. E. (2020) 'Solid lipid nanoparticles for drug delivery: Pharmacological and biopharmaceutical aspects', *Frontiers in Molecular Biosciences*, 7(October), pp. 1–24. doi: 10.3389/fmolb.2020.587997.

Shaath, N. (2010) 'Ultraviolet filters', *Photochemical & Photobiological Sciences: Official Journal of the European Photochemistry Association and the European Society for Photobiology*, 9(4), pp. 464–69. doi: 10.1039/b9pp00174c.

Shakeel, M. *et al.* (2016) 'Toxicity of nano-titanium dioxide (TiO2-NP) through various routes of exposure: A review', *Biological Trace Element Research*, 172(1), pp. 1–36. doi: 10.1007/s12011-015-0550-x.

Shegokar, R. and Müller, R. H. (2010) 'Nanocrystals: Industrially feasible multifunctional formulation technology for poorly soluble actives', *International Journal of Pharmaceutics*, 399(1–2), pp. 129–39. doi: 10.1016/j.ijpharm.2010.07.044.

Shetty, P. K. *et al.* (2015) 'Development and evaluation of sunscreen creams containing morin-encapsulated nanoparticles for enhanced UV radiation protection and antioxidant activity', *International Journal of Nanomedicine*, 10, pp. 6477–91. doi: 10.2147/IJN.S90964.

Simeoni, S. *et al.* (2006) 'Influence of cyclodextrin complexation on the in vitro human skin penetration and retention of the sunscreen agent, oxybenzone', *Journal of Inclusion Phenomena*, 54(3–4), pp. 275–82. doi: 10.1007/s10847-005-9002-0.

Souto, E. B. *et al.* (2022) 'Lipid nanomaterials for targeted delivery of dermocosmetic ingredients: Advances in photoprotection and skin anti-aging', *Nanomaterials* 12(3), p. 377.

Springsteen, A. *et al.* (1999) 'In vitro measurement of sun protection factor of sunscreens by diffuse transmittance 1', *Analytica Chimica* Acta, 380, pp. 155–64.

Srinivasan, M. *et al.* (2005) 'Ferulic acid, a natural protector against carbon tetrachloride-induced toxicity', 19, pp. 491–96. doi: 10.1111/j.1472-8206.2005.00332.x.

Tanner, P. R. (2006) 'Sunscreen product formulation', *Dermatologic Clinics*, pp. 53–62. doi: 10.1016/j.det.2005.09.002.

TGA (2016) 'Australian regulatory guidelines for sunscreens (ARGS)', *International Journal of Pharmaceutics*, 20(January), pp. 1–92.

Tomazelli, L. C. *et al.* (2018) 'SPF enhancement provided by rutin in a multifunctional sunscreen', *International Journal of Pharmaceutics*, 552(1–2), pp. 401–06. doi: 10.1016/j.ijpharm.2018.10.015.

Touloumes, G. J. *et al.* (2020) 'NanoImpact mapping 2D- and 3D-distributions of metal/metal oxide nanoparticles within cleared human ex vivo skin tissues', 17(January). doi: 10.1016/j.impact.2020.100208.

Velasco, M. V. R. *et al.* (2008) 'Broad spectrum bioactive sunscreens', *International Journal of Pharmaceutics*, 363(1–2), pp. 50–57. doi: 10.1016/j.ijpharm.2008.06.031.

Vettor, M. *et al.* (2010) 'Skin absorption studies of octyl-methoxycinnamate loaded poly(D, L-lactide) nanoparticles: Estimation of the UV filter distribution and release behaviour in skin layers', *Journal of Microencapsulation*, 27(3), pp. 253–62. doi: 10.3109/10717540903097770.

Wang, S. Q., Balagula, Y. and Osterwalder, U. (2010) 'Photoprotection: A review of the current and future technologies', *Dermatologic Therapy*, 23(1), pp. 31–47. doi: 10.1111/j.1529-8019.2009.01289.x.

Wang, S. Q. and Tooley, I. R. (2011) 'Photoprotection in the era of nanotechnology', *Seminars in Cutaneous Medicine and Surgery*, 30(4), pp. 210–13. doi: 10.1016/j.sder.2011.07.006.

Wissing, S. A. and Müller, R. H. (2002) 'Solid lipid nanoparticles as carrier for sunscreens: In vitro release and in vivo skin penetration', *Journal of Controlled Release*, 81(3), pp. 225–33. doi: 10.1016/S0168-3659(02)00056-1.

Wu, J. *et al.* (2009) 'Toxicity and penetration of TiO2 nanoparticles in hairless mice and porcine skin after subchronic dermal exposure', *Toxicology Letters*, 191(1), pp. 1–8. doi: 10.1016/j.toxlet.2009.05.020.

Wu, P.-S. *et al.* (2014) 'Effects of the novel poly(methyl methacrylate) (PMMA)-encapsulated organic ultraviolet (UV) filters on the UV absorbance and in vitro sun protection factor (SPF)', *Journal of Photochemistry and Photobiology B: Biology*, 131, pp. 24–30. doi: 10.1016/j.jphotobiol.2014.01.006.

Xia, Q., Saupe, A. and Mu, R. H. (2007) 'Nanostructured lipid carriers as novel carrier for sunscreen formulations', *International Journal of Cosmetic Science*, 29(6), pp. 473–482. doi: 10.1111/j.1468-2494.2007.00410.x.

Zhou, Y. *et al.* (2018) 'Mesoporous silica nanoparticles for drug and gene delivery', *Acta Pharmaceutica Sinica B*, 8(2), pp. 165–77. doi: 10.1016/j.apsb.2018.01.007.

Zielińska, A. *et al.* (2020) 'Polymeric nanoparticles: Production, characterization, toxicology and ecotoxicology', *Molecules*, 25(16), 3731. doi: 10.3390/molecules25163731.

14 Nanoperfumes as a fragrance product

Rajesh Pradhan, Sanskruti Santosh Kharavtekar, Vighnesh Jadhav, Rajeev Taliyan and Sunil Kumar Dubey

CONTENTS

14.1 INTRODUCTION

Cosmetics in the beauty industry have a broad area of coverage, having varied utility. Overall, they confer temporary protection from the environment and aid in providing a better appearance, thus having aesthetic value. In 1961, the term "cosmetic" was given by Raymond Reed (founder of the US society of Cosmetic Chemists). As per the Food Drug and Administration (FDA), cosmetics are

DOI: 10.1201/9781003319146-14

293

defined as "elements (except pure soap) intended to be applied to the human body for cleansing, beautifying, promoting attractiveness, or altering the appearance." Overall, cosmetics are products that enhance cleansing activity, increase beauty, and improve skin appearance. Cosmetic agents are classified according to the definition by Thornfeldt et al. as "skin moisturizers, anti-aging, facial makeup, shampoo, toothpaste, deodorant, hair colour, perfumes and others that are used for beauty enhancement" (Thornfeldt 2005). This article will be focusing on perfumes i.e., nanoperfumes; Perfumes can be defined as a product of fragrance with a pleasant odour when applied to a particular body area and are usually maintained in a fresh condition consisting of oils such as essential or aromatic oils (Cortial et al. 2015; Salvador-Carreno and Chisvert 2003; Salvioni et al. 2021). Contingent upon the concentration of the aromatic compound in perfumes, perfumes are classified as: extrait de parfum (15%–40% aromatic fragrance in 90–95% alcoholic solution), eau de parfum (up to to 15% aromatic fragrance in 90% of alcoholic solution), eau de cologne (2%–8% aromatic fragrance, in 75%–80% alcoholic solution) and also body spray/mist/spritz with 3% of aromatic fragrance called as an au fraiche (Miastkowska et al. 2018; Salvador-Carreno and Chisvert 2003; Sikora et al. 2018). Nowadays, these are gaining emerging importance in everyday life. The fragrances are used to mask the bad odour which positively impacts our mood and provides relief from discomfort if any. They have wide applications in cosmetics and toiletries (Salvador-Carreno and Chisvert 2003). Although perfumes have gained their demand in the market, there are numerous issues with them that persist, thus providing greater opportunities for advancement. Fragrances that are added in perfumes are generally volatile in nature, having a high evaporation rate, poor chemical stability, and being prone to environmental deterioration; leading to loss of aromatic oil and vehicle, shorter half-life, and leaving a high concentrated aroma behind (He, Hu, and Deng 2018). The volatility of the fragrance is one of the crucial factors for its function as a perfume. High volatility affects the longevity of the fragrance. Thus, to tackle this, a controlled release pattern can be employed for obtaining a longer-lasting effect. Ethanol is one of the common solvents that is used for manufacturing perfumes. However, with ethanol, there are several drawbacks such as its high cost, flammability, and tendency to precipitate skin irritation in the case of allergic patients. It is important to find options that will overcome these cons of ethanol. Considering these mentioned limitations of perfume, encapsulation of fragrance in micelle and formulating the alcohol-free perfumes are crucial for sustained release pattern, which will also lead to fragrance longevity and safe solvent-based perfume respectively (Miastkowska and Lasoń 2020; Perinelli et al. 2020; Siddique, Husain, and Singh 2022). Nanotechnology is an emerging development in the field of cosmetics and pharmaceuticals. It has gained a special place in next-generation cosmetics, with an increasing demand for nanoproducts in several markets over the globe (Kaul et al. 2018). EU cosmetic regulations say nanotechnology used in cosmetics should be termed as having the word 'nano' as a prefix when used for developing cosmetic products. Nanotechnology in the field of cosmetics includes nanoemulsion, nanocapsules, nano pigments, liposome-based formulations, nanocrystals, solid lipid nanoparticles, carbon nanotubes, fullerenes, and dendrimers. Developing nanoperfumes can fulfill the research gap that is experienced with conventional perfumes and provide long-lasting properties and overcome the problem of loss of volatile substances (Aziz et al. 2019; Nguyen and Rajendran 2020). This article aims to discuss nanoperfumes, their types, methods of preparation, and excipient used, as well as evaluation of nanoperfumes.

14.2 NANOPERFUMES

Nanoperfumes are fragrant products that are formulated by low and high-energy methods i.e., phase inversion composition and ultrasound, etc., There are several ways to achieve encapsulation of fragrance. "Eau de Toilette" and "Eau de Parfum" were reported to have nanodroplets. Nanoperfumes provide smaller droplet sizes while increasing the interfacial areas, making them perfect for the delivery of fragrances. They have a clear appearance and provide several advantages over conventional perfumes (He, Hu, and Deng 2018; Salvador-Carreno and Chisvert 2003).

14.3 TYPES OF NANOPERFUMES

14.3.1 NANOEMULSION-BASED NANOPERFUMES

While formulating nanoperfumes, ethanol or ethanol and water in combination can be employed, however, this would have certain disadvantages such as drying out quickly over the skin leading to skin irritation, contact eczema, cutaneous intolerance, and inflammation. Ethanol as a solvent for perfumes has several pros, however, due to its disadvantages, there is a need to find an alternative solvent system. An alternative solvent to ethanol can be oil; moreover, oil as a carrier cannot alter or garble the scent of the perfumes. The oil-based perfumes do not dry out quickly on the skin which makes perfume long-lasting. On the other hand, the use of oil as a solvent carrier leads to an oily or greasy appearance over the area applied, as well as leaves the concentrated scent behind. Other alternatives include encapsulation of the fragrance in lipids, polymers, and matrix formulating nanoencapsulation, which leads to improvement of the shelf life of the perfumes and protects it from environmental oxidation. Nanoemulsions are now more widely used in several cosmetics. The standard type of nanoemulsion-based formulations can be formulated as water-based perfumes, consisting of the oil phase, water phase, and surfactant which may or may not contain co-surfactant. The three types of nanoemulsion include o/w nanoemulsion, w/o nanoemulsion, and bicontinuous nanoemulsion. The dispersed phase has a droplet size ranging from 50 nm to 200 nm. Such o/w types of nanoemulsion-based nanoperfumes are thermodynamically stable. Nanoemulsion-based nanoperfumes are formulated by high- and low-energy methods. Ultrasound, high revolution mixers, and high homogenizers are mechanical devices that are categorized under high-energy methods involving mechanical energy. The low-energy method consists of phase inversion composition using chemical energy present in the emulsion ingredients. Nanoemulsion has several applications, such as easy solubilization of hydrophobic moiety, which protects it from oxidation, reduces interfacial tension, and increases the duration of perfume uses (Miastkowska and Lasoń 2020; Sikora et al. 2018).

14.3.2 Solid Lipid Nanoparticle-based Nanoperfumes

SLN-based nanoperfumes have few advantages over nanoemulsion-based nanoperfumes, i.e., these types of nanoperfumes have controlled and sustained release patterns of fragrance which is impractical to be obtained from nanoemulsion. The particle size in SLN-based nanoperfumes is in the range of 50–1000 nm using emulsion and liposomes. SLN mainly consists of a lipoidal layer core in which the fragrant is entrapped, leading to long-lasting, controlled release via entrapment of volatile oil. Two crucial advantages of SLN are that, it is safest in terms of toxicity and can be easily manufactured on large scale. The stability of the SLN is maintained by adding the surfactant, mostly non-ionic surfactant as other surfactants can cause irritation to the skin. SLN-based nanoperfumes can be prepared via multiple methods and principles such as high shear homogenization, high pressure, and cold homogenization, ultrasounds, and emulsification or solvent evaporation, can be employed for the same (Miastkowska and Lasoń 2020). Hot homogenization method eventually follows three essential steps: emulsification, sonication, and cooling, where emulsification includes: (1) Heating lipid and surfactant above the lipid's melting point, (2) Decreasing the droplets' size with the help of ultrasonication, (3) Finally, dispersion of nanoemulsion in water to form SLNs by cooling it to 0°C. After the emulsification method followed by sonication and cooling, the SLNs are prepared from fatty ester, alcohol, hydrocarbon, and triglycerides. Lipids used in the formulation of SLNs-based nanoperfumes include shea butter, Candelilla wax, C10–18 triglycerides, and beeswax. SLNs' properties such as mean average size, PDI, zeta potential, and pH can be in the range of 140–195 nm, 0.25, −10 to −30 mV, and close to 5.0 respectively. Scanning Electron Microscopy (SEM) and fragrance encapsulation in SLNs are important characterization parameters of SLNs-based nanoperfumes (Whitepaper on Nano Technology & AI Application in Fragrance Sector 2020; Nguyen and Rajendran 2020). Figure 14.1a shows SLNs structure and Figure 14.1b shows the preparation method of SLNs.

14.3.3 Non-structured Lipid Carriers-based Nanoperfumes

In the w/o type of nanoemulsion-based nanoperfumes, water evaporates leaving oil enclosed with perfumes on the skin and hindering the evaporation of the perfumes. The NLC-based nanoperfumes are formulated by replacing the oil of o/w emulsion with solid particles, wherein perfumes are enclosed in the solid matrix leading to a slower release of perfumes compared to perfumes' release from nanoemulsion. The release pattern of the perfume from NLC-based nanoperfumes depends on the contents of the perfume, its viscosity, and the NLC structure. Small-sized NLCs are formed based on the ability of surfactant to stabilize the oil/lipids. However excessive use of surfactant may form foam during preparation. NLC are formed via blending the solid lipids with the incompatible liquid lipids. This results in formation of amorphous solids which ideally are in the size range of 10–1000 nm. The preparation method of NLC-based nanoperfumes follows the principle of homogenization at 5°C with different surfactants, lipids, and perfumes enclosed. The loss of perfume due to its volatile nature can be prevented by adding perfume to lipids before

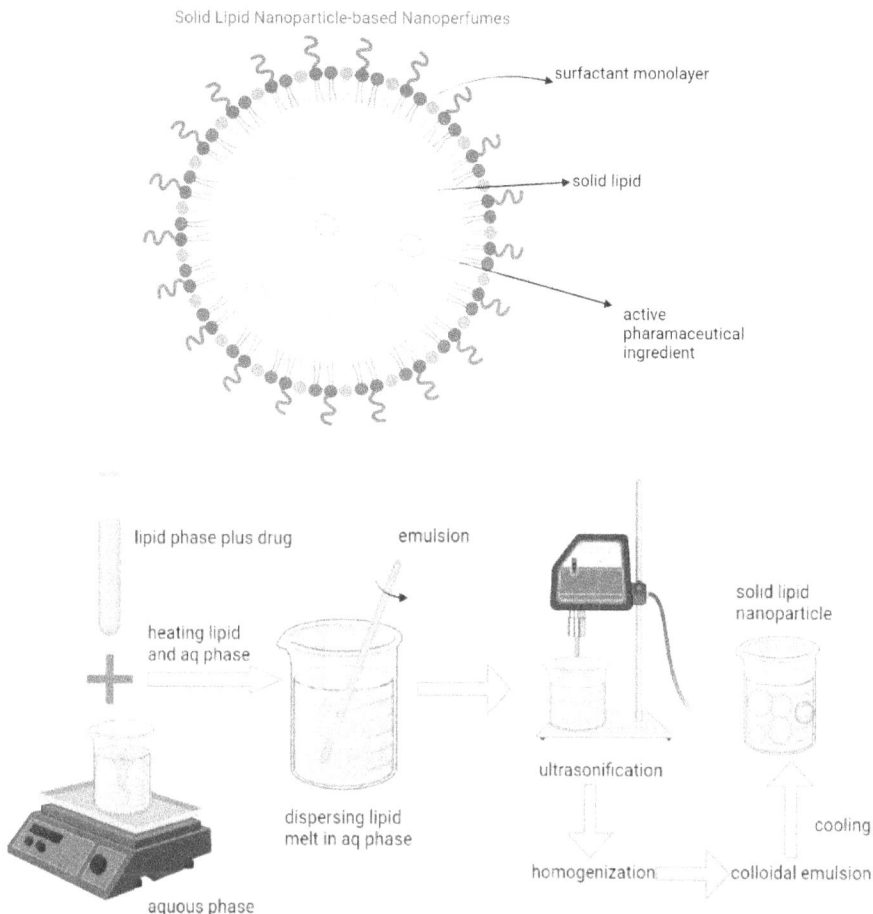

Figure 14.1a SLNs structure

homogenization. The release pattern of perfume from NLC also depends on the type of lipid used and the compatibility of perfume and lipid. In a study, comparing perfumes with two different lipids – Apifil and carnauba wax, concluded that perfumes that were enclosed in Apifil demonstrated controlled release of perfume from NLC. Furthermore, the prolonged fragrance of nanoperfume was achieved by developing positively charged NLC-based nanoperfume as these adhere to the skin (Aiman 2008; Whitepaper on Nano Technology & AI Application in Fragrance Sector 2020).

14.3.4 POLYMERIC-BASED NANOPERFUMES

Another approach to developing nanoperfumes is polymeric-based nanoperfumes, wherein the fragrance is coated in a polymer that protects it from evaporation, and degradation and maintains the controlled release pattern. The polymers used for formulation can be natural, synthetic, and semi-synthetic; broadly categorized as organic and inorganic polymers, either forming a single layer or multilayer core for

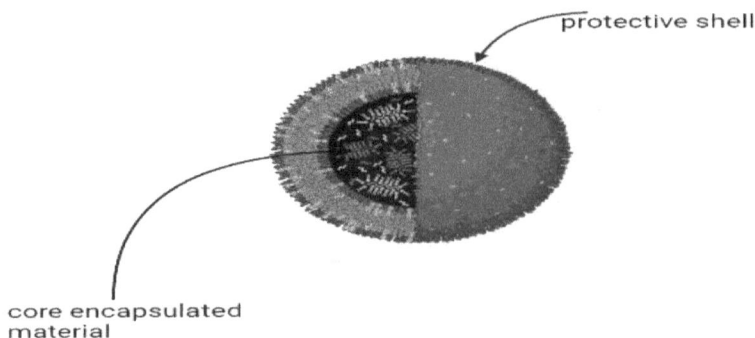

protective shell

core encapsulated
material

FIGURE 14.2 Polymeric nanoparticles

TABLE 14.1
Preparation method of polymeric nanoparticle

Chemical methods	Physicochemical method	Physicomechanical method
situ polymerization	Emulsification	Spray drying
emulsion polymerization	Coacervation	Freeze drying
interfacial polymerization		extrusion

---

nanoencapsulation (Hejmady et al. 2020; Kumar Dubey et al. 2021). Figure 14.2 depicts a polymeric nanoparticle. These polymeric nanoperfumes have several advantages over others, such as ease of formulation. In addition to this, inorganic polymer-based nanoperfume has a unique aroma-releasing pattern. Double-emulsions, like oil-in-water-in-oil (O/W/O) emulsions, are based on the copolymerization principle, a newly developed approach to formulate polymeric-based nanoperfumes by encapsulation of fragrance in the polymer. Cyclodextrin-based fragrance carrier, chitosan-based nanoperfumes, and Poly (lactic-co-glycolic acid) (PLGA) based nanoperfumes are some of the polymeric-based nanoperfumes mentioned in few research articles, where chitosan has the additional advantage of antibacterial property (Lu et al. 2020). Several principle techniques involved in preparation method include three major types shown in Table 14.1.

14.4 EXCIPIENTS USED FOR NANOFORMULATION

Traditionally, alcohol, mainly ethanol, was used as a hydrophobic aromatic solvent but it causes irritation to the skin and causes inflammation. Some people with sensitive skin are prone to allergies due to these ethanol-based perfumes, and to overcome these issues, the alcohol-based solvent phase has been replaced by various aromatic oil phases and perfumed oils. Essential oils procured from several aromatic plants have been conventionally used to prepare dedicated nanoperfumes as they have antimicrobial and antibacterial properties (Jang et al. 2008). Nowadays, alcohol-free perfumes called water-based perfumes are widely used and are in great demand. Three main

TABLE 14.2
Fragrance and excipient used in various nanoperfumes

Fragrance oil phase	Flower	Citrus sinensis L., Lavandula dentata L,
	leaves	Globulus L., Thymus vulgaris L., Mentha piperita L., Satureja hortensis L.,
	rhizomes	Zingiber officinale L., Acorus calamus L.,
	seeds	Carum carvi L., Coriandrum sativum L.,
	fruits	Foeniculum vulgare L., Pimpinella anisum L., Citrus limon L.
	woods	Cinnamomum Cassia Presl., Santalum album L.[1,2]
Aqueous phase	Demineralized water	
Surfactants	Mixture of tween 80, sodium dodecylbenzene sulfonate and hexyl alcohol, tween 80, tween 20, Alkylpolyglucoside, poloxomer 407, Procetyl® AWS, Soy lecithin.	

phases which are combined to form fragrant nanoemulsion, are (1) the oil phase which mainly includes aromatic and fragrant oil, (2) the aqueous phase having demineralized water, and (3) the surfactants, used for dispersion of both these phases, furthermore in order to improve the dispersion, cosurfactant can be added in few instances (Park et al. 2008). The list of various fragrance oil and excipients used is mentioned in Table 14.2.

14.5 PREPARATION METHOD OF NANOPERFUMES

14.5.1 PHASE INVERSION COMPOSITION METHOD (PIC)

The PIC method is a lower-energy, spontaneous emulsification method used to prepare nanoemulsion-based nanoperfumes. The principle relies on transitioning between phases during the emulsification process. Phase transition occurs either by elevation of temperature such that the composition is constant or by altering the composition at a constant temperature. This method can be quickly implicated and has a low cost. In this method, the dispersed phase is thoroughly blended with the continuous phase. Water as solvent is slowly added to the mixture of oil phase containing perfume and surfactant at room temperature with constant stirring, thus forming nanoemulsion-based perfumes (Amin and Das 2019; Gurpreet and Singh 2018). Figure 14.3 shows the PIC preparation method of nanoemulsion.

14.5.2 ULTRASONIC HOMOGENIZATION METHOD

Ultrasonic Homogenization method is high-energy method used to form nano-emulsion-based perfumes. This pre-emulsion is homogenized using an ultrasonic homogenizer. In this method, oil phase (fragrance), aqueous phase, and surfactant are dispersed at room temperature, stirred constantly at 500 rpm for ten minutes with the help of mechanical stirrer, followed by ultrasonic homogenization with 15 W for at least 60 seconds (Buza and Dizdar 2017; Gurpreet and Singh 2018). Figure 14.4 shows about Ultrasonic Homogenization method.

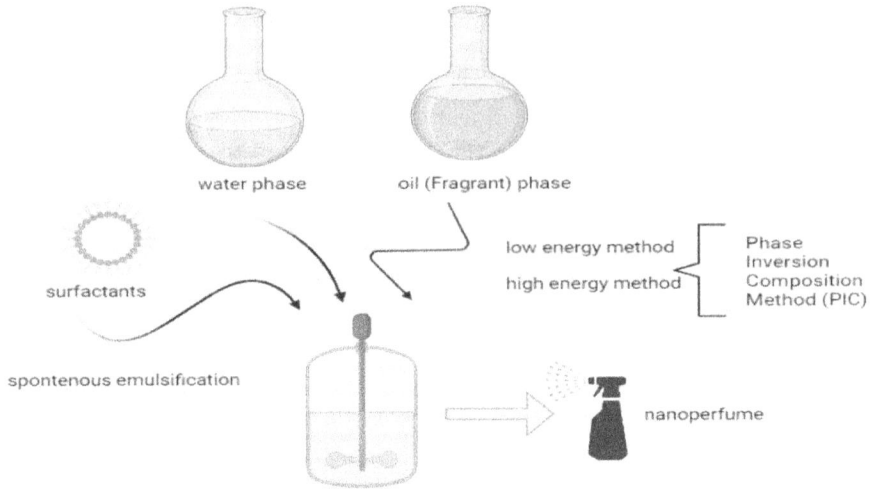

FIGURE 14.3   Phase inversion composition method (PIC)

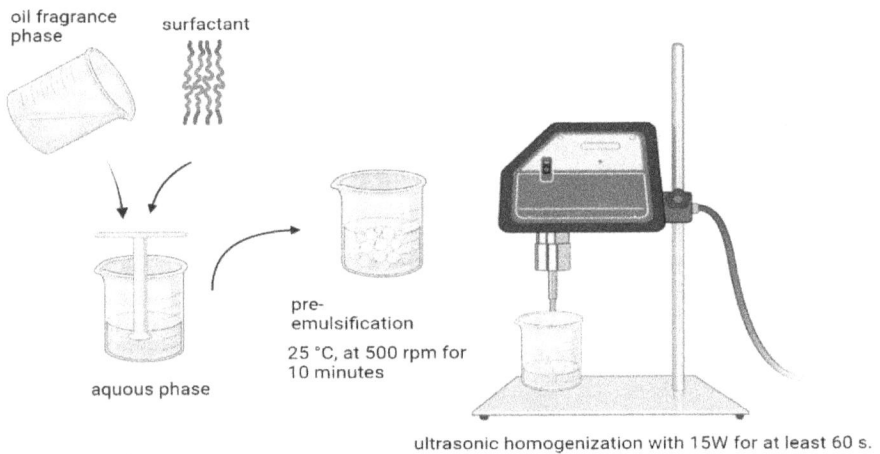

FIGURE 14.4   Ultrasonic homogenization method

14.5.3   ENCAPSULATION

Encapsulation is a widely used method that allows controlled release of the fragrance, thus leading to the protection of fragrant nanoperfumes from evaporation and chemical degradation.

By encapsulating different polymeric materials, it forms a diffusion barrier which helps in enhancing their retention. It basically involves coating or capturing a fragrant molecule using a polymer. Encapsulation of the fragrant molecules would help in preventing contamination and exposure to unfavourable environments. Several features are considered during encapsulation including high-strength shell wall,

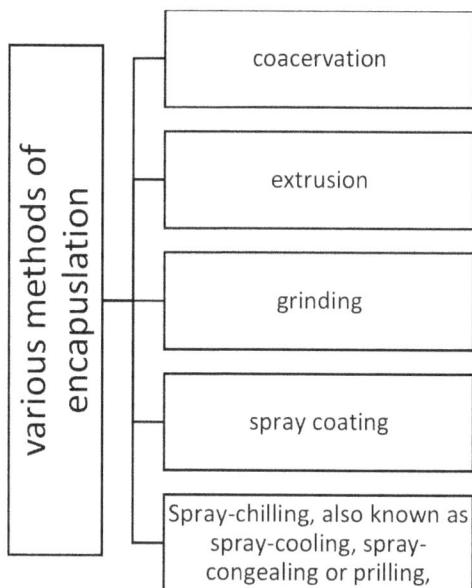

FIGURE 14.5 Encapsulation method

compatibility with aqueous or organic shell cores, ease of use in time-release (or thermal release applications), and simple fabrication. Moreover, along with external shielding, the shelf life of the fragrances could be extended with the help of encapsulation. Herein, the stability of fragrant molecules increases due to the protective nature of the coating material (He, Hu, and Deng 2018; Kaur et al. 2018; Perinelli et al. 2020). Figure 14.5 gives the various methods for encapsulation.

14.5.4 NLC AND SLNS BASED NANOPERFUMES PREPARATION

The preparation methods for both types of nanoperfumes (NLC and SLNs) are almost identical, the difference lies in the use of solid lipids in NLC preparation instead of liquid lipids. The preparation begins with melting the lipid above its melting point and dispersing the fragrance in the lipid melted with a constant stirring at 8000 rpm for 25–30 seconds, further dispersing it in the water phase with surfactant. The obtained pre-emulsion is further homogenized at 800 bars with two cycles of homogenization forming o/w nanoemulsion followed by cooling at room temperature on a water bath set at 15°C (Aiman 2008). The nanoperfumes can be formulated in two different forms, NLC and SLNs depending on the type of lipid used.

14.6 CHARACTERIZATION OF NANOPERFUMES

14.6.1 PARTICLE SIZE AND POLYDISPERSITY INDEX ANALYSIS

Different ranges of particle sizes are measured using the Malvern zeta sizer which follows the principle of dynamic light scattering (DLS) and also it is used to measure the polydispersity index (PDI) which ensures the uniformity in the droplet sizes within the formulation. PDI less than 0.3 indicates monodispersity in the formulation. Polydispersity is the ratio of the standard deviation to the mean droplet size, the higher the PDI, the lower the uniformity of the droplet size in the formulation (Nitheesh et al. 2021).

14.6.2 ZETA POTENTIAL

Zeta potential is an electric charge present on the surface of the particles in nano-perfumes. It is used to predict the dispersion stability of nanoperfumes which in turn depends on the physiochemical property of the drug, polymer, vehicle, presence of surfactant, and electrolytes. Zeta potential is measured using the Zeta Sizer instrument and involves dilution of the sample with distilled water and 0.9% NaCl or surfactant which is used during formulation. The average zeta potential of nanoperfumes at which the formulation is considered stable is ±30 mV (Kaur et al. 2018).

14.6.3 DETERMINATION OF LOADING EFFICIENCY

The loading efficiency of nanoperfumes can be analyzed using a UV spectrophotom-eter at a specific wavelength depending on the lambda max of fragrance. In a par-ticular dilution, a specific amount of nanoperfume is dissolved in an organic solvent followed by ultrasonication leading to the destruction of the nanoparticles, and the extracted fragrance in the organic solvent is analyzed under UV spectrophotometer (Gurpreet and Singh 2018). The fragrance encapsulated is measured by the given formula:

$$\% \, Entrapment\, efficiency = \frac{obtained\, fragrance\, content\, in\, nanoperfume}{amount\, of\, fragrances\, added} \times 100$$

14.6.4 HIGH-PERFORMANCE LIQUID CHROMATOGRAPHY (HPLC) ANALYSIS

Determination of the fragrance content and release from the nanoperfumes can be analyzed using HPLC. This helps in determining the amount of fragrance encap-sulated and the amount that can be released during application. The release rate of fragrance from nanoperfumes can be calculated as follow:

$$\% \, release\, rate = \frac{amount\, of\, fragrance\, released\, from\, nanoperfumes}{amount\, of\, fragrance\, added} \times 100$$

14.6.5 TRANSMISSION ELECTRON MICROSCOPY (TEM)

The morphology and structure of the lipid-based nanoperfumes can be studied using transmission electron microscopy (TEM).

14.6.6 OXIDATIVE STRESS TEST

The oxidative stress test is performed to check the chemical stability and stability against the peroxidation of fragrance. In this test, the peroxide value is been calculated via Wheeler and DGF C-VI 6a method (Fruehwirth et al. 2021) According to this method, the nanoperfumes are added to the isooctane and glacial acetic acid mixture and further allowed to react with potassium iodide. The released iodine by the reaction is determined by titrating it with 0.01 N sodium thiosulfate solution and peroxide value (POV) is calculated. The POV is calculated by formula as mentioned:

$$POV = \frac{(A - B) \times M \times 1000}{2 \times Q}$$

where A is the sodium thiosulfate solution volume consumed, B is sodium thiosulfate solution volume consumed in the blank test, M is the molarity of the sodium thiosulfate solution, and Q is the quantity of the tested sample having an accuracy of ±0.1 mg.

14.7 APPLICATIONS

Nanotechnology in perfumes is a newly developing approach for formulating nanoperfumes with greater advancements and several advantages in various fields such as food industries, cosmetics, and toiletries (Dubey et al. 2022). Furthermore, it can be used to add fragrance to the walls and objects by applying them. Nanoperfumes are reduced particle size molecules enclosing fragrance in polymer, lipid, or other, which allows us to exclude the use of solvents like ethanol via developing nanoemulsion-based nanoperfumes (Lohani et al. 2014). The stability and duration of perfumes can be increased by encapsulation with a polymeric material, which protects the perfumes from oxidation, thereby achieving long-lasting and controlled released pattern nanoperfumes. Nanoperfumes ejectors are nanoperfumes-based applicators that are introduced in the market and have wide application in homes, hospitals, and public places. These nanoperfumes' ejectors make air sterile and also absorb the unpleasant odour if any. Also, some nanoperfumes have been used to alter moods such as anxiety and relieve stress (Vijaya et al. 2020; Wei et al. 2020).

14.8 CONCLUSION

This article highlights nanoperfumes and how these have been overcoming the existing hurdles of formulation and excipients used in perfumes. As cosmetics have been used daily by numerous populations, it is necessary to develop innovative products in the market for consumer satisfaction. Nanotechnology has become a significant pillar for cosmetics in order to convert cosmetics into nanoformulations leading to several advantages. This article also explained the types of nanoperfumes based on their preparation method, followed by their characterization. Nanoemulsion-based nanoperfumes have helped to increase the shelf life of perfume and decrease the deterioration from environmental light and oxygen. Furthermore, formulators have been developing water-based nanoperfume so there is limited use of the irritant chemicals. In addition to this, the SLNs-based nanoperfumes provide the advantages that it makes perfumes long-lasting, and while providing a controlled release. Nanoperfumes have numerous advantages over the normal perfume formulation, making nanoperfumes more demandable in the market. In the future perspective, nanoperfumes will be in higher demand considering their qualities and the way they counteract the issues faced with the perfume formulation.

REFERENCES

Aiman, Hommoss. 2008. "Nanostructured Lipid Carriers (Nlc) in Dermal and Personal." *Freie Universität Berlin*, 1–202.

Amin, Nurul, and Biswajit Das. 2019. "A Review on Formulation and Characterization of Nanoemulsion." *International Journal of Current Pharmaceutical Research* 11 (4): 1–5. doi:10.22159/ijcpr.2019v11i4.34925.

Aziz, Zarith Asyikin Abdul, Hasmida Mohd-Nasir, Akil Ahmad, Siti Hamidah Siti, Wong Lee Peng, Sing Chuong Chuo, Asma Khatoon, Khalid Umar, Asim Ali Yaqoob, and Mohamad Nasir Mohamad Ibrahim. 2019. "Role of Nanotechnology for Design and Development of Cosmeceutical: Application in Makeup and Skin Care." *Frontiers in Chemistry* 7. doi:10.3389/fchem.2019.00739.

Buza, Nermin, and Muamer Dizdar. 2017. "Cmbebih 2017" 62: 317–22. doi:10.1007/978-981-10-4166-2.

Cortial, Angèle, Marc Vocanson, Estelle Loubry, and Stéphanie Briançon. 2015. "Hot Homogenization Process Optimization for Fragrance Encapsulation in Solid Lipid Nanoparticles." *Flavour and Fragrance Journal* 30 (6): 467–77. doi:10.1002/ffj.3259.

Dubey, Sunil Kumar, Anuradha Dey, Gautam Singhvi, Murali Manohar Pandey, Vanshikha Singh, and Prashant Kesharwani. 2022. "Emerging Trends of Nanotechnology in Advanced Cosmetics." *Colloids and Surfaces B: Biointerfaces* 214 (March): 112440. doi:10.1016/j.colsurfb.2022.112440.

Fruehwirth, Sarah, Sandra Egger, Thomas Flecker, Miriam Ressler, Nesrin Firat, and Marc Pignitter. 2021. "Acetone as Indicator of Lipid Oxidation in Stored Margarine." *Antioxidants* 10 (1): 1–17. doi:10.3390/antiox10010059.

Gurpreet, K., and S. K. Singh. 2018. "Review of Nanoemulsion Formulation and Characterization Techniques." *Indian Journal of Pharmaceutical Sciences* 80 (5): 781–89.

He, Lei, Jing Hu, and Weijun Deng. 2018. "Preparation and Application of Flavor and Fragrance Capsules." *Polymer Chemistry* 9 (40): 4926–46. doi:10.1039/c8py00863a.

Hejmady, Siddhanth, Rajesh Pradhan, Amit Alexander, Mukta Agrawal, Gautam Singhvi, Bapi Gorain, Sanjay Tiwari, Prashant Kesharwani, and Sunil Kumar Dubey. 2020. "Recent Advances in Targeted Nanomedicine as Promising Antitumor Therapeutics." *Drug Discovery Today* 25 (12): 2227–44. doi:10.1016/j.drudis.2020.09.031.

Jang, M.H., X.L. Piao, J.M. Kim, S.W. Kwon, and J.H. Park. 2008. "Inhibition of Cholinesterase and Amyloid-&bgr; Aggregation by Resveratrol Oligomers from Vitis Amurensis." *Phytotherapy Research* 22 (4): 544–49. doi:10.1002/ptr.

Kaul, Shreya, Neha Gulati, Deepali Verma, Siddhartha Mukherjee, and Upendra Nagaich. 2018. "Role of Nanotechnology in Cosmeceuticals: A Review of Recent Advances." *Journal of Pharmaceutics* 2018 (March): 1–19. doi:10.1155/2018/3420204.

Kaur, Rajnish, Deepak Kukkar, Sanjeev K. Bhardwaj, Ki Hyun Kim, and Akash Deep. 2018. "Potential Use of Polymers and Their Complexes as Media for Storage and Delivery of Fragrances." *Journal of Controlled Release* 285 (July): 81–95. doi:10.1016/j.jconrel.2018.07.008.

Kumar Dubey, Sunil, Rajesh Pradhan, Siddhanth Hejmady, Gautam Singhvi, Hira Choudhury, Bapi Gorain, and Prashant Kesharwani. 2021. "Emerging Innovations in Nano-Enabled Therapy against Age-Related Macular Degeneration: A Paradigm Shift." *International Journal of Pharmaceutics* 600 (March): 120499. doi:10.1016/j.ijpharm.2021.120499.

Lohani, Alka, Anurag Verma, Himanshi Joshi, Niti Yadav, and Neha Karki. 2014. "Nanotechnology-Based Cosmeceuticals" 2014:843687. doi: 10.1155/2014/843687.

Lu, Zhiguo, Tianlu Zhang, Jun Yang, Jianze Wang, Jie Shen, Xiangyu Wang, Zuobing Xiao, Yunwei Niu, Guiying Liu, and Xin Zhang. 2020. "Effect of Mesoporous Silica Nanoparticles-Based Nano-Fragrance on the Central Nervous System." *Engineering in Life Sciences* 20 (11): 535–40. doi:10.1002/elsc.202000015.

Whitepaper on Nano Technology & AI Application in Fragrance Sector. 2020. Technology Cluster Manager Technology Centre Systems Program (TCSP)Office of DC MSME, Ministry of MSME Miastkowska, Małgorzata, and Elwira Lasoń. 2020. "Water-Based Nanoperfumes." *Nanocosmetics* 173–83. doi:10.1016/b978-0-12-822286-7.00007-3.

Miastkowska, Małgorzata, Elwira Lasoń, Elżbieta Sikora, and Katarzyna Wolińska-Kennard. 2018. "Preparation and Characterization of Water-Based Nano-Perfumes." *Nanomaterials* 8 (12). doi:10.3390/nano8120981.

Nguyen, Tuan Anh, and Susai Rajendran. 2020. "Chapter 23- Current Commercial Nanocosmetic Products." In *Micro and Nano Technologies*, edited by Arun Nanda, Sanju Nanda, Tuan Anh Nguyen, Susai Rajendran, and B T Yassine. *Nanocosmetics Slimani*, 445–53. Elsevier. doi:10.1016/B978-0-12-822286-7.00019-X.

Nitheesh, Yanamandala, Rajesh Pradhan, Siddhant Hejmady, Rajeev Taliyan, Gautam Singhvi, Amit Alexander, Prashant Kesharwani, and Sunil Kumar Dubey. 2021. "Surface Engineered Nanocarriers for the Management of Breast Cancer." *Materials Science and Engineering C* 130 (August): 112441. doi:10.1016/j.msec.2021.112441.

Park, Il Kwon, Junheo N. Kim, Yeon Suk Lee, Sang Gil Lee, Young Joon Ahn, and Sang Chul Shin. 2008. "Toxicity of Plant Essential Oils and Their Components against Lycoriella Ingenua (Diptera: Sciaridae)." *Journal of Economic Entomology* 101 (1): 139–44. doi:10.1603/0022-0493(2008)101[139:TOPEOA]2.0.CO;2.

Perinelli, Diego Romano, Giovanni Filippo Palmieri, Marco Cespi, and Giulia Bonacucina. 2020. "Encapsulation of Flavours and Fragrances into Polymeric Capsules and Cyclodextrins Inclusion Complexes: An Update." *Molecules* 25 (24). doi:10.3390/MOLECULES25245878.

Salvador-Carreno, A., and Chisvert, A. 2003. " Perfumes." *Cosmetics And Toiletries* no. 2002: 36–42.

Salvioni, Lucia, Lucia Morelli, Evelyn Ochoa, Massimo Labra, Luisa Fiandra, Luca Palugan, Davide Prosperi, and Miriam Colombo. 2021. "The Emerging Role of Nanotechnology in Skincare." *Advances in Colloid and Interface Science* 293: 102437. doi:10.1016/j.cis.2021.102437.

Siddique, Jamal Akhter, Fahad Mabood Husain, and Manisha Singh. 2022. *Nanoemulsions: Current Trends in Skin-Care Products. Nanotechnology for the Preparation of Cosmetics Using Plant-Based Extracts*. INC. doi:10.1016/b978-0-12-822967-5.00005-9.

Sikora, Elzbieta, Miastkowska Małgorzata, Katarzyna Wolinska Kennard, and Elwira Lason. 2018. "Nanoemulsions as a Form of Perfumery Products." *Cosmetics* 5 (4): 1–8. doi:10.3390/cosmetics5040063.

Thornfeldt, Carl. 2005. "Cosmeceuticals Containing Herbs: Fact, Fiction, and Future." *Dermatologic Surgery : Official Publication for American Society for Dermatologic Surgery* 31 (7 Pt 2): 873–81. doi:10.1111/j.1524-4725.2005.31734.

Vijaya, N., T. Umamathi, A. Grace Baby, R. Dorothy, Susai Rajendran, J. Arockiaselvi, and Abdulhameed Al-Hashem. 2020. "Nanomaterials in Fragrance Products." *Nanocosmetics* 247–65. doi:10.1016/b978-0-12-822286-7.00012-7.

Wei, Min, Xi Pan, Lin Rong, Aijun Dong, Yunlu He, Xuyan Song, and Junsheng Li. 2020. "Polymer Carriers for Controlled Fragrance Release." *Materials Research Express* 7 (8). doi:10.1088/2053-1591/aba90d.

15 Lipid nanoparticles as a cosmetic delivery system

Ozge Inal, Ulya Badilli, Gulin Amasya and Nilufer Tarimci

CONTENTS

DOI: 10.1201/9781003319146-15

15.1 INTRODUCTION TO NANOTECHNOLOGY

Cosmetics have been used in all civilizations for centuries. Many cosmetic items are produced in conventional dispersion systems such as emulsions and creams due to their ease of use. However, nowadays, the perception of cosmetics differs from person to person and the variation in expectations with the development of cultural and economic conditions have led the cosmetics industry to find new active components and new technologies. These new active ingredients are named *cosmeceuticals* or *dermocosmetics*. Cosmeceuticals are topical agents that are lying in a broad spectrum, between cosmetics and drugs. They can take place in both categories, and they should contain components with specific activity and should deliver these components to the skin to achieve the best performance (Tarimci, 2003; Duarah et al., 2016; Kaul et al., 2018).

Nanotechnology, which was first used as a term by Norio Taniguchi in 1974 finds its place in many areas of life today (Hulla et al., 2015). Nanotechnology has led significant developments in electronics, medical sciences, pharmaceutical sciences, cosmetics, textile, and food industries in the last 30 years. The cosmetics industry was among the first industries to incorporate nanoparticles in its products. Nano-based technology has been used in cosmetic science since 1961 and it takes an important place to enhance the efficacy of cosmetic products (Epstein, 2011; Mihranyan et al., 2012). There are many advantages of nanosized cosmetic delivery systems such as longer-lasting fragrances and perfumes and increased efficiency in anti-aging creams compared to conventional products. In addition, the enhanced stability and controlled release of cosmetic active ingredients (cosmeceuticals) can be obtained with nanoencapsulation of the molecules (Lohani et al., 2014).

15.2 NOVEL NANOCARRIER SYSTEMS FOR COSMECEUTICALS

Novel and innovative cosmetic delivery systems may improve percutaneous penetration of the cosmetic actives as well as their duration into the skin tissue. In addition, they enable their distribution to the targeted skin layer. With nanocarrier systems, the stability and the appearance of the product can be improved and shelf life can be extended. The most important advantage of the transport of cosmetic active substances with nano-delivery systems is gaining controlled release. In this way, localization of substances in the epidermis or dermis can be achieved without reaching the systemic circulation (Tarimci, 2003; Patravale and Mandawgade, 2008; Pradhan et al., 2013). For example, for sunscreen products to be effective as long as possible, they should be kept on the skin, not removed by the effect of water, and at the same time, the substances they contain should not enter the systemic circulation by trans-dermal delivery. On the other hand, the expected effect of anti-aging compounds depends on their ability to reach the living epidermis and dermis. To fulfill the intended performance, products including cosmeceuticals should be encapsulated in

TABLE 15.1

Benefits and examples of controlled release delivery systems

Benefits of delivery systems	Cosmetic examples
Overcoming ingredient incompatibilities	Oil-soluble active ingredients can be prepared into clear and aqueous systems
Protecting from oxidation of active ingredients	Ingredients subject to oxidation, such as vitamin C and vitamin E can be protected from air
Lengthening shelf life of final products	Protection of oxidation of certain active ingredients prevents discoloration of final products
Improving aesthetics	Encapsulation of glycerin reduces tack and allows its incorporation into dry powdery systems
Reducing irritancy of key ingredients	Encapsulation of AHAs
Providing continuous benefits instead of a single, limited exposure	Anti-irritants can soothe skin for sustained periods instead of a single, limited application

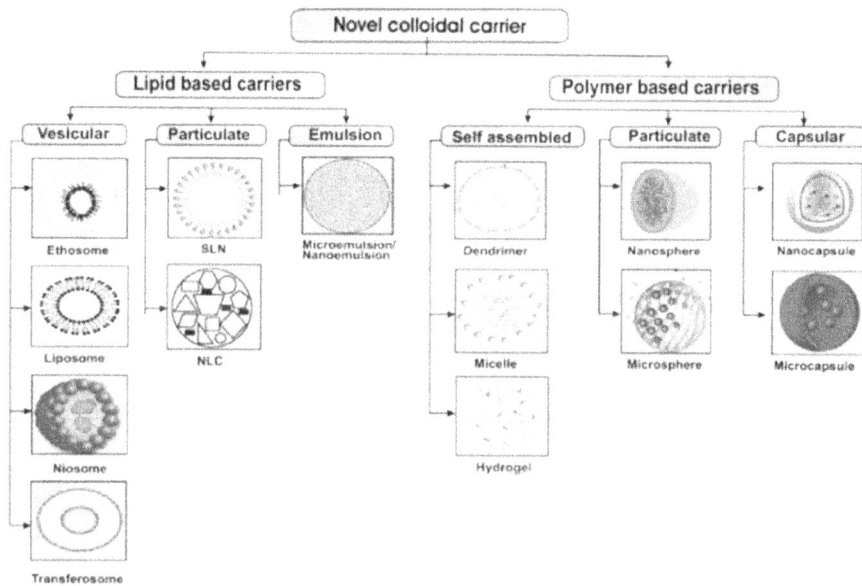

FIGURE 15.1 Nanocarrier colloidal systems (Modified from Moshawih et al., 2019)

a proper nanocarrier delivery system. As indicated in Table 15.1, we can summarize the benefits of the novel controlled release delivery systems (Tarimci, 2011).

The nanocarriers are colloidal systems and they have lipidic or polymeric structures (Figure 15.1). The lipid-based systems can be divided into three broad types:

– Emulsion systems (nanoemulsions)
– Vesicular systems (liposomes, niosomes, and other patented systems)
– Particulate systems (SLN, NLC, and semi-solid lipid nanoparticles)

Nano-based technology gives cosmetic products additional properties related to their enhanced efficacy. For example, sunscreen products containing nano-sized particles of ZnO or TiO_2 offer significantly better performance than those containing micron-sized particles (Smijs and Pavel, 2011). However, the safety assessments of these products should be investigated carefully.

The cosmetics industry is one of the first industries to use nanoparticles in its products (Zhou et al., 2021). The first generation of lipid nanoparticles called solid lipid nanoparticles (SLNs) was developed at the beginning of the 1990s as an alternative carrier to the other nano systems such as liposomes, nanoemulsions, and polymeric nanoparticles. Nowadays, lipid base nanocarriers have been the subject of many studies on topical drug delivery and cosmetics (Souto and Müller, 2008; Puglia et al., 2012). They are submicron colloidal carriers, and their size is between 40 and 1000 nm. Nanocarrier systems have become increasingly important carriers for selectively targeting cosmetic active ingredients to targeted skin layers and cells (Escobar-Chávez et al., 2012; Sengar et al., 2018).

Lipid nanoparticles can be classified into two groups as seen in Figure 15.2.

1 Lipid nanoparticle dispersion systems
   • Solid lipid nanoparticles (SLNs)
   • Nanostructured lipid carriers (NLCs)
2 Semi-solid lipid nanoparticle dispersion systems
   • Semi-solid SLNs
   • Semi-solid NLCs

15.2.1 SOLID LIPID NANOPARTICLES (SLNs)

Conventional dosage forms such as ointments, gels, and creams that are frequently used for dermal drug delivery contain active ingredients in pure form. However, low drug penetration into the skin can frequently occur with high variations. Novel drug delivery systems are being developed extensively in order to overcome these

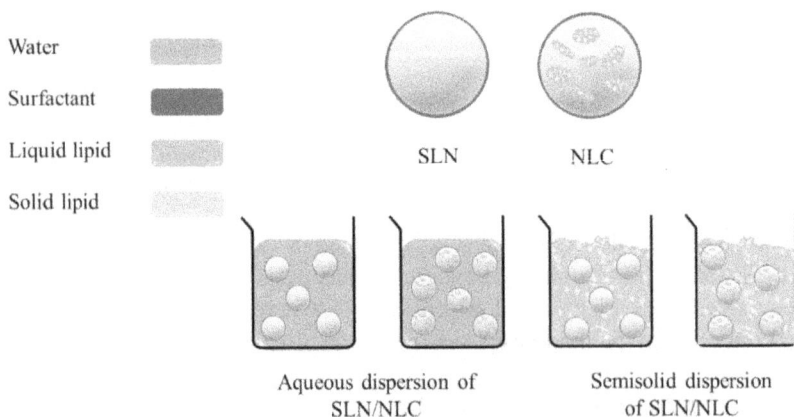

FIGURE 15.2. Different types of lipid nanoparticles

drawbacks. SLNs are nanocarriers formulated as colloidal drug delivery systems as an alternative to liposomes, lipid emulsions, polymeric nanoparticles, and others at the beginning of 1990 (Khater et al., 2021). They play an important role to improve the effectiveness of cosmetics (Zhou et al., 2021). It is known that cosmetic products prepared with lipid nanoparticles are quite capable of protecting the skin against harmful radiations, and they are also very effective products for anti-aging treatment. SLNs are systems formed by replacing the liquid lipid part of oil/water emulsions with lipids that are solid at room and body temperature. Physiologically biocompatible lipids such as triglycerides (tristearin etc.) fatty acids (stearic acid etc.), steroids (cholesterol, etc.), and waxes (cetyl palmitate, etc.) are used as solid lipids. The solid core of SLNs is hydrophobic with a monolayer coating of phospholipids as surfactant and the drug is usually dispersed or dissolved in the core as shown in Figure 15.3.

SLNs can be a successful drug delivery agent for several reasons. Their major advantages can be summarized as follows (Müller et al., 2002; Ramteke et al., 2012; Severino et al., 2012; Badilli et al., 2018; Garcês et al., 2018):

- High drug loading capacity
- Encapsulation of lipophilic and hydrophilic active ingredients
- Enhanced skin hydration with occlusive effect
- Protection of active substance against degradation
- Improving the penetration of active substances into skin layers
- Possibility of drug targeting
- High storage stability
- Suitability for large scale production

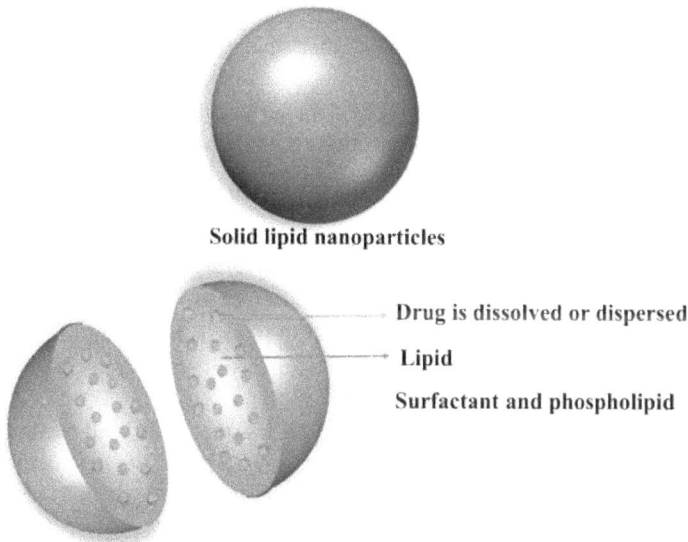

Solid lipid nanoparticles

Drug is dissolved or dispersed

Lipid

Surfactant and phospholipid

FIGURE 15.3 The general structure of SLN loaded with drugs (Mishra et al., 2018)

- Possibility of prolonged or controlled drug release
- Possibility of drying by lyophilization
- Ease in sterilization
- Biocompatible and biodegradable compositional ingredients
- Low toxicity

In addition to these benefits, SLNs also have some disadvantages. Major drawbacks of SLNs are:

- Different colloidal compositions can occur in the dispersion medium.
- Various factors, such as the interaction of drug and lipid melt, and the structure or condition of the lipid matrix, are effective parameters for the encapsulation of drugs in SLNs.
- Particle growth and aggregation may be found during long-term storage.
- Leakage of cosmetic actives from the carrier may occur as a result of the polymorphic transitions during storage.
- SLN dispersions include high water content in their structure (70–90%).

15.2.2 NANOSTRUCTURED LIPID CARRIERS (NLCs)

NLCs were developed as a second-generation technology to overcome the drawbacks of SLNs (Salvi and Pawar, 2019). NLCs are mixtures of solid and liquid lipids and exhibit a non-ideal crystal structure due to this mixture. Thus, they prevent drug expulsion by preventing the perfect crystallization of lipids. NLCs have all the advantages of SLNs. Since the liquid lipid–solid lipid mixtures are used, NLCs have a higher drug loading capacity. In addition, they are more stable than SLNs because of the imperfect solid matrix form of the lipid nanoparticles. It is possible to achieve modulating drug release profile with a semi-solid formulation containing NLC-embedded cosmetic active substances. Another advantage of NLCs over SLNs is less drug leakage during storage.

NLCs can be of three different types depending on the lipid matrix structure: imperfect type, multiple type, and amorphous type (Shah et al., 2015a; Garcês et al., 2018).

- (Type I) – Imperfect type NLCs are prepared with a mixture of solid lipids and small amounts of liquid lipids. This type of NLCs have high drug loading efficiency.
- (Type II) – In contrast, the amount of oily lipids is higher in multiple type NLCs. The lower solubility of lipophilic drugs in solid lipids explains this situation.
- (Type III) – In the amorphous type of NLCs, the solid lipid matrix is not in the crystalline state. This type of NLCs contain additional specific lipids such as hydroxyl octacosanol, isopropyl myristate, and hydroxyl stearate. In this way, drug expulsion can be prevented.

The structural differences between SLNs and NLCs are shown in Figure 15.4.

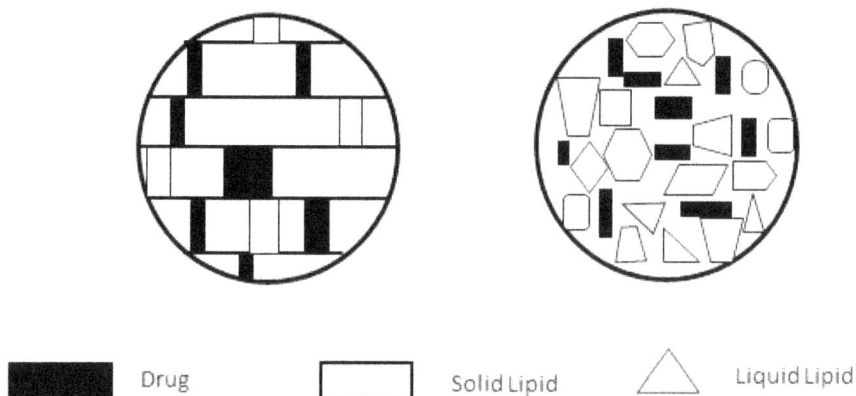

FIGURE 15.4   Structural differences between SLNs and NLCs (Pardeike et al., 2009)

15.2.3   SEMI-SOLID LIPID NANOPARTICLE DISPERSION SYSTEMS

Lipid nanoparticles are in low viscosity aqueous dispersion structure, which is not suitable for cutaneous application. After the production of SLN and NLCs, a second formulation step is required for their application to the skin as a suitable finished product. For this reason, different methods can be applied such as:

1 The SLN/NLC aqueous dispersion can be added to a previously prepared semi-solid base (Müller et al., 2002; Pardeike et al., 2009).
2 As another option, first of all, SLN/NLC is lyophilized and obtained as a dry powder. In the second step, to achieve the desired consistency for cosmetic application, the nanoparticles are incorporated into a conventional semi-solid base (cream or gel) (Souto et al., 2004; Amasya et al., 2019; Amasya et al., 2020).
3 Another method is to add a hydrogel agent such as carbomer or cellulose derivatives to the aqueous phase of the lipid nanoparticle dispersion. In this case, a nano-emulgel formulation is obtained. However, these extra processing steps increase the cost and lengthen the processing time. On the other hand, some problems may be encountered with these methods. For example, creams are heterogeneous systems with a tendency to stability problems, or incompatibility between the ingredients of the semi-solid base and SLN dispersion components can occur (Lippacher et al., 2001, 2002; Badilli et al., 2017).

Semi-solid lipid nanoparticle dispersions were first described by Lippacher and co-workers in 2001 to overcome these drawbacks (Lippacher et al., 2001). SLN formulations that have a semi-solid consistency were developed using a high-pressure homogenization technique and a one-stage production method. Later on, a few studies concerning the one-step production of semi-solid lipid nanoparticle dispersions have been published (Lippacher et al., 2004; Teeranachaideekul et al., 2008; Wu et al., 2011). In the study by Lippacher et al. (2004), it was reported that the particle size of SLNs formulated in a gel did not change for 6 months. In the light of these studies, our research group developed semi-solid SLN and semi-solid NLC formulations as drug and cosmetic delivery systems (Badilli et al., 2015, 2017; Tarimci et al., 2015; Amasya, 2021; Amasya et al., 2021).

As the latest member of the lipid-based nanoparticles, the semi-solid SLN and NLC systems have many advantages. The most important advantage of semi-solid lipid nanoparticle systems is that it allows obtaining a finished product suitable for cutaneous application by creating a semi-solid consistency formula in a single step (Tarimci et al., 2015). Since it does not need to be formulated in a semi-solid dosage form again, it can be produced in a shorter time and more economically. Finally, semi-solid lipid nanoparticle dispersion systems have great importance for dermal applications of drugs and cosmetics.

15.3   INGREDIENTS OF LIPID NANOPARTICLES

The benefits of SLNs and NLCs come from their ingredients. Both natural and synthetic ingredients used in the production of lipid nanoparticles are classified as "Generally Recognized as Safe" (GRAS) due to their low toxicity and well tolerability by the body (Badilli et al., 2018; Dhiman et al., 2021). Besides, the adhesive and occlusive properties that come through lipid ingredients provide a protective film on the skin, therefore hydration can be achieved (Zielińska and Nowak, 2016; Newton and Kaur, 2019). Lipids and surfactants are the main ingredients of lipid nanoparticles. Co-surfactants, preservatives, cryoprotectants, and charge modifiers can also be used in production (Dhiman et al., 2021).

15.3.1   LIPIDS

In case of SLNs, lipids that are solid at body temperature are used to provide a matrix structure. Lipids such as mono, di, and triglycerides, fatty acids, steroids, and waxes are frequently used in the production of SLNs (Badilli et al., 2018; Dhiman et al., 2021).

In different from the SLNs, liquid lipids are also used in the production of NLCs, therefore, the quantity and the stability of the encapsulated active ingredients can be enhanced. The use of liquid lipids provides a melting point relatively less than the solid lipid; however, the lipid matrix is still solid at human body temperature, and the solubility of the active compounds increases in liquid lipids. Either NLCs can be produced with the addition of a small amount of liquid oils (Type I and Type II) or they can be derived from W/O/W type multiple emulsions by the addition of a high amount of liquid lipids into the solid lipids (Type III). The solid lipids used for preparing NLCs are similar to SLNs. Caprylic/capric triglycerides, lauroyl polyoxylglycerides, vitamin E derivatives, oleic acid, soy lecithin, and squalene are some of the liquid lipids used in NLC formulations (Badilli et al., 2018).

The choice of lipid affects formulation parameters such as loading capacity, stability, and release behavior, and it depends on the solubility and partition coefficient of the active ingredient in the lipid, and the crystalline forms of solid lipids. Multiple crystallines in solid lipids can be advantageous in loading efficiency. In addition, the existence of a perfect crystalline lattice structure, due to thermodynamic stability, affects the lipid nanoparticle properties. For example, long fatty acid chains crystallize more slowly than short fatty acid chains. Waxes are physically stable; however, due to high crystallinity, they exhibit higher drug leakage. The β-forms of triglycerides are good examples of stable forms (Shah et al., 2015a). Generally, the use of liquid lipids in NLCs is advantageous in eliminating the lipid crystallinity and polymorphism problems.

15.3.2 Surfactants

Surfactants help in dispersing the melted lipids in the aqueous phase during the production and provide nanodispersion stabilization by either increasing the electrostatic stability (ionic types) or affecting the steric repulsion stability (non-ionic types) after cooling (Shah et al., 2015a; Badilli et al., 2018). In addition, the solubility of hydrophobic molecules in the central solid-lipid core of SLNs is increased in the presence of suitable surfactants. More than one surfactant can be used for the prevention of particle aggregation in the dispersion of lipid nanoparticles (Dhiman et al., 2021). Surfactants can be selected according to their HLB values, effects on the parameters such as lipid modification, particle size reduction, and *in vivo* lipid degradation (Shah et al., 2015a).

One of the most used surfactants is nonionic triblock copolymers (Poloxamer types) consisting of a central hydrophobic polypropylene oxide chain surrounded by two hydrophilic polyethylene oxide chains. Non-ionic polysorbate esters (Tween® types) are also frequently used. Non-ionic surfactants are preferred due to their low toxicity, low irritation, and effectiveness in *in vivo* lipid matrix degradation. Amongst the ionic surfactants, amphoteric types are less toxic. Phospholipids and phosphatidylcholines are amphoteric surfactants, which are commonly used in lipid nanoparticle preparation. Lecithin species, fatty acid esters, sucrose derivatives, polyvinyl alcohol, and bile salts are also among those used (Shah et al., 2015a). Table 15.2 shows lipids and surfactants frequently used in the production of cosmetic SLNs and NLCs.

15.3.3 Other ingredients

Sodium dodecyl sulfate, sodium oleate, sodium glycocholate, or butanol are some examples, which are optionally used as co-surfactant in lipid nanoparticle production. Cryoprotectants such as PVA, PVP, glycine, glucose, lactose, sorbitol, mannitol, gelatine; charge modifiers such as dipalmitoyl phosphatidyl choline, stearyl amine, diacetyl phosphate, and dimyristoyl phosphatidylglycerol can also be used in the production of SLNs (Dhiman et al., 2021).

TABLE 15.2

Frequently used ingredients in the production of cosmetic lipid nanoparticles

Lipids	Surfactants
Triglycerides	**Ionic Surfactants**
Tristearin (Glyceryl tristearate)	Sodium cholate
Tripalmitin (Glyceryl tripalmitate)	Sodium glycocholate
Trilaurin (Glyceryl trilaurate)	Sodium taurocholate
Trimyristin (Glyceryl trimytistate)	Sodium taurodeoxycholate
Tribehenin (Glyceryl tribehenate)	Sodium oleate
Tricaprylin (Glyceryl tricaprilate/caprate)	Sodium dodecyl sulfate
Diglycerides	**Amphoteric Surfactants**
Glyceryl palmitostearate	Phosphatidylcholines (egg/soy)
Glyceryl dibehenate	Hydrogenated phosphatidylcholines (egg/soy)
Monoglycerides	Phospholipids (egg/soy)
Glyceryl oleate	Imidazoline derivatives (Miranol® Ultra C32)
Glyceryl monostearate	**Nonionic Surfactants**
Glyceryl behenate	Polysorbate types (20/80)
Fatty acids	Sorbitan monooleate/ monostearate/tristearate
Stearic acid	Poloxamer 188/407
Palmitic acid	Poloxamine 908
Myristic acid	Brij78/polyoxyethylene
Waxes	Polyglyceryl-3 Methyl glucose distearate
Cetyl ester waxes/Cetyl palmitate	PEG-15 hydroxy stearate
Carnauba	**Co-surfactants**
Beeswax	Tyloxapol
Ceresin	Sodium dodecyl sulfate
Liquid lipids	Sodium oleate
Caprylic and Capric triglycerides	Taurocholate sodium salt
Isopropyl myristate	Sodium glycocholate
Isopropyl palmitate	Butanol
α-tocopherol/Vitamin E	
Oleic acid	
Squalene	
Hydroxystearic acid	
Soya bean oil	

Source: Modified from Shah et al., 2015a.

15.4 PREPARATION METHODS FOR LIPID NANOPARTICLES

The choice of the production method depends on the physicochemical properties and stability of the active ingredient that will be encapsulated and the characteristics of the lipid nanoparticles (Shah et al., 2015b; Zieli´nska and Nowak, 2016). According to their requirements, production methods can be classified into three different groups: (i) high energy requirements such as high-pressure or high-shear/speed homogenization types and supercritical CO_2 extraction methods, (ii) low energy requirements such as microemulsion, double emulsion, membrane contractor methods, (iii) organic solvent requirement such as solvent evaporation, solvent-emulsification-diffusion, solvent injection methods (Dhiman et al., 2021).

Among these methods, hot or cold homogenization with high-pressure and high-speed homogenization methods are most frequently used in cosmetics due to the ease of the scale-up process in large-scale production lines eliminating the possibility of regulatory problems (Souto and Müller, 2007; Pardeike et al, 2009; Abbas et al., 2018; Anderluzzi et al., 2019; Lee and Nam, 2020; Rubiano et al., 2020). Repka and co-workers patented an invention for the continuous pre-emulsion process with the US patent number 20170172937A1 in 2017 for the production of SLNs with HPH (Paliwal et al., 2020). Table 15.3 summarizes the advantages, disadvantages, and general requirements of the methods highly used in the production of lipid nanoparticles in cosmetic delivery.

TABLE 15.3

General advantages, disadvantages, and requirements of frequently used methods in the production of lipid nanoparticles in cosmetics

Method	Advantage	Disadvantage	Requirements
HPH	• Effective in particle size reduction • Suitable for high lipid concentration • Low contamination • Reproducible • Ease of scaling-up • Satisfies with the regulatory aspects	• Energy consuming Hot HPH • Not suitable for thermo-sensitive ingredients • Inappropriate for hydrophilic ingredients Cold HPH • Larger particle size and distribution • Active agent leakage upon storage	• High energy • Strong cavitation forces • Pre-emulsion stage Hot HPH • High temperature Cold HPH • Low temperature • Rapid cooling requires dry ice or liquid nitrogen
High shear Homogenization	• Effective soft and hard particles • Reduced shear stress • Reduced agglomeration • Controllable and repeatable results via ultrasonication • Large amount of surfactant usage is avoided • Suitable for lab-scale	• Physical instability, such as particle growth upon storage • Potential metal contamination • Difficulty in scale-up • High polydispersity • Poor encapsulation efficiency	• High energy • Cavitation forces • Both ultrasonication and high-speed homogenization are required for size reduction. • Temperature over the melting point • Gradual cooling below the crystallization temperature

(Continued)

TABLE 15.3 (Continued)

Method	Advantage	Disadvantage	Requirements
Microemulsion	• No energy needs • Low mechanical stress • Theoretically stable • The microemulsion itself is the low size range	• Low lipid content • Excess water must be removed. • Extremely sensitive to change.	• Controlled environmental • Co-emulsifiers are required
Solvent Emulsification-Evaporation	• Continuous process • Avoids high temperature • Commercially successful method • Suitable for scale-up	• Use of organic solvent • High energy is applied • PDI may be high • Biomolecules may get damaged • Further homogenization by HPH may be needed	• Organic solvent • Homogenizer • Reduced pressure • Rotaevaporator can be required • Co-surfactants can be required
SCF method	• Minimizes the solvent usage • Dry particles are obtained	• Cost ineffective	• Supercritical fluid • Mild pressure

Shah et al. (2015b); Zieliʹnska and Nowak (2016); Newton and Kaur (2019).

15.4.1   High-pressure homogenization (HPH) method

HPH method depends on the production of 100–2000 bar pressure for pushing the liquid/emulsion through a narrow gap in the micron range with a very high velocity (1000 km/h) (Zieliʹnska and Nowak, 2016, Newton and Kaur, 2019; Khater et al., 2021). The mechanical principle of HPH is the particle break- down to nano range with the shear stress and forces through the cavity. Generally, the use of 5–10% (up to 40%) lipid content is reported in the HPH method. HPH is suitable for scale-up processes and eliminates the use of organic solvents. Therefore, it is highly preferred in cosmetic lipid nanoparticle production (Müller et al., 2000; Lippacher et al., 2001; Badilli et al., 2018). There are two types of HPH, hot homogenization and cold homogenization. In both cases, as a first step active ingredient is incorporated into melted lipids by dissolving or dispersing. Figure 15.5 shows the mechanical principles of hot and cold HPH (Svilenov and Tzachev, 2014).

15.4.1.1   Hot homogenization method

In this type of HPH, a temperature 5–10°C higher than the melting point of the lipid and a high-speed homogenizer such as Ultraturrax must be used for producing a pre-emulsion by dispersing the lipid phase into a hot aqueous phase containing

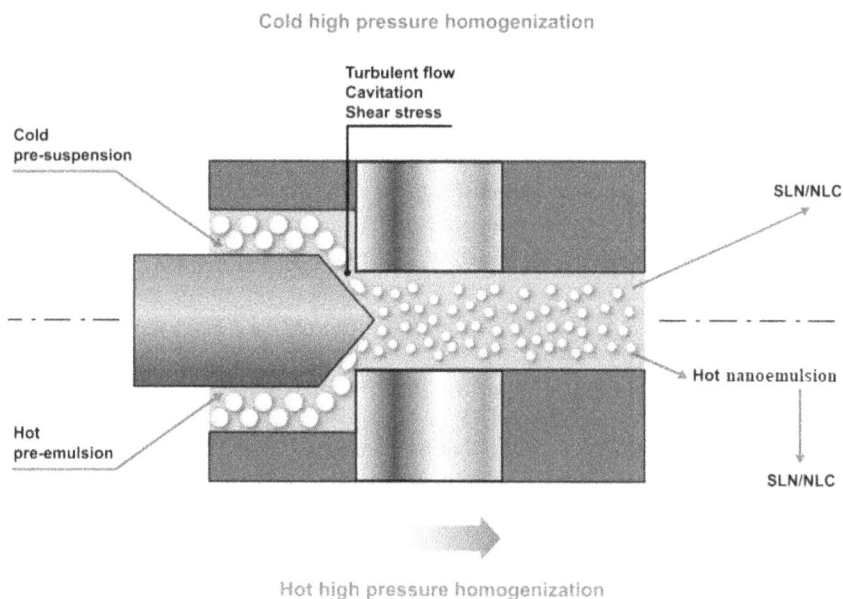

FIGURE 15.5 Mechanism of cold and hot HPH methods (Adopted from Svilenov and Tzachev, 2014)

a surfactant. Later, a piston-gap or a jet-stream homogenizer is used to produce a high shearing force to provide a colloidal emulsion which afterward gradually cooled down to room temperature to provide the formation of nanoparticles. The homogenization stage can be repeated a few times, but in each stage, the temperature increase of approximately 10°C/500 bar must be taken into account (Shah et al., 2015b; Zieliʹnska and Nowak, 2016; Abbas et al., 2018; Badilli et al., 2018; Newton and Kaur, 2019). The lipid-surfactant ratio in pre-emulsion and homogenization parameters such as pressure, temperature, homogenization time, and viscosity highly affects the particle size of lipid nanoparticles (Shah et al., 2015b).

15.4.1.2 Cold homogenization method

Unlike hot homogenization, in cold HPH, lower temperatures are used in melting lipids to ensure a homogeneous lipid mixture. This mixture is rapidly cooled, and the resulting solid lipid mixture is milled to 50–100 μm microparticle size and later dispersed into a cooled surfactant solution prior to homogenization. This method is advantageous for temperature-sensitive active agents and when the active agent is dispersed into the aqueous phase during homogenization (Zieliʹnska and Nowak, 2016; Newton and Kaur, 2019). The choice of lipid is important in method selection. In addition, it is important to comprehensively lower the homogenization temperature to prevent lipid dissolution during the process (Zieliʹnska and Nowak, 2016; Badilli et al., 2018). The particle size produced by this method is larger than hot homogenization and has a wider size distribution (Newton and Kaur, 2019).

15.4.2   HIGH SHEAR HOMOGENIZATION (AND/OR ULTRASONICATION) METHOD

In this method, the melted lipids are emulsified with a high-shear homogenizer in a hot aqueous solution of surfactant that is maintained at the same temperature as the lipid phase. The process is based on the cavitation forces in which a decrease in pressure induces a rapid phase transition (Zieli´nska and Nowak, 2016). To achieve smaller particle size and increase uniformity, further sonication with an ultrasonic homogenizer is required. Lipid nanoparticles are obtained by cooling the hot nano-emulsion (Badilli et al., 2015, 2018; Andreani et al., 2020; Caldas et al., 2021). The increased surface area with the decrease in the particle size/distance affects the rheo-logical properties of the liquid, thus improving the emulsion stability. This method is suitable for the comminution of both soft and hard particles. In terms of ensuring the effectiveness of the method, it is important to provide a narrow polydispersity index to achieve a lower agglomeration (Zieli´nska and Nowak, 2016). The advantages of using an ultrasonic homogenizer are that the equipment has stationary parts, which reduce frictional wear and cleaning time, and that it provides full control over the operational parameters that affect the cavitation (Zieli´nska and Nowak, 2016).

15.4.3   MICROEMULSION METHOD

It is based on rapid crystallization due to the rapid solidification of the microemulsion. The warm microemulsion is dispersed into cold water by continuous stirring and lipid nanoparticles are obtained due to the precipitation of the lipid phase. An oil-in-water microemulsion is prepared from lipids or low-melting fatty acids like stearic acid, emulsifiers such as polysorbate 20, and co-emulsifiers and generally diluted with water at the ratio of 1:25 or 1:50. There is no energy need as the dispersed droplets in the microemulsion phase are readily in the nanometer range. In case of an excess amount of water in comparison with particle content, some need to be removed by using ultra-filtration or lyophilization techniques (Zieli´nska and Nowak, 2016; Badilli et al., 2018; Khater et al., 2021). Rapid lipid crystallization is facilitated with the high-temperature gradient and aggregation is prevented. In the microemulsion method, lipid contents are considerably lower than in the HPH methods. In this method, nanoparticles can be produced with hydrophilic solvents, while larger-sized nanoparticles are produced with lipophilic solvents (Shah et al., 2015b; Zieli´nska and Nowak, 2016; Newton and Kaur, 2019). Shah et al. (2014) also defined the use of microwaves as a heating mechanism in the microemulsion method, in which all ingredients were heated in the same vessel prior to dispersing into cold water.

15.4.4   DOUBLE EMULSION (W/O/W) METHOD

The mechanism depends on the solidification of the emulsion. The main advantages are the low energy input and no need for special equipment; however, the lipid content used is lower in this method (Shah et al., 2015a,b). It is suitable for encapsulating hydro-philic materials into lipid nanoparticles to prevent the drug partition to the outer aqueous phase. Active agents are added into the inner aqueous phase to form a primary water-in-oil emulsion with an organic solvent, which is later emulsified in the external aqueous phase for producing a double (w/o/w) emulsion. Lipid nanoparticles are obtained after the evaporation of the solvent (Amasya et al., 2016; Badilli et al., 2018).

15.4.5 SOLVENT EMULSIFICATION-EVAPORATION (SOLVENT EVAPORATION)

This method depends on dissolving the active agent and lipid phase in an organic solvent, which is immiscible with water. Cyclohexane, chloroform, and dichloromethane are some of the solvents frequently used (Dhiman et al., 2021). A high-speed homogenizer is used for emulsification and the organic solvent is evaporated under reduced pressure such as 40–60 mbar to precipitate the lipids. The average particle size is affected by the concentration of lipids in the organic phase. This method is advantageous for thermo-sensitive materials; however, organic solvent residue can remain (Zielińska and Nowak, 2016).

15.4.6 SOLVENT EMULSIFICATION-DIFFUSION METHOD

SLNs can also be prepared via the solvent emulsification-diffusion method, in which the lipid-containing solvent is added into the co-surfactant containing aqueous phase by dropping. The resultant mixture produced via homogenizers is subjected to rota-evaporation under reduced pressure to form nanoparticles. The advantage of this method is the use of low temperatures; however, organic solvents are used in production. Lipid content, organic phase concentration, emulsifier type, and the HLB value affect the particle size (Newton and Kaur, 2019).

15.4.7 SOLVENT INJECTION METHOD

The solid lipid is dissolved in water-miscible solvents to form a lipid solution that will be rapidly injected into an aqueous phase under continuous mixing and filtering. This aqueous phase may contain a surfactant. The use of pharmaceutically acceptable solvents such as ethanol, isopropyl alcohol, and acetone and the fact that it is a fast production method provides an advantage over other methods using organic solvents (Badilli et al., 2018; Dhiman et al., 2021). Figure 15.6 shows the differences between solvent-based methods (Svilenov and Tzachev, 2014).

15.4.8 SUPERCRITICAL FLUID EXTRACTION (SCF) METHOD

This method depends on using supercritical fluid as the extracting solvent. The most commonly used supercritical fluid is carbon dioxide, which avoids the use of solvents and has an accessible critical point. Moreover, as the method permits the use of mild pressure and temperature conditions, it's environmentally acceptable (Zielińska and Nowak, 2016; Amoabediny et al., 2018). A type of this method known as "Rapid expansion of supercritical solution" (RESS) is used in SLN preparation, in which the fluid is saturated by dissolving the solute. Afterward, the solution is depressurized by passing the solution through a nozzle and allowing the solute to precipitate. Particle gas-saturated solution (PGSS), aerosol solvent extraction solvent, and supercritical fluid extraction of emulsion (SFEE) are some other methods used in the production of lipid nanoparticles.

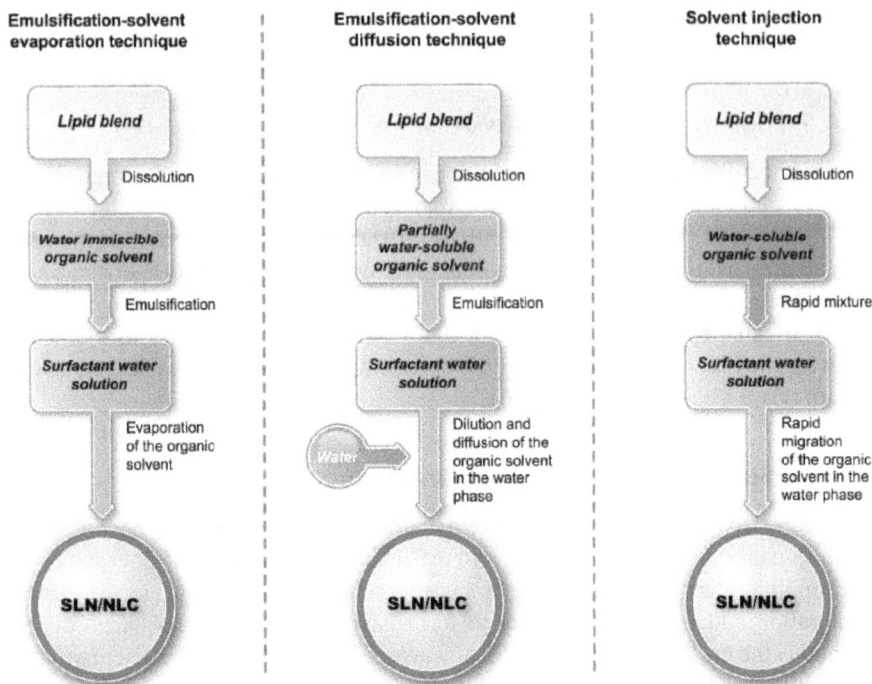

| Emulsification-solvent evaporation technique | Emulsification-solvent diffusion technique | Solvent injection technique |

FIGURE 15.6 Differences between solvent-based methods (Svilenov and Tzachev, 2014)

15.4.9 MEMBRANE CONTRACTOR METHOD

It is an advantageous method as it allows for the mass-scale production of lipid nanoparticles. The method is based on pressing the lipid phase through the membrane pores at a temperature higher than the melting point of the lipids. The lipid droplets formed are removed from the pores by the water phase flowing in a tangential direction with respect to the membrane surface. Lipid nanoparticles are obtained by gradually cooling the mixture to a temperature below the melting point and then to room temperature (Dhiman et al., 2021).

15.4.10 ELECTROSPRAYING METHOD

It is based on the atomization of liquids by electrical forces. As the solution passes through a capillary nozzle, it disperses as small droplets and gains charges due to the electric field. The flow rate of the liquid and the voltage applied effect the particle size. Its advantages are that it is a one-step and reproducible method with low PDI (Badilli et al., 2018).

15.5 CHARACTERIZATION OF LIPID NANOPARTICLES

Nanocosmetics are very up-to-date as they offer great advantages over traditional cosmetic products in providing solutions to skin problems. The unique particle properties reveal their ability to interact with biological cells and enhance the delivery of cosmetic active molecules in different skin layers (Morganti, 2020). Therefore, the particle characteristics of nanocosmetics should be subjected to a series of physical characterization studies, just like drug nanocarrier systems. In general, the physical and biological characterization of lipid nanoparticles as nanocosmetics is important for the quality, efficacy, stability, and safety of the product. The main features that should be focused on for the characterization of lipid nanoparticles can be listed as particle size and distribution, particle morphology, surface properties, and structure, stability, active ingredient release properties, as well as permeability, rheological and occlusive properties (Badilli et al., 2018; Mishra et al., 2018).

Determination of particle size and size distribution is essential to gain information on the colloidal properties of lipid nanoparticles since it was emphasized that the penetration of particles through the skin and skin appendages is directly related to the particle size (Patzelt et al., 2017). Dynamic light scattering (DLS) and electron microscopy are the most widely used methods to describe the particle size and distribution as well as morphological properties of lipid nanoparticles.

In the DLS technique, colloidal dispersion of lipid nanoparticles is used to measure the particle size in terms of hydrodynamic diameter and the polydispersity index of the sample. The method is based on measuring the temporal fluctuations of the scattered laser light by the random multidirectional movement of a particle dispersed in a liquid, known as brown motion (Bhattacharjee, 2016; Stetefeld et al., 2016). Thus, DLS is a fast, reliable, and low-cost measurement tool in which particle sizes from a few nanometers to several micrometers can be determined. The particle size distribution or polydispersity index (PDI) which is expressed numerically between 0 and 1, is closely related to the particle size. The closer the PDI value is to zero, the closer the sizes of the particles in the dispersion are to each other and both particle size and size distribution are directly affected by the lipid nanoparticle composition and preparation method as well. Mehnert and Mader reported that the average particle size increases with higher melting lipids when hot homogenization is used (Mehnert and Mäder, 2001). A similar result was emphasized by Amasya et al. and larger NLC particles were obtained in the presence of Compritol ATO 888 compared to tripalmitin (Amasya et al., 2016). Gordillo-Galeano and Mora-Huertas highlighted that the particle size increases depending on the increase in the total amount of lipid in the dispersion, based on many studies they examined. On the other hand, when the cold method is applied for the high-pressure homogenization process, larger particles and higher PDI values can be obtained compared to the hot homogenization method (Gordillo-Galeano and Mora-Huertas, 2018). While the properties of bulk nanoparticles can be examined by particle size measurement numerically, particles can be observed directly; the image, size, and surface characteristics of a single particle can be visualized through electron microscopy techniques such as transmission electron microscopy (TEM), scanning electron microscopy (SEM), and atomic force microscopy (AFM) (Hallan et al., 2021).

The zeta potential value is another defined particle characteristic, which expresses the certain charge on the surface of the particle. The zeta potential reflects a diffuse electrostatic potential at the electrical double layer surrounding a nanoparticle in solution (Kaszuba et al., 2010). When solid particles are dispersed in the aqueous phase, the particle surfaces become electrically charged. Aggregation is less formed because of the repulsive forces of the charged particles. Therefore, the zeta potential value is important for the stability of colloidal systems during storage. In theory, particles with a high zeta potential either negatively or positively, repel each other and this phenomenon prevents particle aggregation. Hence it can be stated that the zeta potential is an important parameter in the evaluation of the stability of colloidal systems during storage (Patel and Agrawal, 2011). The zeta potential value of the particles either SLN, NLC, or lipid nanoparticle dispersions may affect by lipid ingredients and the cosmetic active molecule encapsulated in the lipid nanoparticles. A study in which the researchers examined the effects of process and formulation parameters on the physicochemical properties of the lipid nanoparticles concluded that the zeta potential values of the SLN formulations were affected by the lipid type. On the other hand, the surfactant also affects the zeta potential (Das et al., 2012). It has been stated by Souto and co-workers that positively or negatively charged particles can interact with skin components in different ways and follow different penetration pathways. So, the zeta potential values of lipid nanoparticles may play a critical role in skin penetration (Souto et al., 2022).

Since the polymorphic form of lipids used in lipid nanoparticles will influence the encapsulation efficiency, the release properties of the cosmetic active molecule, and stability as well, the structural properties of lipid nanoparticles such as crystal structure and the chemical composition are among the basic characterization parameters and should be comprehensively investigated. The differential scanning calorimetry (DSC) technique describes polymorphic transitions and the modification in crystallinity degree of lipid and also the active molecule during the heating and cooling processes of lipid nanoparticle production; while X-ray diffraction (XRD) analysis is used to confirm the crystal structure and the chemical composition at the atomic level (Wu et al., 2011). In both methods, data on the physical and crystallization state of lipids or active substances can be accessed by comparing DSC thermograms or XRD patterns of bulk material with nanoparticles (Das and Chaudhury, 2011). Moreover, spectroscopic methods like Fourier-transform infrared spectroscopy (FTIR) and Raman spectroscopy are frequently used to elucidate the structure of lipids. FTIR, a quantitative analytical method to identify the presence of organic and inorganic compounds in the sample, is frequently employed to detect the possible interaction between drugs and raw materials of lipid nanoparticles. Shifts or diminish in major bands of the bulk lipid, active pharmaceutical ingredient, or other excipients are used to interpret the interaction (Salvi and Pawar, 2019). On the other hand, Raman spectroscopy gives information about the structures and properties of molecules based on their vibrational transitions. Saupe et al. compared SLNs with nanoemulsion and NLC by Raman spectroscopy to investigate the conformational order of hydrocarbon chains and they concluded that the presence of liquid lipid in the structure of NLC did not cause a change in the chain arrangement (Saupe et al., 2006).

The entrapment of active ingredients within the lipid nanoparticle matrix strategy for targeting and transporting the molecules is important for increasing efficacy through percutaneous penetration. Therefore, encapsulation efficiency and loading capacity values are the test parameters that should be examined to determine the amount of cosmetic active ingredient entrapped in the lipid matrix. As mentioned above, while lipid nanoparticles in the form of SLNs and NLCs are obtained as aqueous dispersions at the end of the production processes, semi-solid lipid nanoparticle dispersions have a semi-solid consistency. Detection of the free (un-encapsulated) cosmetic active ingredient in the supernatant by an ultraviolet (UV) spectrophotometer or high-performance liquid chromatography (HPLC) after centrifugation of the lipid nanoparticle dispersion, is called the indirect method. For this method, semi-solid lipid nanoparticles can be dispersed in water or solvent first (Lippacher et al., 2001; Abdel-Salam et al., 2016; Amasya et al., 2021). This is the most frequently used approach to calculate the loading capacity and encapsulation efficiency. On the other hand, as a direct method, the amount of encapsulated active molecules can also be detected by lyophilizing the collected lipid nanoparticles after centrifugation and extracting the loaded active molecule from the dry powder of lipid nanoparticles using solvents. As additional information encapsulation efficiency (EE %) refers to the percentage of active molecules successfully entrapped in the nanoparticle, while the loading capacity (LC %) can be defined as the total amount of entrapped active molecule associated with the total nanoparticle weight (Graverini et al., 2018; Salvi and Pawar, 2019). Besides, how the active ingredient is positioned in the lipid matrix is as essential as the amount of the active ingredient in the lipid matrix. Therefore, Müller and his co-workers defined three different drug incorporation models as a homogeneous matrix, drug-enriched shell, and drug-enriched core, according to the accommodation of the active molecule into the lipid nanoparticle structure (Müller et al., 2000). Both the amount of encapsulated active molecule and the incorporation model of it give information about the *in vitro/in vivo* release properties (Battaglia and Gallarate, 2012). Drug incorporation models of lipid nanoparticles are schematized in Figure 15.7. These models are determined by where the active ingredient is concentrated in the spherical lipid matrix. Burst release is observed in the drug-enriched shell model, where the active substance is concentrated on the particle surface. On the other hand, the extended-release profile can be obtained in the homogeneous matrix where the drug is homogeneously dispersed in the lipid matrix, and in the drug-enriched core model, where the active substance forms a dense core within the lipid nanoparticle. It is known that lipid nanoparticle components and production methods are effective as well as the physicochemical properties of the active ingredient in the incorporation of the active ingredient in the lipid nanoparticle structure (Müller et al., 2002; Üner and Yener, 2007). Investigation of *in vitro* release profile of nanocosmetics is useful for predicting its *in vivo* performance. *In vitro* release studies for nanocosmetics are routinely performed with Franz diffusion cells and the dialysis membrane method.

In addition to the general particle properties mentioned above; there are many advantages that lipid nanoparticles provide in the dermal application. The cosmetic application of lipid nanoparticles is an area of great potential due to their occlusive effect since lipid nanoparticles increase skin hydration through their high occlusive factor. The occlusion of a cosmetic product refers to covering the skin by forming a protective layer on the skin surface. The *in vitro* occlusivity test was first described by Tom De Vringer (De Vringer, 1997). In the application of this test, a beaker filled with a certain weight of water is covered with a membrane or filter paper and sealed carefully to prevent the evaporation of water. When the formulation to be tested is

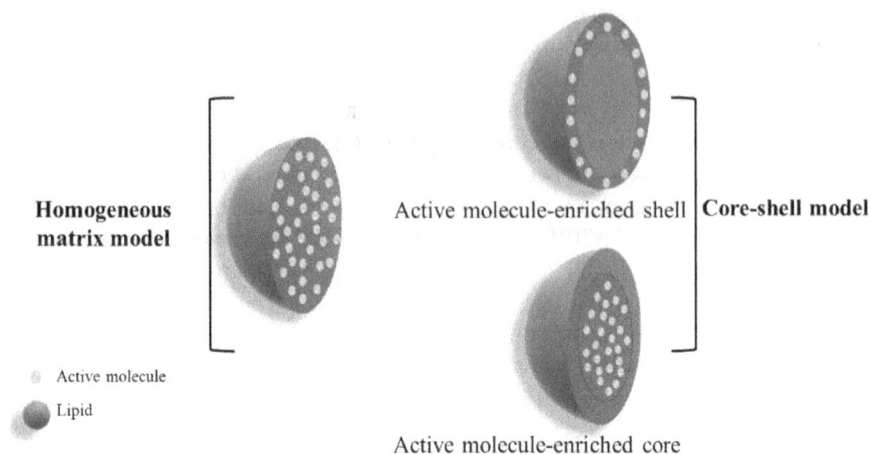

FIGURE 15.7   Drug incorporation models of lipid nanoparticles

applied to the membrane surface without any space, evaporation of the water in the beaker is completely prevented. The occlusivity factor (F) is calculated by detecting the change in the weight of the system depending on time. An experimental setup without any product applied to the membrane is prepared and used as a control group. The F value of each formulation can be calculated with the help of the following equation:

$$F = (A - B) / A \times 100 \tag{15.1}$$

Where A is the water loss of the control and B is the water loss of the formulation. The *in vitro* occlusion test indicates the prevention of water loss. The F value is used to compare formulations and ranges from 0 to 100. Hence the higher the calculated F value, the higher the ability to prevent water loss from the system (De Vringer, 1997; Wissing et al., 2001). In the study of Montenegro et al. in which they examined the correlation between resveratrol-loaded lipid nanocarriers *in vitro* and *in vivo* occlusion factors, they reported that there was a linear relationship between *in vivo* skin hydration and *in vitro* occlusion factor (Montenegro et al., 2017). Therefore, the characterization of *in vitro* occlusive properties of lipid nanoparticles for cosmetic uses can provide information about the effects of lipid nanoparticles on skin hydration without the need for animal and human experiments.

As discussed previously, since SLN and NLC are obtained as aqueous dispersions, a second semi-solid carrier is required for dermal application as well as to have an attractive organoleptic property. Contrarily, both the consistency and the colloidal nature of semi-solid lipid nanoparticles directly turn them into the final product. Therefore, examining the rheological and textural properties of semi-solid lipid nanoparticles; SLNs/NLCs aqueous dispersions, as well as SLNs/ NLCs incorporated in a semi-solid such as cream or gel, is an important assessment criterion in the characterization of lipid nanoparticles. A rheological examination is used in quality control studies of both the intermediate and the finished product and allows prediction of the overall flow behavior, spreadability, and stability of the cosmetic product. A cosmetic product must meet the aesthetic expectations of the consumers while meeting the rheological properties suitable for being taken from the packaging. On the other hand, the texture may be defined as the feel, appearance, or consistency of cosmetic formulation and it is a combination of mechanical, geometrical, and surface attributes of the product (Tai et al., 2014). In many studies, it has been shown that there is a direct correlation between the textural profiles of semi-solid products captured by instrumental devices and their sensory properties. Hence, texture profile analysis (TPA) of cosmetic products is the most preferred method for determining the sensory characteristics of the cosmetic product. Ease of packaging or unpacking, the applicability of semi-solids to the skin surface, consistency, spreadability, or softness can be given as examples of these characteristics, and these properties can be accessed by measuring the parameters such as hardness (firmness), compressibility, cohesiveness, adhesiveness, and elasticity. TPA analysis is mainly used to evaluate the sensory perception of the cosmetic product since the organoleptic properties and the sensory characteristics of the cosmetic product are as important as consumer satisfaction. TPA is a fast and reliable instrumental analysis method in which more than one parameter can be measured at the same time (McMullen et al., 2016; Kowalska et al., 2017; Amasya et al, 2020; César et al., 2020).

According to the World Health Organization (WHO), while penetration is the entry of a substance into a particular layer or structure; permeability is defined as its transport from one layer to a second one (WHO, 2006). As cosmetics are applied to the skin surface daily, the penetration/permeation and therefore the safety of nano-cosmetics in regular use is a big concern. On the other hand, to achieve the desired cosmetic effect, either the active molecule-loaded nanoparticles themselves or encapsulated cosmetic active molecules released from the nanoparticles must pass through the stratum corneum, the barrier layer of the skin. After the barrier property of the stratum corneum is defeated with the help of lipid nanoparticles, the active substance must accumulate in the targeted skin layer without any significant leakage into the systemic circulation and local effect must be achieved (Sengar et al., 2018; Roberts et al., 2021). Therefore, the measurement of skin penetration and permeability of nanocarrier systems, whether for pharmaceutical or cosmetic purposes, should be routinely studied in terms of quality control studies as well as safety and efficacy. Although *in vivo* studies on rodents and human volunteers are being performed for penetration/permeation testing of dermal drug delivery systems, in the case of cosmetics, *ex vivo* and *in vitro* testing are preferred due to recent regulatory changes regarding animal testing. *Ex vivo* studies can be conducted by using excised human cadaver skin or animal skins such as rats, pigs, or monkeys for penetration/permeability testing of dermal nanocarrier systems. But the most current approach is the use of *in vitro*-based non-animal approaches for cosmetic products. Because of the difficulty in obtaining cadaver skin, both ethical concerns and the anatomical differences with animal species have led to the use of synthetic/artificial membranes to characterize the penetration/permeability performance of dermally applied products. Non-lipid-based silicone membranes, the ceramide-derived parallel artificial membrane permeability assay (PAMPA) that can mimic the skin, and the phospholipid vesicle-based permeation assay (PVPA) that can mimic the stratum corneum barrier feature are among the most used artificial in vitro skin models. Many studies have proven to provide a good prediction of in vivo dermal absorption by using synthetic membranes (Casamonti et al., 2019; Zhang et al., 2019; Kim et al., 2020). A vertical or horizontal diffusion cell system is often used alongside a natural or synthetic skin sample to assess drug penetration/permeability through the skin in *in vitro* or *ex vivo* studies. The most classic example of these diffusion cells is the Franz diffusion cell (Hu and He, 2021). The Franz diffusion cell has two chambers separated by a membrane. In the presence of natural or synthetic skin samples, it is possible to determine the amount of substance passing through the skin, flux rates, and layer distribution. At the same time, imaging studies can be carried out with the help of microscope techniques. In particular, the penetration pathway and permeation route can be detected in the presence of fluorescent dye using Confocal Laser Scanning Microscopy. On the other hand, Scanning Electron Microscopy (SEM) and Transmission Electron Microscopy (TEM) are other commonly used imaging techniques.

15.6   SLNS, NLCS, AND SEMI-SOLID LIPID NANOPARTICLES AS COSMETIC NANOCARRIERS

Before examining the exact mechanism of the lipid nanoparticles - skin interaction, it would be useful to mention the structure of the skin and the signs of skin aging. The skin, which consists of three hierarchical layers called epidermis, dermis, and hypodermis from the outside to the inside, is the largest organ of the human body. Therefore, besides having many vital functions; the skin is the most conspicuous part of the body and there is a direct link between human psychology and skin health from the perspective of dermatologists. Having a young and beautiful appearance makes people feel better and has a positive effect on their social behavior. However, skin aging is inevitable due to deterioration of skin functions and structure, both depending on time and environmental factors (Farage et al., 2010). Structural and physiological changes such as dryness, wrinkles, blemishes, loss of elasticity, and coarse-textured skin are the signs of skin aging originating from different layers of the skin. The major cells of the epidermis are keratinocytes, which are produced in the basal layer. Keratinocytes maintain the water balance of the skin and prevent dehydration with the help of keratin they produce and the lipids they secrete. The water content of a healthy epidermis is approximately 20%. During the aging process, the skin loses its ability to retain water and maintain stratum corneum humidity. As a result, the skin surface begins to dry out as the moisture content gradually decreases. However, dryness is not only related to the epidermis but also caused by the moisture loss of the dermis. The melanocytes that arise in the basal layer of the epidermis (deepest part of the epidermis) are responsible for synthesizing the melanin, which is the natural skin pigment. The main task of the melanin pigment is to protect the skin against harmful UV rays. Many pigmentation problems may occur associated with both extrinsic and intrinsic aging processes. In case of hyperpigmentation due to the proliferation of melanocytes and increased amount of melanin, skin whitening or lightening cosmetic agents should reach the deep epidermis. Although there are various types of specialized cells in the dermis, the most common cells are fibroblasts. Fibroblasts are responsible for the synthesis of mainly type I and type III collagen, elastin and fibrillin microfibrils, structural proteoglycan, and hyaluronic acid as well. Therefore, the dermis has a fibrous structure. Collagen is a protein that makes skin cells strong and flexible and collagen fibers in the dermis act as a structure for elastin and hyaluronic acid, which provides skin elasticity and moisture. On the other hand, hair roots originate from the dermis and dermal adipocytes help regenerate the hair follicles and assist wound healing. Mast cells are responsible to secrete vasoactive and proinflammatory mediators, collagen remodeling, and wound healing. Considering the changes in the dermis with aging, dermis thickness decreases due to a time-dependent decrease in cellular turnover and amount of cell number. The deterioration of the cellular integrity of the dermis, the decrease in collagen production, and the decrease in elastin and hyaluronic acid cause a decrease in skin elasticity. Over time, first fine lines and then wrinkles appear (Cernasov, 2009; Nafisi and Maibach, 2018). The deepest layer of the skin, the hypodermis, contains fibroblasts and adipose tissue (fat cells), and the most important function of hypodermis is storing fat or other words storing energy. The most important cosmetic problem observed in the hypodermis and adipose tissue is the enlargement of the fat globules, pushing them towards the dermis and getting stuck between the connective tissue proteins, resulting in the appearance of cellulite. As a result, considering the perception of beauty in today's world, expectations from nanocosmetic products are quite high. Nanocosmetics should play roles such as maintaining homeostasis of the skin, keeping the skin healthy and beautiful, delaying the signs of skin aging, and providing solutions to skin problems, they should be safe in regular use as well. Therefore, cosmetic active substances should reach the relevant skin layers and cells where the aging problem originates.

The unique features of lipid nanoparticles on the skin make them dominant nanocarrier systems in cosmetic applications. The topical application of lipid nanoparticles serves many purposes as promoting their skin penetration and retention, re-enforcement of the lipid barrier, and increase in skin hydration. These properties are closely related to the biocompatible lipid composition of the nanoparticles as well as the particle sizes (Garcês et al., 2018). As the particle size decreases and the surface area increases, the particles gain adhesive properties by the influence of Van der Waal's forces. When lipid nanoparticles are applied to the skin, they form a thin film layer courtesy of their adhesive properties after evaporation of the aqueous phase. Müller and Dingler demonstrated that SLNs form a thin lipid film layer on the skin, and it was stated that an increase in the occlusive effect was obtained as a result of this film layer (Müller and Dingler, 1998; Müller et al., 2002). The pore-free film of lipid nanoparticles has the potential to repair and strengthen the natural barrier of the skin (Aziz et al., 2019). A lipid film layer with hydrophobic nature causes increased skin occlusion. Reduced transepidermal water loss (TEWL) due to the occlusive effect occurs after the application of lipid nanoparticle formulations to the skin, thereby increasing skin hydration (Sala et al., 2018). *In vivo* studies showed that O/W cream formulations containing SLN significantly increased skin hydration compared to conventional O/W cream formulations. Studies also have shown that SLNs have a higher occlusive effect than NLCs, while an increase in occlusive effect is obtained as the particle size of SLNs gets smaller (Souto et al., 2004;

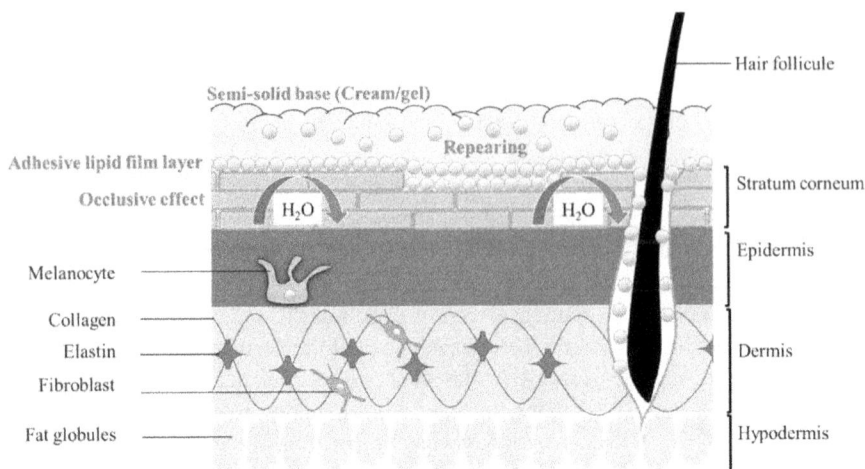

FIGURE 15.8 Schematic illustration of skin structure and the occlusive effect of lipid nanoparticles

Teeranachaideekul et al., 2008; Amasya, 2021). The barrier property of the stratum corneum against exogenous substances may weaken after increased hydration of the skin and as a result of skin hydration, enhanced percutaneous absorption of the active molecule can be achieved. The structure of the skin and the occlusive effect of lipid nanoparticles are schematized in Figure 15.8.

Many researchers listed the advantages of using lipid nanoparticles in the cosmetic field as follows:

Protection: The rigid solid structure of lipid nanoparticles may protect the chemically labile cosmetic active ingredients such as vitamin E and vitamin C, retinol, etc. from light, oxidation, or hydrolysis.

On the other hand, the solid structure of lipid nanoparticles shows the ability to block UV light. Therefore, the lipid nanoparticles themselves can be used as physical UV filters.

Controlling the Release Profile: Since different incorporation models can be achieved depending on the production method and choice of lipid type, extended-release profiles or burst releases can be obtained. Nanocosmetics need to focus on the prolonged release of UV blockers, antioxidants, or perfumes.

Skin Hydration: Even plain lipid nanoparticles show a cosmetic effect against the dry skin problem since water loss and hydration are provided by the occlusive effect.

Skin Targeting: In many studies with lipid nanoparticles, it has been observed that dermal penetration is greater along with the hair follicles. In addition, targeting to different skin layers can be achieved by adjusting the particle size.

15.7   COSMETIC APPLICATIONS OF LIPID NANOPARTICLES

Several researchers have comprehensively evaluated the usage of lipid nanoparticles in the cosmetic field because of their unique and important advantages as a cosmetic delivery system. The examples of the cosmetic applications of lipid nanoparticles during the past ten years were summarized in Table 15.4.

TABLE 15.4
Examples of the cosmetic applications of lipid nanoparticles during the past ten years

Cosmetic application	Cosmetic active substance	LP type	Results	References
Antioxidant	Curcumin	SLN	Improved antioxidant activity	Shrotriya et al., 2018
	Quercetin	NLC	Increased accumulation in epidermis and dermis	Chen-yu et al., 2012
			Enhanced antioxidant effect	
		SLN	Higher skin localization	Bose et al., 2013
		NLC	Highest quercetin amount in the skin with NLC formulation	Bose and Michniak-Kohn, 2013
		SLN		
		SLN	Increased skin permeation of quercetin	Han et al., 2014
	Ellagic acid	NLC	Higher antioxidant activity and lower toxicity	Hallan et al., 2020
	Green tea leaves extract (Epigallocatechin gallate)	SLN	Increased skin penetration of epigallocatechin gallate	Dzulhi et al., 2018
	Alpha-lipoic acid	NLC	Two-step release profile with a burst effect	Lasoń et al., 2017
		NLC	Enhanced photo-stability of alpha-lipoic acid	Wang et al., 2014
	Resveratrol	SLN NLC	Highest occlusive effect and skin hydration with SLN	Montenegro et al., 2017
		SLN NLC	Higher dermis accumulation with NLC	Gokce et al., 2012
	Coenzyme Q10	SLN	Efficient skin targeting of coenzyme Q10 without losing antioxidant activity	Korkmaz et al., 2013
	Vitamin E	NLC	Adequate biocompatibility and non-irritancy for cosmetic application	Eiras et al., 2017
Anticellulite	Caffeine	SLN	Complete lysis of adipocytes in deeper skin layers	Hamishehkar et al., 2015
		SLN	Higher skin permeation	Puglia et al., 2016
		SLN	Improved encapsulation efficiency	Algul et al., 2018
			Controlled caffeine release	

(Continued)

TABLE 15.4 (Continued)

Cosmetic application	Cosmetic active substance	LP type	Results	References
		NLC	Improved skin permeation	Rodrigues et al., 2016
		Semi-solid	Improved permeation into deeper skin layers	Amasya et al., 2021
		NLC		
Depigmentation	Deoxyarbutin	NLC	Higher penetration through Spangler's membrane	Tofani et al., 2016
	Hydroquinone	SLN	Higher skin localization	Ghanbarzadeh et al., 2015
			Enhanced chemical stability of hydroquinone	
		NLC	Enhanced stability of hydroquinone	Wu et al., 2017
	Kojic acid dipalmitate	SLN	Increased kojic acid dipalmitate amount in the epidermis	Mohammadi et al., 2021
	Kojic acid	NLC	Increased tyrosinase inhibitory activity	Khezri et al., 2021
	Trans-Resveratrol	SLN	Enhanced tyrosinase inhibition capacity	Rigon et al., 2016
	MHY498	SLN	Higher skin permeation	Al-Amin et al., 2016
			Efficient inhibition of UVB induced melanogenesis	
	Passion fruit seeds oil	NLC	Greater tyrosinase inhibition activity	Krambeck et al., 2021
			Improved skin penetration	
	Phenylethyl resorcinol	NLC	Significant inhibition of tyrosinase activity	Kim et al., 2017
			Enhanced physicochemical and photostability	
	Curcumin	SLN	Enhanced tyrosinase inhibition activity	Shrotriya et al., 2018
Hair Care	Pueraria mirifica extract	SLN	Increased penetration through skin and hair follicles	Tansathien et al., 2019
	Otoba wax	SLN	High potential of Otoba wax to develop nanocosmetics for hair care	Rubiano et al., 2020
Sunscreen	Bemotrizinol (Tinosorb® S)	NLC	Increased photoprotective activity	Medeiros et al., 2020
	Octylmethoxycinnamate	SLN	Significant reduction in skin penetration and higher prevention of photodegradation by NLC	Puglia et al., 2012
		NLC		

(Continued)

TABLE 15.4 (Continued)

Cosmetic application	Cosmetic active substance	LP type	Results	References
	Ethyl hexyltriazone Diethylamino hydroxybenzoyl hexyl benzoate Bemotrizinol Octylmethoxycinnamate Avobenzone	NLC	Important reduction in skin permeation of sun filters	Puglia et al., 2014
	Octylmethoxycinnamate	SLN	Enhanced photostability of octyl methoxycinnamate	Liu et al., 2015
	Octyl methoxycinnamate Urucum oil	SLN	No reduction in SPF value by replacing 20% of the chemical UV filter with a vegetable oil	Andréo-Filho et al., 2018
	Butyl methoxydibenzoylmethane Octocrylene (Coencapsulation)	SLN NLC	Enhanced photostability and photoprotective capacity Reduced skin penetration	Niculae et al., 2013
	Butyl-methoxydibenzoylmethane Alpha tocopherol (Coencapsulation)	SLN NLC	Higher photoprotective effect with SLNs in comparison with NLCs	Niculae et al., 2014
	Safranal	SLN	Esthetically acceptable and broad-spectrum sunscreen cream	Sanju et al., 2022
	Aloe vera	SLN	Increased photoprotective effect No irritation or hypersensitivity	Rodrigues and Jose, 2020
	Silymarin	SLN	High photoprotective effect based on the in vitro and in vivo SPF values	Netto and Jose, 2018

15.7.1 ANTICELLULITE

Cellulite, a skin alteration characterized by an orange peel-like pitting, is an important esthetic, and cosmetic problem especially for women since most of them have cellulite in different degrees. Cellulite formation is characterized by some structural changes in subcutaneous tissue, which is comprised of adipocytes. Since these white adipose cells store fat, the volume of the cells is enlarged, and the connective tissue is stretched. Increasing the lipolysis by topical application of xanthines, especially caffeine is a widely investigated approach to getting under control the cellulite. In this context, it is important to deliver caffeine effectively into the adipocytes in the subcutaneous layer. Hamishehkar et al. (2015) have prepared a Carbopol® based hydrogel formulation containing caffeine-loaded SLN. The sustained caffeine release was observed in comparison with the hydrogel containing caffeine alone. Higher skin accumulation and lower systemic absorption of caffeine could be obtained with a hydrogel containing SLN. The results of histological studies also showed that complete lysis of adipocytes can be achieved in the deeper skin layers by the application of the SLN formulation of caffeine (Hamishehkar et al., 2015). In another study, caffeine-loaded SLNs were prepared with a high encapsulation efficacy and suitable particle size by Puglia et al. (2016). The cumulative amounts of caffeine permeated through human epidermal membranes were determined for both of the gel formulations containing caffeine-loaded SLNs or caffeine alone. The encapsulation of caffeine in SLNs resulted in a faster permeation than the gel formulation of caffeine over 24 hours (Puglia et al., 2016). Algul et al. (2018) have also prepared an SLN formulation of caffeine and evaluated it *in vitro*. They concluded that SLNs can be evaluated as a potential topical delivery system for caffeine by virtue of improved encapsulation efficiency and controlled release profile (Algul et al., 2018). The potential of NLCs for improving the permeation of caffeine into deeper skin layers was evaluated by Rodrigues et al. (2016). According to the results of *ex vivo* skin permeation studies, the permeation of caffeine could be improved by encapsulation in NLCs that have a particle size below 200 nm (Rodrigues et al., 2016). Amasya et al. (2021) have developed an optimized semi-solid NLC dispersion of caffeine by QbD approach using an artificial neural network (ANN) program. TEM micrographs have revealed that the particles were spherical without any aggregation. The uniform particle size distribution was also demonstrated as a result of the particle size measurements. The *ex vivo* skin distribution profile of semi-solid NLC formulation containing rhodamine B was evaluated by confocal laser scanning microscope and the fluorescence intensity observed in hair follicles and adipose tissue confirmed that the permeation of the hydrophilic molecule into the deep skin layers (Amasya et al., 2021).

15.7.2 DEPIGMENTATION AGENTS

Ghanbarzadeh et al. (2015) evaluated SLNs for improving the stability and increasing the skin localization of hydroquinone, which is a commonly used depigmentation agent. SLN formulations of hydroquinone were prepared with high encapsulation efficiency and increased physicochemical stability. Increased hydroquinone amount localized in the skin could be achieved by encapsulating in SLNs (Ghanbarzadeh et al., 2015). Kojic acid is another tyrosinase enzyme inhibitor, which has a hydrophilic character. Because of its hydrophilic nature, kojic acid has insufficient skin penetration. Kojic acid-loaded NLCs showed prolonged release behavior and higher tyrosinase inhibition activity than pure kojic acid. The accumulation of kojic acid in the skin was also higher than in kojic acid solution (Khezri et al., 2021). In another research, Mohammadi et al. (2021) investigated the SLNs as a topical delivery system for kojic acid dipalmitate. They concluded that the kojic acid dipalmitate amount in the epidermis, which is the target skin layer including melanocytes, could be effectively enhanced (Mohammadi et al., 2021).

15.7.3 ANTIOXIDANTS

The skin is frequently exposed to the environmental stress factors such as UV lights and air pollution. The formation of excess reactive oxygen species can trigger skin disorders and skin aging. Chen-yu et al. (2012) have prepared an NLC formulation of quercetin, a natural flavonoid, which has an important antioxidant property, for dermal application. *Ex vivo* and *in vivo* skin permeation studies revealed that the quercetin amount in epidermis and dermis layers could be increased. The clearance rate to free radicals, i.e., the antioxidant capacity of quercetin-loaded NLCs was found to be higher than quercetin alone (Chen-yu et al., 2012). Dzulhi et al. (2018) have developed SLNs containing green tea leaf extract and evaluated the skin penetration properties. Results of an *ex vivo* skin penetration study showed that the skin penetration of epigallocatechin gallate, which is the main component of green tea extract, could be enhanced by a cream formulation containing SLNs (Dzulhi et al., 2018). In another study, Hallan et al. (2020) demonstrated that the water solubility and antioxidant activity of ellagic acid could be improved by encapsulation in NLCs. Besides, the low toxicity profile could also be achieved by NLC formulation (Hallan et al., 2020).

15.7.4 SUN FILTERS

Lipid nanoparticles have unique properties that provide important advantages for their use in sunscreen products. On the one hand, SLN and NLC act as physical sun filters by themselves. On the other hand, reduced irritation potential and enhanced photostability and photoprotection of organic sun filters can be obtained by encapsulation in lipid nanoparticles (Wissing and Müller, 2003; Lacatusu et al., 2010). The potential of lipid nanoparticles in sunscreen formulations was comprehensively evaluated by several researchers. Puglia et al. (2012) have produced the octyl methoxycinnamate encapsulated SLN and NLC formulations. Higher UV filter loading and smaller particles were obtained with NLC formulation in comparison with SLN. In addition, NLCs were found to be more effective in terms of decreasing skin penetration and preventing the photodegradation of octyl methoxycinnamate. In another study, Puglia et al. (2014) have also investigated the potential use of NLCs and nanoemulsions for five different UV filters; ethyl hexyltriazone, diethylamino hydroxy benzoyl hexyl benzoate, bemotrizinol, octyl methoxycinnamate and avobenzone. As a result of the study, NLC and nanoemulsion showed similar abilities for ensuring photostability of organic UV filters. On the other hand, skin permeation of UV filters could be significantly decreased only by encapsulation in NLC (Puglia et al., 2014). A UV filter, butyl-methoxy dibenzoyl methane, and an antioxidant, alpha-tocopherol, were coencapsulated in SLN and NLC by Niculae et al. (2014). The existence of alpha-tocopherol caused a few decreases in protection factors of formulations in comparison with the SPF values of SLNs and NLCs containing UV filters alone. On the other hand, alpha-tocopherol has an important contribution to the photostability of the UV filter encapsulated lipid nanoparticles. When the effect of the lipid nanoparticle type was evaluated, it was found that the SLN formulations showed higher photoprotection than NLC formulations (Niculae et al., 2014). In a recent study, Medeiros et al. (2020) developed and optimized bemotrizinol, an organic UV filter, loaded NLC by using the Box-Behnken experimental design. They have determined the *in vitro* SPF values and found increased photoprotective activity in comparison with bemotrizinol solution. The researchers have concluded that the enhanced photoprotective activity and decreased irritation potential could be obtained as a result of the synergistic effect between the organic UV filter and NLC (Medeiros et al., 2020). The photoprotective potential of the *Aloe vera*-loaded SLNs incorporated in a cream base was investigated by Rodrigues and Jose (2020). An enhanced photoprotective effect was obtained while irritation or hypersensitivity reactions were not observed with the cream formulation developed (Rodrigues and Jose, 2020).

15.7.5   HAIR CARE

SLN formulation containing *Pueraria mirifica* extract was developed and evaluated by Tansathien et al. (2019) for promoting hair growth. Enhanced penetration of *Pueraria mirifica* extract through the skin and hair follicles can be achieved by SLN formulation compared with the ethanolic solution of the extract. It was estimated that the *Pueraria mirifica* extract encapsulated SLNs could be an effective formulation strategy for inducing hair growth (Tansathien et al., 2019). Otoba wax, which is consisted of mainly lauric and myristic acids has an important potential for promoting hair growth and strengthening. Rubiano et al. (2020) have recently evaluated the usage of Otoba wax for the development of SLN formulation by ultrahigh-pressure homogenization method. SLNs could be produced with high stability without aggregation and cremation. It was concluded that the Otoba wax has promising potential in order to develop SLN formulations for hair care cosmetics.

15.8   COMMERCIAL PRODUCTS OF LIPID NANOPARTICLES IN THE COSMETICS MARKET

Although there is no obvious difference between the cosmetic and pharmaceutical applications of lipid nanoparticles, the time from manufacturing to getting into the market of a cosmetic product is considerably shorter than for pharmaceuticals. Because the regulatory procedure of cosmeceuticals proceeds more easily. Under this circumstance, after the discovery of lipid nanoparticles in the early 1990s, the first commercial product of lipid nanoparticles was launched in the cosmetics market in 2005. This first commercial product is an anti-wrinkle face cream released by Dr. Rimpler under the name of "Cutanova Cream Nano Repair Q10". It contains

Coenzyme Q10, polypeptide, hibiscus extract, ginger extract, and keto sugar. Its anti-aging effect has been proven by clinical trials and the *in vivo* skin hydration test was conducted on healthy volunteers. Significantly increased skin hydration was detected with "Cutanova Cream Nano Repair Q10" in the study conducted in comparison with the classical o/w emulsion (Pardeike et al., 2010) and studies indicated that a significant wrinkle reduction was found after 6 weeks. Currently, there are many lipid-based nano-cosmeceutical products on the market. Some of the commercially available market products containing lipid nanoparticles that have been released so far are listed below in Table 15.5 (Müller, 2007; Saez, 2018; Battaglia, 2019).

TABLE 15.5
Commercially available samples of cosmetic products containing lipid nanoparticles

Product name	Producer	Content	Type
Cutanova-Cream Nanorepair Q10	Dr. Rimpler	Coenzyme Q10, Polypeptide, Hibiscus extract, Ginger extract, Ketosugar	NLC
Intensive Serum Nanorepair Q10		Coenzyme Q10, Polypeptide, Mafane extract	NLC
Cutanova Cream Nanovital Q10		Coenzyme Q10, Polypeptide, Titanium dioxide, Ursolic acid, Oleanolic acid, Sunflower seed extract, Hibiscus extract, Ginger extract, Ketosugar	NLC
NanoLipid Restore CLR	Chemisches Laboratorium Dr. Kurt Richter GmbH	Carnauba wax Black currant seed oil unsaturated fatty acids. (Linoleic acid, alpha- and gamma-linolenic acids)	NLC
NanoLipid Repair CLR		Carnauba wax, black currant seed oil, Manuca oil	NLC
NanoLipid Q10 CLR		Coenzyme Q10, black currant seed oil	NLC
NanoLipid Basic CLR		Caprylic/Capric triglyceride	NLC
SURMER Crème Legère Nano-Protection	Isabelle Lancray	Kukuinut oil, Monoi tiare Tahiti, Pseudopeptide, Milk extract from coconut, Wild indigo, Noni extract	NLC
IOPE Line	Amore Pacific	Coenzyme Q10, Omega-3 omega-6 unsaturated fatty acid	NLC
IOPE Super vital		Coenzyme Q10, Omega-3 omega-6 unsaturated fatty acid	NLC
NLC Deep effect Eye serum	Beate Johnen	Coenzyme Q10, Highly active oligosaccharides, titanium dioxide, Acetyl hexapeptide-3, micronized plant collagen	NLC
NLC Deep effect repair cream		Coenzyme Q10, Highly active oligosaccharides, titanium dioxide	NLC
NLC Deep effect reconstruction cream		Coenzyme Q10, Acetyl hexapeptide-3, Highly active oligosaccharides in a polysaccharide matrix, micronized plant collagen	NLC

(*Continued*)

TABLE 15.5 (Continued)

Product name	Producer	Content	Type
La Prairie	Swiss Cellular White	Glycoproteins, Panax ginseng root extract, Equisetum arvense extract, Camellia sinensis leaf extract, and Viola tricolor extract	NLC
Olivenol Anti Falten Pflegekonzentrat	Dr. Theiss	Olea europaea oil, panthenol, acacia senegal, tocopheryl acetate	NLC
Olivenol Augenpflegebalsam		Olea Europaea oil, prunus amygdalus dulcis oil, hydrolyzed milk protein, tocopheryl acetate, Rhodiola rosea root extract, caffeine	NLC
Regenerations crème Intensive	Scholl	Macadamia ternifolia seed oil, avocado oil, urea, and blackcurrant seed oil	NLC
Nanobase	Yamanouchi	–	SLN

15.9 TOXICITY OF LIPID NANOPARTICLES

In nanocosmetics, the dermal exposure levels are as important as the solubility and particle size in terms of toxicological profile (Newton and Kaur, 2019). In addition, the physicochemical properties of ingredients and impurity influence toxicity. From a toxicological point of view, the main advantage of SLNs and NLCs is their biodegradable and GRAS classified lipid contents. Thus, they show low toxicity and cytotoxicity (Pardeike et al., 2009). In addition, no toxicity specific to this system has been reported. (Dhiman et al., 2021). However, lipid nanoparticles can activate the immune system due to the surfactant ingredients. Therefore, before a product containing lipid nanoparticles can be put on the market, the toxicity profile should also be evaluated by means of cytotoxicity and genotoxicity.

According to the European Commission's Scientific Committee on Consumer Safety (SCCS) guidance, "*The SCCS Notes of Guidance for the Testing of Cosmetic Ingredients and Their Safety Evaluation-2019*", the main safety concerns with nanocosmetics related to the local and systemic exposure of the consumer to the nanomaterials or the harmful effects of these materials. To eliminate the overall health risk to the consumer, the safety assessment of nanomaterials in cosmetics should be evaluated in terms of both local and possible systemic exposures. Therefore, detailed characterization of nanomaterials is gaining importance (SCCS/1611/2019; Dhawan et al., 2020; Souto et al., 2020). FDA also recommends the application of appropriate toxicological testing for nanomaterials (Włodarczyk and Kwarciak-Kozłowska, 2022). Reconstructed human skin models such as EPISkin and EPIDerm for skin irritation, 3T3 NRPT-3T3 fibroblasts neutral red uptake for phototoxicity are examples for dermal testing (SCCS/1611/2019). Committee on the Validation of Alternative Methods (ICCVAM) and the European Centre for the Validation of Alternative Methods (ECVAM) are the bodies working in this direction (Nanda et al, 2016).

15.10 REGULATORY ISSUES RELATED TO LIPID NANOPARTICLES IN TERMS OF CONSUMER SAFETY

Nanomaterials in cosmetic are generally prepared with top-down technologies, which uses physical methods suitable for industrial processes; however, some of these methods require either high energy or solvents (Souto et al., 2020; Włodarczyk and Kwarciak-Kozłowska, 2022). Along with the nanocosmetics safety issues gaining importance the job of regulatory bodies increased.

According to the definition of nanomaterials given in Article 2(k) of EC Regulation No 1223/2009, the production process of lipid nanoparticles corresponds to the term "intentionally manufacture" of nanoscale systems (Selletti, 2020). Regulations related to nanocosmetics are restrictive due to limiting the definition to "intentionally manufacture" and "insoluble or bio-persistent materials" (Rodríguez, 2018).

Article 19 of the EU regulation proposes to include nanoscale material by using the term "nano" on the label to ensure consumer safety regarding such nanocosmetic products placed on the market (Santana et al., 2019). In addition, according to Article 16, nanoproduct notification can be made six months before marketing under the provisions of the "Cosmetic Product Notification Portal (CPNP)" for products that are not subject to regulation according to Annex IV–VI. This statement should provide identification and specifications, safety data, and predictable exposure conditions, such as particle size and physicochemical properties, of nanomaterials within the toxicological profile. The important point here is the notification is necessarily done for the cosmetic product but not for the nanomaterial itself.

Another guidance related to nanotechnology was published by FDA Task Force, under the title of *"Guidance for Industry: Safety of Nanomaterials in Cosmetic Products (FDA-2011-D-D0489)"* for the safety and adulteration of nanomaterials in cosmetics (Nanda et al, 2016). In general, according to FDA, which contains nonbinding recommendations, if a product involves nanotechnology, the size and physical or chemical properties of the material or product in the nano-range are in question even the sizes are outside of the 1–100 nm range (FDA, 2014; Dhawan et al., 2020).

There are some other regulatory bodies and organizations in different countries such as Health Canada (Canada), NICNAS (Australia), NEDO (Japan), EPA (Denmark), TITCK (Turkey) also take initiatives in the direction of regulations related to nanomaterials (Nanda et al., 2016; Włodarczyk and Kwarciak-Kozłowska, 2022).

15.11 CHALLENGES AND FUTURE PERSPECTIVES

Since they are produced from GRAS category and non-animal source materials, lipid nanoparticles are advantageous nanocarriers for effectively carrying the cosmetic active ingredients to deeper skin tissues with less toxicity and high physicochemical stability in comparison with other nanocarrier systems (Khezri et al., 2018). Especially NLCs have gained attraction as cosmetic nanocarriers for their effects on skin hydration, occlusion, and stability enhancement (Ahmed et al., 2020). On the other hand, semi-solid lipid nanoparticles are the most remarkable systems due to their unique advantages such as the semisolid consistency. Besides, the frequently used HPH method in the production of lipid nanoparticles is not only solvent-free but also easier to adapt to industrial scale, therefore suitable for the sustainability approach (Morganti et al., 2020).

Nowadays, the use of lipid nanoparticles as a nanocosmetic delivery system for plant extracts has also attracted attention (Barua et al., 2017; Lacatusu et al., 2020; Araujo et al., 2022).

Despite the positive aspects of lipid nanoparticles, it should be noted that the application of nanosystems in cosmetic formulations raises questions about consumer safety. However, the increase in commercial lipid nanoparticle-containing cosmetic products may offer the advantages of effective and stable formulations through extended shelf life.

REFERENCES

Abbas, H., Kamel, R., El-Sayed, N. 2018. Dermal anti-oxidant, anti-inflammatory and anti-aging effects of Compritol ATO based resveratrol colloidal carriers prepared using mixed surfactants. *International Journal of Pharmaceutical* 541: 37–47. https://doi.org/10.1016/j.ijpharm.2018.01.054

Abdel-Salam, F. S., Elkheshen, S. A., Mahmoud, A. A., Ammar, H. O. 2016. Diflucortolone valerate loaded solid lipid nanoparticles as a semisolid topical delivery system. *Bulletin of Faculty of Pharmacy, Cairo University* 54(1): 1–7.

Ahmed, I. A., Mikail, M. A., Zamakshshari, N., Abdullah, A.-S. H. 2020. Natural anti-aging skincare: Role and potential. *Biogerontology* 21: 293–310. https://doi.org/10.1007/s10522-020-09865-z

Al-Amin, M., Cao, J., Naeem, M., et al. 2016. Increased therapeutic efficacy of a newly synthesized tyrosinase inhibitor by solid lipid nanoparticles in the topical treatment of hyperpigmentation. Drug Design, Development and Therapy 10: 3947–3957.

Algul, D., Duman, G., Ozdemır, S., Turkoz Acar, E., Yener, G. 2018. Preformulation, characterization, and in vitro release studies of caffeine-loaded solid lipid nanoparticles. *Journal of Cosmetic Science* 69: 165–73.

Amasya, G. 2021. A novel formulation strategy for skin occlusion: Semi-solid lipid nanoparticles. *Journal of Research in Pharmacy* 25(4): 388–97.

Amasya, G., Aksu, B., Badilli, U., Onay-Besikci, A., Tarimci, N. 2019. QbD guided early pharmaceutical development study: Production of lipid nanoparticles by high pressure homogenization for skin cancer treatment. *International Journal of Pharmaceutics* 563: 110–21. https://doi.org/10.1016/j.ijpharm.2019.03.056.

Amasya, G., Inal, O., Sengel-Turk, C. T. 2020. SLN enriched hydrogels for dermal application: Full factorial design study to estimate the relationship between composition and mechanical properties *Chemistry and Physics of Lipids* May; 228: 104889. https://doi.org/10.1016/j.chemphyslip.2020.104889.

Amasya, G., Ozturk, C., Aksu, B., Tarimci, N. 2021. QbD based formulation optimization of semi-solid lipid nanoparticles as nano-cosmeceuticals. *Journal of Drug Delivery Science and Technology* 66: 102737.

Amasya, G., Sandri, G., Onay-Besikci, A., Badilli, U., Caramella, C., C Bonferoni, M., Tarimci, N. 2016. Skin localization of lipid nanoparticles (SLN/NLC): Focusing the Influence of formulation parameters. *Current Drug Delivery* 13(7): 1100–10. https://doi.org/10.2174/1567201813666601041305

Amoabediny, G., Haghiralsadat, F., Naderinezhad, S. Helder, M. N., Kharanaghi, E. A., Arough, J. M., Zandieh-Doulabi, B. 2018. Overview of preparation methods of polymeric and lipid-based (noisome, solid lipid, liposome) nanoparticles: A comprehensive review. *International Journal of Polymeric Materials and Polymeric Biomaterials.* 67(6): 383–400. https://doi.org/10.1080/00914037.2017.1332623

Anderluzzi, G., Lou, G., Su, Y., Perrie, Y. 2019. Scalable manufacturing processes for solid lipid nanoparticles. *Pharmaceutical Nanotechnology* 7: 444–59.

Andreani, T., Dias-Ferreira, J., Fangueiro, J. F., et al. 2020. Formulating octyl methoxycinnamate in hybrid lipid-silica nanoparticles: An innovative approach for UV skin protection. *Heliyon* 6: e03831.

Andréo-Filho, N., Bim, A.V.K., Kaneko, T.M., et al. 2018. Development and evaluation of lipid nanoparticles containing natural botanical oil for sun protection: characterization and in vitro and in vivo human skin permeation and toxicity. Skin Pharmacology and Physiology 31: 1–9.

Araujo, A. R. T. S., Rodrigues, M., Mascarenhas-Melo, F., et al. 2022. Chapter 12: New-generation nanotechnology for development of cosmetics using plant extracts. In *Nanotechnology for the Preparation of Cosmetics Using Plant-Based Extracts*, ed. S.H. M. Setapar, A. Ahmad, M. Jawaid, 301–25. Elsevier Inc. https://doi.org/10.1016/B978-0-12-822967-5.00002-3 301

Aziz, Z. A. A., Mohd-Nasir, H., Ahmad, A., Mohd. Setapar, S. H., Peng, W. L., Chuo, S. C., et al. 2019. Role of nanotechnology for design and development of cosmeceutical: Application in makeup and skin care. *Frontiers in Chemistry* 7: 739.

Badilli, U., Gumustas, M., Uslu, B., Ozkan, S. A. 2018. Chapter 9: Lipid-based nanoparticles for dermal drug delivery. In *Organic Materials as Smart Nanocarriers for Drug Delivery*, ed. A. M. Grumezescu. 369–413. Elsevier: William Andrew Publishing. http://dx.doi.org/10.1016/B978-0-12-813663-8.00009-9

Badilli, U., Sengel-Turk, C. T., Amasya, G., Tarimci, N. 2017. Novel drug delivery system for dermal uptake of Etofenamate: Semisolid SLN dispersion. *Current Drug Delivery* 14(3): 386–93. https://doi.org/10.2174/1567201813666160808110245

Badilli, U., Sengel-Turk, C. T., Onay-Besikci, A., Tarimci N. 2015. Development of etofenamate-loaded semisolid SLN dispersions and evaluation of anti-inflammatory activity for topical application. *Current Drug Delivery* 12: 200–09.

Barua, S., Kim, H., Hong, S-C., et al. 2017. Moisturizing effect of serine-loaded solid lipid nanoparticles incorporated in hydrogel bases. *Archieves of Pharmaceutical Research* 40: 250–57. https://doi.org/10.1007/s12272-016-0846-1

Battaglia, L., Gallarate, M. 2012. Lipid nanoparticles: State of the art, new preparation methods and challenges in drug delivery. *Expert Opinion on Drug Delivery* 9(5): 497–508.

Battaglia, L., Ugazio, E. 2019. Lipid nano-and microparticles: An overview of patent-related research. *Journal of Nanomaterials* 2834941.

Bhattacharjee, S. 2016. DLS and zeta potential–What they are and what they are not? *Journal of Controlled Release* 235: 337–51.

Bose, S., Du, Y., Takhistov, P., Michniak-Kohn, B. 2013. Formulation optimization and topical delivery of quercetin from solid lipid based nanosystems. International Journal of Pharmaceutics 441: 56– 66.

Bose, S., Michniak-Kohn, B. 2013. Preparation and characterization of lipid based nanosystems for topical delivery of quercetin. European Journal of Pharmaceutical Sciences 48: 442–452.

Cabello, R. S., Rojo, P. G., Zuluaga, R. 2019. Lessons from the European Regulation 1223 of 2009, on cosmetics: Expectations versus reality. *Nanoethics* 13: 21–35. https://doi.org/10.1007/s11569-019-00335-6

Caldas, A. R., Faria, M. J., Ribeiro, A., Machado, R., Gonçalves, H., Gomes, A. C., Soares G. M. B., Lopez, C. M., Lucio, M. 2021. Avobenzone-loaded and omega-3- enriched lipid formulations for production of UV blocking sunscreen gels and textiles. *Journal of Molecular Liquids* 342: 116965

Casamonti, M., Piazzini, V., Bilia, A. R., Bergonzi, M. C. 2019. Evaluation of skin permeability of resveratrol loaded liposomes and nanostructured lipid carriers using a Skin Mimic Artificial Membrane (skin-PAMPA). *Drug Delivery Letters* 9(2): 134–45.

Cernasov, D. 2009. The design and development of anti-aging formulations. In *Skin Aging Handbook*, ed. N. Dayan, 291–325. William Andrew Publishing.

César, F. C., Maia Campos, P. M. 2020. Influence of vegetable oils in the rheology, texture profile and sensory properties of cosmetic formulations based on organogel. *International Journal of Cosmetic Science* 42(5): 494–500.

Chen-yu, G., Chun-fen, Y., Qi-lu, L., et al. 2012. Development of a quercetin-loaded nanostructured lipid carrier formulation for topical delivery. *International Journal of Pharmaceutics* 430: 292–98.

Das, S., Chaudhury, A. 2011. Recent advances in lipid nanoparticle formulations with solid matrix for oral drug delivery. *AAPS Pharmscitech* 12(1): 62–76.

Das, S., Ng, W. K., Tan, R. B. 2012. Are nanostructured lipid carriers (NLCs) better than solid lipid nanoparticles (SLNs): Development, characterizations and comparative evaluations of clotrimazole-loaded SLNs and NLCs? *European Journal of Pharmaceutical Sciences* 47(1): 139–51.

De Vringer T. Topical preparation containing a suspension of solid lipid particles. U.S. Patent No. 5,667,800. 16 Sep. 1997.

Dhawan, S., Sharma, P., Nanda, S. 2020. Chapter 8. Cosmetic nanoformulations and their intended use. In *Nanocosmetics: Fundamentals, Applications and Toxicity*, ed. A. Nanda, S. Nanda, T.A. Nguyen, S. Rajendran and Y. Slimani, 141–69. Elsevier. https://doi.org/10.1016/B978-0-12-822286-7.00017-6

Dhiman, N., Awasthi, R., Sharma, B., Kharkwal, H., Kulkarni G.T. 2021. Lipid nanoparticles as carriers for bioactive delivery. *Frontiers in Chemistry* 9: 580118.

Duarah, S., Pujari, K., Durai, R. D., Duarah, S., Pujari, K., Durai, R. D., Narayanan V. H. B. 2016. Nanotechnology-based cosmeceuticals: A review. *International Journal of Applied Pharmaceutics* 8 (1): 8–12.

Dzulhi, S., Anwar, E., Nurhayati, T. 2018. Formulation, characterization and in vitro skin penetration of green tea (Camellia sinensis L.) leaves extract-loaded solid lipid nanoparticles. *Journal of Applied Pharmaceutical Science* 8(8): 57–62.

Eiras, F., Amaral, M.H., Silva, R., Martins, E., Sousa Lobo, J.M., Silva, A.C. 2017. Characterization and biocompatibility evaluation of cutaneous formulations containing lipid nanoparticles. International Journal of Pharmaceutics 519: 373–380.

Epstein, H. A. 2011. Nanotechnology in Cosmetic Products. *Skinmed* Mar–Apr; 9(2): 109–10. PMID: 21548515.

Escobar-Chávez, J. J., Díaz-Torres, R., Rodríguez-Cruz I. M., Domínguez-Delgado, C. L. Morales, R. S., Ángeles-Anguiano, E., Melgoza-Contreras, L. M. 2012. Nanocarriers for transdermal drug delivery. *Research and Reports in Transdermal Drug Delivery* 1: 3–17. http://dx.doi.org/10.2147/RRTD.S32621

Farage, M. A., Miller, K. W., Berardesca, E., Maibach, H. I. 2010. Psychological and social implications of aging skin: Normal aging and the effects of cutaneous disease. In *Textbook of Aging Skin*, ed. M. A. Farage, K. M. Miller, H. I. Maibach, 949–57. Springer.

Food and Drug Administration. 2014. *Guidance for Industry: Safety of Nanomaterials in Cosmetic Products*. Washington, DC: FDA.

Garcês, A., Amaral, M. H., Lobo, J. S., Silva, A. C. 2018. Formulations based on solid lipid nanoparticles (SLN) and nanostructured lipid carriers (NLC) for cutaneous use: A review. *European Journal of Pharmaceutical Sciences* 112: 159–67.

Ghanbarzadeh, S., Hariri, R., Kouhsoltani, M., Shokri, J., Javadzadeh, Y., Hamishehkar, H. 2015. Enhanced stability and dermal delivery of hydroquinone using solid lipid nanoparticles. *Colloids and Surfaces B: Biointerfaces* 136: 1004–10.

Gokce, E.H., Korkmaz, E., Dellera, E., Sandri, G., Bonferoni, M.C., Ozer, O. 2012. Resveratrol-loaded solid lipid nanoparticles versus nanostructured lipid carriers: evaluation of antioxidant potential for dermal applications. International Journal of Nanomedicine 7: 1841–1850.

Gordillo-Galeano, A., Mora-Huertas, C. E. 2018. Solid lipid nanoparticles and nanostructured lipid carriers: A review emphasizing on particle structure and drug release. *European Journal of Pharmaceutics and Biopharmaceutics* 133: 285–308.

Graverini, G., Piazzini, V., Landucci, E., Pantano, D., Nardiello, P., Casamenti, F., et al. 2018. Solid lipid nanoparticles for delivery of andrographolide across the blood-brain barrier: In vitro and in vivo evaluation. *Colloids and Surfaces B: Biointerfaces* 161: 302–13.

Hallan, S. S., Sguizzato, M., Esposito, E., Cortesi, R. 2021. Challenges in the physical characterization of lipid nanoparticles. *Pharmaceutics* 13(4): 549.

Hallan, S. S., Sguizzato, M., Pavoni, G., et al. 2020. Ellagic acid containing nanostructured lipid carriers for topical application: A preliminary study. *Molecules* 25: 1449.

Hamishehkar, H., Shokri, J., Fallahi, S., Jahangiri, A., Ghanbarzadeh, S., Kouhsoltani, M. 2015. Histopathological evaluation of caffeine-loaded solid lipid nanoparticles in efficient treatment of cellulite. *Drug Development and Industrial Pharmacy* 41(10): 1640–46.

Han, S.B., Kwon, S.S., Jeong, Y.M., Yu, E.R., Park, S.N. 2014. Physical characterization and in vitro skin permeation of solid lipid nanoparticles for transdermal delivery of quercetin. International Journal of Cosmetic Science 36: 588–597.

Hu, X., He, H. 2021. A review of cosmetic skin delivery. *Journal of Cosmetic Dermatology* 20(7): 2020–30.

Hulla, J. E., Sahu, S. C., Hayes, A. W. 2015. Nanotechnology: History and future. *Human and Experimental Toxicology* 34(12): 1318–21.

Kaszuba, M., Corbett, J., Watson, F. M., Jones, A. 2010. High-concentration zeta potential measurements using light-scattering techniques. *Philosophical Transactions of the Royal Society A: Mathematical, Physical and Engineering Sciences* 368(1927): 4439–51.

Kaul, S., Gulati, N., Verma D., Mukherjee, S., Nagaich, U. 2018. Role of nanotechnology in Cosmeceuticals: A review of recent advances. *Journal of Pharmaceutics* 2018: Article ID 3420204, 1–19. https://doi.org/10.1155/2018/3420204

Khater, D., Nsairat, H., Odeh, F., Saleh, M., Jaber, A., Alshaer, W., Al Bawab, A., Mubarak, M. S. 2021. Design, preparation, and characterization of effective dermal and transdermal lipid nanoparticles: A review. *Cosmetics* 8: 39. https://doi.org/10.3390/cosmetics8020039

Khezri, K., Saeedi, M., Dizaj, S. M. 2018. Application of nanoparticles in percutaneous delivery of active ingredients in cosmetic preparations. *Biomedicine & Pharmacotherapy* 106: 1499–1505.

Khezri, K., Saeedi, M., Morteza-Semnani, K., Akbari, J., Hedayatizadeh-Omran, A. 2021. A promising and effective platform for delivering hydrophilic depigmenting agents in the treatment of cutaneous hyperpigmentation: Kojic acid nanostructured lipid carrier. *Artificial Cells, Nanomedicine, and Biotechnology* 49(1): 38–47. https://doi.org/10.1080/21691401.2020.1865993

Kim, B-S., Na, Y-G., Choi, J-H., et al. 2017. The improvement of skin whitening of phenylethyl resorcinol by nanostructured lipid carriers. Nanomaterials 7: 241.

Kim, M. H., Jeon, Y. E., Kang, S., Lee, J. Y., Lee, K. W., Kim, K. T., et al. 2020. Lipid nanoparticles for enhancing the physicochemical stability and topical skin delivery of orobol. *Pharmaceutics* 12(9): 845.

Korkmaz, E., Gokce, E.H., Ozer, O. 2013. Development and evaluation of coenzyme Q10 loaded solid lipid nanoparticle hydrogel for enhanced dermal delivery. Acta Pharmaceutica 63: 517–529.

Kowalska, M., Wozniak, M., Pazdzior, M. 2017. Assessment of the sensory and moisturizing properties of emulsions with hemp oil. *Acta Polytechnica Hungarica* 14(8): 183–95.

Krambeck, K., Silva, V., Silva, R., et al. 2021. Design and characterization of nanostructured lipid carriers (NLC) and nanostructured lipid carrier-based hydrogels containing passiflora edulis seeds oil. *International Journal of Pharmaceutics* 600: 120444.

Lacatusu, I., Badea, N., Murariu, A., Bojin, D., Meghea, A. 2010. Effect of UV sunscreens loaded in solid lipid nanoparticles: A combined SPF assay and photostability. *Molecular Crystals and Liquid Crystals* 523(1): 247/[819]–259/[831].

Lacatusu, I., Istrati, D., Bordei, N., Popescu, M., Seciu, A. M., Panteli, N. M., Badea, N. 2020. Synergism of plant extract and vegetable oils-based lipid nanocarriers: Emerging trends in development of advanced cosmetic prototype products. *Materials Science & Engineering C* 108: 110412. https://doi.org/10.1016/j.msec.2019.110412

Lasoń, E., Sikora, E., Miastkowska, M., Socha, P., Ogonowski, J. 2017. NLC delivery systems for alpha lipoic acid: Physicochemical characteristics and release study. Colloids and Surfaces A 532: 57–62.

Lee, Y-J., Nam, G.-W. 2020. Sunscreen boosting effect by solid lipid nanoparticles-loaded fucoxanthin formulation. *Cosmetics* 7: 14. https://doi.org/10.3390/cosmetics7010014

Lippacher, A., Müller, R. H., Mäder, K. 2001. Preparation of semisolid drug carriers for topical application based on solid lipid nanoparticles. *International Journal of Pharmaceutics* 214(1–2): 9–12.

Lippacher, A., Müller, R. H., Mäder, K. 2002. Semisolid SLN™ dispersions for topical application: Influence of formulation and production parameters on viscoelastic properties. *European Journal of Pharmaceutics and Biopharmaceutics* 53(2): 155–60. https://doi.org/10.1016/S0939-6411(01)00233-8

Lippacher, A., Müller, R. H., Mäder, K. 2004. Liquid and semisolid SLN™ dispersions for topical application: Rheological characterization. *European Journal of Pharmaceutics and Biopharmaceutics* 58(3): 561–67. https://doi.org/10.1016/j.ejpb.2004.04.009

Liu, X-H., Liang, X-Z., Fang, X., Zhang, W.P. 2015. Preparation and evaluation of novel octylmethoxycinnamate-loaded solid lipid nanoparticles. International Journal of Cosmetic Science 37: 446–453.

Lohani, I., Verma, A., Joshi, H., Yadav, N., Karki, N. 2014. Nanotechnology-based cosmeceuticals. *International Scholarly Research Notices.* 2014: Article ID 843687, 14. https://doi.org/10.1155/2014/843687

McMullen, R. L., Gorcea, M., Chen, S. 2016. Emulsions and their characterization by texture profile analysis. In *Handbook of Formulating Dermal Applications*, ed. N. Dayan, 131–55. Scrivener Publishing.

Medeiros, T. S., Moreira, L. M. C. C., Oliveira, T. M. T., et al. 2020. Bemotrizinol-loaded carnauba wax-based nanostructured lipid carriers for sunscreen: Optimization, characterization, and in vitro evaluation. *AAPS PharmSciTech* 21: 288.

Mehnert, W., Mäder, K. 2001. Solid lipid nanoparticles: Production, characterization and applications. *Advanced Drug Delivery Reviews* 64: 83–101.

Mihranyan, A., Ferraz, N., Strømme, M. 2012. Current status and future prospects of nanotechnology in cosmetics. *Progress in Materials Science* 57: 875–910.

Mishra, V., Bansal, K. K., Verma, A., Yadav, N., Thakur, S., Sudhakar, K., et al. 2018. Solid lipid nanoparticles: Emerging colloidal nano drug delivery systems. *Pharmaceutics* 10(4): 191.

Mohammadi, F., Giti, R., Meibodi, M. N., Ranjbar, A. M., Bazooband, A. R., Ramezani, V. 2021. Preparation and evaluation of kojic acid dipalmitate solid lipid nanoparticles. *Journal of Drug Delivery Science and Technology* 61: 102183.

Montenegro, L., Parenti, C., Turnaturi, R., Pasquinucci, L. 2017. Resveratrol-loaded lipid nanocarriers: Correlation between in vitro occlusion factor and in vivo skin hydrating effect. *Pharmaceutics* 9: 58.

Morganti, P. 2020. Chapter 1: Nanocosmetics: An introduction. In *Nanocosmetics: Fundamentals, Applications and Toxicity*, ed. A. Nanda, S. Nanda, T. A. Nguyen, S. Rajendran, Y. Slimani, 3–16. Elsevier. https://doi.org/10.1016/B978-0-12-822286-7.00001-2

Morganti, P., Chen, H. D., Morganti, G. 2020. Chapter 24: Nanocosmetics: Future perspective. In *Nanocosmetics: Fundamentals, Applications and Toxicity*, ed. A. Nanda, S. Nanda, T.A. Nguyen, S. Rajendran, Y. Slimani, 455–481. Elsevier. https://doi.org/10.1016/B978-0-12-822286-7.00020-6

Moshawih, S., Rabiatul Basria S. M. N. Mydin, Kalakotla, S., Jarrar. Q. B. 2019. Potential application of resveratrol in nanocarriers against cancer: Overview and future trends. *Journal of Drug Delivery Science and Technology* 53(1): 101187. https://doi.org/10.1016/j.jddst.2019.101187.

Müller, R. H., Dingler, A. 1998. The next generation after the liposomes: solid lipid nanoparticles (SLN, Lipopearls) as dermal carrier in cosmetics. *Eurocosmetics* 7(8): 19–26.

Müller, R. H., Mäder, K., Gohla, S. 2000. Solid lipid nanoparticles (SLN) for controlled drug delivery–a review of the state of the art. *European Journal of Pharmaceutics and Biopharmaceutics* 50(1): 161–77.

Müller, R. H., Petersen, R. D., Hommoss, A., Pardeike, J. 2007. Nanostructured lipid carriers (NLC) in cosmetic dermal products. *Advanced Drug Delivery Reviews* 59(6): 522–30.

Müller, R. H., Radtke, M., Wissing, S. A. 2002. Solid lipid nanoparticles (SLN) and nanostructured lipid carriers (NLC) in cosmetic and dermatological preparations. *Advanced Drug Delivery Reviews* 54: 131–55.

Nafisi, S., Maibach, H. I. 2018. Skin penetration of nanoparticles. In *Emerging Nanotechnologies in Immunology*, ed. R. Shegokar, E. B. Souto, 47–88. Elsevier.

Nanda, S., Nanda, A., Lohan, S., Kaur, R., Singh, B. 2016. Chapter 3: Nanocosmetics: Performance enhancement and safety assurance. In *Nanobiomaterials in Galenic Formulations and Cosmetics,* ed. A. M. Grumezescu, 47–67. Elsevier: William Andrew Publishing. http://dx.doi.org/10.1016/B978-0-323-42868-2.00003-6

Netto, G., Jose, J. 2018. Development, characterization, and evaluation of sunscreen cream containing solid lipid nanoparticles of silymarin. Journal of Cosmetic Dermatology 17: 1073–1083.

Newton, A. M. J., Kaur, S. 2019. Chapter 9: Solid lipid nanoparticles for skin and drug delivery: Methods of preparation and characterization techniques and applications. In *Nanoarchitectonics in Biomedicine,* ed. A. M. Grumezescu, 295–334. Elsevier: William Andrew Publishing. https://doi.org/10.1016/B978-0-12-816200-2.00015-3

Niculae, G., Badea, N., Meghea, A., Oprea, O., Lacatusu, I. 2013. Coencapsulation of butyl-methoxydibenzoylmethane and octocrylene into lipid nanocarriers: UV performance, Photostability and in vitro Release. Photochemistry and Photobiology 89: 1085–1094.

Niculae, G., Lacatusu, I., Bors, A., Stan, R. 2014. Photostability enhancement by encapsulation of a-tocopherol into lipid-based nanoparticles loaded with a UV filter. *Comptes Rendus Chimie* 17: 1028–33.

Paliwal, R., Paliwal, S. R., Kenwat, R., Kurmi, B. D., Sahu M. K. 2020. Solid lipid nanoparticles: A review on recent perspectives and patents. In *Expert Opinion on Therapeutic Patents.* Taylor and Francis. https://doi.org/10.1080/13543776.2020.1720649

Pardeike, J., Hommoss, A., Müller, R. H. 2009. Lipid nanoparticles (SLN, NLC) in cosmetic and pharmaceutical dermal products. *International Journal of Pharmaceutical Sciences* 366: 170–84.

Pardeike, J., Schwabe, K., Müller, R. H. 2010. Influence of nanostructured lipid carriers (NLC) on the physical properties of the Cutanova Nanorepair Q10 cream and the in vivo skin hydration effect. *International Journal of Pharmaceutics* 396(1–2): 166–73.

Patel, V. R., Agrawal, Y. K. 2011. Nanosuspension: An approach to enhance solubility of drugs. *Journal of Advanced Pharmaceutical Technology & Research* 2(2): 81.

Patravale, V. B., Mandawgade, S. D. 2008. Novel cosmetic delivery systems: An application update. *International Journal of Cosmetic Sci*ence Feb; 30(1): 19–33.

Patzelt, A., Mak, W. C., Jung, S., Knorr, F., Meinke, M. C., Richter, H., et al. 2017. Do nanoparticles have a future in dermal drug delivery? *Journal of Controlled Release* 246: 174–82.

Pradhan, M., Singh, D., Singh, M. R. 2013. Novel colloidal carriers for psoriasis: Current issues, mechanistic insight and novel delivery approaches. *Journal of Controlled Release* 170(3): 380–95.

Puglia, C., Bonina, F., Rizza, L., et al. 2012. Lipid nanoparticles as carrier for octyl-methoxycinnamate: In vitro percutaneous absorption and photostability studies. *Journal of Pharmaceutical Sciences* 101(1): 301–11.

Puglia, C., Damiani, E., Offerta, A., et al. 2014. Evaluation of nanostructured lipid carriers (NLC) and nanoemulsions as carriers for UV-filters: Characterization, in vitro penetration and photostability studies. *European Journal of Pharmaceutical Sciences* 51: 211–17.

Puglia, C., Offerta, A., Tirendi, G. G., et al. 2016. Design of solid lipid nanoparticles for caffeine topical administration. *Drug Delivery* 23(1): 36–40.

Ramteke, K. H., Joshi, S. A., Dhole, S. N. 2012. Solid lipid nanoparticle: A review. *IOSR Journal of Pharmacy* 2(6): 34–44. https://doi.org/10.9790/3013-26103444

Rigon, R.B., Fachinetti, N., Severino, P., Santana, M.H.A., Chorilli, M. 2016. Skin delivery and in vitro biological evaluation of trans-resveratrol-loaded solid lipid nanoparticles for skin disorder therapies. Molecules 21: 116.

Roberts, M. S., Cheruvu, H. S., Mangion, S. E., Alinaghi, A., Benson, H. A., Mohammed, Y., et al. 2021. Topical drug delivery: History, percutaneous absorption, and product development. *Advanced Drug Delivery Reviews* 177: 113929.

Rodrigues, F., Alves, A. C., Nunes, C., et al. 2016. Permeation of topically applied caffeine from a food by—product in cosmetic formulations: Is nanoscale in vitro approach an option? *International Journal of Pharmaceutics* 513: 496–503.

Rodrigues, L. R., Jose, J. 2020. Exploring the photo protective potential of solid lipid nanoparticle-based sunscreen cream containing Aloe vera. *Environmental Science and Pollution Research* 27: 20876–88.

Rodríguez, H. 2018. Nanotechnology and risk governance in the European Union: The constitution of safety in highly promoted and contested innovation areas. *Nanoethics* 12: 5–26. https://doi.org/10.1007/s11569-017-0296-3

Rubiano, S., Echeverri, J. D., Salamanca, C. H. 2020. Solid lipid nanoparticles (SLNs) with potential as cosmetic hair formulations made from Otoba wax and ultrahigh pressure homogenization. *Cosmetics* 7(2): 42. https://doi.org/10.3390/cosmetics 7020042

Saez, V., Souza, I. D. L., Mansur, C. R. E. 2018. Lipid nanoparticles (SLN & NLC) for delivery of vitamin E: A comprehensive review. *International Journal of Cosmetic Science* 40(2): 103–16.

Sala, M., Diab, R., Elaissari, A., Fessi, H. 2018. Lipid nanocarriers as skin drug delivery systems: Properties, mechanisms of skin interactions and medical applications. *International Journal of Pharmaceutics* 535(1–2): 1–17.

Salvi, V. R., Pawar, P. 2019. Nanostructured lipid carriers (NLC) system: A novel drug targeting carrier. *Journal of Drug Delivery Science and Technology* 51: 255–67.

Sanju, N., Vineet, M., Kumud, M. 2022. Development and evaluation of a broad spectrum polyherbal sunscreen formulation using solid lipid nanoparticles of safranal. Journal of Cosmetic Dermatology 00:1–14.

Saupe, A., Gordon, K. C., Rades, T. 2006. Structural investigations on nanoemulsions, solid lipid nanoparticles and nanostructured lipid carriers by cryo-field emission scanning electron microscopy and Raman spectroscopy. *International Journal of Pharmaceutics* 314(1): 56–62.

Scientific Committee on Consumer Safety. 2019. Guidance on The Safety Assessment of Nanomaterials in Cosmetics (no SCCS/1611/19). SCCS, Brussels.

Selletti, S. 2020. Chapter 20: Current legal frameworks and consumer protection in nanocosmetics. In *Nanocosmetics: Fundamentals, Applications and Toxicity,* ed. A. Nanda, S. Nanda, T.A. Nguyen, S. Rajendran, Y. Slimani, 393–403. Elsevier. https://doi.org/10.1016/B978-0-12-822286-7.00016-4

Sengar, V., Jyoti, K., Jain, U. K., Katare, O. P., Chandra, R., Madan, J. 2018. Lipid nanoparticles for topical and transdermal delivery of pharmaceuticals and cosmeceuticals: A glorious victory. In *Lipid Nanocarriers for Drug Targeting,* ed. A. M. Grumezescu, 413–436. William Andrew Publishing.

Severino, P., Andreani, T., Macedo, A. S., Fangueiro, J. F., Santana, M. H., Silva, A. M., Souto, E. B. 2012. Current state-of-art and new trends on lipid nanoparticles (SLN and NLC) for oral drug delivery. *Journal of Drug Delivery* 2012: 750891. https://doi.org/10.1155/2012/750891.

Shah, R., Eldridge, D., Palombo, E., Harding, I. 2015a. Chapter 2. Composition and structure. In *Lipid Nanoparticles: Production, Characterization and Stability,* 11–22. Springer.

Shah, R., Eldridge, D., Palombo, E., Harding, I. 2015b. Chapter 3. Production techniques. In *Lipid Nanoparticles: Production, Characterization and Stability,* 23–44. Springer.

Shah, R., Malherbe, F., Eldridge, D., Palombo, E., Harding, I. 2014. Physicochemical characterization of solid lipid nanoparticles (SLNs) prepared by a novel microemulsion technique. *Journal of Colloid and Interface Science* 428: 286–94.

Shrotriya, S., Ranpise, N., Satpute, P., Vidhate, B. 2018. Skin targeting of curcumin solid lipid nanoparticles-engrossed topical gel for the treatment of pigmentation and irritant contact dermatitis. Artificial Cells, Nanomedicine, and Biotechnology 46(7): 1471–1482.

Smijs, T. G., Pavel, S. 2011. Titanium dioxide and zinc oxide nanoparticles in sunscreens: Focus on their safety and effectiveness, *Nanotechnology, Science and Applications* 4: 95–112.

Souto, E. B., Fangueiro, J. F., Fernandes, A. R., et al. 2022. Physicochemical and biopharmaceutical aspects influencing skin permeation and role of SLN and NLC for skin drug delivery. *Heliyon* 8: e08938.

Souto, E. B., Müller, R. H. 2007. Chapter 14. Lipid nanoparticles (Solid lipid nanoparticles and nanostructured lipid carriers) for cosmetic, dermal, and transdermal applications. In *Nanoparticulate Drug Delivery Systems,* ed. D. Thassu, M. Deelers, Y. Pathak, 213–33. Informa Healtcare.

Souto, E. B., Müller, R. H. 2008. Cosmetic features and applications of lipid nanoparticles (SLN, NLC). *International Journal of Cosmetic Science* 30: 157–65.

Souto, E. B., Wissing, S. A., Barbosa, C. M., Müller, R. H. 2004. Development of a controlled release formulation based on SLN and NLC for topical clotrimazole delivery. *International Journal of Pharmaceutics* 278(1): 71–77.

Stetefeld, J., McKenna, S. A., Patel, T. R. 2016. Dynamic light scattering: A practical guide and applications in biomedical sciences. *Biophysical Reviews* 8(4): 409–27.

Svilenov, H., Tzachev, C. 2014. Chapter 8. Solid lipid nanoparticles – A promising drug delivery system. In *Nanomedicine,* ed. A. Seifalian, A. de Mel, D. M. Kalaskar, 188–213. One Central Press.

Tai, A., Bianchini, R., Jachowicz, J. 2014. Texture analysis of cosmetic/pharmaceutical raw materials and formulations. *International Journal of Cosmetic Science* 36(4): 291–304.

Tansathien, K., Nuntharatanapon, N., Jaewjira, S., Pizon, J. R. L., Opanasopit, P., Rangsimawong, W. 2019. Solid lipid nanoparticles containing Pueraria mirifica ethanolic extract for hair growth promotion. *Key Engineering Materials* 819: 175–80.

Tarimci, N. 2003. New cosmetic delivery systems. In *COSMODERM III and 5th International Cosmetic Symposium (Icos) PROCEEDİNGS: Skin Care and Aesthetics in the Millenium,* ed. Y. Yazan, 140–150, *İstanbul -TURKEY.*

Tarimci, N. 2011. Chapter 7: Cyclodextrins in the cosmetic field. In *Cyclodextrins in Pharmaceutics, Cosmetics and Biomedicine: Current and Future Biomedical Applications,* ed. E. Bilemsoy, 131–44. John Wiley & Sons.

Tarimci, N., Sengel Turk, C. T., Badilli, U. 2015. A novel approach for topical delivery of SLNs: Semisolid SLNs. LAP LAMBERT Academic Publishing 2015-09-07. ISBN 13: 9783659779015.

Teeranachaideekul, V., Boonme, P., Souto, E. B., Müller, R. H., Junyaprasert, V. B. 2008. Influence of oil content on physicochemical properties and skin distribution of Nile red-loaded NLC. *Journal of Controlled Release* 128(2): 134–41.

Tofani, R.P., Sumirtapura, Y.C., Darijan, S.T. 2016. Formulation, characterisation, and in vitro skin diffusion of nanostructured lipid carriers for deoxyarbutin compared to a nano-emulsion and conventional cream. Scientia Pharmaceutica 84: 634–645.

Üner, M., Yener, G. 2007. Importance of solid lipid nanoparticles (SLN) in various administration routes and future perspectives. *International Journal of Nanomedicine* 2: 289–300.

Wang, J., Tang, J., Zhou, X., Xia, Q. 2014. Physicochemical characterization, identification and improved photo-stability of alpha-lipoic acid-loaded nanostructured lipid carrier. Drug Development and Industrial Pharmacy 40(2).

Wissing, S., Lippacher, A., Müller, R. 2001. Investigations on the occlusive properties of solid lipid nanoparticles (SLN). *Journal of Cosmetic Science* 52(5): 313–24.

Wissing, S. A., Müller, R. H. 2003. Cosmetic applications for solid lipid nanoparticles (SLN). *International Journal of Pharmaceutics* 254: 65–68.

Włodarczyk, R., Kwarciak-Kozłowska, A. 2021. Nanoparticles from the cosmetics and medical industries in legal and environmental aspects. *Sustainability* 13: 5805. https://doi.org/10.3390/su13115805

World Health Organization. 2006. Dermal absorption. World Health Organization. https://apps.who.int/iris/handle/10665/43542

Wu, L., Zhang, J., Watanabe, W. 2011. Physical and chemical stability of drug nanoparticles. *Advanced Drug Delivery Reviews* 63(6): 456–69.

Wu, P-S., Lin, C-H., Kuo, Y-C., Lin, C-C. 2017. Formulation and characterization of hydroquinone nanostructured lipid carriers by homogenization emulsification method. Journal of Nanomaterials Article ID 3282693.

Zhang, Y., Lane, M. E., Hadgraft, J., Heinrich, M., Chen, T., Lian, G., et al. 2019. A comparison of the in vitro permeation of niacinamide in mammalian skin and in the Parallel Artificial Membrane Permeation Assay (PAMPA) model. *International Journal of Pharmaceutics* 556: 142–49.

Zhou, H., Luo, D., Chen, D., et al. 2021. Current advances of nanocarrier technology-based active cosmetic ingredients for beauty applications. *Clinical, Cosmetic and Investigational Dermatology* 14: 867.

Zieli´nska, A., Nowak, I. 2016. Chapter 10. Solid lipid nanoparticles and nanostructured lipid carriers as novel carriers for cosmetic ingredients. In *Nanobiomaterials in Galenic Formulations and Cosmetics*, ed. A. M. Grumezescu, 231–55. Elsevier. http://dx.doi.org/10.1016/B978-0-323-42868-2.00010-3

16 Regulatory aspects, recent legal contexts, consumer protection and future perspectives of nanocosmetics

Anuradha Dey and Sunil Kumar Dubey

CONTENTS

16.1 INTRODUCTION

The question that often emerges, given that the cosmetic industry is ever expanding, is, 'Should more stringency be imposed for regulating the cosmetic products that are being newly launched along with policing the products that are already in the market?' Another question, which arises, is, 'Is there a necessity to have an independent review board for cosmetic products?' Several other questions that arise in the case

DOI: 10.1201/9781003319146-16

349

of cosmetic products are: Should there be guidelines mandating the requirement of proving the efficacy of each ingredient that is being incorporated in the preparation? The existence of a rating system that classifies efficacy, for example, the sun protection factor (SPF) helps in gauging the efficacy of sunscreens. Nanomaterials are defined as substances having sizes falling in the range of 1–100 nm. The cosmetic market is less regulated in comparison to pharmaceutical products. However, with the continued addition of numerous products every year by multi-national cosmetic giants, domestic cosmetic companies, small businesses, etc., whose products are not tracked comprehensively in a unified manner, it is becoming essential to regulate the same in order to avoid repercussions such as no track history of cosmetic efficacy, no consumer feedback, no categorization of adverse effects or side effects. Given the diverse categories of nanostructures that are used in making cosmetics such as liposomes (Rapalli et al. 2021), nanostructured lipid carriers (Rapalli, Kaul, et al. 2020; Waghule et al. 2020), solid lipid nanoparticles, nanoemulsions, polymeric nanoparticles, inorganic nanoparticles (Bapat et al. 2020), nanocrystals (Rapalli, Waghule, et al. 2020), dendrimers (Dubey et al. 2021), etc., all of which have varied formulation procedures and characteristic properties, regulating them is essential. An increase in the cases of adverse effects and rising questions regarding the safety of using nanomaterials makes it essential to have stringent regulations in place.

16.2   REGULATORY SCENARIO AND ASSOCIATED LEGAL CONTEXTS OF NANOCOSMETICS IN LEADING ECONOMIES

The United States, European Union, India, China, Japan, etc. have their individual norms and practices when it comes to the cosmetic market. Each of them has its own regulations to which they adhere. The International Cooperation on Cosmetic Regulation (ICCR) is an example of an international body that, as the name suggests, works toward unifying the rules and regulations pertaining to the use of nanomaterials in cosmetics. The USA, European Union, Canada, and Japan are the core members of ICCR. The 4th Annual meeting of the ICCR was where the Joint Working Group (JWG) was constituted for examining the different safety approaches for validating cosmetics. A report generated from the discussions focussed on outlining the different parameters that are to be evaluated whilst determining the safety of nanomaterials used in cosmetics, such as the physicochemical parameters concerning nanomaterials, the avenues of exposure to them, the hazardous implications of such substances, their identification and quantification, etc. (Gupta et al. 2022). The following section discusses briefly the regulatory scenario of the US and European markets.

16.2.1   NANOCOMETICS IN THE UNITED STATES MARKET

In the United States, it was in the year 2007 that the Nanotechnology Task Force (NTF) was assigned the responsibility of putting forth its recommendations on how to regulate nanotechnology-based cosmetic products. The NTF assimilated data on the adequacy of using a particular nanomaterial in a cosmetic product along with gathering information on the long-term safety of the same. In the same year, the "Working

Party on Manufactured Nanomaterials" instituted by the OECD (Organization for Economic Cooperation and Development) launched a "Testing Program" for testing all the nanomaterials. Post this, in 2011, after incorporating the recommendations given by NTF, the Food and Drugs Administration (FDA), published a guidance document for the cosmetic industry titled, "Guidance for Industry: Safety of Nanomaterials in Cosmetic Products," having docket no. FDA-2011-D-0489. This document served to provide guidance on the use of nanomaterials in the cosmetic industry and food products as well. It establishes the FDA's framework of incorporating nanotechnology-based principles into cosmetics, the parameters for technically assessing the variety of nanoproducts that are being generated. The guidance document urges to assess each product based on its specifications, particularly the biological and mechanical effects that it exhibits. The document lays emphasis on two important features, the parameters measured that are associated with the particle size and the properties/phenomena that are influenced by the size. The same guidance document has a Federal Register notice: 79 FR 36532 associated with it which is concerned with the safety evaluation of nanomaterials used in cosmetic formulations. This section provides details on the route of exposure, the absorption of the cosmetic agent, and its toxicity evaluation. The suggestion about studying dermal absorption, skin irritation, and the corrosive nature of the formulation using pig skin or available alternatives of reconstructed human skin (Example: EpiSkin® and Epiderm®, etc.) are discussed. The FDA, again in 2014, released another guidance document titled, "Considering whether an FDA regulated product involves the application of nanotechnology", having the docket number FDA-2010-D-0530, which is pertaining to regulating all products (both pharmaceutical drugs and cosmetics) which are of the nano-size form (Dhawan, Sharma, and Nanda 2020).

In general, cosmetic products in the U.S. market do not need pre-market FDA approval, only the color additives that are being used need approval. The responsibility of ensuring that the launched cosmetic product is safe lies with the company itself ("Cosmetic Products | FDA" 2021). The FDA does not mandate the cosmetic manufacturer to provide the names of the nanomaterials used in the label as that may make the consumers skeptical about purchasing the product (Jeswani et al. 2019). However, the products that are being sold on U.S. soil have to comply with the laws of two major Acts, the FD&C Act and the FPL Act which are the Federal Food, Drug and Cosmetic Act and Fair Packaging and Labeling Act respectively. Subchapter-VI (FD&C Act) is concerning cosmetics and the Sections 361–364 are concerned with cosmetic products that are either adulterated or misbranded along with outlining scenarios of exemptions from the regulations ("FD&C Act Chapter VI: Cosmetics | FDA" 2021). Chapter 39 (FPL Act) lays down the framework of what qualifies to be correct labeling and describes what falls under deceptive/wrong labeling (Commissioner, n.d.).

The FDA and Personal Care Products Council (PCPC) jointly initiated and established the Voluntary Cosmetic Registration Program (VCRP) which serves to be a platform for gathering details on all the materials that are used for manufacturing cosmetics and keeps a record of the adverse effects or side effects of the same. The manufacturers can voluntarily report the above data on this platform (Dubey, Dey, et al. 2022). Via the VCRP platform, the FDA can update the cosmetic manufacturers about the nanomaterials which have been found to have risks associated with their use. However, in general, the FDA limits the use of certain chemicals beyond a certain concentration, such as zinc oxide can be used only up to 25% ("CFR - Code of Federal Regulations Title 21" 2022). The Cosmetic Ingredient Review Expert Panel (CIR) has the responsibility of assessing all the chemical ingredients that are used in cosmetics.

The Food and Drug Administration (FDA) is mostly reliant on the cosmetic industry to self-regulate itself, as the manufacturers need to ensure the safety of their products in order for consumer acceptance to be high. However, the governing bodies FDA, Cosmetic Ingredient Review Expert Panel (CIR), and Research Institute for Fragrance Materials (RIFM) have no jurisdictional authority to enforce the withdrawal of the unsafe product from the market. The FDA is trying to gain authority via the Personal Care Products Safety Act (PCPSA) to recall products wherever necessary along with bringing into practice a mandatory system for reporting adverse events and an annual safety analysis of all products (Kwa, Welty, and Xu 2017). The American Contact Dermatitis Society has the Contact Allergen Management Program (CAMP) in place which is a repository of all allergic reactions that have been reported by patients upon the use of cosmetic/personal care products.

16.2.2 NANOCOSMETICS IN THE EUROPEAN MARKET

The European Commission (EC) is the governing body that is responsible for approving cosmetics in the European market. The primary "Cosmetic Products Regulation (CPR) No. 1223/2009" is the key legislation for regulating cosmetics. Prior to this, the EC directive 76/768/EC was in effect which had been instituted in 1976. According to the CPR, a nanomaterial is defined as "An insoluble or biopersistent and intentionally manufactured material with one or more external dimensions, or an internal structure, on the scale from 1 to 100 nm".

Article 16 of the current European legislation is concerned with all the nanomaterials that are used in cosmetics apart from nano-sized coloring agents, UV filters, and preservatives. The directives pertaining to the above-mentioned three constituents are discussed in Article 14 of the legislation. The Cosmetics Products Notification Portal (CPNP) is the platform on which every approved product has to be listed. This happens after each product is subjected to scrutiny by an individual, appointed by the European Union who is responsible for overseeing the approval of that particular product. Articles 3, 4, and 13 deal with specifying the responsibilities of this appointed person. This appointed person tasked with this responsibility has to finish the process six months before the launch date of the product. The documentation filed has to have certain details such as the IUPAC names of the nanomaterial(s) in the product, its size, its approximate quantity present in the product, its physicochemical properties, details on its toxicological assessment and safety data (For example skin irritation/sensitization, eye irritation test data). Article 13(1-f) requires notifying the products which contain nanomaterials and Article 19 lays down rules for labelling of nanomaterials. The manufacturer is required to submit all the information at the CPNP portal and thus "notify" the EC about its nano-product.

Another body, the Scientific Committee on Consumer Safety (SCCS), as the name suggests, is responsible for providing surety that the constituents in cosmetic products are safe. In October 2019, the SCCS released a guidance document titled, "Guidance on the Safety Assessment of Nanomaterials in Cosmetics", and again in March 2021 released another revised version of the "SCCS Notes Of Guidance for the Testing of Cosmetic Ingredients and their Safety Evaluation" (11th revision), both of each are crucial documents for guiding the development of nanocosmetics that are safe ("Scientific Committee on Consumer Safety SCCS the SCCS Notes of Guidance for the Testing of Cosmetic Ingredients and Their Safety Evaluation 11th Revision," n.d.). The EC has often taken the assistance of the SCCS in assessing the safety of nanomaterials and made the final reports publically available as well. The EC considers three nanomaterials as UV filters, namely zinc oxide, tris-biphenyl triazine, and titanium oxide. Nanoform of carbon black is also considered a cosmetic colorant. Article 16-Section 10a has been recently updated on June 15, 2017, to include a catalog of all the nanomaterials that are used in cosmetic preparation. Broadly the catalog is divided into three sections, i.e. preservatives, UV filters, and colorants, and information on each nanomaterial belonging to any of the three categories is provided. The information includes their names, route of exposure, details on rinsing off or leaving the product, etc. On November 21, 2019, the latest version of the catalog was updated again ("DocsRoom - European Commission" 2022). The SCCP also provides details on in vitro methods that have been validated for assessing cosmetic products so that testing on animals can be avoided as much as possible.

Recommendation documents lay out the proposal for the use of nanomaterials for developing nanoproducts but are not legally binding or compulsory in nature.

The first paragraph of the Article 16 of the European Legislation should mention all the cosmetic ingredients that are a 'nano-material' by nature and are used in the product. The Article 16, at the bare minimum, should give the following information:

1. IDENTIFICATION of the nanomaterial such as its IUPAC name along with its brief descriptors as mentioned in the Annexure II-VI
2. SPECIFICATIONS of the nanomaterial such as its dimension, shape, physical properties and chemical properties
3. ESTIMATION of the approximate quantity of the nanomaterial that is present in the cosmetic product and how much of it is going to be used per year
4. TOXICITY data pertaining to the nanomaterial
5. SAFETY data of the nanomaterial relating to the category of cosmetic it belongs to
6. EXPOSURE conditions that are foreseeable for the nanomaterial formulated

The European legislation has the right to make amendments to the list of information that needs to be provided based on the scientific progression of the studies. Moreover, if the EC has any safety related concerns about the product, it can ask the SCCS to provide their opinion on the same. The SCCS generally provides its opinion within six months from the time when the request is raised by the EC. If this data provided by the SCCS stands to be unsatisfactory, then the EC asks the 'appointed person responsible' for further data. Once this additional information is provided by the appointed person within the fixed time frame, the SCCS considers it and then again shares its opinion. The final information/opinion that has been concluded by the

EC is made available publically. The entire procedure in instances of the above-mentioned safety concerns are laid out in the Section 4 of Article 16. Based upon the data provided by the SCCS, if the EC concludes that there are risks associated with human health, then, it can consequently make amendments to the Annexure II and III of the EC regulation 1223/2009 which enlists the substances whose use is prohibited in the formulation of cosmetics.

Further, Article 19 lays down the regulations pertaining to labelling such as the nominal content that is present in the formulation, the name of the person who is responsible, precautions to be taken when using the product, the function/use of the product, the batch number and expiry date of the product and a list of ingredients that are present in it. The labelling should be done in a manner that the above-mentioned writings are legible and visible clearly. This article makes it obligatory to enlist all the names of the ingredients which are present in the nanoform using the suffix 'nano' present in brackets following the INCI name.

Article 10 of this EC regulation makes it compulsory to provide a safety assessment report according to details mentioned in Annex I of the EC regulation. The report generally comprises two parts, Part A contains information about the safety of the product and Part B contains information about the safety assessment that has been conducted. Part A contains information that assists in identifying and quantifying the potential hazards which may arise at different steps, be it from the raw materials or from any step of the manufacturing process or during packaging. The report should also contain the toxicological profile of all the substances present in it and the respective assessment conducted for each of them, their side effects and changes in toxicity profile which is dependent upon the particle size. The cosmetic safety report prepared hereunder by the person responsible should be updated from time to time based on the additional information that is obtained even after the product is placed on the market. As nanomaterials are undergoing continuous evaluation and progress, regular update regarding their safety is a time-bound and sensitive matter.

16.3   CONSUMER PROTECTION, ASSOCIATED HEALTH RISKS, AND ENVIRONMENTAL CONCERNS IN RELATION TO NANOCOSMETICS

Nanoparticles have the potential of causing cytotoxicity and pose to be a consequent risk to human health which is dependent upon various factors such as the physicochemical properties of the nanomaterials, the quantity of the nanomaterial present, the route of administration, the duration of exposure to the nanomaterial, etc. The three main routes of nanomaterial entry are by either inhaling or ingesting or through the skin. The National Institute of Occupational Health and Safety (NIOHS) states that the maximum exposure to nanomaterials generally occurs by inhaling the airborne nanoparticles. This can especially happen during the manufacturing process and thus it stands to be a major health hazard to the people who are involved in various steps of its production (Schulte et al. 2008; "Engineered Nanoparticles - Current Knowledge about OHS Risks and Prevention Measures" 2022). Ingestion, most likely can occur from mouth to hand and travel to vital body organs. Topically, nanomaterials can travel deeper into skin layers and pose to be a risk. Studies have shown that nanomaterials have travelled to the deeper layers of pig skin within 24 hours of exposure (Ryman-Rasmussen, Riviere, and Monteiro-Riviere 2006). The US Government Accountability Office (GAO) has shown that upon application of sunscreen on damaged skin, the nanomaterials present in it can travel from it into our body and lead to serious side effects (Schulte et al. 2008). Given the small size of nanoparticles, they can easily traverse across the membranes in the body and enter the cells and tissues wherein they may elicit effects that may lead to cell death (Wani et al. 2011). When the materials are nano-sized, their fundamental properties are altered and different from that of the larger-sized particles of the same. Usually, the biological and chemical reactivity are enhanced in the nanoform owing to high surface area to volume ratio. Nanoparticles being more chemically reactive in nature lead to increased formation of reactive oxygen species (ROS) and free radicals. These species are majorly responsible for the cytotoxicity brought about by the nanoparticles as they lead to oxidative stress which may subsequently lead to damage of the cells, membranes, and DNA. Damages incurred at the cellular and DNA level can lead to development of malignancy as well (Ahamed 2014). Nanomaterials have the potency of precipitating toxicity in various systems be it pulmonary or cardiovascular or reticuloendothelial or neurological systems.

Neonatal safety of nanocosmetic products is another area of major concern. Passage of nanomaterials across the placental barrier or the endometrium or the yolk sac or to the fetus can lead to damage of the placenta or neonatal development, fetal deformities, and other functional disorders in the newborn (Yang, Yi, and Bing 2014).

They accumulate in the body, leading to development of oxidative stress, formation of ROS, autophagy, and even disruption in the functions of lysosome, etc. (Stern, Adiseshaiah, and Crist 2012). Nanoparticles are capable of escaping biological barriers and interfering with the normal reactions in biological systems.

Efforts are concerted toward utilizing nanomaterials as tools for improving the quality of air/water, managing waste effectively, reducing emissions of greenhouse gases, treating chemical effluents effectively, etc. However, nanomaterials also have a downside associated with their use, upon being discharged into water/air/soil systems in definite quantities, they can pose to be a huge environmental risk. As per investigations conducted by the US Government Accountability Office (GAO), and by the University of Toledo, etc. some nanomaterials that have antibacterial activity may hamper the normal activities of the microbes that are present in water-treatment plants and thus render the water treatment ineffective (Raj et al. 2012). Nanomaterials at high concentrations can interrupt the metabolic functions of the microbe which diminishes their role of maintaining balance in various biogeochemical cycles and consequently leads to imbalance in the nutrient cycle. Titanium oxide (TiO_2) nanoparticles that are used for disinfecting and also decomposing pollutants may have the potential to influence other photochemical reactions in the atmosphere. Chronic exposure to TiO_2 can also damage marine life, especially algae species and daphnids, consequently leading to disturbing the marine ecological balance (Hund-Rinke and Simon 2006).

Consolidating the risks associated with nanomaterials, both externally and internally, is yet to be comprehensively mapped for all the nanomaterial-based products that are manufactured. Controlling the synthesis of nanomaterials is yet to be implemented. Adopting principles of green nanotechnology, ensuring biosynthesis of nanomaterials from pre-existing materials, and eliminating bioaccumulation of nanoparticles are essential steps that are needed to be incorporated.

16.4 FUTURE PERSPECTIVES

Cosmetics are no longer limited within the boundaries of having the sole purpose of enhancing the aesthetic appeal but are having broader uses, repositioning themselves as therapeutic agents along with having nutritional significance, such that they promote the overall health of the skin/hair/nail/face as per their scope of application. The market share of nanotechnology-based cosmetic products was approximately valued at 460 billion USD in 2014 which rose to approximately 675 billion USD in 2020, thus depicting the exponential growth that it has happened and is still undergoing. There are several innovations underway pertaining to cosmetics. Moving beyond the regular creams, lotions, gels, etc. to advanced nanosystems to adopting newer materials and platforms for better results is an ongoing journey. A few prominent scopes and facets of advancing cosmetics and future outlooks are discussed in the following section.

16.4.1 EXPLORING NANOCOSMETICS USING BLOCK COPOLYMERS

Block copolymers stand to be one of the most promising nanoparticles for developing cosmetics in the coming years. Currently, there are no cosmetic products that use them. Most of the developments are still in the stages of exploratory research. The two-block copolymers that have been studied most extensively are polyesters and poly-ethylene oxide, which are hydrophobic and hydrophilic in nature respectively. A CAS number of REACH registration has not yet been allocated to them as of yet. Poly(ethylene glycol)-block-poly

(e-caprolactone) methyl ether (PCL-b-PEG), Poly(ethylene glycol)-block-polylactide methyl

Ether (PLA-b-PEG), and Poly(ethylene glycol) methyl ether-block-poly(L-lactide-co-glycolide) (PLGA-b-PEG) are the three commercially available block copolymers from Sigma-Aldrich. One of the earliest reports of topical application of block copolymers was PCL-b-PEG loaded with minoxidil onto guinea pig skin. In this study, the micellar block copolymers exhibited high skin permeation. Dermatological formulations require penetration across the skin and herein the delivery of minoxidil proved to be quite promising. In a study by Bourezg et al., nanoparticles of poly(dihexyl lactide)-block-poly(ethylene glycol) loaded with several antifungal agents were more efficacious in their activity than commercially available oil-in-water cream Pevaryl® (Laredj-Bourezg et al. 2015). In a study comparing the amount of Vitamin A being absorbed across the pig skin from different formulations, it was found that the formulation made using block copolymers gave one and two orders of magnitude higher absorption than polysorbate 80 surfactant micelles and oil solution respectively. Moreover, a differentiating characteristic was also noted, that the PLA-b-PEG nanoparticles showed higher absorption than PCL-b-PEG nanoparticles (Hansen 2007). Block copolymers having lengthier hydrophobic blocks of amphiphilic nature have a self-assembling property and closely resemble phospholipid liposomes. They are known as 'Polymersomes' (Rastogi, Anand, and Koul 2009). Block copolymers are also used as stabilizers for preparing nanoemulsions as they get adsorbed onto the surfaces of the oil droplets (Laredj et al. 2012). Drug or cosmetic actives can be loaded onto the block copolymer nanoparticles and oil droplets of the emulsion simultaneously, thus making provision for double-level skin delivery along with prolonged effect. It has been found that block copolymers achieve faster delivery than emulsion droplets (Laredj-Bourezg et al. 2014; Laredj-Bourezg et al. 2017). They stand to be promising carriers for the upcoming generation of nanocosmetics.

16.4.2 EXPLORING PHOTONIC NANO-CHITIN OR NANO-LIGNIN-BASED NATURAL COLORS

Several consumers prefer using nanocosmetics that do not have any artificial coloring agents. The natural phenomenon of the scattering of light by the chitin-photonic-nanocrystals in combination with pigments gives rise to the bright colors present in the wings of butterflies and several birds. Conceptualizing and developing high-technological nanotech approaches for utilizing this principle for coloring cosmetics will be an advanced innovation (Piszter et al. 2011; Elbaz et al. 2017). Understanding and elucidating these complex mechanisms wherein chitin and lignin can be used as natural colors for mimicking the same approach as seen with colors present in a peacock's tail, etc. will be path-breaking.

16.4.3 SUSTAINABLE WASTE MANAGEMENT AS THE SOURCE OF NANOCOSMETIC MATERIAL

Each year approximately 3 billion tonnes of food waste is produced worldwide, which stands to be a great environmental problem as it becomes a challenge to manage this waste effectively, leaving behind a very high carbon footprint. Rerouting this waste to serve as raw material for producing a variety of biodegradable materials is a viable alternative. Other sources are agricultural waste, animal kingdom by-products, sugarcane, other plant biomass, etc. Chitin is a major constituent of the exoskeleton of numerous invertebrates, for example, crustaceans. It is also obtained from the by-products of fisheries. Lignin on the other hand is obtained majorly from plant-biomass and food and agricultural sources (Morganti et al. 2019). Around 75% of the total body weight of crustaceans ends up as waste material that stands to be an environmental concern as these wastes are either dumped back into the sea or allowed to spoil or buried in landfills, all of which are not sustainable alternatives. Extracting chitin from the shells and utilizing the same is a preferred option (Hamed, Özogul, and Regenstein 2016). Lignin is another abundant macromolecule available at our disposal. Every year, approximately 100 million tonnes of lignin is produced. It should be noted that lignin constitutes 15–25% of plant biomass, and is naturally available as a renewable source. Manufacturing facial wipes, baby wipes, diapers, face masks, etc. can be done using these. The field of nanocosmetics thus has a huge scope of using these as raw materials for formulating sustainable natural cosmetics.

16.4.4 Upcoming generation of smart carriers, bio-fibers, and bio-skin

Polymeric materials that are biodegradable in nature, when made into non-woven tissues, serve to be promising carriers for application in the field of biomedicine and cosmetics. Vast options of biodegradable polyester and biopolymer molecules fall under this category. For example, cellulose-based, chitin-based, lignin-based, silk-based, and starch-based compounds and their respective derivatives are the diverse options available amongst polysaccharides. Amongst biopolyesters, polylactic acid (PLA) and polyhydroxyalkanoate (PHA) are the major ones (Morganti, Coltelli, and Serena 2018).

Adopting techniques like spray-drying to make chitin nanofibrils and nano-lignins that are encapsulating actives is a feasible strategy. Once these nanoparticles are formed, they can be transferred onto the nanofiber that is used for making the nonwoven tissue scaffolds and delivering them to the skin. This innovative approach has the potential of directing the growth of skin cells, delivery of nutrients, the release of waste, oxygen, etc. (Rancan, Blume-Peytavi, and Vogt 2014; Morganti et al. 2016). They can elicit a variety of pharmacological actions, working as immunomodulators or anti-inflammatory agents or anti-aging agents (reducing fine lines, spots, and wrinkles) or mediating repair of tissues or inhibiting the growth of bacteria, etc., all of which lead to healthier skin (Rancan, Blume-Peytavi, and Vogt 2014). Developing smart tissues by combining nano-chitin and nano-lignin together by the gelation method was done in various studies (Desai 2016). These nanofibers can also be used for making clothing materials that have antibacterial properties by manufacturing them by incorporating silver, copper, etc. Skin cancer, which is one of the most common forms of cancer affecting many can be curtailed by making garments incorporating nanofibers that have zinc oxide and titanium oxide in their structure that can ward off UV rays. Apart from this, nanofibers and nonwovens have a huge scope of expansion in making industrial and healthcare filtration products and air ventilation systems, both for heating and cooling. These nanomaterials are used for manufacturing fuel cells, batteries, capacitors, etc. As nanomaterials have the capability of migrating and entering the deep tissues which may precipitate unknown/unwanted reactions, it becomes an absolute necessity to ascertain the biocompatibility and safety of the nano-chitin/lignin that are used, presenting their effectiveness and absence of any side effects such as the development of allergies, etc. (Morganti, Coltelli, and Serena 2018). Sustainably developing chitin nanofibrils and their use thereof has been patented as well ("US8552164B2- Spray-Dried Chitin Nanofibrils, Method for Production and Uses Thereof - Google Patents" 2022; "US8383157B2- Preparation of Chitin and Derivatives Thereof for Cosmetic and Therapeutic Use - Google Patents" 2022). Some studies which substantiate the same, for example, hyaluronan-loaded chitin nanofibrils for anti-aging action were determined to be safe both in vitro (using human keratinocytes and fibroblasts) and in vivo (Morganti et al. 2014).

The above-discussed biomaterials can be used for generating scaffolds of tissues which can lead to restoration of the skin microenvironment, improvement in the delivery of cosmetic actives, and also serve as depots mediating slow-release for long-lasting effects (Morganti et al. 2017; Morganti, Coltelli, and Serena 2018). The production of these smart carriers is possible without using chemical emulsifiers/

preservatives/coloring agents/ fragrances, etc. thus offering safer, completely natural alternatives (Morganti et al. 2016). Thus bio-based nanoparticles and nanosheets offer increased loading, enhanced delivery, more safety, and less or no toxicity as even the metabolic products of these sources are biological products (Shikhman et al. 2009). These scaffolds, when properly designed structure-wise with the right actives, have the potential of replacing damaged or diseased skin, thus conferring aesthetic improvement (Parisi et al. 2018). They offer the flexibility of modifying their physicochemical properties to exhibit controlled release of the actives loaded onto their nanofiber structures. Innovative development of PolyBioskin® in Europe is a noteworthy example wherein bio-nanofibers were used for fabricating biodegradable nonwoven tissues that have utility in making facial masks/tissues, wipes, wound-healing, etc. (Morganti et al. 2017; Morganti, Coltelli, and Serena 2018). Using electrospinning technology or centrifugal spinning technology properties of the material can be modulated to mimic skin's extracellular matrix. Also, based on the consumer group, be it aged or middle-aged or young or people with diseased skin, these carriers and the ingredients chosen can be modified (Morganti 2020; Morganti, Chen, and Gianluca 2020).

16.4.5 EMERGING COLD PLASMA THERAPY AS A PLATFORM FOR COSMETIC THERAPY

Cold atmospheric plasma (CAP) therapy is a new and upcoming physical treatment methodology for dermatological conditions. It has already proved to be useful in wound-healing and cancer therapy (Dubey, Parab, et al. 2022). Gases such as nitrogen or noble gases or normal air are ionized to a plasma state and are in the temperature range of 30–100°C. It mediates skin resurfacing which is much less invasive in nature than plastic surgery. Plasma treatment changes the permeability of skin, which can facilitate the passage of exogenous cosmetic/drug actives (Busco et al. 2020). CAP influences skin-cell proliferation and oxygen levels and affects cellular signaling pathways, skin vascularity, and extracellular matrix properties. Thus, depicting the vast scope of skin modulation at different levels. CAP also has the ability to modify pH and kill bacteria. Nascent stages of research with few successes are underway that are aiming to use CAP for anti-wrinkling, skin-lightening, and skin-tightening effects along with reducing other signs of aging ("Plasma: Groundbreaking Skin Rejuvenation" 2022 ; "Plasma Skin Rejuvenation | The Solent Medi Spa" 2022). Understanding of the precise mechanism involved and effects at cellular and molecular levels are yet to be understood clearly, and thus it is an area that requires further exploration. CAP therapy has huge scope in regenerating skin and promoting skin wellness (Busco et al. 2020).

16.4.6 MARINE NANOCOSMETICS

Approximately 70% of the earth's surface is covered by oceans and marine species amount to more than 50% of biodiversity on the planet, thus serving to be a rich bounty of novel compounds yet to be explored (Aneiros and Garateix 2004). Material ranging from polysaccharides to vitamins to minerals to poly-unsaturated fatty acids to several enzymes and peptides are found in marine sources (Kim 2014).

Carbohydrates are the most abundant organic compound and form an integral part of biogeochemical cycles. Marine carbohydrates are present as sulphated or non-sulphated polysaccharides, which are a rich source of several nutrients. Seaweeds are a rich source of sulphated polysaccharides that have diverse utility, working as antioxidants, blood-coagulating agents, anti-cancer agents, and antiviral agents, thus carving out use in pharmaceutical and cosmetic industries. Brown algae can regulate sebum production along with having anti-inflammatory properties, along with working as conditioning agents (Yuan and Walsh 2006; Fitton, Irhimeh, and Falk 2007).

Phlorotannins are polyphenolic compounds found majorly in brown algae that have huge potential in cosmetics. Marine carotenoids are potential cosmetics as they can be enzymatically converted to form different vitamins, for example, Vitamin A (retinol) which is extensively used in cosmetics (Kim 2014). Fucosterol from algae has high free radical scavenging activity. Marine collagen stands to be a good substitute for bovine collagen that has been in use for a long time, it has good antioxidant and skin-repairing properties. Other prominent marine polysaccharides are agar, galactan, chitin, carrageenan, fucoidan, alginate, etc. Chitin nanofibrils facilitate the maintenance of epidermal homeostasis, neutralization of free radicals, and improvement in skin granular density (Ito et al. 2014). Oligomers of chitosan have been shown to boost fibroblast production, and inhibit metalloproteinase and have excellent water-absorbing properties, rendering them a good moisturizing agent (Zhang and Kim 2009). Fucoidan, obtained from brown algae and carrageenan from red algae are good moisturizing and texture-improving agents that are used in novel cosmetics (Swatschek et al. 2002). Matrix metalloproteinase-1 (MMP-1) are involved in the photoaging process. Treatment with fucoidan has shown to decrease the expression of MMP-1 whilst increasing the expression of type-1 procollagen mRNA and also keep skin elastic fibers elastic, thus having the potential of being developed into anti-photoaging cosmetics (Aneiros and Garateix 2004). Carrageenan and alginate have good gelling properties and are used in cosmetics (30ruco). Employing nanotechnological principles for exploring the potential of marine-based nanocosmetics is yet to reach fruition.

16.5 CONCLUSION

With the coming of age nanotechnological advances and their consequent incorporation into almost all areas of research, even cosmetics, have led to significant improvement in their overall performance. It has made possible the use of different organic and inorganic substances and enhanced the activity brought forth by their nanosized forms. The most used materials in cosmetics are nano-sized titanium oxide, zinc oxide, and gold and silver particles. The use of nanoparticles has been explored in developing cosmetics spanning skin-care, hair-care, nail care, protection against UV rays, dust, pollution, deodorants, antiperspirants, etc. the extent of interaction taking place between the nanomaterial and our cells or cellular processes is a major factor in assessing the effects the said nanomaterial has on our bodies. This is of critical importance and hence having guidelines in place for keeping a check on their use is an utmost necessity. Each guideline has a battery of tests mentioned that are to be undertaken for establishing safety. However, it should be kept in mind that cosmetics applied onto the skin, barely pass the skin barrier and have less contact time; moreover, the results obtained from toxicity assays are dependent upon the route of exposure and the dose undertaken. Thus, results from such studies should not be extrapolated to cosmetic use without careful consideration.

REFERENCES

Ahamed, Niyas. 2014. "Ecotoxicity Concert of Nano Zero-Valent Iron Particles – A Review." *Journal of Critical Reviews* 1 (1): 36–39.

Aneiros, Abel, and Anoland Garateix. 2004. "Bioactive Peptides from Marine Sources: Pharmacological Properties and Isolation Procedures." *Journal of Chromatography. B, Analytical Technologies in the Biomedical and Life Sciences* 803 (1): 41–53. doi:10.1016/j.jchromb.2003.11.005.

Bapat, Ranjeet A, Tanay V Chaubal, Suyog Dharmadhikari, Anshad Mohamed Abdulla, Prachi Bapat, Amit Alexander, Sunil K Dubey, and Prashant Kesharwani. 2020. "Recent Advances of Gold Nanoparticles as Biomaterial in Dentistry." *International Journal of Pharmaceutics* 586: 119596. doi:10.1016/j.ijpharm.2020.119596.

Busco, Giovanni, Eric Robert, Nadira Chettouh-Hammas, Jean-Michel Pouvesle, and Catherine Grillon. 2020. "The Emerging Potential of Cold Atmospheric Plasma in Skin Biology." *Free Radical Biology and Medicine* 161: 290–304. doi:10.1016/j. freeradbiomed.2020.10.004.

"CFR – Code of Federal Regulations Title 21." 2022. Accessed August 7. https://www. accessdata.fda.gov/scripts/cdrh/cfdocs/cfcfr/CFRSearch.cfm?fr=352.50.

Commissioner, Office of the. n.d. "Laws Enforced by FDA - Fair Packaging and Labeling Act." Office of the Commissioner.

"Cosmetic Products | FDA." 2021. Accessed October 26. https://www.fda.gov/cosmetics/ cosmetic-products-ingredients/cosmetic-products.

Desai, Kashappa Goud. 2016. "Chitosan Nanoparticles Prepared by Ionotropic Gelation: An Overview of Recent Advances." *Critical Reviews in Therapeutic Drug Carrier Systems* 33 (2): 107–58. doi:10.1615/CritRevTherDrugCarrierSyst.2016014850.

Dhawan, Surbhi, Pragya Sharma, and Sanju Nanda. 2020. "Cosmetic Nanoformulations and Their Intended Use." In, 141–69. doi:10.1016/B978-0-12-822286-7.00017-6.

"DocsRoom - European Commission." 2022. Accessed September 9. https://ec.europa.eu/ docsroom/documents/38284.

Dubey, Sunil Kumar, Anuradha Dey, Gautam Singhvi, Murali Manohar Pandey, Vanshikha Singh, and Prashant Kesharwani. 2022. "Emerging Trends of Nanotechnology in Advanced Cosmetics." *Colloids and Surfaces B: Biointerfaces*, 112440. doi:10.1016/j. colsurfb.2022.112440.

Dubey, Sunil Kumar, Maithili Kali, Siddhanth Hejmady, Ranendra Narayan Saha, Amit Alexander, and Prashant Kesharwani. 2021. "Recent Advances of Dendrimers as Multifunctional Nano-Carriers to Combat Breast Cancer." *European Journal of Pharmaceutical Sciences : Official Journal of the European Federation for Pharmaceutical Sciences* 164 (September): 105890. doi:10.1016/j.ejps.2021.105890.

Dubey, Sunil Kumar, Shraddha Parab, Amit Alexander, Mukta Agrawal, Vaishnav Pavan Kumar Achalla, Udit Narayan Pal, Murali Monohar Pandey, and Prashant Kesharwani. 2022. "Cold Atmospheric Plasma Therapy in Wound Healing." *Process Biochemistry* 112: 112–23. doi:10.1016/j.procbio.2021.11.017.

Elbaz, Abdelrahman, Jie Lu, Bingbing Gao, Fuyin Zheng, Zhongde Mu, Yuanjin Zhao, and Zhongze Gu. 2017. "Chitin-Based Anisotropic Nanostructures of Butterfly Wings for Regulating Cells Orientation." *Polymers.* doi:10.3390/polym9090386.

"Engineered Nanoparticles – Current Knowledge about OHS Risks and Prevention Measures." 2022. Accessed September 23. https://www.irsst.qc.ca/en/publications-tools/publication/i/100529/n/engineered-nanoparticles-current-knowledge-about-occupational-health-and-safety-risks-and-prevention-measures-second-edition-r-656.

"FD&C Act Chapter VI: Cosmetics | FDA." 2021. Accessed October 26. https://www.fda.gov/regulatory-information/federal-food-drug-and-cosmetic-act-fdc-act/fdc-act-chapter-vi-cosmetics.

Fitton, J Helen, Mohammad Irhimeh, and Nick Falk. 2007. "Macroalgal Fucoidan Extracts: A New Opportunity for Marine Cosmetics." *Cosmetics and Toiletries* 122 (8): 55.

Gupta, Vaibhav, Sradhanjali Mohapatra, Harshita Mishra, Uzma Farooq, Keshav Kumar, Mohammad Javed Ansari, Mohammed F Aldawsari, Ahmed S Alalaiwe, Mohd Aamir Mirza, and Zeenat Iqbal. 2022. "Nanotechnology in Cosmetics and Cosmeceuticals-A Review of Latest Advancements." *Gels (Basel, Switzerland)* 8 (3). doi:10.3390/gels8030173.

Hamed, Imen, Fatih Özogul, and Joe M Regenstein. 2016. "Industrial Applications of Crustacean By-Products (Chitin, Chitosan, and Chitooligosaccharides): A Review." *Trends in Food Science & Technology* 48: 40–50. doi:10.1016/j.tifs.2015.11.007.

Hansen, Charles M. 2007. "Methods of Characterization - Surfaces." In *Hansen Solubility Parameters: A Users Handbook, Second Edition*, January, 113–23. CRC Press. doi:10.1201/9781420006834/HANSEN-SOLUBILITY-PARAMETERS-CHARLES-HANSEN.

Hund-Rinke, Kerstin, and Markus Simon. 2006. "Ecotoxic Effect of Photocatalytic Active Nanoparticles (TiO2) on Algae and Daphnids." *Environmental Science and Pollution Research International* 13 (4): 225–32. doi:10.1065/espr2006.06.311.

Ito, Ikuko, Tomohiro Osaki, Shinsuke Ifuku, Hiroyuki Saimoto, Yoshimori Takamori, Seiji Kurozumi, Tomohiro Imagawa, et al. 2014. "Evaluation of the Effects of Chitin Nanofibrils on Skin Function Using Skin Models." *Carbohydrate Polymers* 101 (January): 464–70. doi:10.1016/j.carbpol.2013.09.074.

Jeswani, Gunjan, Swarnali Paul, Lipika Chablani, and Ajazuddin Ajaz. 2019. "Safety and Toxicity Counts of Nanocosmetics." In 299–335. doi:10.1007/978-3-030-16573-4_14.

Kim, Se-Kwon. 2014. "Marine Cosmeceuticals." *Journal of Cosmetic Dermatology* 13 (1): 56–67. doi:10.1111/jocd.12057.

Kwa, Michael, Leah J Welty, and Shuai Xu. 2017. "Adverse Events Reported to the US Food and Drug Administration for Cosmetics and Personal Care Products." *JAMA Internal Medicine* 177 (8): 1202–4. doi:10.1001/jamainternmed.2017.2762.

Laredj, Faiza, Yves Chevalier, Olivier Boyron, and Marie-Alexandrine Bolzinger. 2012. "Emulsions Stabilized with Organic Solid Particles." *Colloids and Surfaces A: Physicochemical and Engineering Aspects* 413 (November): 252–259.

Laredj-Bourezg, Faiza, Marie-Alexandrine Bolzinger, Jocelyne Pelletier, and Yves Chevalier. 2017. "Pickering Emulsions Stabilized by Biodegradable Block Copolymer Micelles for Controlled Topical Drug Delivery." *International Journal of Pharmaceutics* 531 (1): 134–42. doi:10.1016/j.ijpharm.2017.08.065.

Laredj-Bourezg, Faiza, Marie-Alexandrine Bolzinger, Jocelyne Pelletier, Marie-Rose Rovere, Batoule Smatti, and Yves Chevalier. 2014. "Pickering Emulsions Stabilised by Biodegradable Particles Offer a Double Level of Controlled Delivery of Hydrophobic Drugs." In *Advances in Dermatological Sciences*, 143–56. The Royal Society of Chemistry. doi:10.1039/9781849734639-00143.

Laredj-Bourezg, Faiza, Marie-Alexandrine Bolzinger, Jocelyne Pelletier, Jean-Pierre Valour, Marie-Rose Rovère, Batoule Smatti, and Yves Chevalier. 2015. "Skin Delivery by Block Copolymer Nanoparticles (Block Copolymer Micelles)." *International Journal of Pharmaceutics* 496 (2): 1034–46. doi:10.1016/j.ijpharm.2015.11.031.

Morganti, Pierfrancesco. 2020. "Chapter 1- Nanocosmetics: An Introduction." In *Micro and Nano Technologies*, edited by Arun Nanda, Sanju Nanda, Tuan Anh Nguyen, Susai Rajendran, and Yassine B T - Nanocosmetics Slimani, 3–16. Elsevier. doi:10.1016/B978-0-12-822286-7.00001-2.

Morganti, Pierfrancesco, Hong-Duo Chen, Xing-Hua Gao, Morganti Gianluca, and Febo D. 2019. "Chitin & Lignin: Turning Food Waste into Cosmeceuticals." *Journal of Clinical and Cosmetic Dermatology* 3 (January). doi:10.16966/2576-2826.135.

Morganti, Pierfrancesco, Hong-Duo Chen, and Morganti Gianluca. 2020. "Nanocosmetics: Future Perspective." In, 455–81. doi:10.1016/B978-0-12-822286-7.00020-6.

Morganti, Pierfrancesco, Maria Coltelli, and Danti Serena. 2018. "Biobased Tissues for Innovative Cosmetic Products: Polybioskin as an EU Research Project." *Global Journal of Nanomedicine* 3 (January): 1–5. doi:10.19080/GJN.2018.03.555620.

Morganti, Pierfrancesco, Maria Coltelli, Danti Serena, and Elodie Bugnicourt. 2017. "The Skin: Goal of the EU Polybioskin Project." *Global Research Journal of Pharmacy and Pharmacology* 2 (October): 7–13.

Morganti, Pierfrancesco, Marco Palombo, Francesco Carezzi, Maria L Nunziata, Gianluca Morganti, Maria Cardillo, and Angelo Chianese. 2016. "Green Nanotechnology Serving the Bioeconomy: Natural Beauty Masks to Save the Environment." *Cosmetics*. doi:10.3390/cosmetics3040041.

Morganti, Pierfrancesco, Marco Palombo, Galina Tishchenko, Vladimir E Yudin, Fabrizio Guarneri, Maria Cardillo, Paola Del Ciotto, Francesco Carezzi, Gianluca Morganti, and Giuseppe Fabrizi. 2014. "Chitin-Hyaluronan Nanoparticles: A Multifunctional Carrier to Deliver Anti-Aging Active Ingredients through the Skin." *Cosmetics*. doi:10.3390/cosmetics1030140.

Parisi, Ludovica, Andrea Toffoli, Giulia Ghiacci, and Guido M Macaluso. 2018. "Tailoring the Interface of Biomaterials to Design Effective Scaffolds." *Journal of Functional Biomaterials*. doi:10.3390/jfb9030050.

Piszter, Gábor, Krisztián Kertész, Zofia Vértesy, Zsolt Bálint, and László Péter Biró. 2011. "Color Based Discrimination of Chitin–Air Nanocomposites in Butterfly Scales and Their Role in Conspecific Recognition." *Analytical Methods* 3 (1): 78–83. doi:10.1039/C0AY00410C.

"Plasma: Groundbreaking Skin Rejuvenation." 2022. Accessed August 7. https://blog.theahomebeauty.com/plasma-groundbreaking-skin-rejuvenation.

"Plasma Skin Rejuvenation | The Solent Medi Spa." 2022. Accessed August 7. https://www.thesolentmedispa.co.uk/plasma-skin-rejuvenation/.

Raj, Shoma Jose, US Sumod, and M Sabitha. 2012. "Nanotechnology in Cosmetics: Opportunities and Challenges." *Journal of Pharmacy And Bioallied Sciences* 4 (3): 186. doi:10.4103/0975-7406.99016.

Rancan, Fiorenza, Ulrike Blume-Peytavi, and Annika Vogt. 2014. "Utilization of Biodegradable Polymeric Materials as Delivery Agents in Dermatology." *Clinical, Cosmetic and Investigational Dermatology* 7: 23–34. doi:10.2147/CCID.S39559.

Rapalli, Vamshi Krishna, Saswata Banerjee, Shahid Khan, Prabhat Nath Jha, Gaurav Gupta, Kamal Dua, Md Saquib Hasnain, Amit Kumar Nayak, Sunil Kumar Dubey, and Gautam Singhvi. 2021. "QbD-Driven Formulation Development and Evaluation of Topical Hydrogel Containing Ketoconazole Loaded Cubosomes." *Materials Science and Engineering: C* 119: 111548. doi:10.1016/j.msec.2020.111548.

Rapalli, Vamshi Krishna, Vedhant Kaul, Tejashree Waghule, Srividya Gorantla, Swati Sharma, Aniruddha Roy, Sunil Kumar Dubey, and Gautam Singhvi. 2020. "Curcumin Loaded Nanostructured Lipid Carriers for Enhanced Skin Retained Topical Delivery: Optimization, Scale-up, in-Vitro Characterization and Assessment of Ex-Vivo Skin Deposition." *European Journal of Pharmaceutical Sciences : Official Journal of the European Federation for Pharmaceutical Sciences* 152 (September): 105438. doi:10.1016/j.ejps.2020.105438.

Rapalli, Vamshi Krishna, Tejashree Waghule, Neha Hans, Arisha Mahmood, Srividya Gorantla, Sunil Kumar Dubey, and Gautam Singhvi. 2020. "Insights of Lyotropic Liquid Crystals in Topical Drug Delivery for Targeting Various Skin Disorders." *Journal of Molecular Liquids* 315: 113771. doi:10.1016/j.molliq.2020.113771.

Rastogi, Rachna, Sneh Anand, and Veena Koul. 2009. "Flexible Polymerosomes-an Alternative Vehicle for Topical Delivery." *Colloids and Surfaces. B, Biointerfaces* 72 (1): 161–66. doi:10.1016/j.colsurfb.2009.03.022.

Ryman-Rasmussen, Jessica P, Jim E Riviere, and Nancy A Monteiro-Riviere. 2006. "Penetration of Intact Skin by Quantum Dots with Diverse Physicochemical Properties." *Toxicological Sciences : An Official Journal of the Society of Toxicology* 91 (1): 159–65. doi:10.1093/toxsci/kfj122.

Schulte, Paul, Charles Geraci, Ralph Zumwalde, Mark Hoover, and Eileen Kuempel. 2008. "Occupational Risk Management of Engineered Nanoparticles." *Journal of Occupational and Environmental Hygiene* 5 (4): 239–49. doi:10.1080/15459620801907840.

"Scientific Committee on Consumer Safety SCCS the SCCS Notes of Guidance for the Testing of Cosmetic Ingredients and Their Safety Evaluation 11th Revision." n.d.

Shikhman, A R, D C Brinson, J Valbracht, and M K Lotz. 2009. "Differential Metabolic Effects of Glucosamine and N-Acetylglucosamine in Human Articular Chondrocytes." *Osteoarthritis and Cartilage* 17 (8): 1022–28. doi:10.1016/j.joca.2009.03.004.

Stern, Stephan T, Pavan P Adiseshaiah, and Rachael M Crist. 2012. "Autophagy and Lysosomal Dysfunction as Emerging Mechanisms of Nanomaterial Toxicity." *Particle and Fibre Toxicology* 9 (June): 20. doi:10.1186/1743-8977-9-20.

Swatschek, Dieter, Wolfgang Schatton, Josef Kellermann, Werner E G Müller, and Jörg Kreuter. 2002. "Marine Sponge Collagen: Isolation, Characterization and Effects on the Skin Parameters Surface-PH, Moisture and Sebum." *European Journal of Pharmaceutics and Biopharmaceutics : Official Journal of Arbeitsgemeinschaft Fur Pharmazeutische Verfahrenstechnik e.V* 53 (1): 107–13. doi:10.1016/s0939-6411(01)00192-8.

"US8383157B2- Preparation of Chitin and Derivatives Thereof for Cosmetic and Therapeutic Use - Google Patents." 2022. Accessed August 6. https://patents.google.com/patent/US8383157B2/en.

"US8552164B2- Spray-Dried Chitin Nanofibrils, Method for Production and Uses Thereof - Google Patents." 2022. Accessed August 6. https://patents.google.com/patent/US8552164.

Waghule, Tejashree, Vamshi Krishna Rapalli, Srividya Gorantla, Ranendra Narayan Saha, Sunil Kumar Dubey, Anu Puri, and Gautam Singhvi. 2020. "Nanostructured Lipid Carriers as Potential Drug Delivery Systems for Skin Disorders." *Current Pharmaceutical Design* 26 (36): 4569–79. doi:10.2174/1381612826666200614175236.

Wani, Mohmmad Younus, Mohd Ali Hashim, Firdosa Nabi, and Maqsood Ahmad Malik. 2011. "Nanotoxicity: Dimensional and Morphological Concerns." In *Advances in Physical Chemistry* 2011, Edited by Vicki H Grassian, 450912. Hindawi Publishing Corporation. doi:10.1155/2011/450912.

Yang, Li, Zhang Yi, and Yan Bing. 2014. "Nanotoxicity Overview: Nano-Threat to Susceptible Populations." *International Journal of Molecular Sciences.* doi:10.3390/ijms15033671.

Yuan, Yvonne V, and Natalie A Walsh. 2006. "Antioxidant and Antiproliferative Activities of Extracts from a Variety of Edible Seaweeds." *Food and Chemical Toxicology : An International Journal Published for the British Industrial Biological Research Association* 44 (7): 1144–50. doi:10.1016/j.fct.2006.02.002.

Zhang, Chen, and Se-Kwon Kim. 2009. "Matrix Metalloproteinase Inhibitors (MMPIs) from Marine Natural Products: The Current Situation and Future Prospects." *Marine Drugs* 7 (2): 71–84. doi:10.3390/md7020071.

17 Safety assessment of nanocosmetics

M.P. Vinardell and M. Mitjans

CONTENTS

17.1 INTRODUCTION

According to European regulations, cosmetic products are substances or mixtures that come in contact with the external parts of the body, such as the epidermis, hair, nails, lips, etc., in order, to clean, perfume, and/or protect them. The US Food and Drug Administration (FDA) defines a cosmetic as a product (excluding pure soap) intended to be applied to the human body for cleansing, beautifying, promoting attractiveness, or altering appearance. Cosmetics imported into the US must comply with the same FDA laws and regulations as those for the products manufactured domestically (https://www.fda.gov/industry/regulated-products/cosmetics-overview#cosmetic).

When these products contain nanomaterials, they are called nanocosmetics. The first nanocosmetic introduced into the market was the Capture line from Christian Dior in 1986. There has been an increase in the number of products containing nanomaterials since then (Fytianos et al., 2020).

The cosmetic industry uses nanotechnology because these materials enhance certain properties such as color, transparency, solubility, and protection. In many cases, the use of nanomaterials allows the penetration of some ingredients into the skin. Some of the advantages of using nanomaterials in cosmetics are presented in Figure 17.1.

One of the advantages of nanomaterials is that they can stay in suspension longer than their corresponding non-nanomaterial, conferring great stability to the suspension and avoiding phase inversion. The optical activity and transparency of

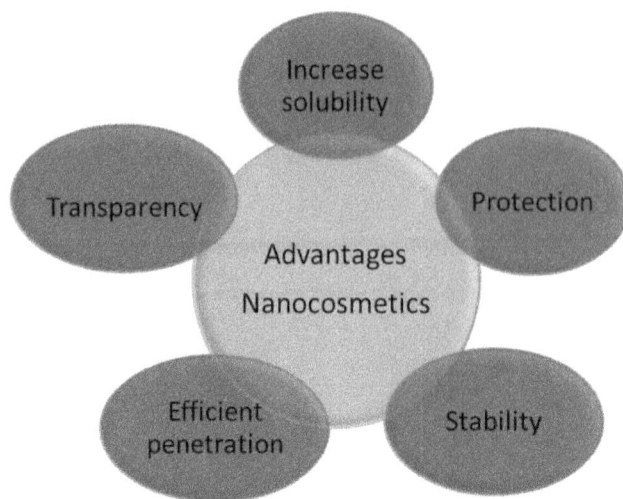

FIGURE 17.1   Advantages of nanocosmetics

nanomaterials give a better appearance to cosmetic products. For instance, zinc oxide (ZnO) and titanium oxide (TiO$_2$) give a white color to sunscreens, but in their nanoforms, they have no such effect, which is preferred by consumers (Kim et al., 2021).

The most common use of nanomaterials in cosmetics is in skincare products and as UV filters (e.g., ZnO and TiO$_2$ nanoparticles) in sunscreens. Different types of nanomaterials are used in nanocosmetics such as liposomes and niosomes, which allow the incorporation of different ingredients. Similar micelles are used to solubilize poorly soluble ingredients. Other forms are nanocapsules, nanoemulsions, solid lipid nanoparticles, nanocrystals, and, more recently, cubosomes (Effiong et al., 2020).

17.2   TOXICITY CONCERNS

There has been an increase in the use of nanocosmetics in recent years, which has led to an increase in concerns about their potential toxicity.

A search in the PubMed database using the term "nanomaterials and cosmetics" retrieved the articles shown in Figure 17.2, illustrating the evolution in the number of studies. The first article in the scientific literature on the use of nanomaterials in cosmetics was in 2000. Thereafter, there was a progressive increase, especially between 2005 and 2010. In the subsequent years, the number of papers stabilized. When the word "toxicity" was included in the search term, the first article was from 2004, followed by a progressive increase in the number of papers that also stabilized. In both searches, there was a decrease in the number of articles in 2021, which can be attributed to the decrease in research as a consequence of the global COVID-19 pandemic.

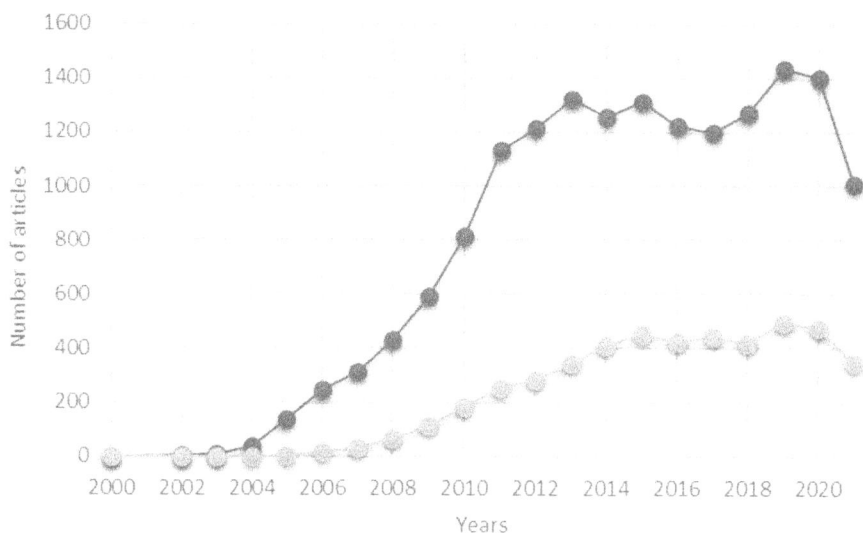

FIGURE 17.2 Number of articles in PubMed since 2000, with those obtained with the search term "nanomaterials and cosmetics" shown in black and those obtained with the search term "nanomaterials and cosmetics and toxicity" shown in gray

17.3   NANOCOSMETIC RISK

The nanomaterials present in nanocosmetics can enter the body by inhalation, ingestion, or skin penetration. The inhalation of nanomaterials can happen with sprayed products. The small size of the nanoparticles allows access to the alveoli, where the nanoparticles can interact with the respiratory epithelium to cause serious toxicity to the respiratory system, such as that seen with ZnO and TiO_2 nanoparticles (Kim et al., 2017). The interaction of nanoparticles with the olfactory epithelium can lead to their absorption and spread to the olfactory bulb in the central nervous system (Garcia et al., 2015). This is one of the more significant concerns related to nanomaterials.

Ingestion occurs with cosmetics applied to the oral area and also accidentally for other cosmetics. The nanomaterials present in the gut are poorly absorbed, such as the TiO_2 nanoparticles. Studies in vivo and in vitro have demonstrated no significant absorption (Jones et al., 2015).

A variety of cosmetics containing nanoparticles are on the market that can penetrate the skin after their application to it. In general, nanomaterials have the potential for systemic exposure as a result of increased skin absorption because of their small size. However, it has been reported that the dermal absorption of ZnO and TiO_2 nanoparticles is very low (Kim et al., 2017).

The skin is the largest organ of the body. The transport of nanoparticles through the skin is related to the nature and physicochemical properties of the nanoparticles and vehicles, as well as the conditions of the skin. In healthy skin, the percentage of nanoparticles absorbed is very small (Senzui et al., 2010). The principal concerns are for damaged skin. However, experiments both in vivo and in vitro have shown that TiO_2 nanoparticles cannot penetrate through damaged skin or intact skin, suggesting that these nanoparticles should be safe when applied to the skin (Xie et al., 2015). One study showed that the TiO_2 and ZnO present in sunscreen formulations displayed slightly enhanced penetration through UVB-damaged skin, but no transdermal absorption was detected. Therefore, these nanoparticles are not absorbed into the bloodstream (Monteiro-Riviere et al., 2011). Some studies have suggested that the capacity of nanomaterials, such as liposomes, to penetrate through the skin depends on their flexibility (Zeb et al., 2019).

17.4 SAFETY REGULATIONS FOR NANOCOSMETICS

In Europe, the term "nanomaterial" has had a legal definition from the European Commission (EC) since 2011: https://eur-lex.europa.eu/LexUriServ/LexUriServ.do?uri=OJ:L:2011:275:0038:0040:en:PDF.

According to the recommendations of the EC, a "nanomaterial" includes a natural or synthetic active or non-active agent ('particle') that can be unbounded or present in 'aggregation' or as an 'agglomerate'. Around 50% or more of the particles must have at least one dimension in the range of 1–100 nm. This definition is currently under review.

Cosmetic products have to be registered on the cosmetic products notification portal (CPNP) by manufacturers, importers, or third persons assigned by the European Union (EU) [https://ec.europa.eu/growth/sectors/cosmetics/cosmetic-product-notification-portal_es]. The CPNP is a free-of-charge online notification system created for the implementation of Regulation (EC) No. 1223/2009 for cosmetic products. When a product has been registered on the CPNP, there is no need for any further notification at the national level within the EU. Complete details about the product must be mentioned in the notification, such as whether nanomaterials are present in the product, their identification, and the probable exposure conditions. Furthermore, all the cosmetic products with nanomaterials, apart from UV filters, colorants, preservatives, and products that are not controlled by the regulation, are required to follow an additional procedure. According to Article 16(3), a particular notice is to be submitted to the CPNP by the responsible persons six months before a product comes onto the market. A risk assessment can be planned by the Scientific Committee on Consumer Safety (SCCS) at the request of the EC if there are some doubts about the safety of the nanomaterials.

In the US, the FDA is responsible for the regulation of a wide range of products, including cosmetics. Cosmetic product manufacturers must ensure that the product is not adulterated. Except for color additives and ingredients that are prohibited or restricted from use in cosmetics, a manufacturer may use any ingredient including nanomaterials in the formulation of a cosmetic, provided that the use of the ingredient does not cause the cosmetic to be adulterated (section 361 of the Food Drug & Cosmetic Act (21 U.S.C. 361) (http://uscode.house.gov/view.xhtml?req=granuleid:USC-prelim-title21-section361&num=0&edition=prelim). Manufacturers or distributors are responsible for obtaining all the data needed to substantiate the safety of their products before introducing them onto the marketplace. To help in the evaluation of the safety of cosmetics, the *Cosmetic Ingredient Review* was established in 1976 by the industry trade association (then the Cosmetic,

Toiletry and Fragrance Association, now the Personal Care Products Council) with the support of the FDA and the Consumer Federation of America. The Expert Panel for the Cosmetic Ingredient Safety and the review process are independent from the Council and the cosmetic industry (https://www.cir-safety.org/about).

17.5 GUIDANCE TO STUDY THE TOXICITY OF NANOMATERIALS IN COSMETICS

There are different guidelines to study the toxicity of nanomaterials. The FDA has published a set of guidelines for the industry (FDA, 2014). This guidance describes how to perform the safety assessment of this type of cosmetic ingredient. In this sense, the characterization of nanomaterials forms an integral part of the safety assessment, which includes a proper identification of the chemical composition and any impurities, as well as the structure and configuration of the nanomaterial. The stability of the nanomaterial under testing conditions and in a formulation under the intended conditions of use should be determined. Appropriate analytical methods suitable for the specific nanomaterial and the cosmetic product formulation should be applied.

In toxicology studies, it is important to consider the applicability of the test methods, considering that the potential agglomeration and aggregation of particles can affect their toxicity. Therefore, the studies should be modified, as needed, with respect to factors such as solvents and dosing formulations, solubility, the agglomeration and aggregation of the particles, and the stability conditions associated with the cosmetic product containing the nanomaterials.

The FDA recommends the assessment of the physicochemical characteristics, agglomeration, and size distribution of the nanomaterials under the conditions of the toxicity assays; and in the final product, impurities, potential routes of exposure to the nanomaterials, potential for nanoparticle aggregation, and agglomeration; dosimetry for in vitro and in vivo toxicology studies, in vitro and in vivo toxicological data on the nanomaterial ingredients, and their impurities, dermal penetration, potential inhalation, irritation (to the skin and eyes), sensitization, and mutagenicity/genotoxicity (FDA, 2014).

In Europe, the SCCS is responsible for the evaluation of the safety of cosmetics (https://ec.europa.eu/health/scientific-committees/scientific-committee-consumer-safety-sccs_en). Their recommendations concerning nanomaterials were published in 2012 as a guidance on the safety assessment of nanomaterials in cosmetics (SCCS/1484/12), which has since been updated with new guidance (SCCS/1611/19). In this guidance, the safety assessment of cosmetic nanomaterial ingredients should consider exposure and the characteristics of these ingredients at the nano-scale. As presented in Figure 17.3, the guidance recommends a detailed physicochemical characterization of nanomaterials in view of the potential changes in the properties of these materials at the nanoscale. The guidance provides important parameters and methods that should be performed for the identification and characterization of nanomaterials intended for use in cosmetic products. Similar to the guidance of the FDA, the exposure assessment and the identification of potential exposure routes form the first critical decision point in the overall safety assessment.

The animal testing ban in the European Union has meant that all studies assessing cosmetic nanomaterial safety since 2013 have been performed without animals,

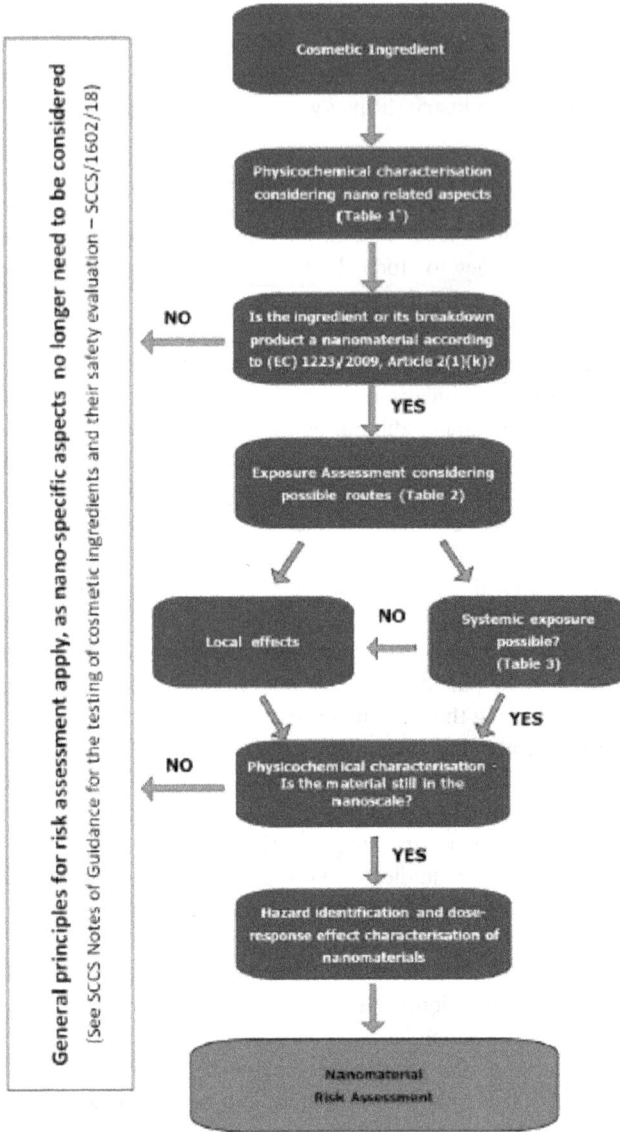

FIGURE 17.3 Schematic outline for the safety assessment of nanomaterials in cosmetics (from SCCS/1611/19)

differing from the requisites of the FDA. For the estimation of systemic exposure based on in vitro models, there are different physiologically based pharmacokinetic (PBPK) models developed especially for some nanomaterials (Utembe et al., 2020).

In 2021, the SCCS published scientific advice on the safety of nanomaterials in cosmetics with a corrigendum in response to the EC requesting a description of the specific concerns identified for certain nanomaterials, such as colloidal silver (nano), styrene/acrylates copolymer (nano), sodium styrene/acrylates copolymer (nano), silica, hydrated silica, and silica surface modified with alkyl silylates (nano form) (SCCS/1618/20). The inconclusive opinions about these nanomaterials were a consequence of the lack of information supplied to the SCCS for their evaluation.

The SCCS concluded that certain aspects of the nanomaterials constituted a concern when used in cosmetic products. These included physicochemical aspects

Descriptor	Question	Answer[a] (score)		
		Yes (3)	No (0)	? (1)
Physico-chemical properties[b] (max 12 pts)	Indication of low or no dissolution or degradation rate in physiologically relevant media?			
	Indication of reactivity? E.g. due to surface area, type of chemical, surface treatment.			
	Indication of release of toxic ions or molecules?			
	Indication that the nanomaterial is persistent and rigid, i.e. a High Aspect Ratio Nanoparticle (HARN)[c] ?			
Hazard (max 12 pts)	Is the chemical itself a substance of very high concern, relating to human health hazard[d] ?			
	Indication of mutagenicity/carcinogenicity (of the material)?			
	Indication of immunotoxicity (of the material)?			
	Indication of other toxicity (of the material)?			
Kinetics (max 12 pts)	Indication of absorption?			
	Indication of distribution to brain or reproductive organs?			
	Indication of accumulation in any tissue?			
	Indication of change in kinetic profile compared to non-nano situation?			
Exposure [e] (max 12 pts)	Products used or likely to be used much or in many products and/or by wide population?			
	Is exposure of sensitive subgroups anticipated? (e.g. babies or elderly people)			
	Is exposure likely to occur frequently (more than a few incidental times)?			
	Is there potential for nanomaterial exposure likely, based on the product use description?			
	Total marks			
		x 3	x 0	x 1
	Sub-score		0	
	Total score			

[a] An indication for a specific physicochemical property, hazard, (toxico)kinetic behaviour or exposure is sufficient to attribute the maximum score of 3. Unknown (= ?) can also be interpreted as 'maybe', in case the indications are weak.
[b] Take into account that outer layers may not be stable and therefore consider changes in surface properties.
[c] HARN = a material that has a diameter < 100 nm and a length many times greater than its diameter (aspect ratio greater than 3 or 5:1), as defined by ECHA (2017) [11].
[d] Reference to ZZS list: http://www.rivm.nl/rvs/Stoffenlijsten/Zeer_Zorgwekkende_Stoffen, only substances on this list that relate to human health hazards are considered
[e] Restricted to exposure of consumers.

FIGURE 17.4 Scoring system proposed by Brand et al. (2019), with key questions to assess the selected signal for prioritization on the potential risk of nanomaterials to human health (from SCCS/1618/20)

related to their very small dimensions, solubility and persistence in the body, their chemical nature and toxicity, their morphology (especially those with a needle shape or rigid long fibers), their surface characteristics determining their reactivity, and their surface modifications that can change the biokinetic behavior of a nanomaterial. There was also concern regarding exposure and the potential accumulation of the nanomaterial in the body.

The SCCS adapted the scoring system proposed by Brand et al. (2019) for food to cosmetic ingredients, which considers key aspects of nanomaterials that can be a signal for their level of risk (Figure 17.4). This should be combined with expert judgment to assign an arbitrary score for prioritization on the basis of potential risk to human health.

17.6   NANOMATERIALS EVALUATED BY THE SCCS

Table 17.1 shows the different nanomaterials evaluated by the SCCS. Among the nanomaterials evaluated was the UV filter bis-(diethylaminohydroxybenzoyl benzoyl) piperazine. The conclusion of the evaluation was based on the available data showing a very low dermal and oral absorption, thereby indicating that it is unlikely to have systemic effects and is safe when used as a UV filter in dermally applied cosmetic products up to a maximum concentration of 10%. However, there are concerns regarding its use in sprayable formulations as it has been shown to have inflammatory effects in the lungs of rats after acute inhalation (SCCS/1634/21). A similar observation has been made with TiO_2 used as a UV filter in sprayable formulations. The SCCS concluded that there was not enough data to draw a conclusion about the safety of this type of formulation and more evaluations should be performed to avoid any harmful effects to consumers (SCCS/1583/17).

The SCCS has evaluated different nanomaterials used in cosmetics that have raised concerns, such as copper nanoparticles that have been linked to possible systemic uptake and a consequent accumulation in certain organs (Tang et al., 2018; Xu et al., 2018). Moreover, the mutagenicity of copper nanoparticles has been reported in the TA98 and TA100 *Salmonella typhimurium* strains, in micronucleus tests and in comet assays with different cell lines (Alarifi et al., 2013; Sadiq et al., 2015). Other nanomaterials raising concerns include platinum nanoparticles, which are absorbed by the respiratory and digestive tracts and can also penetrate the skin (Mauro et al., 2015). Mutagenic effects have been reported for platinum nanoparticles (Nejdl et al., 2017).

Gold nanoparticles have been demonstrated to penetrate through the epidermis, reaching the dermis and hypodermis (Mahmoud et al., 2018; Raju et al., 2018). The penetration of gold nanoparticles depends on their size and shape (Gupta and Rai, 2017) and is enhanced when functionalized with ethylene glycol (Hsiao et al., 2019). Hair follicles are one of the routes for the penetration of gold nanoparticles, as demonstrated recently (Friedman et al., 2021). Most studies, both in vivo and in vitro, have demonstrated genotoxic effects of gold nanoparticles (George et al., 2017). However, other studies (Wang et al., 2020) have not found these effects. Hence, more studies are needed. Particle size plays an important role in determining the genotoxicity of gold nanoparticles (Xia et al., 2017).

TABLE 17.1
SCCS opinions of nanomaterials for different application

Reference	Date	Nanomaterial	Uses	Conclusions
SCCS/1649/23	March 2023	Fullerenes, Hydroxylated Fullerenes and hydrated forms of Hydroxylated Fullerenes	Not provided	The SCCS has not been able to conclude on the safety of fullerenes and (hydrated) hydroxylated forms of fullerenes due to a number of uncertainties and data gaps in regard to physicochemical, toxicokinetic and toxicological aspects.
SCCS/1648/22	March 2023	Hydroxyapatite	Toothpaste, mouthpaste	The SCCS considers hydroxyapatite (nano) safe when used at concentrations up to 10% in toothpaste, and up to 0.465% in mouthwash.
SCCS/1634/21	October 2021	HAA299 *Bis-(Diethylaminohydroxybenzoyl Benzoyl) Piperazine*	UV-filter	The SCCS considers that HAA299 (nano) is safe when used as UV-filter in dermally-applied cosmetic products up to a maximum concentration of 10% because it is a practically insoluble material, with very low dermal and oral absorption.
SCCS/1630/21	June 2021	*Platinum, Colloidal Platinum, Acetyl tetrapeptide-17 Colloidal Platinum*	Leave-on cosmetics	The SCCS is of the opinion that it is not possible to carry out a safety assessment of any of the notified platinum nanomaterials due to limited or missing essential information.
SCCS/1629/21	June 2021	*Gold and Colloidal Gold, Gold Thioethylamino Hyaluronic Acid and Acetyl heptapeptide-9 Colloidal gold*	Leave-on cosmetics	The SCCS is of the opinion that it is not possible to carry out a safety assessment of any of the notified gold nanomaterials due to limited or missing essential information.
SCCS/1624/20	March 2021	*Hydroxyapatite*	Leave-on and rinse-off dermal and oral cosmetic products	The SCCS cannot conclude on the safety of the hydroxyapatite composed of rod–shaped nanoparticles for use in oral-care cosmetic products at the maximum concentrations and specifications given in the Opinion

(Continued)

Table 17.1 (Continued)

Reference	Date	Nanomaterial	Uses	Conclusions
SCCS/1621/20	March 2021	*Copper and Colloidal Copper*	Leave-on and rinse-off dermal and oral cosmetic products	The SCCS has considered all the information provided by the Notifiers and is of the opinion that it is not possible to carry out safety assessment of the nanomaterials Copper and Colloidal Copper due to the limited or missing essential information.
SCCS/1618/20	January 2021 **Corrigendum March 2021**	Scientific Advice on the safety of nanomaterials in cosmetics	-	The SCCS has identified certain aspects of nanomaterials that constitute a basis for concern over safety to consumers when used in cosmetic products. These include: - Physicochemical aspects relating to very small particles; solubility/persistence; chemical nature and toxicity of the nanomaterial; physical/ morphological features of the particles; surface chemistry and surface characteristics (surface modifications/coatings); - Exposure aspects relating to: the frequency and the amounts used and potential accumulation in the body; - Other aspects relating to: novel properties, activity or function, and specific concern arising from the type of application.
SCCS/1606/19	June 2019 Corrigendum December 2019	Solubility of Synthetic Amorphous Silica (SAS)	-	Discussion on the solubility of SAS. The SCCS has noted that the protocols used for solubility tests have a strong influence on the solubility of SAS materials.
SCCS/1596/18	October 2018	*Colloidal Silver*	Antimicrobial agent in cosmetics including toothpastes and skin care products with a maximum concentration limit of 1%.	Due to a number of major data gaps, the SCCS is not in the position to draw a conclusion on the safety of colloidal silver in nano form when used in oral and dermal cosmetic products.

(*Continued*)

Table 17.1 (Continued)

Reference	Date	Nanomaterial	Uses	Conclusions
SCCS/1595/18	June 2018	*Styrene/acrylates copolymer and Sodium styrene/Acrylates copolymer*	Slimming formulation. Anti-aging cosmetics in milks, emulsions, creams, lotions, solutions.	The SCCS cannot conclude on the safety of any of the three styrene/acrylate copolymer nano-entities submitted by the Applicants. The data submitted are insufficient to evaluate possible toxicity
SCCS/1583/17	Last version January 2018	*Titanium Dioxide in spray*	UV filter	The SCCS has concluded that the information provided is insufficient to allow assessment of the safety of the use of nano-TiO2 in spray applications that could lead to exposure of the consumer's lungs.
SCCS/1580/16	November 2016 Final version March 2017 **Corrigendum June 2018**	*Cetyl Phosphate, Manganese Dioxide and Triethoxycaprylylsilane as coatings for Titanium Dioxide (nano) used as UV-filter in dermally-applied cosmetic products*	UV-Filter in dermally applied cosmetic products	In view of the above discussion, which indicates a general lack of dermal absorption and low general toxicity of nano-forms of titanium dioxide, the SCCS considers that the use of the three TiO2 nanomaterials (A, B, C), coated with either cetyl phosphate, manganese dioxide or triethoxycaprylylsilane, can be considered safe for use in cosmetic products intended for application on healthy, intact or sunburnt skin
SCCS/1489/12	September 2012 Revision December 2012	*Zinc oxide*	UV filter	On the basis of available evidence that the use of ZnO nanoparticles with the characteristics as indicated in the dossier, at a concentration up to 25% as a UV-filter in sunscreens, can be considered not to pose a risk of adverse effects in humans after dermal application. This does not apply to other applications that might lead to inhalation exposure to ZnO nanoparticles (such as sprayable products)

17.7   IN VITRO METHODS TO STUDY NANOMATERIAL TOXICITY

The principal difficulty of using classical in vitro methods to assess toxicity is that these methods have not been validated for nanomaterials. Moreover, nanomaterials can interact with the medium and interfere with the conventional in vitro cytotoxic assays, presenting false positives and false negatives (Guadagnini et al., 2015; Mitjans et al., 2018). Therefore, it is necessary to modify these methodologies to adapt them to nanomaterials (Gu et al., 2019). In this sense, the SCCS accepts data obtained with these adapted methods, discarding the information obtained from studies that have not been modified. Table 17.2 shows the most relevant in vitro methods in the opinions of the SCCS for different nanomaterials and the relevant conclusion of the opinion related to each method.

There are different in vitro studies to assess the toxicity of materials and nanomaterials at the local or systemic level.

When a material contacts the skin or eyes, it can induce irritation. The methods applied to evaluate these effects have been validated and adopted by the OECD, involving the use of reconstructed human epidermis (OECD, 2021a) or reconstructed human corneal epithelium (OECD, 2019), respectively. Recently, a different method using another reconstructed human epidermis was validated, with the objective of it being introduced in the corresponding OECD guidelines (Han et al., 2020).

Materials contacting the skin can induce sensitization, which should be evaluated. The in chemico and in vitro tests for this evaluation cover specific key events within the adverse outcome pathway (AOP) for skin sensitization, which is a sequence of events from the underlying molecular events to the adverse outcomes in the whole organism (OECD, 2012). The in chemico tests are the direct peptide reactivity assay (DPRA), amino acid derivative reactivity assay (ADRA), and kinetic direct peptide reactivity assay (kDPRA). These methods are based on in chemico covalent bindings to proteins and correspond to the first event in the sensitization pathway (OECD, 2021b). The assays determining this event in keratinocyte activation are the ARE-Nrf2 luciferase KeratinoSens™ and ARE-Nrf2 luciferase LuSens test methods, which use luciferase to assess the Nrf2-mediated activation of antioxidant response element (ARE)-dependent genes (OECD, 2018a). Other methods include the human Cell Line Activation Test (h-CLAT), which measures the expression of the specific cell surface markers linked to dendritic cell (DC) maturation (i.e., CD86 and CD54) using flow cytometry; the U937 cell line activation test (U-SENSTM), which measures the expression of the specific cell surface marker CD86; and the Interleukin-8 Reporter Gene Assay (IL-8 Luc assay), which measures changes in a cytokine linked to the activation of DCs by measuring IL-8 mRNA induction. These assays investigate the activation of DCs (OECD, 2018b). All these assays should always be considered in combination and/or with other information.

A nanomaterial can be genotoxic if it induces DNA damage or alterations in genes. There are two more relevant in vitro methods to study the genotoxicity of cosmetic ingredients. These include the micronucleus assay and the chromosome aberration assay using mammalian cells, but adapted to nanomaterials to avoid possible interferences (Li et al., 2017). Other methods such as the Ames test using bacteria cannot be applied to nanomaterials because the process of nanomaterial penetration into bacteria is unclear (Doak et al., 2012). However, the results on this are inconsistent

TABLE 17.2

In vitro methods used to assay the safety of nanomaterials in cosmetics evaluated by the SCCS

Nanomaterial	SCCS opinion	Toxicity study	In vitro method	Conclusion
HAA299 (nano)	SCCS/1634/21	Dermal absorption	OECD TG 428 human skin	Systemic availability after dermal application is very low
		Skin irritation	Irritation test Reconstructed Human Epidermis (RhE) according to OECD 439	Not irritant
		Eye corrosion	Bovine Corneal Opacity and Permeability Test (BCOP Test) OECD Guideline no. 437	Not corrosive
		Eye irritation	EpiOcularTM Eye Irritation Test OECD Guideline no. 492	The test findings do not rule out an eye irritation potential from the HAA299 (nano) itself.
Platinum (nano), Colloidal Platinum (nano) and Acetyl tetrapeptide-17 Colloidal Platinum (nano)	SCCS/1630/21	Eye irritation	Het-Cam	Not irritant
Gold (nano), Colloidal Gold (nano), Gold Thioethylamino	SCCS/1629/21	Skin irritation	Reconstructed human epidermis ('EpiSkin') OECD 439	Not irritant
Hyaluronic Acid (nano) and Acetyl heptapeptide-9		Eye irritation	Reconstructed human epidermis ('Skinethic') Reconstructed human Cornea-like Epithelium (tissues) OECD 492	No information about potential interferences
Colloidal gold (nano)		Skin sensitisation	ARE-Nrf2 luciferase ('KeratinoSens') OECD 442D Human cell line activation test – hCLAT. OECD 442E	There is very limited experience with the testing of nanoparticles in the in-vitro assays.
		Intestinal absorption	In vitro permeability across Caco-2 cell monolayers	It has not been validated for nanomaterials
Hydroxiapatite	SCCS/1624/20	Oral irritation	SkinEthic reconstructed Human Oral Epithelium)	Not irritant
		Genotoxicity	Micronucleus test	No conclusion, not enough information

in the literature and a recent study demonstrated the efficacy of the Ames test in detecting the mutagenicity of metal nanoparticles using the liver S9 fraction for bioactivation and a preincubation procedure (Pan, 2021).

Another important aspect to consider when performing in vitro methods is the dosimetry used. Conventional chemicals are expressed based on weight or volume units. However, for nanomaterials, this can be expressed as a function of area or per cell (Huk et al., 2015). This aspect is also considered by the SCCS in their evaluations.

17.8 CONCLUSIONS

Despite the reluctance of some consumers to use cosmetics containing nanomaterials, the nanocosmetics available on the market are safe as their risk has been carefully assessed. In Europe, the SCCS has evaluated the nanoparticle ingredients used in cosmetics. There is guidance available from the SCCS and FDA on the evaluations that should be performed to assess the safety of nanocosmetics. The guidance describes the principal requirements for nanomaterials, such as their characteristics and the safety evaluation process.

REFERENCES

Alarifi, S., Ali, D., Verma, A., et al. 2013. Cytotoxicity and Genotoxicity of Copper Oxide Nanoparticles in Human Skin Keratinocytes Cells. *Int J Toxicol.* 32:296–307. https://journals.sagepub.com/doi/pdf/10.1177/1091581813487563

Brand, W., van Kesteren, P.C.E., Oomen, A.G. 2019. Potential health risks of nanomaterials in food: a methodology to identify signals and prioritise risks [Mogelijke gezondheidsrisico's van nanomaterialen in voedsel: een methode om risico's te signaleren en te prioriteren], RIVM letter report 2019–0191, www.rivm.nl/bibliotheek/rapporten/2019–0191.pdf.

Doak, S.H., Manshian, B., Jenkins, G.J., Singh, N. 2012. In vitro genotoxicity testing strategy for nanomaterials and the adaptation of current OECD guidelines. *Mutat Res.* 745(1–2):104–111.

Effiong, D.E., Uwah, T.O., Jumbo, E.U. et al. 2020. Nanotechnology in Cosmetics: Basics, Current Trends and Safety Concerns—A Review. Advances in Nanoparticles , 9, 1–22. https://www.scirp.org/journal/paperinformation.aspx?paperid=97148.

FDA. 2014. Guidance for Industry: safety of nanomaterials in cosmetic products. https://www.fda.gov/media/83957/download

Friedman, N., Dagan, A., Elia, J., et al. 2021. Physical properties of gold nanoparticles affect skin penetration via hair follicles. *Nanomedicine.* 36:102414

Fytianos, G., Rahdar, A., Kyzas, G.Z. 2020. Nanomaterials in Cosmetics: Recent Updates. *Nanomaterials (Basel)*10(5):979. https://www.ncbi.nlm.nih.gov/pmc/articles/PMC7279536/pdf/nanomaterials-10-00979.pdf

Garcia, G.J., Schroeter, J.D., Kimbell, J.S. 2015. Olfactory deposition of inhaled nanoparticles in humans. *Inhal Toxicol.* 27(8):394–403. https://www.ncbi.nlm.nih.gov/pmc/articles/PMC4745908/pdf/nihms755292.pdf

George, J.M., Magogotya, M., Vetten, M.A., et al. 2017. From the Cover: An Investigation of the Genotoxicity and Interference of Gold Nanoparticles in Commonly Used In Vitro Mutagenicity and Genotoxicity Assays. *Toxicol. Sci.* 156(1):149–166. https://academic.oup.com/toxsci/article/156/1/149/2930558?login=false

Gu, Q., Cuevas, E., Ali, S.F., et al. 2019. An Alternative In Vitro Method for Examining Nanoparticle-Induced Cytotoxicity. *Int J Toxicol.* 38(5):385–394. https://journals.sagepub.com/doi/pdf/10.1177/1091581819859267

Guadagnini, R., Halamoda Kenzaoui, B., Walker, L., et al. 2015. Toxicity screenings of nanomaterials: challenges due to interference with assay processes and components of classic in vitro tests. *Nanotoxicology.* 9(1):13–24.

Gupta, R., Rai, B. 2017. Effect of Size and Surface Charge of Gold Nanoparticles on their Skin Permeability: A Molecular Dynamics Study. *Scientific Reports.* 7:45292.

Hsiao, P.F., Tsai, H.C., Peng, S., et al. 2019. Transdermal delivery of poly(ethylene glycol)-co-oleylamine modified gold nanoparticles: Effect of size and shape. *Materials Chem. Phys.* 224:22–28. https://www.ncbi.nlm.nih.gov/pmc/articles/PMC5507526/pdf/pone.0180798.pdf

Huk, A., Collins, A.R, El Yamani, N., et al. 2015. Critical factors to be considered when testing nanomaterials for genotoxicity with the comet assay. *Mutagenesis.* 30(1):85–88. https://academic.oup.com/mutage/article/30/1/85/1002750?login=false.

Han, J., Kim, S., Lee, S.H. et al. 2020. Me-too validation study for in vitro skin irritation test with a reconstructed human epidermis model, KeraSkin™ for OECD test guideline 439. *Reg. Toxicol. Pharmacol.* 117: 104725

Jones, K., Morton, J., Smith, I., et al. 2015. Human in vivo and in vitro studies on gastrointestinal absorption of titanium dioxide nanoparticles. *Toxicol Lett.* 233(2):95–101.

Kim, K., Kim, Y.W., Lim, S.K., et al. 2017. Risk assessment of zinc oxide, a cosmetic ingredient used as a UV filter of sunscreens. *J. Toxicol. Environ. Health B.* 20 (3):155–182.

Kim, K.B., Kwack, S, J., Lee, J.Y., et al. 2021. Current opinion on risk assessment of cosmetic. *J. Toxicol. Environ. Health B.* 24(4):137–161.

Li, Y., Doak, S.H., Yan, J., et al. 2017. Factors affecting the in vitro micronucleus assay for evaluation of nanomaterials. *Mutagenesis.* 32(1):151–159.

Mahmoud, N.N., Harfouche, M., Alkilany, A.M., et al. 2018. Synchrotron-based X-ray fluorescence study of gold nanorods and skin elements distribution into excised human skin layers. *Colloids and Surfaces B: Biointerfaces* 165:118–126.

Mauro, M., Crosera, M., Bianco, C., et al. 2015. Permeation of platinum and rhodium nanoparticles through intact and damaged human skin. *J. Nanopart. Res.* 17:253.

Mitjans, M., Nogueira-Librelotto, D.R., Vinardell, M.P. Nanotoxicity In Vitro: Limitations of the Main Cytotoxicity Assays. In: Nanotoxicology. Toxicity Evaluation, Risk Assessment and Management Ed. ByVineet Kumar, Nandita Dasgupta, Shivendu Ranjan, 2018, pp 172–192.

Monteiro-Riviere, N.A., Wiench, K., Landsiedel, R., et al. 2011. Safety evaluation of sunscreen formulations containing titanium dioxide and zinc oxide nanoparticles in UVB sunburned skin: an in vitro and in vivo study. *Toxicol Sci.* 123(1):264–80.

Nejdl, L., Kudr, J., Moulick, A., et al. 2017. Platinum nanoparticles induce damage to DNA and inhibit DNA replication. *PLoS ONE.* 12(7):e0180798.

OECD 2012. The Adverse Outcome Pathway for Skin Sensitisation Initiated by Covalent Binding to Proteins. Part 1: Scientific Evidence. Series on Testing and Assessment No. 16 http://www.oecd.org/officialdocuments/publicdisplaydocumentpdf/?cote=ENV/JM/MONO(2012) 10/PART1&docLanguage=En. (accessed February 10, 2022).

OECD 2018a. Test No. 442D: In Vitro Skin Sensitisation: ARE-Nrf2 Luciferase Test Method, OECD Guidelines for the Testing of Chemicals, Section 4, OECD Publishing, Paris, https://doi.org/10.1787/9789264229822-en. (accessed January 20, 2022).

OECD 2018b. Test No. 442E: In Vitro Skin Sensitisation: In Vitro Skin Sensitisation assays addressing the Key Event on activation of dendritic cells on the Adverse Outcome Pathway for Skin Sensitisation, OECD Guidelines for the Testing of Chemicals, Section 4, OECD Publishing, Paris, https://doi.org/10.1787/9789264264359-en. (accessed January 20, 2022).

OECD 2019. Test No. 492: Reconstructed human Cornea-like Epithelium (RhCE) test method for identifying chemicals not requiring classification and labelling for eye irritation or serious eye damage, OECD Guidelines for the Testing of Chemicals, Section 4, OECD Publishing, Paris, https://doi.org/10.1787/9789264242548-en.

OECD 2021a. Test No. 439: In Vitro Skin Irritation: Reconstructed Human Epidermis Test Method, OECD Guidelines for the Testing of Chemicals, Section 4, OECD. (accessed February 14, 2022). Publishing, Paris, https://doi.org/10.1787/9789264242845-en. (accessed February 14, 2022).

OECD 2021b. Test No. 442C: In Chemico Skin Sensitisation: Assays addressing the Adverse Outcome Pathway key event on covalent binding to proteins, OECD Guidelines for the Testing of Chemicals, Section 4, OECD Publishing, Paris, https://doi. org/10.1787/9789264229709-en. (accessed January 20, 2022).

Pan ,X. 2021. Mutagenicity Evaluation of Nanoparticles by the Ames Assay. *Methods Mol Biol.* 2326:275–285.

Raju, G., Katiyar, N., Vadukumpully, S., et al. 2018. Penetration of gold nanoparticles across the stratum corneum layer of thick-Skin. *J. Dermatol. Sci.* 89(2):146–154.

Sadiq, R., Khan, Q.M., Mobeen, A., et al. 2015. In vitro toxicological assessment of iron oxide, aluminium oxide and copper nanoparticles in prokaryotic and eukaryotic cell types. *Drug Chem. Toxicol.* 38(2):152–161.

SCCS (Scientific Committee on Consumer Safety) (2012). Guidance on the Safety Assessment of Nanomaterials in Cosmetics. SCCS/1484/12. https://ec.europa.eu/health/system/ files/2020-10/sccs_o_233_0.pdf

SCCS (Scientific Committee on Consumer Safety), Guidance on the Safety Assessment of Nanomaterials in Cosmetics, 30–31 October 2019, SCCS/1611/19 https://ec.europa.eu/ health/sites/default/files/scientific_committees/consumer_safety/docs/sccs_o_233.pdf.

SCCS (Scientific Committee on Consumer Safety), Opinion on Titanium Dioxide (nano form) as UV-filter in sprays, preliminary version of 7 March 2017, final version of 19 January 2018, SCCS/1583/17. http://publications.europa.eu/resource/cellar/ b635a200-38cd-11e9-8d04-01aa75ed71a1.0001.01/DOC_1

SCCS (Scientific Committee on Consumer Safety), Scientific advice on the safety of nanoma-terials in cosmetics, preliminary version of 6 October 2020, final version of 8 January 2021, SCCS/1618/20, Corrigendum of 8 March 2021. https://ec.europa.eu/health/sys-tem/files/2021-08/sccs_o_239_0.pdf

SCCS (Scientific Committee on Consumer Safety) Opinion on HAA299 (nano) 2021. SCCS/1634/21. https://ec.europa.eu/health/sites/default/files/scientific_committees/ consumer_safety/docs/sccs_o_256.pdf

SCCS (Scientific Committee on Consumer Safety), Opinion on Hydroxyapatite (nano), pre-liminary version 4 January 2023, final version 21–22 March 2023, SCCS/1648/22. https://health.ec.europa.eu/publications/hydroxyapatite-nano-0_en.

SCCS (Scientific Committee on Consumer Safety), Opinion on 2 Fullerenes, Hydroxylated Fullerenes and hydrated forms of Hydroxylated Fullerenes (nano), 3 preliminary version of 21–22 March 2023, SCCS/1649/23. https://health.ec.europa.eu/publications/fuller-enes-hydroxylated-fullerenes-and-hydrated-forms-hydroxylated-fullerenes-nano_en

Senzui, M., Tamura, T., Miura, K. et al. 2010. Study on penetration of titanium dioxide (TiO2) nanoparticles into intact and damaged skin in vitro. *J. Toxicol. Sci.* 35(1):107–113.

Tang, H., Xu, M., Zhou, X., et al. 2018. Acute toxicity and biodistribution of different sized copper nano-particles in rats after oral administration. *Mater Sci Eng C Mater Biol Appl.* 93:649–663.

Utembe W, Clewell H, Sanabria N, et al. 2020. Current Approaches and Techniques in Physiologically Based Pharmacokinetic (PBPK) Modelling of Nanomaterials. *Nano-materials (Basel).* 10(7):1267.

Wang, Y., Zhang, H., Shi, L., et al. 2020. A focus on the genotoxicity of gold nanopar-ticles. *Nanomedicine (Lond).* 15(4):319–323. https://www.futuremedicine.com/doi/ epub/10.2217/nnm-2019–0364

Xia, Q., Li, H., Liu, Y., et al. 2017. The effect of particle size on the genotoxicity of gold nanoparticles. *J. Biomed. Mater. Res. A.* 105(3):710–719.

Xie G, Lu W, Lu D. 2015. Penetration of titanium dioxide nanoparticles through slightly dam-aged skin in vitro and in vivo. *J Appl Biomater Funct Mater.* 13(4):e356–61.

Xu, M., Tang, H., Zhou, X., et al. 2018. Effects and mechanisms of sub-chronic exposure to copper nanoparticles on renal cytochrome P450 enzymes in rats. *Environ Toxicol Pharmacol.* 63:135–146.

Zeb, A., Arif, S.T., Malik, M. et al. 2019. Potential of nanoparticulate carriers for improved drug delivery via skin. *J. Pharm. Investig.* 49:485–517.

18 Current trends and marketed products in nanocosmetics

Anuradha Dey and Sunil Kumar Dubey

CONTENTS

18.1 INTRODUCTION

Employing nanomaterials and nanostructured compounds for developing cosmetic formulations has been gaining momentum in the research industry along with a continued expansion of its applications. Nano-sized versions of several materials provide significant advantages in comparison to their micro/macro forms. They have a high surface area-to-volume ratio, have electromagnetic properties, increased biological activity, higher penetrability, etc. which makes them promising carriers. The term 'nanomaterial' is designated to substances having dimensions less than 100 nm. Reducing the size of existing materials to nanoform or utilizing materials that itself exist in nanoform is requisite for developing the concerned formulations.

DOI: 10.1201/9781003319146-18

Nanoparticles are one of the most industry-relevant materials which have been gaining huge acceptance, especially over the last few decades given that they are of diverse types, confer a high extent of tuning capacity pertaining to various properties, formulation techniques, etc., and can also be functionally modified based on the requirement (Vance et al. 2015; Heiligtag and Niederberger 2013). These flexibilities and avenues for continued exploration make nanocosmetics an area of immense research interest.

18.2   CURRENT TRENDS IN NANOCOSMETICS

One of the major areas of nanotechnology-based applications in cosmetics is the development of skincare products. Commercial cosmetic products developed by utilizing nanoparticles have unique characteristics, gaining the upper hand in comparison to their bulk/molecular forms that have been used in existing products (Dubey et al. 2022). For example, inorganic nanoparticulate form of TiO_2 serves as UV filters whereas its micro/macro form does not (Vinod and Jelinek 2019; Smijs and Pavel 2011). The nanocosmetic industry is undergoing continuous development and the major trends in the industry as of today are discussed herein.

18.2.1   POPULARITY OF CLAIM-SPECIFIC PRODUCTS WITH SCIENTIFIC BACKING AND ENCOURAGEMENT OF CRITICAL FEEDBACK

Ideally, skincare products exert their action in such a way that the natural functions of the skin are not hindered; rather the inherent functioning of the skin is supported by the activity of the applied cosmetic agent. These agents serve to help in regulating the moisture content or act as a protective shield or help in overcoming the gaps wherein the skin itself is failing to carry out its functions along with serving as a cosmetic remedy to acne, blemishes, dark spots, etc. (Rapalli, Waghule, et al. 2020). The vast plethora of products that are available in the market makes it necessary for the current industry players to launch their products that have specific claims along with substantial data backing those claims. With an increase in consumer awareness, people have a more skeptical attitude, such that they prefer those products which provide proof of their efficacy. The incorporation of nanotechnology-based approaches has widened the scope of applicability and combinations of products thereof that have different mechanisms to exert their action. For example, several liposomes that are being developed are made of biocompatible lipids and acids that mimic the biochemical properties of the epidermal layers of the skin, hence they not only act as carriers of the cosmetic agents but also act as an active principle independently. Products fall within different categories based on their area of application, be it as aesthetic appearance enhancers such as make-up products, foundations, concealers, mascara, bronzers, highlighters, etc., or nourishing agents such as lotions, moisturizers, creams, gels, etc. or exfoliators or skin-repairing agents or products conferring protection such as UV-blockers, cleansers, etc. or products mediating anti-aging, skin whitening, anti-hyperpigmentation, collagen-boosting actions, etc. In general, nanocosmetic products have particular activities associated with themselves that they claim to fulfill.

For example, the Rovisome ACE Plus by Evonik claims anti-wrinkling and skin revitalizing action ("ROVISOME® ACE Plus (Antioxidant) by Evonik - Technical Datasheet" 2022), Royal Jelly Lift Concentrate by Jafra Cosmetics claims to restore skin radiance and smoothness along with reducing wrinkles and fine lines within just 14 days of application (Singh and Sharma 2016) are few examples. Sharing clinical results and taking unbiased feedback on their cosmetic products are beneficial in ensuring a longer lifetime of the product in the market.

18.2.2 MARCHING FORWARD WITH SMARTLIPIDS: THIRD GENERATION OF LIPID CARRIERS

The first generation of lipid carriers, solid lipid nanoparticles (SLNs), were formulated incorporating the advantages of two nanosystems: liposomes and polymeric nanoparticles. Liposomes are fluid lipid systems that have high regulatory acceptance and are well tolerated along with being easy to scale up. Polymeric nanoparticles on the other hand have a solid matrix of particulate systems conferring chemical protection along with the controlled release. SLNs are manufactured by high-pressure homogenization approach, similar to that of manufacturing liposomes and emulsions, however at much higher temperatures. In the initial steps, the solid lipids are melted and the actives are incorporated to form a macroemulsion which is then made to pass through a high-pressure homogenizer to form nanoemulsions. In the later steps, the cooling of these nanoemulsions leads to the formation of solid lipids. The disadvantage faced with SLNs was that they had a lower capacity for loading actives and also had the problem of the expulsion of actives. The nanostructured lipid carriers (NLCs) were the second generation of lipid carriers that served to circumvent these challenges (Waghule, Gorantla, et al. 2020). NLCs consist of a blend of solid and liquid lipids wherein the liquid lipids have the capability of solubilizing higher amounts of actives, thus improving the drug/active loading capacity (Waghule, Rapalli, et al. 2020; Rapalli, Kaul, Gorantla, et al. 2020). For example, the capacity of loading retinol increased from 2% to 5% in SLN to NLC (Jenning and Gohla 2001). However, the challenge with NLCs is that the oil form present in them accelerates the process of polymorphic transitioning (Ding, Pyo, and Müller 2017). This propelled the search for more stable systems, leading to the inception of the third generation of 'SmartLipids'. SmartLipids are a polydispersion of many lipids, mono, di, and triglycerides with fatty acids (Müller, Shegokar, and Keck 2011; Ding, Pyo, and Müller 2017), both in solid and lipid form. This latest generation of lipids exhibited a higher increase in the capacity of loading retinol, increasing from 5% in NLCs to 5% in SmartLipids (Olechowski, Müller, and Pyo 2019). SmartLipids confer several advantages such as having high adherence to the skin, acting as stronger protective barriers, more chemical stability, repairing skin defects, hydrating the skin, warding off pollutants, and facilitating increased absorption of the cosmetic actives. They are the upcoming lipid nanocarriers which are being used for developing cosmetic formulations. In a study by Kopke et al., phenyl ethyl resorcinol loaded SmartLipids were developed for working as skin brightening agents (Köpke, Müller, and Pyo 2019).

18.2.3   A SHIFT FROM ANTI-AGING TO ANTI-POLLU-AGING COSMETICS

Nanocosmetics working based on the principle of addressing the signs of aging, by either diminishing their appearance or slowing down their manifestation have been in the market since the late 1980s with numerous products catering to the demands of wide demographics. However, in recent years, the trend has been to develop cosmetics that can curtail the harmful effects that pollutants have on our skin. Skin, being the largest organ of our body faces maximum exposure to the environment. Fine dust particles, exhaust gases, smog, and smoke comprise the major sources of environmental pollutants which not only harm our respiratory system but also interact with our skin. These are the underlying causes that accelerate the process of skin aging. Many of the pollutants in these sources are organic compounds or heavy metals which get adsorbed onto the skin surface leading to skin irritation or inflammation or redness etc. Soot particles in the air clog our pores, lead to skin pigmentation, etc. Many of these pollutants cause oxidative stress leading to more damage and accelerated aging. In a study by Krutmann et al., it was found that women living in high pollution areas had 20% more pigmentation and spots on their skin (Vierkötter et al. 2010). The growing interest in this arena especially lies in Asian countries given that pollution is a major concern there, but in recent times European countries have also started focusing on the same. Several cosmetic giants like Nivea, Lancome, etc. have launched their anti-pollution line of products. Urban Skin Protect by Nivea, Germany, Hydro Effect Serum by Iavera, Germany, and City Miracle Crème by Lancome, France, is some of the leading anti-pollution-based nanocosmetics that are present in the market. Nanocosmetics have a huge scope of development in the arena of anti-pollu-aging agents which can function as protective skin barriers, and deliver cosmetic actives that have anti-oxidant activity along with actives that mediate skin hydration, and skin lightening effects.

18.2.4   STRENGTHENED MARKETING STRATEGY

With the digital revolution, global cosmetic leaders have strengthened their advertising and marketing gimmicks. Utilizing all the available platforms, be it social media platforms, billboards, conducting touted pre-launch launch and post-launch events, organizing competitions, signing on celebrities/models, etc., are the strategies being undertaken by companies for ensuring the success of products. This has especially increased in recent years as companies now have direct access to the consumers via these platforms. For a nanocosmetic product post its launch, to be successful and create hold in the market, it is necessary to have a strong marketing community that can assertively imprint in the mind of the customers the benefits/effects of their products. All companies have aggressive marketing departments which have a huge role in making sure that the product is a success. Highlighting the superior efficacy or enhanced performance or quicker effects or novel application or a combination of either property needs to be conveyed to the consumers shortly and crisply. Numerous global cosmetic leaders have bountiful resources to expend for developing novel cosmetic products utilizing nanotechnology to ensure top-notch marketing for the same. Marketing and research go hand-in-hand as only the marketing team which is in the field are best at perceiving the wants of the customers and relay the same to the research and development teams for aligning their works based on the same. Thus, this approach of undertaking research based on the trends in consumer preferences is the best way to ensure that the products being launched in the market prove to be successful in terms of generating returns (Sakamoto et al. 2017).

18.2.5 EXPANDING HORIZONS: COSMOTHERAPY AND COSMECEUTICALS

Since the coining of the term "Cosmeceutical" by Albert Kligman in the year 1984, which as the name indicates is a hybrid between "cosmetics" and "pharmaceuticals", the field of cosmeceuticals has come a long way, blurring the boundaries between cosmetic applications and pharmaceutical effects (Kligman 2000). They encompass products that contain constituents that have therapeutic and nutritive value, promoting healthy skin or hair, or nails. They include all formulation types, be it creams or gels or lotions or ointments or liquids or ingestible pills, and functional foods as well. These constituents may comprise vitamins or minerals or natural herbal actives, actives from marine sources, antioxidants, etc. (Ruocco et al. 2016; Ahmad and Ahsan 2020; Kim 2014). "Cosmotherapy", as per Javed Ahmad, does not entail ingestible or nutrient-containing products (Ahmad 2021). Cosmeceuticals or cosmotherapeutics have wide-ranging applications. Acne is one of the most common skin conditions, affecting a majority of individuals. Developing lipid nanoparticles for the delivery of actives that address acne have proven to be successful in overcoming the toxicity or side-effect-related concerns that are commonplace with most existing anti-acne formulations (Date, Naik, and Nagarsenker 2006). In a study by Munster et al. SLNs loaded with RU-58841 myristate, the prodrug form of a nonsteroidal antiandrogen was seen to be successfully taken up by the follicles and bring about anti-acne and anti-aging effects on the skin. Moreover, the SLN formulation brought about its action by forming a local depot on the skin that was limited to the upper layers and did not lead to systemic absorptions (Münster et al. 2005). In a study by Souto et al., the lipid nanoparticles loaded with clotrimazole showed prolonged release and good efficacy in treating mycoses (Souto and Müller 2005, 2006, 2007). Similar results were obtained with econazole nitrate-loaded SLNs (Sanna et al. 2007; Passerini et al. 2009) and miconazole nitrate-loaded SLNs to name a few (Bhalekar et al. 2009). Similarly, terconazole-loaded niosomes were found to be more efficacious in treating napkin candidiasis than the suspension form of the same (Mosallam et al. 2021). Water Bank by Amore Pacific is a rich moisturizer having a lot of vitamins. AcneWorx, a novasome preparation of salicylic acid, marketed by Dermworx, is effective in acne treatment (Singh, Malviya, and Sharma 2011). Calcipotriol and methotrexate-loaded NLCs were found to be safe and effective in treating psoriasis (Lin et al. 2010; Rapalli et al. 2018). Tacrolimus-loaded lipid nanoparticles exhibited enhanced drug release, reduced toxicity, enhanced skin permeation, and prolonged accumulation, all leading to better treatment of atopic dermatitis (Pople and Singh 2010). Another example of cosmotherapeutic application is that of minoxidil which is a common constituent of anti-hair fall therapies. Mostly minoxidil preparations have high alcohol content which leads to dryness, itching, and reddening of the scalp. An NLC-based minoxidil preparation developed by Silva et al. proved to be safer and more effective than the alcoholic formulations (Silva et al. 2009). All these are a few examples that are a testament to the blurring lines of cosmetics and therapeutics/pharmaceuticals.

18.3   MARKETED NANOCOSMETIC PRODUCTS

The cosmetic products that have been developed using nanotechnology can be classified into several classes depending upon the type of nanoformulation that has been developed. Currently, there are numerous nanocosmetics in the market that have different cosmetic actives incorporated into their systems and bring about diverse actions as per their claims. A few of the most important classes of nanocosmetics that have several products in the market have been discussed hereunder.

18.3.1   LIPOSOMES

Sphere-shaped lipid vesicles are known as liposomes which have an aqueous core in the center that is surrounded by a phospholipid bilayer. This structure enables the loading of both hydrophilic and lipophilic drugs. Liposomes confer multiple advantages to the system, they increase the solubility of the actives, exhibit enhanced therapeutic action, are more biocompatible in nature, and provide an avenue for controlled release, whilst also protecting the system from being degraded externally (Kaur and Agrawal 2007; Sriraman and Torchilin 2014). Moreover, liposomes are biocompatible and biodegradable along with having site-specificity in action (Akbarzadeh et al. 2013). Capture (Dior) was the first liposome-nanotechnology-based product to be launched in the market way back in 1986. Liposomes are formulated using lipids such as phosphatidylethanolamine, ceramides, phosphatidylcholine, and cholesterol, mimicking the lipids found on our epidermal layers for easing delivery. There are several liposomal cosmetic products in the market that have diverse use, spanning skin care, hair care, etc. Liposomes make it possible to load plant actives into them, for example, Daeses Lifting Cream® is made of jojoba and fruit extracts and Derma Stemness Reviving Serum® is a combination of hyaluronic acid with stem cells of argan plant. With extensive research pertaining to exploiting liposomes, now there are several classes of liposomes having unique properties of their own. Examples of such systems are niosomes, novasomes, transferosomes, ethosomes, cubosomes, etc. Table 18.1 lists some of the prominent marketed liposomal nanocosmetics.

18.3.2   LIPID NANOPARTICLES

Solid lipid nanoparticles (SLNs) were the first generation of lipid nanoparticles wherein a solid lipid is present inside an aqueous core (Prabhakaran, Sathali, Ekambaram, and Priyanka 2012). The central lipid matrix has made it possible to load hydrophilic or lipophilic or even poorly water-soluble actives into it. Upon application, SLNs generally form an occlusive layer on the skin, thus hindering transepidermal water loss (Arora, Agarwal, and Rayasa 2012). SLNs of paraffin are present in Nanobase® and marketed by Yamanouchi which keeps the skin hydrated. Allure Body Crème by Chanel contains a mixture of lipids, a few of them being squalene,

TABLE 18.1
Marketed liposome based nanocosmetic preparations

Nanocosmetic product	Prominent use/marketing claims	Marketed by
Capture Totale	Anti-aging, skin firming	Dior
Liposome Face and Neck Lotion	Anti-aging, skin hydration, and repair	Clinicians Complex
Ageless Facelift Cream	Diminish fine lines, wrinkles, decrease pore size	I-Wen Naturals
Moisture Liposome: Eye Cream/ Face Cream	Moisturizing, brightening, and firming skin around eyes	Decorte
Rovisome ACE Plus	anti-aging and antioxidant action	Evonik
Royal Jelly Lift Concentrate	Restore moisture and skin radiance	Jafra Cosmetics
Bio Performance Liposome	Skin hydration, nourishment, and age-defying effect	Dead Sea Premier
Derma Stemness Reviving Serum	Oil-free serum for skin rejuvenation	Kaya Skin Clinic
Acnel Lotion N	Restoring essential fatty acids, anti-acne treatment, skin hydration	Dermaviduals
Isocell MAP	Anti-wrinkle, anti-irritant action, moisturizing effect	Lucas Meyer
Acglicolic Classic Crema Hidratante SPF 15	Anti-aging, skin hydration, renewing skin	Sesderma
Daeses Lifting Cream	Instant skin lift, skin firming agent	Sesderma
Rovisome® Q10 NG	antioxidant and radical scavenger activity	Evonik
Liposome Concentrate	Skin calming, firming, softening, smoothening effect	Russell Organics
C-Vit	Illuminating the face, brightening complexion, minimizing signs of aging	Sesderma
Lumessence Eye Cream	Moisturizing, brightening, and firming skin around eyes	Aubrey Organics
Azelac Ru Serum	Clarifying blemishes, reducing their size and improving skin appearance	Sesderma
Rehydrating Liposome Day Creme	Anti-inflammatory and anti-oxidant properties	Kerstin Florian
Advanced Night Repair Protective Recovery Complex	Reduce multiple signs of aging having long-lasting effects	Estee Lauder
Seskavel Mulberry Anti-Hair Loss Foam	Stimulate hair growth by preventing and stopping hair fall	Sesderma
Clearly It!® Complexion Mist	Facial toner for balancing sebum production	Kara Vita
Fillderma Lips	Lip volumizer	Sesderma
Spectral DNC-N Hair Loss	Hair loss treatment	DS Laboratories,Inc.

tocopheryl acetate and linalool, etc. along with hexyl cinnamal and limonene, etc. SLNs provide an option for the release kinetics to be either controlled or immediate; however, SLNs have quite a few disadvantages associated with them: expulsion of actives, low-loading capacity, long-term stability, etc., thus making it necessary to improve upon them. Expulsion of the actives and reduced loading capacity of actives occurs as SLNs tend to undergo re-crystallization and polymorphic changes during storage (Yoon, Park, and Yoon 2013; Wissing and Müller 2003).

Following this, NLCs are the second generation of lipid carriers whose formulation process was driven by the goal to overcome the limitations of SLNs. NLCs have an improved structure, that is, it is unstructured in nature with an imperfect crystal structure which increases their loading capacity (Ghasemiyeh and Mohammadi-Samani 2018). The matrix core comprises both liquid lipids and solid lipids which confer several advantages such as preventing the expulsion of contents along with increased stability, enhanced entrapping efficiency, and higher loading of actives (Rapalli, Kaul, Waghule, et al. 2020). During storage, crystallization-associated changes are minimized due to the use of liquid lipids in the formulation. Moreover, NLCs exhibit more controllable release profiles when compared to SLNs (Shidhaye et al. 2008). NanoRepair Q10 is one of the first NLC products to be launched by Dr. Rimpler GmbH, Germany, for anti-aging activity having Coenzyme-Q10, hibiscus, and ginger extracts in it (Müller et al. 2007). Olivenol by Dr. Theiss contains senega extracts, milk proteins, almond extracts, caffeine, etc. NLCs are suitable for making creams and body lotions, for example, the SURMER crème by Isabelle Lancray is a mixture of wheat protein and Kukuinut oil; Phyto NLC Active Cell Repair by Sireh Emas has olive oil as the base which is loaded with cucumber, lemon, and curcumin extracts. Entrapping volatile oils and plant bioactives in the lipid matrix and ensuring their controlled release onto the skin has made NLCs popular versatile carriers for cosmetic agents. NLCs are more adhesive in nature compared to other lipid systems (Schäfer-Korting, Mehnert, and Korting 2007). Table 18.2 lists some of the prominent marketed NLCs and SLNs.

18.3.3 NIOSOME

These are self-assembling liposomal vehicles comprising a bilayer of non-ionic surfactant which imparts improved stability and higher entrapment efficiency than liposomes. Several types of non-ionic surfactants can be used for making niosomes, be it alkyl amides, spans, tweens, sorbitan esters, brij, etc. (Kazi et al. 2010). As per the feasibility of production, niosomes are easier to scale up and have lower manufacturing costs than liposomes (Ge et al. 2019; Abdelkader, Alani, and Alany 2014). Niosomes have lesser systemic absorption and are mostly limited to the skin surface for meditating their action and hence are suitable carriers for cosmetic delivery (Lasic 1998). In the 1970s, the cosmetic conglomerate L'Oreal was the first to develop niosomes and acquired a patent for the same (Nasir, Sl, and Kaur 2012). They have a chain of products under the brand name Lancome that is based on niosome technology. The Anti-age Response Cream by Simply Man-match is a niosomal preparation having pomegranate seed oil, avocado oil, and ginseng as its major constituents for anti-aging action. Similarly, the Mayu Base Cream by Laon cosmetics also contains ginseng extracts with saponins for anti-aging effects. Niosomal preparations also have proteins as their constituents, for example, the Mayu Base Cream mentioned earlier has ribonucleotides and Hair Repair by Identik has adenosine along with yeast extracts as its components. Table 18.3 lists examples of marketed niosome preparations.

TABLE 18.2
Marketed SLN and NLC-based nanocosmetic preparations

Nanocosmetic product	Nanoparticle type	Prominent use	Marketed by
Iope Supervital Extra Moist Eye Cream	NLC	Firming and anti-winkling of eye area skin	Amore Pacific
NLC deep effect repair cream	NLC	Extra moisturization and skin softening	Beate Johnen
Cutanova Nano Repair Q10	NLC	Anti-aging and anti-oxidant Skin energizer	Dr. Rimpler GmbH
Phyto NLC Active Cell Repair	NLC	Anti-aging, sebum control, and anti-pigmentation	Sireh Emas
Swiss Cellular White Illuminating Eye Cream	NLC	Diminish dark circles and moisturize the eye area	La Prairie
Regenerations Cream Intensive Ampoules	NLC	Recovery from skin photoaging. Imparting luminosity and firmness to skin	Endocare
NanoLipid Repair CLR	NLC	Skin regeneration and revitalization	Chemisches Laboratorium (Dr. Kurt Richter)
SURMER Crème	NLC	Skin regeneration and moisturization	Isabelle Lancray
Olivenol: Anti Falten	NLC	Inhibit collagen breakdown, reduce the depth of wrinkles	Dr.Theiss (Medipharma Cosmetics)
Nanobase	SLN	Long lasting anti-dryness	Yamanouchi
Allure Body Cream	SLN	Lightweight lotion imparting airy smooth texture	Chanel

TABLE 18.3
Marketed niosome based nanocosmetic preparations

Nanocosmetic product	Prominent use	Marketed by
Anti-Age Response Cream	Anti-aging, moisturizing, and skin protection	Simply Man-Match
Masque Floral Repair	In-depth nourishment of hair	Identik
Newlight serum	Skin whitening	Guinot
Mayu Niosome Base Cream	Anti-aging and skin repair	Laon Cosmetics

18.3.4 INORGANIC NANOPARTICLE-BASED COSMETICS

Inorganic nanoparticles are cosmetic preparations that use metals and metalloids as their constituents. Cosmetic preparations incorporating gold and silver particles have been popular for many decades and now their nanoforms are also widely marketed. Properties such as anti-aging, skin rejuvenation, working as skin-firming agents, etc. are attributed to gold and silver. They also improve blood circulation along with having anti-microbial properties. There are numerous gold nanoparticle-based cosmetic products on the market, for example, the NanoGold Energizing Cream by Chantecaille has 24-carat gold particles fused with algae extracts and vitamins C and E; the Nano Gold Anti-aging Lifting serum by Nuvoderm has coffee, aloe vera, pumpkin seed and coconut extracts with gold particles for anti-aging effects. Several other products are listed in Table 18.4. Nanosilver-based products treat photo-damaged skin, work as a preservative, and are generally used in anti-aging, nailcare, and dental products (Pereira et al. 2014; Mota et al. 2020). Silver nanoparticles along with TiO_2 are used for preparing deodorants with anti-bacterial properties ("US4659560A - Deodorant Compositions - Google Patents" 2021). 99.99% nanosilver is present in a colloidal form in the Cosil whitening Mask marketed by Natural Korea. They also have variants that incorporate aloe vera or collagen extracts into them. Silver also works as an exfoliator, for example, Cleanser Silver is made of nanosilver and sericin (silk-fiber) acting as a pore cleanser. The NanoGold 24-Hour Cream has both gold and silver nanoparticles which have anti-bacterial action.

TABLE 18.4
Marketed nanocosmetic preparations containing inorganic nanoparticles

Nanocosmetic product	Nanoparticle type	Prominent use	Marketed by
NanoGold Energizing Cream	Gold	Pure botanicals and 24-karat gold for anti-aging and moisturization	Chantecaille
NanoGold 24-Hour Cream	Gold and silver	Antibacterial, cleansing, and skin balancing	Joyona International
Leorex hypoallergenic	Silicon dioxide	Dislodge dead cells, improve skin complexion	GlobalMed Technologies
Nano Gold Anti-Aging Lifting Serum	Gold	Reverse and reduce signs of aging	Nuvoderm
Nano Gold BB Cream SPF 50 PA+++	Gold	Triple function: skin whitening, UV protection, and anti-wrinkling	Tony Moly
Cosil Whitening Mask	Silver	Absorbing skin contaminants and skin brightening	Natural Korea
Nouriva Repair Moisturizing Cream		Moisturizing itchy, dry, scaly and rough skin	Ferndale Laboratories, Inc.

(Continued)

TABLE 18.4 (Continued)

Nanocosmetic product	Nanoparticle type	Prominent use	Marketed by
24-K Nano Ultra Silk Serum	Gold	Light weight waterless solution for repairing and moisturizing skin	Orogold
Renergie	Avobenzone (3%), octisalate (5%), octocrylene (7%)	Anti-aging	Lancome
Cleanser Silver	Silver	Make-up cleanser	Nano Cyclic
DiorSnow Pure UV Base SPF 50	Octinoxate, TiO_2, benzophenone and octocrylene	Anti-hyperpigmentation and sun protection	Dior
Nano Gold & Silk Day Cream	Gold	Skin care against premature aging	LR Zeitgard
Cosil Whitening Mask	Silver and plant extracts	Absorbing skin contaminants and skin brightening	Natural Korea
Colorstay Natural Tan 24 hrs	Zinc oxide, TiO_2	Absorbent and anti-caking	Revlon
NanoCare Gold	Gold and silver	Cavity disinfectant	Dental NanoTechnology
Sunforgettable® Total Protection™ Brush on shield 50	Zinc oxide (22.5%) and Titanium oxide (22.5%)	All mineral skin-shield against UV rays	ColoreScince
Nanorama—Nano Gold Mask Pack	Gold	Moisture retention and anti-aging	Lexon Nanotech
24 K Gold Gel Cream	Gold	Impart glow and shine to skin	O3+
NanoWorks ShineLuxe	Mica	Hair polish	Pureology
SkinLights highlighter and Bronzer	Topaz	Light reflective powder for multidimensional glow	Revlon
Zinclear® (ZinClear IM®, ZinClear XP)	Zinc oxide	Skin moisturizer and broad-spectrum UV protection	Antaria
Nano-In Hand and Nail Moisturizing Serum and Foot Moisturizing Serum	Zinc oxide	Nail and foot moisturizer	Nano-Infinity Nanotech
Micro Silk White Lotus Intensive Lotion Mask	Silver	Skin pH balance and hydration	Joyona International
Nanowax	polymer-resin hybrid	Color protection from heat	Pureology
Soltan facial sun defence cream	Titanium oxide	Long-lasting UV protection	Boots (Optisol)

18.3.5 Nanospheres and Nanocapsules

Nanospheres and nanocapsules are the two major forms of polymeric nanoparticles. In nanospheres, the loaded drug/cosmetic active is present and dispersed throughout the polymer matrix, whereas in nanocapsules, as the name implies, the drug/cosmetic active is present encapsulated inside the core from which gradual release takes place (Aziz et al. 2019). Varying parameters such as the nature and type of polymer used, the formulation technique adopted, the physiochemical characteristics of the drug/cosmetic active, etc. along with the objective to be achieved, widen the scope of making different types of polymeric nanoparticles. Both natural and synthetic polymers are used for making nanospheres, for example, the natural ones are derived from starch or chitin or algae or albumin, etc. and the synthetic ones are poly-acrylates/caprolactone, PLGA, etc. Nanosphere containing hyaluronic acid formulated with chitosan as polymer has shown efficacy as an anti-wrinkling agent (Chaouat et al. 2017); minoxidil-chitosan nanospheres have shown enhanced efficacy in hair-fall rescue (Matos et al. 2015). Nanocapsules on the other hand are feasible systems for delivering cosmetic actives/ drugs that are prone to degrade by triggers such as light/pH/enzymes etc. Thus, encapsulating them provides the required protection and also facilitates controlled release kinetics (El-Say and El-Sawy 2017; Frank et al. 2015). Nanospheres have utility in several categories of products, for examples, the Lip Tender (Kara Vita) containing argireline, willow herb, green tea extracts, etc. is used for moisturizing lips; the moisturizing sunscreen SPF29 by Dermazone Solutions provides UV protection; and the Celazome' Eye Treat contains sea butter, olive oil, vitamin E, etc. for under eye care. Double dose in a box, nanocapsules by Dr Brandt contains grapeseed and green tea extracts along with sesame oil and lavender oil which mediate anti-aging actions. The Primordiale Optimum Lip by Lancome contains nanocapsules of Vitamin E and gatulline for preventing wrinkling of lip skin. The portfolio of Super Aqua Skin creams by Enprani contains blue algae and avocado extracts for skin moisturization. Several other examples of polymeric nanoparticles are listed in Table 18.5.

18.3.6 Nanoemulsions

Nanoemulsions are liquid-in-liquid biphasic dispersion systems, being either oil-in-water (O/W) or water-in-oil (W/O) or having multi-layers such as O/W/O or W/O/W. Appearance wise they are either translucent or transparent in nature, are kinetically and thermodynamically very stable, and have droplet size in the 50–200 nm range. Nanoemulsion systems are widely used for making creams, lotions, gels, shampoos, etc. as they yield homogenized products with high loading capacity for actives along with ensuring better skin permeations for efficacy. When used in skincare products, nanoemulsions reduce water loss by forming a layer on the skin (Sonneville-Aubrun, Yukuyama, and Pizzino 2018). Kemira and L'Oreal are leading companies that have acquired nanoemulsions-based patents (Ashaolu 2021; Aziz et al. 2019). Agera Nano EyeLift and Nano-Lipobelle™ DN CoQ10 are anti-aging nanoemulsions preparations. Bepanthol Ultra Facial Protect Cream by Bayer Health Care contains ceramides, vitamins, and glycerine, etc. for moisturization and protection from

TABLE 18.5

Marketed nanocosmetic preparations containing polymeric nanoparticles

Nanocosmetic product	Category	Prominent use	Marketed by
Nanosphere Plus Stem cell	Nanosphere	Preservation and protection of skin cells	Derma Swiss
Moisturizing sunscreen SPF29	Nanosphere	UV Protection and moisturizer	Dermazone Solutions
Lip Tender	Nanosphere	Lip moisturizer	Kara Vita
Ultra Moisturizing Day Cream	Nanosphere	Skin moisturizer	Hydralane Paris
Enlighten me	Nanosphere	Skin whitening	Kara Vita
Filler Intense Cream	Nanosphere	Anti-aging, flawless smooth skin	CellAct Switzerland
CelazomeEye Treat	Nanosphere	Under eye cream	Celazome New Zealand Limited
Soleil Express Protection Instant Cooling Sun Spritz SPF 15	Nanocapsule	Cooling spray for UV protection	Lancome
Hydra Zen Cream	Nanocapsule	Anti-stress and 24-hour moisturization	Lancome
Super aqua skin cream line	Nanocapsule	Skin brightening, anti-aging	Enprani
Primordiale Optimum Lip	Nanocapsule	Lip moisturizer	Lancome
Hydra Flash Bronzer Daily Face moisturizer	Nanocapsule	Anti-aging	Lancome
Nano vita C	Nanocapsule	Anti-aging	Eccos
Double dose in a box	Nanocapsule	Rapid wrinkle eraser	Dr. Brandt

pollution. Vita-Herb Nano-Vital Skin Toner contains a mixture of medicinal agents niacinamide and adenosine with aloe vera, ginseng, and other plant extracts which mediates skincare for acne-prone skin along with providing moisturization and anti-aging benefits. Several other marketed products are listed in Table 18.6.

18.3.7 OTHER MARKETED NANOCOSMETICS

Apart from liposomes and niosomes that have been previously discussed, there are other vesicular systems that have characteristic properties that make them unique carriers. Novasomes are made of two to seven bilipid layers and have an amorphous core in the center. IGI Laboratories have proprietary rights to a novasome encapsulation technology known as NOVAVAX (Mosallam et al. 2021). Another two vesicular forms are ethosomes which are phospholipid vesicles having a high ethanol content and cubosomes having a cuboidal honeycomb structure (Verma and Pathak 2010; Kaul et al. 2018; Rapalli et al. 2021). Cosmetic giants, L'Oreal, Nivea, and P&G are developing cubosomal cosmetics (Singh 2021). The new upcoming

TABLE 18.6

Marketed nanoemulsions based nanocosmetic preparations

Nanocosmetic product	Prominent use	Marketed by
Nano-Lipobelle™ DN CoQ10 oA	Skin rejuvenation by promoting mitochondrial activity and collagen production	MiBelle Biochemistry
Phyto-Endorphin Hand Cream	Diminish age spots and wrinkles onhand	Rhonda Allison
Coco Mademoiselle Fresh Moisture Mist	Moisturization	Chanel
Bepanthol Ultra Facial Protect Cream	Anti-aging; Anti-pollutant; Moisturization	Bayer Health Care
Vital A-VC	Skin restoration and revitalization	Marie Louise
Agera Nano EyeLift	Anti-aging	Agera
Vita-Herb Nano-Vital Skin Toner, lotion, aqua cream	Anti-wrinkling and skin whitening	Vitacos Cosmetics
Hyaluronic Acid & Nanoemulsion Intensive Hydration Toner	Skin hydration	Coni Beauty
Calming Alcohol Free Nanoemulsion	Antipollution, Moisturizer	Chanel
Red Vine Hair Sun Protection Serum	UV protection for hair	Korres

polymersomes are made of block-copolymers. Nanofibers are less than 100 nm in diameter size, nanomaterial strands that have a high surface area and a very porous nature which imparts high loading capacity. Either synthetic (PVA or PVP etc.) or natural (silk or collagen or chitosan etc.) are used for making nanofibers. They are mostly used for making face masks, cleansing agents, skin-healing agents, etc. Successful enhanced delivery of vitamins (A and E) was achieved using nanofibers in comparison to their film forms (Taepaiboon, Rungsardthong, and Supaphol 2007). Dendrimers are tree-like branched-shaped nanostructures conferring high loading capacity owing to their increased surface area (Surekha et al. 2021; Dubey et al. 2021). Unilever has acquired patent rights on hydroxyl group functionalized polyester-derived dendrimers for developing sprays, gel, and lotion-based products ("CA2383521A1- Hydroxyl-Functionalised Dendritic Macromolecules for Treating Hair - Google Patents" 2022). Nanocrystals are compact crystals of the actives themselves. The Nanovital Vitanics is a marketed nanocrystal preparation containing niacinamide and ascorbic acid with a mixture of herbal extracts for skin moisturization (Kaul et al. 2018). Carbon-based nanocosmetics are of a few types as well. Carbon nanotubes are hollow tube-like structures; fullerenes are cage-like structures comprising carbon rings. LipoFullerene, made from olive oil and contains squalane has been shown to have anti-oxidant and anti-wrinkle properties clinically (Kato et al. 2010). Some of the marketed preparations are listed below in Table 18.7.

TABLE 18.7

Marketed nanocosmetic preparations of miscellaneous nanoparticles

Nanocosmetic product	Category	Prominent use	Marketed by
Water Bank	Novasome	Moisturization	Pacific
Hydroviton.CR Liquid Normalizing Soap	Nanosomes	Exfoliating agent; Spot reduction; Pore cleanser	DS Laboratories, Inc.
Genie Sparkle Bottles	Nanovitamins	Bronzer/highlighter	ColoreScience
C-60 Night Cream	Fullerenes	Moisturization and anti-oxidant effect	Zelens
White Out+	Fullerenes	Treating eye puffiness and dark circles	Sircuit Skin Cosmeceuticals Inc.
Nanovital Vitanics	Nanocrystal	Anti-aging and moisturization	Vitacos Cosmetics
Viterol. A 16%	Nanosomes	Anti-aging	DS Laboratories, Inc.
Nano Skin Tech Tennis Player Sun and Wind Protection	Nanocomplexes and UV chromophores	UV Protection	Bionova
Ketac Nano	Nano ionomer	Restoring teeth, filling defects	3M Science
AcneWORX	Novasome	Acne treatment, spot reduction, and moisturization	DermWORX

18.4　CONCLUSION

Developing functional nanocosmetics is an ever-growing field of interest that is quite dynamic in nature, with new combinations being explored by several researchers continuously. The numerous nanotech-based cosmetic products are a testament to this. Post strict preliminary screening, using cosmetically active ingredients that are effective, being both of synthetic and natural origin, are used for developing several products, some of which finally reach the market. Nanocosmetics are yet to capture the mass of the consumer base given that their production is expensive compared to conventional cosmetics. However, increasing awareness about the enhanced efficacy makes them more sought-after products, thus hinting at the scope of expansion they are yet to undergo, thus providing incentive for more research in this arena.

REFERENCES

Abdelkader, Hamdy, Adam W G Alani, and Raid G Alany. 2014. "Recent Advances in Non-Ionic Surfactant Vesicles (Niosomes): Self-Assembly, Fabrication, Characterization, Drug Delivery Applications and Limitations." 21 (2): 87–100. doi:10.3109/10717544. 2013.838077.

Ahmad, Anas, and Haseeb Ahsan. 2020. "Lipid-Based Formulations in Cosmeceuticals and Biopharmaceuticals." *Biomedical Dermatology* 4 (1): 12. doi:10.1186/s41702-020-00062-9.

Ahmad, Javed. 2021. "Lipid Nanoparticles Based Cosmetics with Potential Application in Alleviating Skin Disorders." *Cosmetics*. doi:10.3390/cosmetics8030084.

Akbarzadeh, Abolfazl, Rogaie Rezaei-Sadabady, Soodabeh Davaran, Sang Woo Joo, Nosratollah Zarghami, Younes Hanifehpour, Mohammad Samiei, Mohammad Kouhi, and Kazem Nejati-Koshki. 2013. "Liposome: Classification, Preparation, and Applications." *Nanoscale Research Letters* 8 (1): 102. doi:10.1186/1556-276X-8-102.

Arora, Nageen, Shilpi Agarwal, and Murthy Rayasa. 2012. "Latest Technology Advances in Cosmaceuticals." *International Journal of Pharmaceutical Sciences and Drug Research* 4 (July): 168–182.

Ashaolu, Tolulope Joshua. 2021. "Nanoemulsions for Health, Food, and Cosmetics: A Review." *Environmental Chemistry Letters* 19 (4): 3381–95. doi:10.1007/S10311-021-01216-9.

Aziz, Zarith Asyikin Abdul, Hasmida Mohd-Nasir, Akil Ahmad, Siti Hamidah Mohd Setapar, Wong Lee Peng, Sing Chuong Chuo, Asma Khatoon, Khalid Umar, Asim Ali Yaqoob, and Mohamad Nasir Mohamad Ibrahim. 2019. "Role of Nanotechnology for Design and Development of Cosmeceutical: Application in Makeup and Skin Care." *Frontiers in Chemistry* 7 (November): 739. doi:10.3389/fchem.2019.00739.

Bhalekar, Mangesh R, Varsha Pokharkar, Ashwini Madgulkar, Nilam Patil, and Nilkanth Patil. 2009. "Preparation and Evaluation of Miconazole Nitrate-Loaded Solid Lipid Nanoparticles for Topical Delivery." *AAPS PharmSciTech* 10 (1): 289–96. doi:10.1208/s12249-009-9199-0.

"CA2383521A1- Hydroxyl-Functionalised Dendritic Macromolecules for Treating Hair - Google Patents." 2022. Accessed August 2. https://patents.google.com/patent/CA2383521A1/en.

Chaouat, C, Stéphane Balayssac, Myriam Malet-Martino, F Belaubre, Emmanuel Questel, A M Schmitt, Steph Poigny, S Franceschi, and E Perez. 2017. "Green Microparticles Based on a Chitosan/Lactobionic Acid/Linoleic Acid Association. Characterization and Evaluation as a New Carrier System for Cosmetics." *Journal of Microencapsulation* 34 (March): 1–21. doi:10.1080/02652048.2017.1311956.

Date, A A, B Naik, and M S Nagarsenker. 2006. "Novel Drug Delivery Systems: Potential in Improving Topical Delivery of Antiacne Agents." *Skin Pharmacology and Physiology* 19 (1): 2–16. doi:10.1159/000089138.

Ding, Y, S M Pyo, and R H Müller. 2017. "SmartLipids(®) as Third Solid Lipid Nanoparticle Generation - Stabilization of Retinol for Dermal Application." *Die Pharmazie* 72 (12): 728–35. doi:10.1691/ph.2017.7016.

Dubey, Sunil Kumar, Anuradha Dey, Gautam Singhvi, Murali Manohar Pandey, Vanshikha Singh, and Prashant Kesharwani. 2022. "Emerging Trends of Nanotechnology in Advanced Cosmetics." *Colloids and Surfaces B: Biointerfaces*, 112440. doi:10.1016/j.colsurfb.2022.112440.

Dubey, Sunil Kumar, Maithili Kali, Siddhanth Hejmady, Ranendra Narayan Saha, Amit Alexander, and Prashant Kesharwani. 2021. "Recent Advances of Dendrimers as Multifunctional Nano-Carriers to Combat Breast Cancer." *European Journal of Pharmaceutical Sciences: Official Journal of the European Federation for Pharmaceutical Sciences* 164 (September): 105890. doi:10.1016/j.ejps.2021.105890.

El-Say, Khalid M, and Hossam S El-Sawy. 2017. "Polymeric Nanoparticles: Promising Platform for Drug Delivery." *International Journal of Pharmaceutics* 528 (1–2): 675–91. doi:10.1016/j.ijpharm.2017.06.052.

Frank, Luiza A., Renata V. Contri, Ruy C. R. Beck, Adriana R. Pohlmann, and Silvia S. Guterres. 2015. "Improving Drug Biological Effects by Encapsulation into Polymeric Nanocapsules." *Wiley Interdisciplinary Reviews: Nanomedicine and Nanobiotechnology* 7 (5): 623–39. doi:10.1002/WNAN.1334.

Ge, Xuemei, Minyan Wei, Suna He, and Wei-En Yuan. 2019. "Advances of Non-Ionic Surfactant Vesicles (Niosomes) and Their Application in Drug Delivery." *Pharmaceutics* 11 (2): 55. doi:10.3390/pharmaceutics11020055.

Ghasemiyeh, Parisa, and Soliman Mohammadi-Samani. 2018. "Solid Lipid Nanoparticles and Nanostructured Lipid Carriers as Novel Drug Delivery Systems: Applications, Advantages and Disadvantages." *Research in Pharmaceutical Sciences* 13 (4): 288–303. doi:10.4103/1735-5362.235156.

Heiligtag, Florian J, and Markus Niederberger. 2013. "The Fascinating World of Nanoparticle Research." *Materials Today* 16 (7): 262–71. doi:10.1016/j.mattod.2013.07.004.

Jenning, V, and S H Gohla. 2001. "Encapsulation of Retinoids in Solid Lipid Nanoparticles (SLN)." *Journal of Microencapsulation* 18 (2): 149–58. doi:10.1080/02652040010000361.

Kato, Shinya, Hikam Taira, Hisae Aoshima, Yasukazu Saitoh, and Nobuhiko Miwa. 2010. "Clinical Evaluation of Fullerene-C 60 Dissolved in Squalane for Anti-Wrinkle Cosmetics." *Journal of Nanoscience and Nanotechnology* 10 (10): 6769–74. doi:10.1166/JNN.2010.3053.

Kaul, Shreya, Neha Gulati, Deepali Verma, Siddhartha Mukherjee, and Upendra Nagaich. 2018. "Role of Nanotechnology in Cosmeceuticals: A Review of Recent Advances." *Journal of Pharmaceutics* 2018 (March): 1–19. doi:10.1155/2018/3420204.

Kaur, I P, and R Agrawal. 2007. "Nanotechnology: A New Paradigm in Cosmeceuticals." *Recent Patents on Drug Delivery & Formulation* 1 (2): 171–82. doi:10.2174/187221107780831888.

Kazi, Karim Masud, Asim Sattwa Mandal, Nikhil Biswas, Arijit Guha, Sugata Chatterjee, Mamata Behera, and Ketousetuo Kuotsu. 2010. "Niosome: A Future of Targeted Drug Delivery Systems." *Journal of Advanced Pharmaceutical Technology & Research* 1 (4): 374–80. doi:10.4103/0110-5558.76435.

Kim, Se-Kwon. 2014. "Marine Cosmeceuticals." *Journal of Cosmetic Dermatology* 13 (1): 56–67. doi:10.1111/jocd.12057.

Kligman, Albert M. 2000. "Cosmetics: A Dermatologist Looks to the Future: Promises and Problems." *Dermatologic Clinics* 18 (4): 699–709. doi:10.1016/S0733-8635(05)70221-7.

Köpke, Daniel, Rainer H Müller, and Sung Min Pyo. 2019. "Phenylethyl Resorcinol SmartLipids for Skin Brightening – Increased Loading & Chemical Stability." *European Journal of Pharmaceutical Sciences* 137: 104992. doi:10.1016/j.ejps.2019.104992.

Lasic, D D. 1998. "Novel Applications of Liposomes." *Trends in Biotechnology* 16 (7): 307–21. doi:10.1016/S0167-7799(98)01220-7.

Lin, Yin-Ku, Zih-Rou Huang, Rou-Zi Zhuo, and Jia-You Fang. 2010. "Combination of Calcipotriol and Methotrexate in Nanostructured Lipid Carriers for Topical Delivery." *International Journal of Nanomedicine* 5 (March): 117–28. doi:10.2147/ijn.s9155.

Matos, Breno Noronha, Thaiene Avila Reis, Taís Gratieri, and Guilherme Martins Gelfuso. 2015. "Chitosan Nanoparticles for Targeting and Sustaining Minoxidil Sulphate Delivery to Hair Follicles." *International Journal of Biological Macromolecules* 75 (April): 225–29. doi:10.1016/j.ijbiomac.2015.01.036.

Mosallam, Shaimaa, Maha H Ragaie, Noha H Moftah, Ahmed Hassen Elshafeey, and Aly Ahmed Abdelbary. 2021. "Use of Novasomes as a Vesicular Carrier for Improving the Topical Delivery of Terconazole: In Vitro Characterization, In Vivo Assessment and Exploratory Clinical Experimentation." *International Journal of Nanomedicine* 16: 119. doi:10.2147/IJN.S287383.

Mota, Ana Henriques, Alexandra Sousa, Mariana Figueira, and Mariana Amaral. 2020. "Natural-Based Consumer Health Nanoproducts: Medicines, Cosmetics, and Food Supplements Mucoadhesion and Drug Delivery View Project European Network of Bioadhesion Expertise (ENBA) View Project." Elsevier. https://doi.org/10.1016/B978-0-12-816787-8.00019-3.

Müller, Rainer H, R D Petersen, A Hommoss, and J Pardeike. 2007. "Nanostructured Lipid Carriers (NLC) in Cosmetic Dermal Products." *Advanced Drug Delivery Reviews* 59 (6): 522–30. doi:10.1016/J.ADDR.2007.04.012.

Müller, Rainer H, Ranjita Shegokar, and Cornelia M Keck. 2011. "20 Years of Lipid Nanoparticles (SLN and NLC): Present State of Development and Industrial Applications." *Current Drug Discovery Technologies* 8 (3): 207–27. doi:10.2174/157016311796799062.

Münster, U, C Nakamura, A Haberland, K Jores, W Mehnert, S Rummel, M Schaller, et al. 2005. "RU 58841-Myristate--Prodrug Development for Topical Treatment of Acne and Androgenetic Alopecia." *Die Pharmazie* 60 (1): 8–12.

Nasir, Ali, Harikumar Sl, and Amanpreet Kaur. 2012. "Niosomes: An Excellent Tool for Drug Delivery." *International Journal of Research in Pharmacy and Chemistry* 2 (January): 479–487.

Olechowski, Florence, Rainer H Müller, and Sung Min Pyo. 2019. "BergaCare SmartLipids: Commercial Lipophilic Active Concentrates for Improved Performance of Dermal Products." *Beilstein Journal of Nanotechnology* 10: 2152–62. doi:10.3762/bjnano.10.208.

Passerini, Nadia, Elisabetta Gavini, Beatrice Albertini, Giovanna Rassu, Marcello Di Sabatino, Vanna Sanna, Paolo Giunchedi, and Lorenzo Rodriguez. 2009. "Evaluation of Solid Lipid Microparticles Produced by Spray Congealing for Topical Application of Econazole Nitrate." *Journal of Pharmacy and Pharmacology* 61 (5): 559–67. doi:10.1211/jpp.61.05.0003.

Pereira, Leonel, Nicolina Dias, Juliana Carvalho, Sara Fernandes, Cledir Santos, and Nelson Lima. 2014. "Synthesis, Characterisation and Antifungal Activity of Chemically and Fungal-Produced Silver Nanoparticles against Trichophyton Rubrum." *Journal of Applied Microbiology* 117 (September). doi:10.1111/jam.12652.

Pople, Pallavi V, and Kamalinder K Singh. 2010. "Targeting Tacrolimus to Deeper Layers of Skin with Improved Safety for Treatment of Atopic Dermatitis." *International Journal of Pharmaceutics* 398 (1): 165–78. doi:10.1016/j.ijpharm.2010.07.008.

Prabhakaran, Ekambaram, Sathali A Hasan, and Priyanka Karunanidhi. 2012. "Solid Lipid Nanoparticles: A Review." *Scientific Reviews and Chemical Communications* 2 (1): 80–102.

Rapalli, Vamshi Krishna, Saswata Banerjee, Shahid Khan, Prabhat Nath Jha, Gaurav Gupta, Kamal Dua, Md Saquib Hasnain, Amit Kumar Nayak, Sunil Kumar Dubey, and Gautam Singhvi. 2021. "QbD-Driven Formulation Development and Evaluation of Topical Hydrogel Containing Ketoconazole Loaded Cubosomes." *Materials Science and Engineering: C* 119: 111548. doi:10.1016/j.msec.2020.111548.

Rapalli, Vamshi Krishna, Vedhant Kaul, Srividya Gorantla, Tejashree Waghule, Sunil Kumar Dubey, Murali Monohar Pandey, and Gautam Singhvi. 2020. "UV Spectrophotometric Method for Characterization of Curcumin Loaded Nanostructured Lipid Nanocarriers in Simulated Conditions: Method Development, in-Vitro and Ex-Vivo Applications in Topical Delivery." *Spectrochimica Acta. Part A, Molecular and Biomolecular Spectroscopy* 224 (January): 117392. doi:10.1016/j.saa.2019.117392.

Rapalli, Vamshi Krishna, Vedhant Kaul, Tejashree Waghule, Srividya Gorantla, Swati Sharma, Aniruddha Roy, Sunil Kumar Dubey, and Gautam Singhvi. 2020. "Curcumin Loaded Nanostructured Lipid Carriers for Enhanced Skin Retained Topical Delivery: Optimization, Scale-Up, In-Vitro Characterization and Assessment of Ex-Vivo Skin Deposition." *European Journal of Pharmaceutical Sciences : Official Journal of the European Federation for Pharmaceutical Sciences* 152 (September): 105438. doi:10.1016/j.ejps.2020.105438.

Rapalli, Vamshi Krishna, Gautam Singhvi, Sunil Kumar Dubey, Gaurav Gupta, Dinesh Kumar Chellappan, and Kamal Dua. 2018. "Emerging Landscape in Psoriasis Management: From Topical Application to Targeting Biomolecules." *Biomedicine & Pharmacotherapy* 106: 707–13. doi:10.1016/j.biopha.2018.06.136.

Rapalli, Vamshi Krishna, Tejashree Waghule, Neha Hans, Arisha Mahmood, Srividya Gorantla, Sunil Kumar Dubey, and Gautam Singhvi. 2020. "Insights of Lyotropic Liquid Crystals in Topical Drug Delivery for Targeting Various Skin Disorders." *Journal of Molecular Liquids* 315: 113771. doi:10.1016/j.molliq.2020.113771.

"ROVISOME® ACE Plus (Antioxidant) by Evonik - Technical Datasheet." 2022. Accessed August 2. https://cosmetics.specialchem.com/product/i-evonik-rovisome-ace-plus.

Ruocco, Nadia, Susan Costantini, Stefano Guariniello, and Maria Costantini. 2016. "Polysaccharides from the Marine Environment with Pharmacological, Cosmeceutical and Nutraceutical Potential." *Molecules (Basel, Switzerland)* 21 (5). doi:10.3390/molecules21050551.

Sakamoto, Kazutami, Howard Lochhead, Howard Maibach, and Yuji Yamashita. 2017. *Cosmetic Science and Technology: Theoretical Principles and Applications*. Amsterdam: Elsevier.

Sanna, Vanna, Elisabetta Gavini, Massimo Cossu, Giovanna Rassu, and Paolo Giunchedi. 2007. "Solid Lipid Nanoparticles (SLN) as Carriers for the Topical Delivery of Econazole Nitrate: In-Vitro Characterization, Ex-Vivo and In-Vivo Studies." *Journal of Pharmacy and Pharmacology* 59 (8): 1057–64. doi:10.1211/jpp.59.8.0002.

Schäfer-Korting, Monika, Wolfgang Mehnert, and Hans-Christian Korting. 2007. "Lipid Nanoparticles for Improved Topical Application of Drugs for Skin Diseases." *Advanced Drug Delivery Reviews* 59 (6): 427–43. doi:10.1016/j.addr.2007.04.006.

Shidhaye, S S, Reshma Vaidya, Sagar Sutar, Arati Patwardhan, and V J Kadam. 2008. "Solid Lipid Nanoparticles and Nanostructured Lipid Carriers--Innovative Generations of Solid Lipid Carriers." *Current Drug Delivery* 5 (4): 324–31. doi:10.2174/156720108785915087.

Silva, A C, D Santos, D C Ferreira, and E B Souto. 2009. "Minoxidil-Loaded Nanostructured Lipid Carriers (NLC): Characterization and Rheological Behaviour of Topical Formulations." *Die Pharmazie* 64 (3): 177–82.

Singh, Anupama, Rishabha Malviya, and Pramod Sharma. 2011. "Novasome-A Breakthrough in Pharmaceutical Technology a Review Article." *Advances in Biological Research* 5 (January): 184–89.

Singh, Archana. 2021. "Carbon Nanofiber in Cosmetics." In, 341–63. doi:10.1002/9781119769149.ch14.

Singh, Thakur Gurjeet, and Neha Sharma. 2016. "Chapter 7- Nanobiomaterials in Cosmetics: Current Status and Future Prospects." In, edited by B T Alexandru Mihai, *Nanobiomaterials in Galenic Formulations and Cosmetics Grumezescu*, 149–74. William Andrew Publishing. doi:10.1016/B978-0-323-42868-2.00007-3.

Smijs, Threes G, and Stanislav Pavel. 2011. "Titanium Dioxide and Zinc Oxide Nanoparticles in Sunscreens: Focus on Their Safety and Effectiveness." *Nanotechnology, Science and Applications* 4 (October): 95–112. doi:10.2147/NSA.S19419.

Sonneville-Aubrun, Odile, Megumi N. Yukuyama, and Aldo Pizzino. 2018. "Application of Nanoemulsions in Cosmetics." *Nanoemulsions: Formulation, Applications, and Characterization*, 435–75. Academic Press. doi:10.1016/B978-0-12-811838-2.00014-X.

Souto, E B, and R H Müller. 2005. "SLN and NLC for Topical Delivery of Ketoconazole." *Journal of Microencapsulation* 22 (5): 501–10. doi:10.1080/02652040500162436.

———. 2006. "The Use of SLN and NLC as Topical Particulate Carriers for Imidazole Antifungal Agents." *Die Pharmazie* 61 (5): 431–37.

———. 2007. "Rheological and in Vitro Release Behaviour of Clotrimazole-Containing Aqueous SLN Dispersions and Commercial Creams." *Die Pharmazie* 62 (7): 505–9.

Sriraman, Shravan Kumar, and Vladimir P. Torchilin. 2014. "Recent Advances with Liposomes as Drug Carriers." *Advanced Biomaterials and Biodevices* 9781118773635 (July): 79–119. doi:10.1002/9781118774052.CH3.

Surekha, Bhavya, Naga Sreenu Kommana, Sunil Kumar Dubey, A V Pavan Kumar, Rahul Shukla, and Prashant Kesharwani. 2021. "PAMAM Dendrimer as a Talented Multifunctional Biomimetic Nanocarrier for Cancer Diagnosis and Therapy." *Colloids and Surfaces B: Biointerfaces* 204: 111837. doi:10.1016/j.colsurfb.2021.111837.

Taepaiboon, Pattama, Uracha Rungsardthong, and Pitt Supaphol. 2007. "Vitamin-Loaded Electrospun Cellulose Acetate Nanofiber Mats as Transdermal and Dermal Therapeutic Agents of Vitamin A Acid and Vitamin E." *European Journal of Pharmaceutics and Biopharmaceutics : Official Journal of Arbeitsgemeinschaft Fur Pharmazeutische Verfahrenstechnik e.V* 67 (2): 387–97. doi:10.1016/j.ejpb.2007.03.018.

"US4659560A - Deodorant Compositions - Google Patents." 2021. Accessed October 27. https://patents.google.com/patent/US4659560A/en.

Vance, Marina E, Todd Kuiken, Eric P Vejerano, Sean P McGinnis, Michael F Jr Hochella, David Rejeski, and Matthew S Hull. 2015. "Nanotechnology in the Real World: Redeveloping the Nanomaterial Consumer Products Inventory." *Beilstein Journal of Nanotechnology* 6: 1769–80. doi:10.3762/bjnano.6.181.

Verma, Poonam, and K Pathak. 2010. "Therapeutic and Cosmeceutical Potential of Ethosomes: An Overview." *Journal of Advanced Pharmaceutical Technology & Research* 1 (3): 274. doi:10.4103/0110-5558.72415.

Vierkötter, Andrea, Tamara Schikowski, Ulrich Ranft, Dorothea Sugiri, Mary Matsui, Ursula Krämer, and Jean Krutmann. 2010. "Airborne Particle Exposure and Extrinsic Skin Aging." *The Journal of Investigative Dermatology* 130 (12): 2719–26. doi:10.1038/jid.2010.204.

Vinod, T P, and Raz Jelinek. 2019. "Inorganic Nanoparticles in Cosmetics BT - Nanocosmetics: From Ideas to Products." In, edited by Jean Cornier, Cornelia M Keck, and Marcel Van de Voorde, 29–46. Cham: Springer International Publishing. doi:10.1007/978-3-030-16573-4_3.

Waghule, Tejashree, Srividya Gorantla, Vamshi Krishna Rapalli, Pranav Shah, Sunil Kumar Dubey, Ranendra Narayan Saha, and Gautam Singhvi. 2020. "Emerging Trends in Topical Delivery of Curcumin Through Lipid Nanocarriers: Effectiveness in Skin Disorders." *AAPS PharmSciTech* 21 (7): 284. doi:10.1208/s12249-020-01831-9.

Waghule, Tejashree, Vamshi Krishna Rapalli, Srividya Gorantla, Ranendra Narayan Saha, Sunil Kumar Dubey, Anu Puri, and Gautam Singhvi. 2020. "Nanostructured Lipid Carriers as Potential Drug Delivery Systems for Skin Disorders." *Current Pharmaceutical Design* 26 (36): 4569–79. doi:10.2174/1381612826666200614175236.

Wissing, Sylvia A, and Rainer H Müller. 2003. "Cosmetic Applications for Solid Lipid Nanoparticles (SLN)." *International Journal of Pharmaceutics* 254 (1): 65–68. doi:10.1016/S0378-5173(02)00684-1.

Yoon, Goo, Jin Woo Park, and In-Soo Yoon. 2013. "Solid Lipid Nanoparticles (SLNs) and Nanostructured Lipid Carriers (NLCs): Recent Advances in Drug Delivery." *Journal of Pharmaceutical Investigation* 43 (5): 353–62. doi:10.1007/s40005-013-0087-y.

Index

Note: **Bold** page numbers refer to tables and *italic* page numbers refer to figures.

For Product Safety Concerns and Information please contact our EU
representative GPSR@taylorandfrancis.com
Taylor & Francis Verlag GmbH, Kaufingerstraße 24, 80331 München, Germany